The Parvoviruses

THE VIRUSES

Series Editors
HEINZ FRAENKEL-CONRAT, *University of California*
Berkeley, California
ROBERT R. WAGNER, *University of Virginia School of Medicine*
Charlottesville, Virginia

THE HERPESVIRUSES, Volumes 1, 2, 3, and 4
Edited by Bernard Roizman

THE REOVIRIDAE
Edited by Wolfgang K. Joklik

THE PARVOVIRUSES
Edited by Kenneth I. Berns

The Parvoviruses

Edited by

KENNETH I. BERNS
University of Florida
Gainesville, Florida

PLENUM PRESS • NEW YORK AND LONDON

Library of Congress Cataloging in Publication Data

Main entry under title:

The Parvoviruses.

(The Viruses)
Includes bibliographical references and index.
1. Parvoviruses. I. Berns, Kenneth I., 1938– II. Series. [DNLM: 1. Parvoviridae. QW 165.5.P3 P2765]
QR408.P37 1983 576′.6484 83-19181
ISBN 978-1-4684-8014-6 ISBN 978-1-4684-8012-2 (eBook)
DOI 10.1007/978-1-4684-8012-2

Softcover reprint of the hardcover 1st edition 1984
A Division of Plenum Publishing Corporation
233 Spring Street, New York, N.Y. 10013

Contributors

Kenneth I. Berns, Department of Immunology and Medical Microbiology, College of Medicine, University of Florida, Gainesville, Florida 32610

Neil R. Blacklow, Division of Infectious Diseases, University of Massachusetts Medical Center, Worcester, Massachusetts 01605

Barrie J. Carter, Laboratory of Cell Biology and Genetics, National Institute for Arthritis, Diabetes, and Digestive and Kidney Diseases, National Institutes of Health, Bethesda, Maryland 20205

George Cukor, Division of Infectious Diseases, University of Massachusetts Medical Center, Worcester, Massachusetts 01605

William W. Hauswirth, Department of Immunology and Medical Microbiology, College of Medicine, University of Florida, Gainesville, Florida 32610

M. David Hoggan, Laboratory of Viral Diseases, National Institute of Allergy and Infectious Diseases, National Institutes of Health, Bethesda, Maryland 20205

F. Brent Johnson, Department of Microbiology, Brigham Young University, Provo, Utah 84601

Catherine A. Laughlin, Section on Macromolecular Genetics, Laboratory of Cell Biology and Genetics, National Institute for Arthritis, Diabetes and Digestive and Kidney Diseases, National Institutes of Health, Bethesda, Maryland 20205

Carol J. Marcus-Sekura, Section on Macromolecular Genetics, Laboratory of Cell Biology and Genetics, National Institute for Arthritis, Diabetes, and Digestive and Kidney Diseases, National Institutes of Health, Bethesda, Maryland 20205

Günter Siegl, Institute for Hygiene and Medical Microbiology, University of Bern, CH 3010 Bern, Switzerland

Preface

The Parvoviridae have been of increasing interest to reseachers in the past decade. Their small size and simple structure have made them amenable to detailed physiochemical analysis, and from this work relatively detailed information has resulted that has signficantly increased our understanding of the biology of these viruses. It has become clear that the Parvoviridae are of interest not only for their own sake, but also because their relative simplicity renders them useful probes in the study of the biology of host cells and of other DNA viruses with which they interact. The *Dependovirus* genus, for instance, contains the defective adeno-associated viruses (AAV), which require a coinfection with either an adenovirus or a herpesvirus for productive multiplication. Studies of AAV, therefore, necessarily impinge on our understanding of the control of macromolecular synthesis by the helper virus. Similarly AAV has been reported to inhibit the oncogenicity of both adeno- and herpesviruses and has been used as a probe of mechanism in these instances as well. Finally, AAV establishes latent infections *in vivo* and is the only mammalian DNA virus where a comparable model system has been established in cell culture. This system has allowed study of the mechanism of latent infection at the molecular level.

In some ways the popularity of parvoviruses as a subject for research has suffered because AAV has never been associated with disease and because no human autonomous parvovirus had been characterized until very recently. This consideration has altered in recent years for several reasons. In addition to the discovery of a human autonomous parvovirus associated with disease, there has been a growing realization that latent infection by AAV may actually provide some protection to the host by inhibition of helper adeno- and herpesvirus infections. Finally canine parvovirus has also increased awareness of the propensity of the family for changes in host range associated with the potential of lethal epidemics.

This volume is an effort to summarize in coherent fashion the progress and excitement in this field.

Kenneth I. Berns

Contents

Chapter 3

Adeno-Associated Virus Defectiveness and the Nature of the Adenovirus Helper Function

Barrie J. Carter and Catherine A. Laughlin

Chapter 4

Autonomous Parvovirus DNA Structure and Replication

William W. Hauswirth

Chapter 5

Parvovirus Transcription

Barrie J. Carter, Catherine A. Laughlin, and Carol J. Marcus-Sekura

Chapter 6

Variant and Defective Interfering Parvoviruses

Barrie J. Carter

Chapter 7

Parvovirus Proteins

F. Brent Johnson

Chapter 8

Biology of Pathogenicity of Autonomous Parvoviruses

Günter Siegl

CHAPTER 1

Adeno-Associated Virus DNA Structure and Replication

Kenneth I. Berns and William W. Hauswirth

I. AAV DNA STRUCTURE

The primary structure of adeno-associated virus (AAV) DNA has been pivotal in understanding its replication. The nucleotide sequence arrangement in virion DNA and the structure of replicative forms have suggested rather directly an experimentally verifiable model for replication. Thus, as each detail of genome structure became known, the mechanism of replication became better defined. For this reason we have chosen to describe experiments leading to our present ideas of AAV DNA structure and replication in more or less historical order.

The strandedness of the AAV genome was not easily established for AAV. This was because in the initial studies plus and minus strands annealed during isolation of virion DNA. However, a careful comparison of AAV particle size and DNA content with those of well-characterized bacteriophage and autonomous parvovirus particles (Crawford *et al.*, 1969) revealed a potential twofold excess of AAV DNA per virion. This suggested that AAV might not contain a duplex genome. To experimentally prove this contention, Rose *et al.* (1969) mixed virions containing DNA substituted with bromodeoxyuridine (BUdR) with normal density virions, isolated duplex DNA, and showed that 50% of the molecules were of hybrid density. Thus, the complementary strands in the duplex DNA examined after purification must have come from different virions.

KENNETH I. BERNS and WILLIAM W. HAUSWIRTH • Department of Immunology and Medical Microbiology, College of Medicine, University of Florida, Gainesville, Florida 32610.

Because nondenaturing conditions had been used in the extraction procedure, the virions must have contained single-stranded DNA. This conclusion was directly confirmed by isolating only single-stranded DNA from virions under nondenaturing and nonrenaturing conditions (Mayor *et al.*, 1969; Berns and Rose, 1970). Finally, based on the observation that BUdR-substituted AAV DNA strands were separable (Berns and Rose, 1970; Rose and Koczot, 1971) because the thymidine content in plus (21%) and minus (27%) strands differed, virions containing each strand polarity were separated on density gradients (Berns and Adler, 1972).

Establishing the strandedness of the AAV genome resolved only one of several unusual structural features. In addition to the predominant linear duplex of unit length found under normal isolation conditions, duplex unit length circles and duplex dimeric linear forms were also detected by electron microscopy and gel electrophoresis (Gerry *et al.*, 1973). After limited digestion of the 3′ termini of duplex molecules with exonuclease III followed by annealing (Gerry *et al.*, 1973) circular duplex molecules were formed, indicating the presence of natural terminal sequence repetitions. Circle formation was most efficient when 1–3% of the genome had been digested from each 3′ end, hence the length of the natural terminal repeat was within these limits. Additionally, inverted terminal repetitions were also present because the single-stranded form of the DNA could form hydrogen-bond-stabilized single-strand circles (Kozcot *et al.*, 1973). The size of the short duplex "panhandles" holding the circles intact (Berns and Kelly, 1974) was the same as that estimated for the natural repeat regions. In later studies it was shown that any AAV DNA terminal restriction fragment less than 100 nucleotides long was identical from either molecular end (Spear *et al.*, 1977). Thus, the termini of AAV DNA exhibit the properties of both natural and inverted repeated nucleotide sequences. The only type of sequence possessing both properties is a DNA palindrome (Gerry *et al.*, 1973). One property of palindromic terminal sequences is that, as single strands, they will fold over to self-hybridize creating hairpinned ends. Such ends should be protected from digestion with single strand-specific DNases whereas internal regions should be accessible. By using 3′ or 5′ ^{32}P terminally labeled molecules this property was demonstrated by AAV DNA (Fife *et al.*, 1977). Because only one radiolabeled nucleotide was added to each terminus the hairpinned sequence or palindrome must begin at or very near the molecular termini to protect the label. Also, the accessibility of ends for enzymatic labeling meant that proteins or unusual chemistry did not exist at the termini.

Finally, the exact structure of the AAV DNA termini was resolved by nucleotide sequencing (Fig. 1) (Berns *et al.*, 1979; Lusby *et al.*, 1980). The inverted terminal repetition is 145 bases long with the first 125 bases forming a large palindrome. Within this palindrome the first 41 bases pair perfectly with bases 85–125. Between bases 42 to 84, two smaller palindromic sequences exist such that bases 42 to 62 and bases 64 to 84

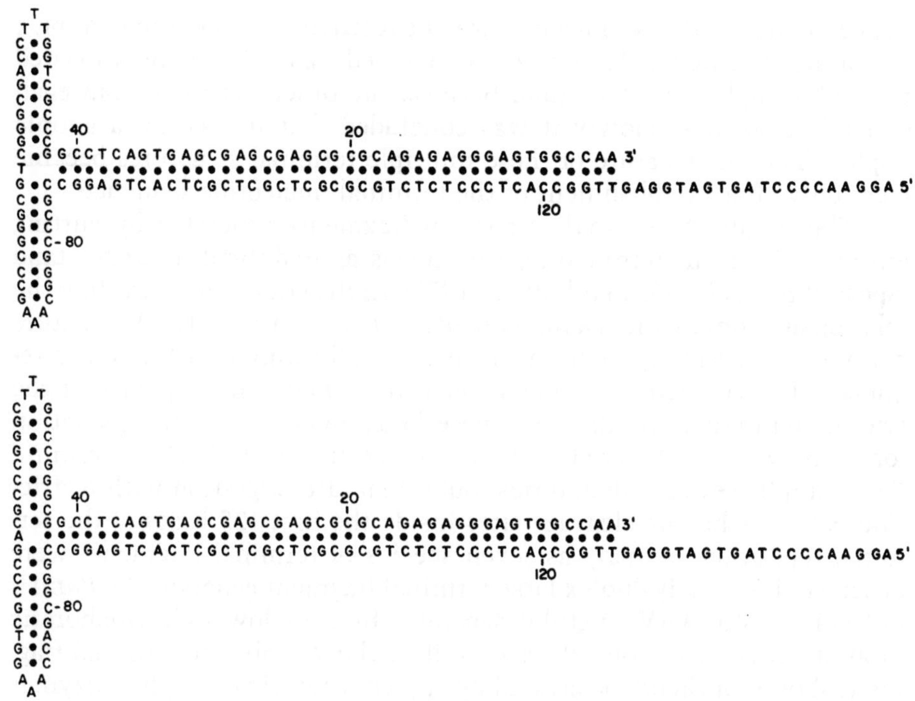

FIGURE 1. Nucleotide sequence of the 145 base terminal repeat in AAV DNA shown in the two possible orientations, in its maximally base paired configuration.

hairpin perfectly with small single-stranded loop regions of three adenine or three thymine residues at the turn-around point. One unpaired nucleotide, residue 63, separates these two internal palindromes. The maximally base paired form of this sequence is consistent with all the properties previously deduced for the termini of the AAV genome. Additionally, a limited terminal sequence heterogeneity in which the 5′ terminal sequence was 5′-TTGGCC in 35% of the molecules, TGGCC in 50%, and GGCC in 15% was reported by Fife *et al.* (1977).

A number of unusual properties of AAV DNA are related to the terminal nucleotide sequence. In particular it was noted early that the relationships among terminal restriction fragments are complex. Initial mapping of restriction enzyme sites showed that terminal restriction fragments which extend beyond the terminal repetition always appear as two species from either molecular end by gel electrophoresis (Berns *et al.*, 1975a; Carter and Khoury, 1975; Carter *et al.*, 1975; de la Maza and Carter, 1976). Pairs of fragments created at the left molecular end by *Bam*HI (Spear *et al.*, 1977; Berns *et al.*, 1979) or at the right end by *Taq*I (Berns *et al.*, 1979) can be interconverted by denaturing and renaturing either member of the pair. In these studies it was noted that one species migrated in accordance with its molecular weight relative to other simple duplex restriction fragments. Thus it was concluded that this fragment existed in a fully duplex conformation. Because the other fragment from each end migrated more slowly it was concluded that it was not a simple duplex. The exact nature of these multiple terminal structures became clear only after establishment of the terminal nucleotide sequence.

The origin of terminal restriction fragments generated by cutting entirely within the terminal repetition was more difficult to understand (Spear *et al.*, 1977; Denhardt *et al.*, 1976). In the course of unraveling the relationship among them a fundamental property of AAV DNA structure became clear, namely, that within the overall terminal palindromic sequence (1–125) there is an inversion which affects the sequence of the two small internal palindromes. These data provided an exact explanation for all previously observed structural properties of AAV DNA termini. For example, several laboratories found that after digestion with restriction enzymes having cleavage sites inside the first 125 bases of the terminal repetition as many as seven species of terminal fragments were observed. If the fully duplex large terminal fragment generated by *Bam*HI or *Taq*I (10% of AAV length) is separated from its lower electrophoretic mobility counterpart on gels as described above, only one terminal fragment should in theory be created by any enzyme. However, for enzymes cutting between nucleotides 41 and 85 (*Hpa*II, *Sma*I, and *Bgl*I) two terminal fragments from each end were seen (Spear *et al.*, 1977; Lusby *et al.*, 1980). The *Hpa*II fragments were 43 and 47 base pairs long, those for *Sma*I 44 and 80 base airs, while the *Bgl*I fragments were 46 and 79 base pairs. Referring to the terminal sequence in Fig. 1 the simplest explanation for these results is evident if it is assumed that the 125 base pair palindrome exists with equal probability in two mutually inverted orientations. By the nature of DNA palindromes the two orientations are equivalent. However, the internal regions (bases 42–84) are not part of the overall palindrome and will therefore not be identical in the two orientations. Hence, restriction enzymes cutting within this region will generate two different terminal fragments from such a set of ends. It can be seen that the sizes of terminal fragment pairs predicted by this model correspond exactly to the experimental results.

Finally, an inconsistency was noted between the numbers of HpaII and SmaI sites between bases 42 and 84 and the number of fragments actually found. By comparing the number and position of restriction sites mapped by partial digestion of DNA ^{32}P-labeled at one end using the technique of Smith and Birnstiel (1976) with restriction fragments from a complete digest, it was noted that for both *Hpa*II (Berns *et al.*, 1979; Spear *et al.*, 1977) and *Sma*I (Lusby *et al.*, 1980) exactly twice as many sites existed as were needed to account for the fragments mapping to this region. This provided additional evidence for sequence inversion because the positions of these sites are exactly consistent with the two sequences shown in Fig. 1.

The nucleotide sequence provided ultimate proof for two mutually inverted terminal sequence populations. For this purpose it was necessary to find a restriction fragment which contains the entire sequence altered by the inversion. Such a fragment will not be affected by the inversion because the cleavage sites are at identical positions in both orientations by virtue of the fact that they are in the larger overall palindrome. In this way the overall sequences of the two small internal plaindromes were determined. The *Hae*III fragment extending from nucleotide 41 to 85 is such a fragment. To confirm by sequencing that the inversion exists, a comparison was made of sequences of the terminal and penultimate *Sma*I fragments which either begin or end within the small internal palindromic sequences affected by the inversion (Lusby *et al.*, 1980).

Taken together these data provide a rational explanation for the unusual properties of AAV DNA termini, including (1) duplex dimeric molecules formed through hydrogen-bonded termini, (2) single-strand circles also formed through hydrogen-bonded termini, (3) single-stranded DNA with hairpinned ends, in which the terminal nucleotide is protected from single-strand-specific nuclease digestion, (4) multiple structures for the termini of duplex DNA, and (5) two sets of restriction sites for restriction enzymes cutting between nucleotides 42 and 84.

One of the most intriguing features of the structure of the AAV genome is the "T" shape formed by hairpinning of the 3′ terminus of the first 125 nucleotides of the inverted terminal repetition in AAV DNA. The significance of a "T" rather than a simple hairpin is not known, but there are two possible explanations worth consideration. The first is that the proteins involved in initiation of DNA synthesis require not only an appropriate primary sequence but also specific secondary and tertiary structures within the genome terminus in much the way that tRNA tertiary structure is important to function. The distance between the arms of the "T" which may contain protein binding sites and the putative site for initiation of DNA synthesis in the "T" stem is approximately 13.6 nm or well within the physical domains of many proteins. A second possible explanation, which does not exclude the first, relates to the hypercycle theory of Eigen and his colleagues (Eigen and Schuster, 1978). This theory suggests that the secondary and tertiary structures in self-complementary polynucleotides are the natural end product of relatively unconstrained sequence evolution. Regardless of the original sequence, base substitutions will normally occur, and those that promote increased base pairing are favored, especially if a GC base pair is formed. When this theory is applied to an 80-base-pair polyribonucleotide, the possible structures converge to a common structure, regardless of the original sequence. The "common" structure formed is that of tRNA. Extension of the theory to longer polyribonucleotide chain lengths yields a structure very similar to that observed for the product synthesized by bacteriophage Qβ replicase in the absence of template (Sumper and Luce, 1975). The theory has not been extended to DNA but is conceivable that the "T"-

shaped structure observed at the termini of parvovirus genomes would be the most likely structure for a polydeoxyribonucleotide of chain length 115–125 rather than a more simple hairpin loop. We note that the arms of the "T" are particularly G-C rich as would be predicted. Thus evolutionary and protein recognition constraints may have combined to define the terminal structure of AAV DNA.

The *in vivo* structure of the AAV genome structure remains unknown. Because virions contain only one single-stranded copy per particle, single-stranded circular forms seen with purified DNA are perhaps the favored structures. These panhandled circles are attractive because they are entirely consistent with the early steps proposed for DNA replication (see below). In this regard it would be critical to know what role AAV capsid proteins may play in stabilizing a particular structure. Despite the fact that AAV proteins have not been shown to have a specific affinity for AAV DNA, in view of the potential role of an AAV gene product in replication (see below), DNA–capsid-protein interactions are entirely reasonable.

II. REPLICATION OF AAV

There are multiple potential fates for AAV DNA during cell infection; the exact fate depends on the character of the host cell and helper virus. However, early events are rather independent of the cell type or presence of helper virus. AAV adsorbs irreversibly to KB cells in suspension within 1–2 hr postinfection and AAV DNA can be detected in the nucleus soon thereafter (Rose and Koczot, 1972). Uncoating takes place coincident with or following nuclear penetration because sensitivity of input AAV DNA to DNase can be detected only in the nucleus (Berns and Adler, unpublished observations). Helper virus is unnecessary for either penetration or uncoating, and it is at this point in the infection cycle, after the AAV genome has entered the nucleus, that the fate of the AAV genome can vary.

In the presence of a helper virus the DNA may be copied, transcribed, its mRNA translated, and infectious virions assembled. In this same system, normal DNA replication may begin but lead to the generation of defective genomes. Several classes of such abortive DNA may also be packaged. If no helper virus is present, AAV DNA does not replicate. However, the AAV genome may establish a stable relationship with the host by integrating into its chromosomes where replication proceeds along with that of chromosomal DNA (see Chaper 2). Integrated viral DNA may then be rescued and replicated if the latently infected cell is subsequently coinfected with a normal helper virus (Hoggan *et al.*, 1972; Berns *et al.*, 1975b; Cheung *et al.*, 1980; Handa *et al.*, 1977a). Finally, recent data suggest that the early steps in DNA replication (synthesis of the complementary strand) may occur in the absence of helper functions

(Bohenzky *et al.*, unpublished data) and this may be related to the integration process. These fates for the AAV genome represent essentially the full spectrum of replicative processes for most linear DNA-containing viruses. Each process will now be examined separately.

A. Productive DNA Replication

Locating the origin of DNA synthesis is often pivotal to an understanding of the overall mechanism of replication. By determining the gradient of label introduced in a pulse shorter than the time required for one round of replication in mature length progeny molecules it is possible to map both the origin and termination of DNA replication for simple viral genomes (Danna and Nathans, 1972). To this end Hauswirth and Berns (1977, 1978) analyzed the genome pattern of pulsed radiolabel incorporation into mature AAV DNA by measuring the activity of each restriction fragment in unit length molecules. Because the most label will reside in that portion of the genome replicated last, it was possible to conclude that AAV DNA replication terminated at or very near each molecular end. However, it was not possible from these data to decide between (1) a undirectional mechanism in which replication started at each molecular end and proceeded by strand displacement to the other end or (2) bidirectional replication starting at an internal genome position. To distinguish between these alternatives, strands of most *Hpa*II restriction fragments from the above pulse label experiment were separated and, on the basis of knowing the polarity of each strand of every fragment, radiolabel incorporation profiles for each complementary strand were constructed. Both resulting profiles were unimodal, with the most label at each 3′ end. This result is not consistent with bidirectional replication in which single strand gradients would remain bimodal. Thus a unidirectional replication mechanism starting at or near the 3′ end of the parental strand was indicated.

One mechanistic problem for a replicating linear DNA genome is how to produce progeny strands without losing 5′ terminal sequences during each round of replication. In the context of the Cavalier-Smith (1974) model the palindromic structure of AAV DNA provides a way to circumvent this problem. It was first necessary to prove that AAV DNA actually utilizes the hairpinning property of the termini during replication. Straus *et al.* (1976b) and Handa *et al.* (1976), using temperature-sensitive, replication-defective Ad as a helper, showed that replicating pools of AAV DNA contained dimer length concatemeric single strands. Because pulsed radiolabel could be chased through this form to unit length single strands in several hours Straus *et al.* (1976b) concluded that these molecules were true replicative intermediates. Such intermediates were also shown to be composed of strands of opposite polarity because they would self-anneal creating nearly unit length duplex molecules. A hairpinned sequence on one end of each molecule was deduced because digestion with single strand-specific DNase converted all these molecules to unit length on alkaline sucrose gradients. Indeed, Straus *et al.* (1976b) concluded that the Cavalier-Smith model could best explain their data. Hauswirth and Berns (1979) repeated and extended this study using wild-type Ad type 2 as helper. About 10% of the labeled AAV DNA sedimented at the duplex dimer position after a 5–10 min pulse. Of these molecules about one-third sedimented in alkali as dimer length single strands, as noted earlier. The remainder were presumed to be unit length duplex molecules held together by hydrogen-bonded termini as seen earlier by Gerry *et al.* (1973). The genome gradient of pulse label was identical to that of unit length molecules, and this label could be chased into shorter DNA. The large preexisting pool of labeled unit length molecules precluded a firm conclusion that dimeric concatemers were true precursors to unit length strands. However, other data strongly favor this view. All internal restriction fragments of dimeric concatemers were found to exist in the identical molarity as in virion DNA. However, some differences in terminal fragments could be seen, although no evidence for tandemly repeated terminal sequences was found. The right terminal *Taq*I fragment was found to exist in two forms. In addition to the normal form, a form 60 base pairs shorter than normal was found. Restriction analysis, S1 nuclease digestion, and hydroxylapatite binding showed that this shortened terminal fragment was created by hairpinning the terminal palindromic sequences. Straus *et al.* (1978) reported shortened restriction fragments from both ends of replicating molecules which also self-annealed as if hairpinned. Thus both molecular ends of the AAV genome appeared to be hairpinned at some point during replication. This suggested a functional role for the AAV terminal palindromic sequences. In summary, these studies located the replication origin, the polarity of replication, and the major replicative intermediate. These data led directly to a replication model (see below). However, one problem remained: a large fraction of a pulse-labeled precursor was incorporated into neither unit nor dimer length AAV DNA. The relevance of these forms to AAV replication will be dealt with in Section II B and Chapter 9.

Relative to the mechanism of DNA replication, it is important to know whether the orientation of the terminal repetition at one molecular end defines the orientation at the other end in the same molecule. Lusby *et al.* (1981) have answered this question by separating complementary strands of whole AAV DNA and allowing each polarity strand to self-anneal, forming hydrogen-bond-stabilized single-strand circles. By then digesting with restriction enzymes which make their terminal cuts between nucleotides 42–84, it was possible to distinguish between fragments arising from normal double helical structures and those arising from an aberrant double helical structure. It was found that approximately equal fractions of terminal fragments had fully duplex and alternative duplex structures. This suggested that equal proportions of individual strands have terminal repetitions of either the same polarity or opposite polarity. Therefore the orientation of one terminal repetition does not define the orientation of the other on the same molecule.

B. Abortive DNA Replication

The products of abortive DNA replication and their biological consequences are the primary subject of Chapter 9. They will be briefly discussed here insofar as their structure and function has led to a better understanding of the sequence requirements for AAV replication. The relicating pool of AAV DNA in infected cells contains short DNA as well as unit and dimeric length AAV DNA molecules. About 35% of a 5–10-min pulse label was incorporated into such short DNA molecules (Straus *et al.*, 1976b; Hauswirth and Berns, 1977; Hauswirth and Berns, 1978). Sedimentation on neutral gradients showed they ranged from nearly full length to about 5% AAV genome length. The modal size was about one-third unit length, and by hybridization experiments the majority were complementary to AAV DNA and about half were single stranded (Hauswirth and Berns, 1979). All duplex DNA less than unit length is AAV DNA by restriction fragment analysis and all sequences represented mapped to within 25% of either molecular end. Restriction fragments near each molecular end predominated (Hauswirth and Berns, 1978; 1979). Further analysis suggested that one end of each molecule contained the normal complement of terminal restriction fragments (a normal AAV DNA end) while the other end was hairpinned at an internal genome site. Pulse-chase experiments showed that these molecules were relatively stable products of replication and were not elongated to unit length or dimeric molecules. Hence AAV DNA replication may lead to abortive products with a very specific set of structural properties. Assuming that they are not the products of recombination, it should be readily possible to generate them from any model for AAV DNA replication because they constitute such a considerable fraction of newly replicated DNA.

Abortive replication does not lead to fully functional genomes, but they appear to contain all sequence determinants required for efficient encapsidation. As such they may play a role in several biological effects of AAV. Light density (1.32 to 1.41 g/cm^3) AAV virions containing various amounts of DNA have been reported (Torikai *et al.*, 1970; Hoggan, 1971; Johnson *et al.*, 1975; de la Maza and Carter, 1980). Recently these particles have been characterized (de la Maza and Carter, 1980) and were found to contain the normal complement of protein, but the AAV DNA ranged from 3% to full genome length. All molecules contained apparently normal termini and most contained both normal termini with internal deletions of varying size. This was not the major variant form deduced earlier for the intracellular products of abortive replication (Hauswirth and Berns, 1979). However, we note that about 50% of the intracellular DNA which incorporated label during a brief pulse and was shorter than unit length was single stranded. No further analysis was made on these molecules except to note that most hybridized to AAV DNA sequences. One could hypothesize that encapsidated products of abortive replication came from this class of molecules. It is interesting to note that an essentially identical DNA structure was deduced for the DNA in one class of defective MVM virions (Faust and Ward, 1979). Another class of MVM particles contained hairpinned duplex molecules identical to the defective duplex AAV molecules described above. It is uncertain why autonomous and defective parvoviruses encapsidate specific classes of defective genomes, but template differences particularly at the genome ends may play a role. AAV may require both terminal sequences for encapsidation because strands are symmetrically copied while MVM may require only its 5′ sequence. If so the AAV encapsidation process would select only those molecules with the two normal terminal sequences, regardless of the completeness of the internal sequences. Tattersall and Ward (1980) have suggested that strands of parvovirus DNA may be encapsidated as they are displaced from the duplex replicative intermediate. This suggests that aberrant AAV DNA molecules hairpinned at internal genome positions may not present the appropriately exposed progeny 5′ end to begin encapsidation. Regardless of the exact details it is clear from these studies that if DNA sequence recognition occurs during virion assembly, it must reside within a few percent from either molecular end, probably within the terminal repetition itself.

AAV virions containing defective genomes have been implicated in a number of biological interference processes. They are extensively reviewed in Chapter 9 and are of interest here because their effects appear to be exerted mainly at the level of DNA replication. Light density particles (1.35 and 1.32 g/cm^3) containing internally deleted genomes are not infectious but will inhibit normal AAV (1.41 g/cm^3) replication (Laughlin *et al.*, 1979). Interestingly, although the yield of both infectious particles and unit length DNA were reduced in the presence of either class of light particle, replication of helper adenovirus was unaffected. This suggests that interference affects a step in AAV replication which is distinct from any step in Ad replication. Inhibition could be reduced by preirradiation with UV light, suggesting that the variant DNA, and not capsid protein, was the inhibitory agent. The simplest explanation is competition by defective AAV DNA terminal sequences for a limiting replication factor because only terminal genome regions are common to all light density particles. Presumably, defective genomes inhibit by competing for a viral or host protein involved in initiation of replication. If this view is correct, then aberrant genomes can undergo at least one early step in their replication or at a minimum bind to one initiation factor. But aberrant genomes do not appear to contain all determinants for replication because they are only synthesized when intact genomes are also present (Laughlin *et al.*, 1979). This may suggest that an AAV gene product is required for AAV DNA replication. However, the question of whether aberrant DNA can be replicated must remain unanswered because all such DNA could result from defective replication of normal genomes and this data may simply mean that defective genomes lack a *cis*-function required for replication. Other studies, however (see below), suggest that in fact an AAV protein(s) may be required, which leaves open the possibility of aberrant AAV DNA replication.

C. Replication in the Absence of Helper Functions

It is unclear whether any steps in the synthesis of AAV DNA take place without helper virus coded or induced proteins. The possibility that some early events could take place in the absence of helper virus is not entirely unreasonable because several key factors involved in initiation of replication are present, including the host DNA polymerase and all nucleotide sequence requirements. If an AAV coded protein serves a replicative function, it must either be a capsid protein or an undetected early AAV gene product. To test for replication without helper, Bohenzky and Berns (unpublished) isolated ^{32}P-labeled AAV virions containing DNA substituted with BUdR. After infection of KB cells with these virions in the absence of helper virus, hybrid density AAV DNA was recovered. The shift from heavy to hybrid density suggested one round of replication. This was not due to repair synthesis because the hybrid density material was duplex DNA whose strands were covalently linked at one end or the other, the expected initial product of the replication process. Also, if repair synthesis had occurred it should have produced a continuous density range of DNA, not predominately hybrid density material as observed. It is clear that these molecules strongly resembled normal AAV replicative intermediates, but more characterization is necessary. Because no light density progeny strands could be detected in these experiments, it appears that AAV DNA strands may serve as primer-templates for conversion to duplex replicative intermediates without helper virus, but additional replicative steps leading to progeny strands cannot ensue. The nature of the block for progeny strand synthesis is unknown but it must be related to helper virus functions, or host functions induced or altered by helper virus coinfection.

Without helper virus, AAV can establish a latent infection in human cell lines. The details of this process are reviewed in Chapter 2 (see also Berns and Hauswirth, 1979) and it suffices here to state that 3–6 copies of AAV DNA were found covalently associated with the chromosomal DNA in cloned latently infected cells. There is no clear evidence that replicative events were required for this integration event. However, it seems likely that replication, either prior to or in the course of integration, would be required to insert a single-stranded DNA into a double-stranded

DNA. Less likely would be annealing of plus and minus strands from different virons prior to integration.

The requirements for AAV DNA replication could be more precisely defined if a cell-free system for synthesizing AAV specific DNA were available. Nuclei isolated from AAV type I infected human embryonic kidney (HEK) cells coinfected with the replication-defective, temperature-sensitive mutant of adenovirus type 31 (A31*ts*13) at the nonpermissive temperature, were able to synthesize AAV specific DNA (Handa and Shimojo, 1977b). AAV products were sized on alkaline sucrose gradients and shown to vary from short DNA segments to DNA longer than unit length. Subsequently, a nuclei-free 23S complex from H5*ts*125 infected cells was isolated and the AAV-1 products partially characterized. AAV-1 DNA sedimented on neutral gradients as unit length but on alkaline gradients all material was only 8–10 S (Handa and Shimojo, 1977a). The possibility of repair synthesis was not ruled out in either study nor were hairpinned AAV termini searched for. Nevertheless, these studies do represent efforts at developing an *in vitro* replicating system for AAV and deserve continued development.

D. Replication of Cloned AAV DNA

The intact AAV genome has been inserted into the bacterial plasmid pBR322 and cloned in *Escherichia coli*. Upon transfection of the recombinant plasmid into human cells in conjunction with adenovirus coinfection, the AAV genome is excised from the plasmid, replicated, and infectious virions produced (Samulski *et al.*, 1982). The development of this system opens the way to detailed analysis of the effects of specific modifications of the genome on the process of AAV DNA replication. Already the system has provided support for the current model of replication and has revealed an additional unsuspected property of AAV DNA replication that is undoubtedly a consequence of the inverted terminal repetition.

Cloning of the AAV genome was accomplished by inserting the linear duplex form of the DNA which had been tailed at the 3′ termini with dC into the *Pst*I site of pBR322 which had been tailed at the 3′ termini with dG. Covalent sealing of the annealed AAV-pBR322 recombinant molecules occurred *in vivo* after transformation of *E. coli*. For several reasons there was a low yield of recombinant clones containing the intact AAV genome. Two major problems reflect the special structure of the inverted repetition.

(1) Because there are two terminal sequences as a consequence of the inversion which occurs during replication, only $\frac{1}{4}$ of the duplex AAV DNA molecules formed during *in vitro* annealing have perfect base pairing at both ends (i.e., form a Watson–Crick-type double helix at the ter-

mini). Thus repair processes which take place after transformation lead to extensive deletions in regions of imperfect base pairing.

(2) Even after covalent linkage of the AAV sequences to pBR322 there is a tendency for deletions to occur near the termini of the AAV insert. This reflects the palindromic nature of the terminal repetition. There is evidence from other systems that palindromic sequences are preferentially deleted from circular plasmids (Lilley, 1981). Such sequences are thought to be deleterious because of topological constraints imposed on the replication of covalently closed circular DNAs. While these constraints made the cloning of the AAV genome in an intact form laborious, the existence of clones with varying deletions in the terminal repetition has proven useful for beginning the analysis of the structural requirements for AAV DNA replication.

The initial result obtained from the replication of the cloned AAV DNA was confirmation that inversion of the terminal palindromic sequences occurs during AAV DNA replication. Clearly in a clone there can be only one sequence orientation at either end of the AAV sequence. In the case of the clone studied both terminal sequences were in the same orientation as determined by restriction enzyme analysis. Analysis of DNA isolated from virions produced after rescue of the cloned genome revealed that now both orientations were present at both ends of the DNA. Thus the existence of the two sequences in the inverted terminal repetition is a result of the replication process and not a consequence of the existence of two or more mutually independent populations of AAV DNA molecules.

As stated above most of the recombinant clones characterized contained deletions at one or both ends of the AAV insert. Most of the deletions were confined to the terminal 125 nucleotide stretch comprising the palindromic sequence. Surprisingly, AAV genomes from several of the clones with a significant deletion of the palindromic sequence at only one end of the AAV sequence were able to be rescued, replicated, and packaged. Detailed analysis of one of these clones revealed that there was a deletion of 111 nucleotides from the left end and 7 nucleotides from the right terminus. However, similar analysis of the DNA contained within the rescued virions showed that the deletions had been repaired during the rescue process so that the terminal sequences were now intact and indistinguishable from wild-type DNA (Samulski *et al.*, 1983). The mechanism by which repair takes place is not known but is presumed to be dependent upon the existence of the inverted terminal repetition. Experimental support for this assumption is cited below. It seems likely that the 7 base deletion at the right end could be repaired by hairpin formation at the initiation of DNA synthesis. The first 7 bases inserted would effect repair of this lesion. A mechanism for repair of the more extensive deletion at the left end is more problematic. One possibility could involve the proposed single-stranded circular structural intermediate (form IV) depicted in Fig. 2, which is stabilized by base pairing

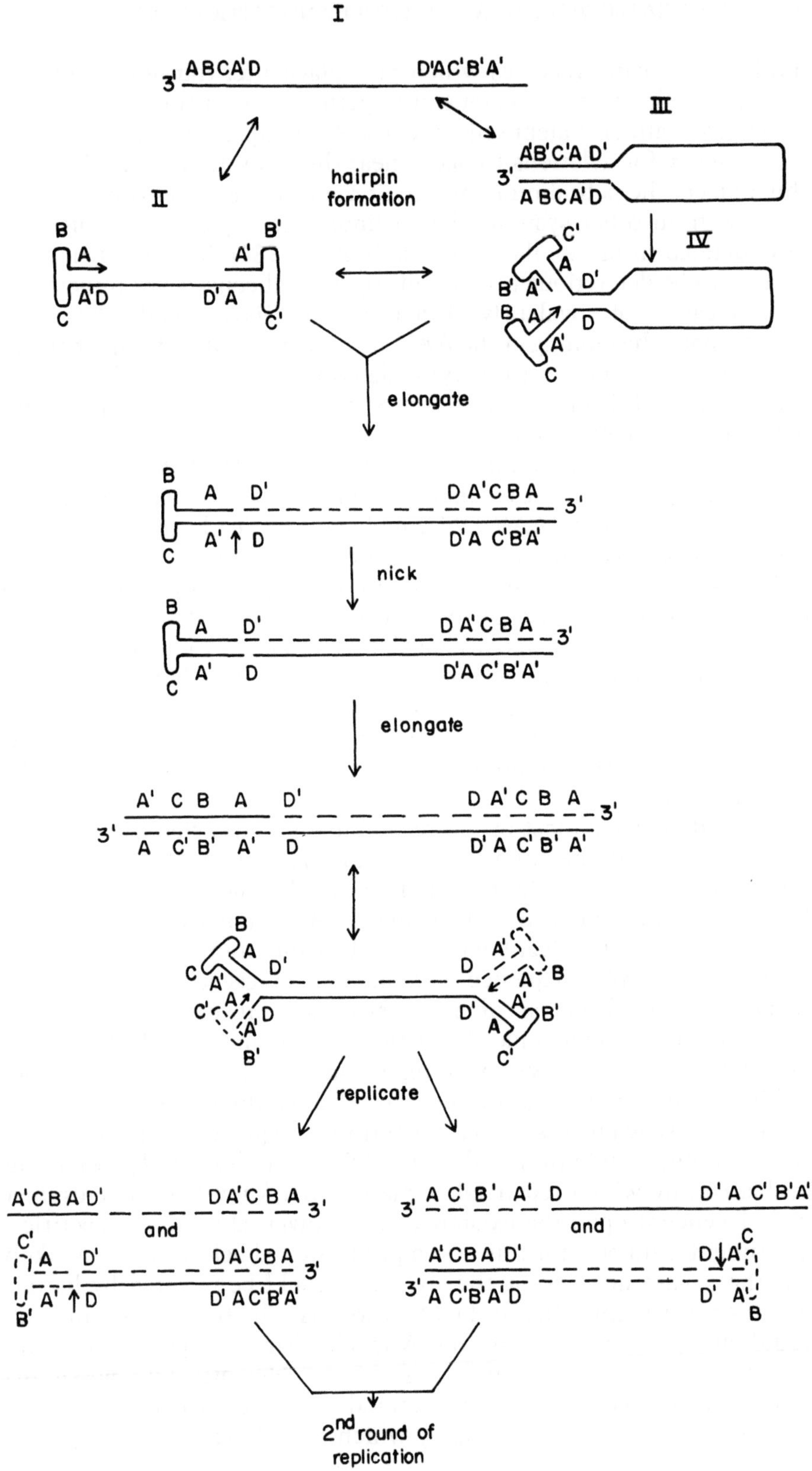
I
3' ABCA'D D'AC'B'A'
II
hairpin
formation
III
3' A'B'C'A D'
ABCA'D
IV
elongate
nick
elongate
replicate
and
and
2nd round of
replication

between the inverted terminal repeats. The 32 bases remaining in the left terminal repeat would be enough to stabilize the structure and if they were at the 3′ end of the polynucleotide chain could act as a primer for repair synthesis of the terminal gap, thus filling in the deleted sequence and leading to an intact AAV DNA molecule which could then go on to replicate normally. Similar rescues of terminal deletions have also been observed in the case of Ad and have also been attributed to the existence of an inverted terminal repetition (Stowe, personal communication).

The ability of the AAV DNA replication process to rescue deletions occurring at one end of the genome means that in order to study the effects of a modification of the sequence within the inverted terminal repetition, the same sequence must be modified at both ends of the genome and may, even then, have to occur beyond base 41 so that it cannot be repaired by simple hairpin formation and repair synthesis. Study on one such mutant has already begun. The deleted clone described above has 111 bases deleted from the left end. *Sma*I cleaves AAV DNA only within the inverted repeat at two closely spaced sites 11 bases apart. These sites are missing from the left end of this clone. The clone was cleaved with *Sma*I and the resulting long linear fragment which was missing the 11 base pairs between the two *Sma*I sites was circularized and recloned. As a result the 11 base sequence in the middle of the terminal repeat was now absent from both ends of the cloned AAV DNA. When this recombinant plasmid was transfected into human cells the AAV DNA was rescued and replicated. However, there was an almost complete inhibition of the production of unit length monomers equivalent in length to mature AAV virion DNA. Thus, the deletion did not appear to inhibit AAV DNA synthesis *per se* but did seem to have a profound effect on the processing and/or packaging of the replicative intermediates. As anticipated there was also a profound decrease in the production of intact virions.

The final observation to date using cloned AAV DNA is that those clones with significant deletions in the terminal repetition at both ends are not active in terms of DNA replication, supporting the notion expressed above that rescue of a deletion at one terminus is dependent upon the existence of an intact sequence at the other end of the genome.

Specific deletions within the internal unique sequences have also been created. Of interest is a mutant with the sequence between 50 and 90 map units deleted. This mutant is not rescued even through the termini are intact. These data would be in accord with the idea that an AAV gene product is required in terms of complete AAV DNA replication, and

←

FIGURE 2. Model for AAV DNA replication. The terminal sequence domains represented by primed letters are the complements of those represented by unprimed letters. Details of the sequence domains and replicative steps are discussed in the text.

that one possible function that such a product could subserve would be to cleave the terminal hairpin structure.

This system promises to be extremely useful in the elucidation of structural and functional requirements for AAV DNA replication.

III. A MODEL FOR AAV DNA REPLICATION

To replicate linear DNA without circular or concatemeric intermediates requires a precise set of structural and functional properties as detailed by Cavalier-Smith (1974). Replication of AAV DNA should serve as a model of how a small, linear viral DNA manages this and may, more generally, apply to the ends of any duplex DNA. In considering the details of AAV DNA replication it is striking that the properties demanded by the Cavalier-Smith model correlate so well with the experimental features found for AAV DNA. Thus, our model for replication is essentially an adaptation of the original proposal of Cavalier-Smith and corresponds to that suggested by Straus *et al.* (1976) upon finding hairpinned intermediates.

A scheme for AAV DNA replication is shown in Fig. 2. Each step of the model will be detailed as it applies to AAV. We will make particular note of experimentally observed features as well as those processes yet to be demonstrated. The 145 base terminal repeats, in terms of sequence domains, are abbreviated so that A corresponds to the terminal 41 base sequence, B and C represent bases 42–62 and bases 64–84, respectively, each composed of a palindromic sequence, A′ represents bases 85–125 which are the inverted complement of A, and D represents bases 126–145, the internal portion of the inverted terminal repeat. The inverted complement of any sequence domain is represented by a primed letter.

It is possible to begin with any combination of terminal sequences on an AAV DNA strand. We will arbitrarily start with an AAV DNA strand whose terminal sequences are the inverted complements of each other and will demonstrate how the terminal sequence at the 3′ end of one strand can be inverted independently of the 5′ sequence on that strand during replication. The identical replication mechanism still pertains whether or not the starting genome contains terminal sequences which are perfect inverted complements (e.g., 3′ABCA′D–D′AC′B′A′). This is a necessary property of the model because both types of terminal sequence arrangements are known to exist in the AAV genome with about equal probability (Lusby *et al.*, 1981) and both types can be generated after rescue and replication of AAV containing one type of terminal sequence (Samulski *et al.*, 1982).

The first step in AAV DNA replication is formation of an appropriate template primer. Palindromic terminal sequence arrangements in all intact AAV strands will spontaneously form 3′ and 5′ terminal hairpinned structures. The template primer for the first round of replication may be either a simple 3′ hairpin or the 3′-hairpinned form of the single-stranded circular structure (forms III and IV). Several considerations lead us to favor structure IV. First, the terminal structure of this molecule is formally identical to that of the duplex RF (see below). Therefore, protein-DNA interactions required for recognizing the origin of replication should be identical whether or not a replication complex is using an infecting single-stranded AAV genome or a duplex RF. Second, it will be shown that defective genomes are most directly synthesized if template ends are close together. Third, this structure, a single-stranded circle held together by duplex ends, has been directly observed in the electron microscope (Berns and Kelly, 1974). Admittedly, within the limits of resolution, alternative hydrogen-bonded structures cannot be *a priori* disregarded. The fraction of AAV genomes which will spontaneously anneal to form either structure III or IV *in vivo* remain undetermined. In considering the overall thermodynamic stability, regions D′ and D are hydrogen bonded in structures III and IV and they are not in structure II. Hence III and IV are formally more stable by the free energy of these extra hydrogen bonds than structure II in the absence of external factors such as DNA binding proteins. Structures III and IV are almost equally probable on thermodynamic grounds. On a kinetic basis, however, each palindromic terminal sequence in I will more rapidly anneal to itself to create structure II rather than anneal to the other termini as in structure III. Structure II, once formed, should spontaneously convert to IV. For these reasons, we favor structure IV as the most likely primer-template to initiate replication. By these arguments the sequence of annealing would be I, II, IV. Clearly other factors such as the ionic strength and protein–DNA interaction within the viral capsid and after uncoating in the infected cell nucleus may well define the final equilibrium between alternative hairpinned structures.

After formation of a 3′ hairpinned terminus self-primed DNA synthesis may begin. Analogous to autonomous parvovirus DNA (Bourguignon *et al.*, 1976) bacterial polymerases will synthesize *in vitro* a complementary strand using the AAV DNA as a template-primer (Riva, Hauswirth, and Berns, unpublished). Therefore, hairpinned strands must display the required 3′-hydroxyl group and primer-template geometry. Elongation of the template strand from its 3′ end to create a hairpinned duplex molecule has been shown using pulsed label analysis (Hauswirth and Berns, 1979). These molecules are found in replicating pools and are shown to be made by initiating and terminating DNA synthesis at their ends.

Completion of a progeny strand requires creation of a single strand cut exactly opposite the original 3′ parental terminus and the re-elongation of the parental strand at this point. The net result is displacement and inversion of the original parental 3′ terminal sequence. This effects a transfer of the parental 3′ terminus to the 5′ end of the progeny strand. An enzymatic activity which can create the requisite nick prior to parental strand elongation has not yet been documented. Similar DNA nicking activities can be demonstrated in uninfected human cell lines (Buller and Rose, 1978). However, the site-specific nature of eukaryotic DNA nicking is undocumented and it remains a pivotal feature of our model for parvovirus DNA replication. Finally, elongation of the parental strand from the 3′—OH at the nicked site can, in theory, be carried out by the normal repair functions of the host cell.

The infecting strand has now been converted to a duplex RF, but no free progeny strands have yet been created. Replicative forms having both fully duplex and the nicked hairpin terminal structures have also been seen in pulse label studies (Hauswirth and Berns, 1979) and it was noted that these molecules were the precursors to progeny strands. About two-thirds of unit length duplex material was not covalently hairpinned, and thus was presumably nicked RF or fully duplex DNA. This is consistent with the finding that both normal and hairpinned terminal restriction fragments are detectable from this pool. Thus, only about one-third of the replicative intermediates are nicked or intact hairpinned duplex structures while the majority have been elongated to full duplex copies of the AAV genome.

The next step is production of progeny strands. This requires the identical priming event that intiated replication of the infecting viral strand. However, in contrast to hairpin formation of a single strand, the ends of the replicative intermediate are now base paired to their complementary strand and must separate before hairpinning. There is no obvious way for this to happen spontaneously and we therefore postulate the intervention of specific DNA–protein interactions to melt out the duplex ends. A number of single strand-specific DNA binding proteins are known to lower melting temperatures of duplex DNA particularly at their ends. This would facilitate the melting of the AAV DNA RF. Once melted out and self-annealed, the double hairpin ends of the RF are identical to structure IV in their primer-template properties. Thus, the analogous elongation at one or the other 3′ end of each such duplex intermediate leads to direct displacement of completed progeny strands.

The overall effect of one round of replication on the terminal sequence according to this model is (1) replacement of the parental 3′ palindromic sequence (ABCA′) by its inverted complement (AC′B′A′) while leaving the 5′ sequence intact, and (2) inversion of the parental 3′ palindromic sequence and transfer to the 5′ end of the progeny strand. It is evident that if the parental strand now in the RF were to undergo another round of replication, its 3′ end would again invert and the original 3′ sequence would be regenerated. In summary, through the first two replication cycles, the 3′ terminal sequence on the parental strand would be 3′-ABCA′ then 3′AC′B′A′ and then again 3′-ABCA′ while all 5′ terminal sequences would remain 5′-A′B′C′A. For the first progeny strand the cycle would proceed 3′-ABCA′ then 3′-AC′B′A′ and then again 3′-ABCA′. However, it should be noted that the 5′-terminal sequence is now 5′-A′CBA for all progeny strands. Therefore, this model generates two classes of 3′ and two classes of 5′ terminal sequences regardless of the original parental sequence arrangement. It is also evident that a total of four types of strands will be generated. There will be two of each coding polarity and, for each coding polarity pair, one arrangement of terminal sequences can base pair perfectly with themselves, e.g., 3′-AC′B′A′–ABCA′ and the other cannot, e.g., 3′AC′B′A′–AC′B′A′. Thus, consistent with all experimental evidence, two populations of mutually inverted terminal sequences at each genome end are generated by this model.

The initiation of DNA synthesis in the replicative intermediate should not have a strand preference because the sequence surrounding this point is independent of strand polarity. Therefore, all four strand types should be produced in approximately equal numbers, as found by Lusby *et al.* (1981), who demonstrated that equivalent numbers of AAV strands can form perfect and imperfect terminal duplex structures to stabilize single-strand circles.

Encapsidation of progeny strands is formally the last step in replication. There is essentially nothing known about this process, but, as discussed above, all sequence determinants for encapsidation must reside in or near the terminal repetitions because deleted AAV DNA molecules lacking almost all sequences except the terminal repetitions appear to be efficiently packaged. AAV progeny strands may be encapsidated like autonomous parvovirus DNA, which Tattersall and Ward (1976) have suggested is packaged during displacement from the duplex RF. However, at least some progeny strand AAV DNA must escape this sort of immediate packaging to account for production of one class of defective genomes (see below).

In summary the model postulates a undirectional, displacement synthesis of progeny strands from a duplex RF. In this sense it is similar to that proposed for the autonomous parvoviruses (Chapter 4). Unlike the autonomous parvoviruses, however, the symmetry of terminal AAV sequences leads naturally to equal production of both polarity strands rather than preferential synthesis of one polarity strand. The self-priming property and synthesis of exclusively leading strands is consistent with the lack of evidence for RNA priming or discontinuous synthesis.

The genome products of abortive replication by AAV fall into four structural classes: (1) linear single-stranded molecules containing functional terminal regions and internal sequence deletions, (2) duplex versions of structure [1], (3) linear duplex molecules with one fully duplex terminal region and more internal genome sequences linked by a hairpin at the internal "turn-around" point, and (4) structures similar to structure [3] but containing imperfectly base paired terminal regions. The essence of several models previously proposed for generating defective AAV genomes is that either a template strand switch or jump occurring during otherwise normal replication will produce these classes of aberrant replication products. A complete discussion of the replicative process leading to defective genomes is presented in Chapter 9. In general defective genomes made in this way necessitate no extra factors because all experimentally documented defective genomes arise from normally replicating molecules without postulating new structures.

IV. HOST AND VIRAL PROTEINS REQUIRED FOR AAV DNA REPLICATION

To fully understand AAV DNA replication it is necessary to define viral and helper functions involved in the replication of AAV DNA. Although this issue is the central focus of Chapter 8, a brief summary of the situation as it relates to AAV, helper virus, and host cell function is appropriate here. The requirement for helper virus coinfection complicates the picture because three general classes of proteins may be potentially involved: host proteins, helper virus proteins, and AAV proteins. Additionally, neither eukaryotic cellular nor helper virus DNA replication are well understood in terms of specific functional proteins. Thus progress in understanding AAV replication is necessarily tied to progress in cell and animal virus replication in general. A final problem is that *in vitro* analysis of host or viral gene products may not accurately depict the *in vivo* situation because of viral modification of host cell activities. The general problem of viral DNA replication is obviously not trivial at the level of understanding specific protein function, but the simple structure and small size of the AAV genome make it an attractive model system. Significantly, most, if not all, nucleotide sequence determinants for DNA replication reside within the terminal repetitions of AAV DNA, which will make analysis of AAV mutants particularly interesting. We will begin by reviewing data pertaining to host functions, then helper virus functions, and finally to potential roles for AAV proteins themselves.

AAV has little, if any, host range limitation *in vitro* relative to its species of origin. That is AAV will infect across species lines (at least in cell culture) provided a suitably infectious helper virus is available. For example, human AAV will replicate productively in canine cells if canine adenovirus is coinfected as helper (Casto *et al.*, 1967). Not all host–helper–AAV combinations have been tested, but the implication is clear that the host range of AAV is probably defined not by the host cell but by the existence of an infectious homologous helper adenovirus. Thus the ubiquity of adenoviruses defines the host range of AAV. This suggests that if host cell gene products are required directly for AAV replication they are common components of many species of animal cells. AAV replication may utilize host DNA polymerase, although no presently available reports clearly implicate a specific host polymerase in AAV DNA replication. Other host cell proteins, such as DNA nucleases and binding proteins are also reasonable candidates for host cell helper functions.

The most effective hosts for helper virus infection are actively replicating cells. Thus they are also best for AAV. To test the dependency of AAV replication on host cell status it would be interesting to know how effectively AAV can establish a latent infection (i.e., cryptic infection in the absence of a helper virus) in resting cells relative to rapidly dividing cells. This may be a feasible experiment because the possibility exists that some limited replication of the AAV genome may occur in the absence of helper virus. Hence the metabolic activity of a cell may well be critical for this sort of helper-independent AAV replication. Under normal circumstances, however, helper virus gene expression is an absolute requirement for productive AAV infection. This fact was first demonstrated by the delay always observed between the appearance of early helper virus proteins and AAV DNA synthesis (Rose and Koczot, 1971; Blacklow *et al.*, 1967; Carter *et al.*, 1973). If AAV and adenovirus are coinfected there is a 3–4 hr delay in AAV DNA synthesis relative to the situation where AAV infects cells actively producing early adenovirus proteins. Thus there is a rather immediate requirement for early adenovirus protein(s) for AAV DNA synthesis.

The precise helper virus functions utilized during AAV replication remain unknown. This is primarily because of our imprecise knowledge of the function of adenovirus proteins. Nevertheless, the increased availability of mutant adenoviruses whose protein defects are understood at the molecular level coupled with modern molecular techniques has provided a number of insights. A reasonable starting assumption is that adenovirus gene products necessary for AAV DNA replication might serve the analogous replicative function for adenovirus DNA. Thus, some adenovirus mutants defective in their own replication might be expected to also be defective in helping AAV DNA replication. Unfortunately all such early temperature-sensitive mutants (*ts*) of adenovirus yield measurable AAV DNA synthesis upon coinfection. In this respect the most completely analyzed system has been the human adenoviruses. Thus far, all temperature-sensitive mutants tested will support at least detectable AAV DNA replication at the nonpermissive temperature. These include early mutants unable to replicate their own DNA (H5*ts*125 and H5*ts*149) as well as late mutants defective in virus particle maturation (H5*ts*116, H5*ts*135, H5*ts*142, H5*ts*147) (Handa *et al.*, 1975; Straus *et al.*, 1976; Mayor and Young, 1978; Ostrove and Berns, 1980). Recent work suggests that only early adenovirus functions may be required for AAV replication. Richardson *et al.* (1980) showed that microinjection of Ad2 early mRNA into Vero cells supported complete AAV replication. In this system no late adenovirus proteins were detectable. Thus early functions alone appear sufficient for complete AAV replication.

Ostrove and Berns (1980) have tested several adenovirus host range (*hr*) mutants for their ability both to help AAV DNA replication and to rescue AAV from latently infected cells. Host range mutants will normally replicate only in adenovirus-transformed cells which contain all or part of the T-antigen region (early region 1) to complement the viral defect. Adenovirus *hr*3, defective in its own DNA synthesis, was unable to support AAV synthesis except in complementing adenovirus-transformed cells. The defect maps to adenovirus early region 1a (E1a, 0 to 4.4 map units) (Frost and Williams, 1978; Berk *et al.*, 1979). The *hr*3 protein defect is not well characterized because all proteins known to be coded in this region are present in *hr*3-infected cells although in reduced amounts (Lassam *et al.*, 1978, 1979). A second adenovirus mutant, *hr*6 (defect mapping between 4.4 and 9.0 units, E1b), supports both its own and AAV DNA replication without complementation. Interestingly, this mutant will not support rescue and replication of AAV DNA from latently infected cells. Thus a second adenovirus function, one not required for helping replicate free AAV DNA, is required for the rescue of the AAV genome from the latent state. The defect in *hr*6 appears to be due to a nonfunctional 58K T-antigen (Lassam *et al.*, 1979; Graham *et al.*, 1978). An analogous situation appears to exists for the excision of integrated polyoma virus DNA which requires a functional polyoma A gene product (Basilico *et al.*, 1979). Based on their results with *hr*6, Ostrove and Berns (1980) suggested that a similar adenovirus-coded protein requirement may exist for excision of integrated AAV DNA. If so it is curious that an Ad gene product would provide such a specific function in AAV latency and not, as far as is known, provide the analogous function for itself.

Other studies using mutants of adenovirus types 2 and 5 have shown that expression of multiple early regions are necessary for AAV replication. Myers *et al.* (1980) showed that H5*ts*125 is a very inefficient helper of AAV at the nonpermissive temperature. The defect in this mutant maps to a gene for a 75K DNA binding protein required for elongation of nascent DNA (Ginsberg *et al.*, 1976; Frost and Williams, 1978; Van der Vliet, 1975). Its exact role relative to AAV DNA replication remains uncertain because it appears to also affect splicing of both adenovirus (Klessig and Chow, 1980) and AAV mRNAs (Myers *et al.*, 1980). Specifically, although normal amounts of a precursor to 2.3-kb AAV RNA accumulate, smaller spliced mRNAs do not. It was subsequently shown that this 2.3-kb RNA codes for all three AAV capsid proteins by *in vitro* translation. This suggested that the defect on Ad5*ts*125 affected a gene product involved in processing or translating AAV mRNA. Interestingly, in a cell line of monkey origin (AGMK) in which only early adenovirus protein synthesis occurs, including the 75K protein,' no infectious AAV is detectable but replicative intermediates accumulate (Buller *et al.*, 1979). It was suggested that these results and the fact that progeny AAV DNA is reduced 5- to 50-fold with H5*ts*125 in permissive human cells (Myers *et al.*, 1980) are most consistent with a role for an AAV protein in progeny strand synthesis from the RF. If true, the temperature-sensitive 72K protein of H5*ts*125 would exert its effect on AAV replication at the level of mRNA processing.

Janik *et al.* (1981) further defined the adenovirus genome regions necessary for helping AAV replication by transfecting restriction fragments of adenovirus types 2 and 5 into an adenovirus type 5 transformed cell line expressing only early regions 1a and 1b. They found that three fragments containing VAI RNA, E2 and E4 needed to be simultaneously transfected to observe AAV. These studies suggest that multiple early regions of adenovirus are required for the helper effect. However, the relationship between the proteins encoded by these regions in affecting Ad and ultimately AAV gene expression remained unresolved.

Recently Richardson and Westphal (1981) have begun to unravel the interrelationship between early adenovirus gene products and their role in AAV replication. By microinjecting various combinations of either adenovirus type 2 DNA restriction fragments or the corresponding mRNAs into the nuclei of AAV-infected cells an ordered array of adenovirus early gene regulation was deduced. The order proceeds, E1 → E2a → E4. Because injection of E4 mRNA alone was sufficient for AAV DNA and capsid production, it was concluded that E4 gene expression alone provides the only necessary helper function for AAV. Thus E1 and E2A participate in a regulatory cascade which, for the purposes of AAV replication, result in expression of the E4 gene. In these studies a role for VAI was not found, but its gene product may have a role in regulating the quantitative level of AAV production.

In addition to adenovirus, many classes of herpesviruses (HSV) serve as at least partial helpers for AAV (Atchison, 1970; Blacklow *et al.*, 1975; Dolin and Rabson, 1973; Boucher *et al.*, 1971; Johnson *et al.*, 1972). Additionally, cells transformed by HSV type II will also support AAV replication (Blacklow, 1975). Although these early reports found apparently normal AAV DNA, RNA, and protein made when herpesvirus was the helper, the level of infectious particles was very low. Recently, however, Buller *et al.* (1981) reported that both HSV types I and II are complete helpers for AAV replication and find levels of infectious virus close to these found with adenovirus as helper. Unlike adenovirus mutants, several replication-defective mutants of HSV type I also support AAV DNA and protein synthesis (Handa *et al.*, 1977b; Drake *et al.*, 1974). The nature of the defect (or inhibition) may not therefore be at the level of DNA replication but more likely at some later step, perhaps in virus assembly. It is interesting that AAV is able to interchangeably utilize either adenovirus or herpesvirus gene products for its replication, and this fact underlines the resourcefulness of AAV in its dependency on two of the most widespread animal viruses. In this sense, the helper-virus-dependent steps of AAV DNA replication may be shared by many DNA-containing animal viruses.

Recent data indirectly suggest that AAV-coded capsid proteins may play a role in its DNA replication. When adenovirus H5*ts*125 was used as helper at nonpermissive temperatures a large reduction in AAV infectious particle production and capsid protein synthesis was seen (Myers *et al.*, 1980). The amount of replicative form AAV DNA (covalently linked plus and minus strands) labeled under these conditions was unchanged. However, the amount of unit-length single-stranded AAV DNA was greatly reduced (Myers and Carter, 1981). A similar pattern of normal levels of replicative intermediates but few progeny strands was noted with wild-type helper adenovirus if the protein synthesis inhibitor L-canavanine was present (Buller and Rose, 1978). L-Canavanine has been shown to inhibit AAV capsid synthesis, and it was therefore argued that an AAV protein (presumably a capsid protein or some combination of capsid proteins) is required for effective synthesis of progeny single strands from RF molecules. Interestingly a *ts* mutant of the autonomous parvovirus H-1 exhibits the identical phenotype (Rhode, 1974). The model of Tattersall and Ward (1976) in which the 5′ ends of nascent progeny strands are bound to capsid protein prior to complete virus assembly is consistent with these observations. Thus capsids may have two roles, one in packaging progeny strands and another in processing progeny strand synthesis from RF molecules. In the absence of efficient capsid production progeny strands might either not form at all or, once formed, they might be converted back to RF structures. This view is supported by a report that defective AAV genomes are only replicated in the presence of both helper virus and intact AAV (Laughlin *et al.*, 1979). Cheung and Berns (unpublished) have further tested this hypothesis by transfecting *Eco*RI restriction fragments into HeLa cells. They found that any terminal fragment or any combination of ligated terminal fragments which maintained both normal terminal structures but which lacked some portion of the internal coding region would not be replicated if just helper adenovirus were added. However, if both helper and normal AAV DNA were added, these transfected fragments were replicated. Thus an AAV gene product also appeared to be required for AAV DNA synthesis.

In summary, the identity and role of host and viral proteins in AAV DNA synthesis is beginning to be understood. Presently there is experimental evidence favoring a complex system in which cellular proteins (whether normal or virally induced is uncertain), at least the E4 protein of helper adenovirus, and an AAV-coded function all play a role.

V. CONCLUSION

In summary a reasonably complete picture of AAV DNA replication in terms of structural intermediates and proteins involved is emerging. The majority of this view was derived from a detailed characterization of the mature virion genome. Any model for AAV DNA replication must be able to explain the nucleotide sequence arrangements found in the mature genome because it is the natural end product. The model presented meets this qualification and is further supported by data obtained from studies of replicative intermediates and defective viral genomes isolated from infected cells. It remains to develop adequate *in vitro* systems to be able to study in depth the biochemistry involved in AAV DNA replication. The most promising development in this regard has been cloning of the AAV genome into a bacterial plasmid. It should now be possible to construct site-specific mutations to determine precisely which AAV gene functions are necessary for replication and which AAV DNA sequences are *cis*-functioning for replication.

Recent elucidation of an adenovirus gene function necessary for AAV replication bodes well for understanding the helper virus function. However, the identity of several proteins involved in AAV DNA replication remains elusive. In normal lytic infections host cell DNA polymerase is clearly used for elongation of AAV DNA. If HSV is used as the helper virus it is interesting to note that replicative complexes isolated from AAV infected cells contain HSV DNA polymerase. This potential promiscuity in using DNA polymerases is interesting. Another protein predicted by the models to be involved in AAV replication is a site-specific endonuclease needed to cleave the terminal hairpin structure. Whether this function resides in one of the capsid proteins or a nonstructural protein or is even virus coded at all remains to be determined. It may be equivalent to the transfunctional *rep* gene described by Rhode (1976) for the autonomous parvovirus H1. There may well be additional host-coded proteins involved in the replication complex, particularly factors used in the initiation process if one can argue by analogy with studies on single-stranded DNA phages.

The data so far available suggest that the general scheme first proposed by Cavalier-Smith (1974) is almost totally applicable to AAV DNA replication. A major attraction of the parvovirus system is its small genome size. This has allowed detailed structural analysis with direct implication for its mode of replication. A significant unanswered question is how far the picture of AAV replication can be extended to other viral and cellular eukaryotic systems. It should be emphasized that the Cavalier-Smith model was developed for cellular chromosomal DNA, and in this respect, the possibilities are exciting. One caveat is that the data and models derived for the parvoviruses may be only applicable to events at the termini of chromosomal DNA molecules. It is encouraging to note that the general property of a nucleotide sequence at the ends of linear DNA genomes with the potential of self-base-pairing by hairpin formation has already been described for HSV (Roizman, 1980) and vaccinia virus (Geshelin and Berns, 1974). The latter virus has a linear duplex genome in which the complementary strands are covalently cross-linked at both ends by apparent hairpin structures. Additionally Forte and Fangman (1979) have described similar structures at the termini of yeast chromosomal DNA. Thus the possibility of validly extending the AAV model to other systems is promising.

ACKNOWLEDGMENTS. Some of the investigations cited in this manuscript were supported by National Science Foundation research grant PCM 79-09354 and National Institutes of Health research grant 1 RO1 AI16326.

REFERENCES

Astell, C.R., Smith, M., Chow, M.D., and Ward, D.C., 1979, Structure of the 3′ hairpin termini of four rodent parvovirus genomes: Nucleotide sequence homology at the origins of DNA replication, *Cell* **17:**691.

Atchison, R.W., 1970, The role of herpesviruses in adenovirus-associated virus replication *in vitro, Virology* **42:**155.

Atchison, R.W., Casto, B.C., and Hammon, W. McD., 1965, Adenovirus-associated defective virus particles, *Science* **194:**754.

Basilico, C., Gattoni, S., Zouzias, D., and Valle, G.D., 1979, Loss of integrated viral DNA sequences in polyoma-transformed cells is associated with an active viral A function, *Cell* **17:**645.

Berk, A.J., and Sharp, P.A., 1977, Sizing and mapping of early adenovirus mRNAs by gel electrophoresis of S1 endonuclease digested hybrids, *Cell* **12:**721.

Berk, A.J., Lee, F., Harrison, T., Williams, J., and Sharp, P.A., 1979, Pre-early Ad5 gene product regulates synthesis of early viral mRNAs, *Cell* **17:**935.

Berns, K.I., 1974, Molecular biology of adeno-associated viruses, *Curr. Topics Microbiol. Immunol.* **65:**1.

Berns, K.I., and Adler, S., 1972, Separation of two types of adeno-associated virus particles containing complementary polynucleotide chains, *J. Virol.* **9:**394.

Berns, K.I., and Hauswirth, W.W., 1979, Adeno-associated virus, *Adv. Virus Res.* **25:**407.

Berns, K.I., and Kelly, T.J., Jr., 1974, Visualization of the inverted terminal repetition in adeno-associated virus DNA, *J. Mol. Biol.* **82:**267.

Berns, K.I., and Rose, J.A., 1970, Evidence for a single-stranded adenovirus-associated virus genome: Isolation and separation of complementary single strands, *J. Virol.* **5:**693.

Berns, K.I., Kort, J., Fife, K.H., Grogan, W.E., and Spear, I., 1975a, Study of the fine structure of adeno-associated virus DNA with bacterial endonucleases, *J. Virol.* **16:**712.

Berns, K.I., Pinkerton, T.C., Thomas, G.F., and Hoggan, M.D., 1975b, Detection of adeno-associated virus-specific nucleotide sequences in DNA isolated from latently infected Detroit 6 cells, *Virology* **68:**556.

Berns, K.I., Hauswirth, W.W., Fife, K.H., and Spear, I.S., 1978, Terminal structure of adeno-associated virus particles, in: *Replication of Mammalian Parvoviruses* (D.C. Ward and P. Tattersall, eds.), p. 179, Cold Spring Harbor, New York.

Berns, K.I., Hauswirth, W.W., Fife, K.H., and Lusby, E., 1979, Adeno-associated virus DNA replication, *Cold Spring Harbor Symp. Quant. Biol.* **43:**781.

Blacklow, N.R., 1975, Potentiation of an adenovirus-associated virus by herpes simplex virus type 2 transformed cells, *J. Natl. Can. Inst.* **54:**241.

Blacklow, N.R., Hoggan, M.D., and Rowe, W.P., 1967, Immunofluorescent studies of the potentiation of an adenovirus-associated virus by adenovirus 7, *J. Exp. Med.* **125:**755.

Blacklow, N.R., Hoggan, M.D., and McClanahan, M.S., 1970, Adenovirus-associated viruses: Enhancement by human herpesviruses, *Proc. Soc. Exp. Biol. Med.* **134:**952.

Boucher, D.W., Melnick, J.L., and Mayor, H.D., 1971, Non-encapsidated infectious DNA of adeno-satellite virus in cells coinfected with herpesvirus, *Science* **173:**1243.

Bourguignon, G.J., Tattersall, P.J., and Ward, D.C., 1976, DNA of a minute virus of mice: Self-priming, nonpermuted, single-stranded genome with a 5′-terminal hairpin duplex, *J. Virol.* **20:**290.

Bratosin, S., Laub, D., Tal, J., and Aloni, Y., 1979, Mechanism for circularization of linear DNAs: Circular parvovirus MVM DNA is formed by a noose sliding in a lasso-like DNA structure, *Proc. Natl. Acad. Sci. USA* **76:**4289.

Buller, R.M.L., and Rose, J.A., 1978, Characterization of adeno-associated virus-induced polypeptides in KB cells, *J. Virol.* **25**:331.

Buller, R.M.L., Straus, S.E., and Rose, J.A., 1979, Mechanism of host restriction of adeno-associated virus replication in African Green Monkey kidney cells, *J. Gen. Virol.* **43**:663.

Buller, R.M.L., Janik, J.E., Sebring, E.D., and Rose, J.A., 1981, Herpes simplex virus types 1 and 2 help adeno-associated virus replication, *J. Virol.* **40**:241.

Carter, B.J., and Khoury, G., 1975, Specific cleavage of adenovirus-associated virus DNA by restriction endonuclease R. EcoR1: Characterization of cleavage products, *Virology* **63**:523.

Carter, B.J., Koczot, F.J., Garrison, J., Rose, J.A., and Dolin, R., 1973, Separate helper functions provided by adenovirus for adenovirus-associated virus multiplication, *Nature New Biol.* **244**:71.

Carter, B.J., Khoury, G., and Denhardt, D.T., 1975, Physical map and strand polarity of specific fragments of adenovirus-associated virus DNA produced by endonuclease R. EcoR1, *J. Virol.* **16**:559.

Casto, B.C., Atchison, R.W., and Hammon, W. McD., 1967, Studies on the relationship between adeno-associated virus type I (AAV-1) and adenoviruses. I. Replication of AAV in certain cell cultures and its effects on helper adenovirus, *Virology* **32**:52.

Cavalier-Smith, T., 1974, Palindromic base sequences and replication of eukaryotic chromosome ends, *Nature* **250**:467.

Cheung, A.K.M., Hoggan, M.D., Hauswirth, W.W., and Berns, K.I., 1980, Integration of the adeno-associated virus genome into cellular DNA in latently infected human Detroit 6 cells, *J. Virol.* **33**:738.

Chow, M.B., and Ward, D.C., 1978, Comparison of the terminal nucleotide structure in the DNA of nondefective parvoviruses. in: *Replication of Mammalian Parvoviruses* (D.C. Ward and P. Tattersall, eds.), p. 205, Cold Spring Harbor, New York.

Chow, L.T., Gelinas, R.E., Broker, T.R., and Roberts, R.J., 1977, An amazing sequence arrangement at the 5′ ends of adenovirus 2 messenger RNA, *Cell* **12**:1.

Crawford, L.V., 1966, A minute virus of mice, *Virology* **29**:605.

Crawford, L.V., Follet, E.A.C., Burdon, M.G., and McGeoch, D.J., 1969, The DNA of a minute virus of mice, *J. Gen. Virol.* **4**:37.

Danna, K.J., and Nathans, D., 1972, Bidirectional replication of simian virus 40 DNA, *Proc. Natl. Acad. Sci. USA* **69**:309.

de la Maza, L.M., and Carter, B.J., 1976, Cleavage of adeno-associated virus DNA with SalI, PstI and HaeII restriction endonucleases, *Nucleic Acids Res.* **3**:2605.

de la Maza, L.M., and Carter, B.J., 1978, DNA structure of incomplete adeno-associated virus particles. in: *Replication of Mammalian Parvoviruses* (D. Ward and P. Tattersall, eds.), p. 193, Cold Spring Harbor, New York.

de la Maza, L.M., and Carter, B.J., 1980, Molecular structure of adeno-associated virus variant DNA, *J. Biol. Chem.* **255**:3194.

Denhardt, D.T., Eisenberg, G., Bartok, K., and Carter, B.J., 1976, Multiple structures of adeno-associated virus DNA: Analysis of terminally-labeled molecules with endonuclease R. HaeIII, *J. Virol.* **18**:672.

Dolin, R., and Rabson, A.S., 1973, Herpesvirus saimiri: Enhancement of adenovirus-associated virus, *J. Natl. Can. Inst.* **50**:205.

Drake, S., Schaffer, P.A., Esparza, J., and Mayor, H.D., 1974, Complementation of adeno-associated satellite viral antigens and infectious DNA by temperature-sensitive mutants of herpes simplex virus, *Virology* **60**:230.

Eigen, M., and Schuster, P., 1978, The hypercycle, a principle of natural self organization. Part C, the realistic hypercycle, *Naturwissenschaften* **65**:341.

Ensinger, M.J., and Ginsberg, H.S., 1972, Selection and preliminary characterization of temperature-sensitive mutants of type 5 adenovirus, *J. Virol.* **10**:328.

Faust, E.A., and Ward, D.C., 1979, Incomplete genomes of the parvovirus minute virus of mice: Selective conservation of genome termini, including the origin for DNA replication, *J. Virol.* **72**:131.

Fife, K.H., Berns, K.I., and Murray, K., 1977, Structure and nucleotide sequence of the terminal regions of adeno-associated virus DNA, *Virology* **78**:475.

Forte, M.A., and Fangman, W.L., 1979, Yeast chromosomal DNA molecules have strands which are cross-linked at their termini, *Chromosoma* **72**:131.

Frost, E., and Williams, J., 1978, Mapping temperature-sensitive and host-range mutants of adenovirus type 5 by marker rescue, *Virology* **91**:39.

Gerry, H.W., Kelly, T.J., Jr., and Berns, K.I., 1973, Arrangement of nucleotide sequences in adeno-associated virus DNA, *J. Mol. Biol.* **79**:207.

Geshelin, P., and Berns, K.I., 1974, Characterization and localization of the naturally occurring cross-links in vaccinia virus DNA, *J. Mol. Biol.* **88**:785.

Graham, F.L., Harrison, T., and Williams, J., 1978, Defective transforming capacity of adenovirus type 5 host-range mutants, *Virology* **86**:10.

Handa, H., and Shimojo, H., 1977a, Isolation of the viral DNA replication complex from adeno-associated virus type I-infected cells, *J. Virol.* **24**:444.

Handa, H., and Shimojo, H., 1977b, Viral DNA synthesis *in vitro* with nuclei isolated from adenovirus-associated virus type-I infected cells, *Virology* **77**:424.

Handa, H., Shiroki, K., and Shimojo, H., 1975, Complementation of adeno-associated virus growth with temperature-sensitive mutants of human adenovirus types 12 and 5, *J. Gen. Virol.* **29**:239.

Handa, H., Shimojo, H., and Yamaguchi, K., 1976, Multiplication of adeno-associated virus type I in cells coinfected with a temperature-sensitive mutant of human adenovirus type 31, *Virology* **74**:1.

Handa, H., Shiroki, K., and Shimojo, HJ., 1977a, Establishment and characterization of KB cell lines latently infected with adeno-associated virus type I, *Virology* **82**:84.

Handa, H., Shiroki, K., and Shimojo, H., 1977b, Helper factor(s) for growth of adeno-associated virus in cells transformed by adenovirus 12, *Proc. Natl. Acad. Sci. USA* **74**:4508.

Handa, H., Shiroki, K., and Shimojo, H., 1978, Expression of helper function for adeno-associated virus in adenovirus-transformed cells. in: *Replication of Mammalian Parvoviruses* (D.C. Ward and P. Tattersall, eds.), p. 99, Cold Spring Harbor, New York.

Harrap, K.A., and Payne, C.C. The structural properties and identification of insect viruses, *Adv. Virus Res.* **25**:273.

Hauswirth, W.W., and Berns, K.I., 1977, Origin and termination of adeno-associated virus DNA replication, *Virology* **79**:488.

Hauswirth, W.W., and Berns, K.I., 1978, Initiation and termination of adeno-associated virus DNA replication. in: *Replication of Mammalian Parvoviruses* (D.C. Ward and P. Tattersall, eds.), p. 257, Cold Spring Harbor, New York.

Hauswirth, W.W., and Berns, K.I., 1979, Adeno-associated virus DNA replication: Non unit-length molecules, *Virology* **93**:57.

Hayward, G.S., Bujard, H., and Gunther, M., 1978, Three distinct replicative forms of Kilham-rat-virus DNA. in: *Replication of Mammalian Parvoviruses* (D.C. Ward and P. Tattersall, eds.), p. 327, Cold Spring Harbor, New York.

Hirt, B., 1977, Selective extraction of polyoma DNA from mouse cell cultures, *J. Mol. Biol.* **26**:365.

Hoggan, M.D., 1970, Adenovirus-associated viruses, *Prog. Med. Virol.* **12**:112.

Hoggan, M.D., 1971, in: *Comparative Virology* (K. Maramarosch and E. Kurstak, eds.), p. 49, Academic Press, New York.

Hoggan, M.D., Blacklow, N.R., and Rowe, W.P., 1966, Studies of small DNA viruses found in various adenovirus preparations: Physical, biological, and immunological characteristics, *Proc. Natl. Acad. Sci. USA* **55**:1457.

Hoggan, M.D., Thomas, G.F., and Johnson, F.B., 1972, Continuous "carriage" of the adenovirus-associated virus genome in cell cultures in the absence of helper adenoviruses. in: *The Proceedings of the Fourth Lepetit Colloquium*, p. 243, Cocoyac, Mexico, North-Holland Publ., Amsterdam.

Johnson, F.B., Blacklow, N.R., and Hoggan, M.D., 1972, Immunological reactivity of antisera prepared against the sodium dodecyl sulfate-treated structural polypeptides of adenovirus-associated virus, *J. Virol.* **9**:1017.

Johnson, F.B., Whitaker, C.W., and Hoggan, M.D., 1975, Structural polypeptides of adeno-associated virus top component, *Virology* **65:**196.

Klessig, D.F., and Chow, L.T., 1980, Incomplete splicing and deficient accumulation of fiber messenger RNA in monkey cells infected by human adenovirus type 2, *J. Mol. Biol.* **139:**221.

Kollek, R., Tseng, B.Y., and Goulian, M., 1980, Studies on the effect of inhibitors of DNA polymerase on synthesis of parvovirus H-1 DNA, *J. Supramolec. Struct. Suppl.* **4:**332.

Kozcot, F.J., Carter, B.J., Garon, C.F., and Rose, J.A., 1973, Self-complementarity of terminal sequences within plus or minus strands of adenovirus-associated virus DNA, *Proc. Natl. Acad. Sci. USA* **70:**215.

Lassam, N.J., Bayley, S.T., and Graham, F.L., 1978, Synthesis of DNA, late polypeptides, and infectious virus by host-range mutants of adenovirus 5 in nonpermissive cells, *Virology* **87:**463.

Lassam, N.J., Bayley, S.T., and Graham, F.L., 1979, Tumor antigens of human Ad5 in transformed cells and in cells infected with transformation-defective host-range mutants, *Cell* **18:**781.

Laughlin, C.A., Myers, M.W., Risin, D.L., and Carter, B.J., 1979, Defective-interfering particles of the human parvovirus adeno-associated virus, *Virology* **94:**162.

Li, A.T., Lavelle, G.C., and Tennant, R.W., 1978, DNA replication of Kilham rat virus: Characterization of intracellular forms of viral DNA extracted by guanidine hydrochloride. in: *Replication of Mammalian Parvoviruses* (D.C. Ward and P. Tattersall, eds.), p. 341, Cold Spring Harbor, New York.

Lusby, E., Fife, K.H., and Berns, K.I., 1980, Nucleotide sequence of the inverted terminal repetition in adeno-associated virus DNA, *J. Virol.* **34:**402.

Lusby, E., Bohenzky, R., and Berns, K.I., 1981, The inverted terminal repetition in adeno-associated virus DNA: Independence of orientation at either end of the genome, *J. Virol.* **37:**1083.

Mayor, H.D., and Young, J.F., 1978, Complementation of adeno-associated virus by temperature-sensitive mutants of human adenovirus and herpesvirus. in: *Replication of Mammalian Parvoviruses* (D.C. Ward and P. Tattersall, eds.), p. 109, Cold Spring Harbor, New York.

Mayor, H.D., Torikai, K., Melnick, J., and Mandel, M., 1969, Plus and minus single-stranded DNA separately encapsidated in adeno-associated satellite virions, *Science* **166:**1280.

Muller, H.P., Gautschi, M., and Sigel, G., 1978, Defective particles of parvovirus LuIII. in: *Replication of Mammalian Parvoviruses* (D.C. Ward and P. Tattersall, eds.), p. 231, Cold Spring Harbor, New York.

Myers, M.W., and Carter, B.J., 1981, Adeno-associated virus replication. The effect of L-canavanine or a helper virus mutation on accumulation of viral capsids and progeny single-stranded DNA, *J. Biol. Chem.* **256:**567.

Myers, M.W., Laughlin, C.A., Jay, F.T., and Carter, B.J., 1980, Adenovirus helper function for growth of adeno-associated virus: Effect of temperature sensitive mutants in adenovirus early gene region 2, *J. Virol.* **35:**65.

Ostrove, J.M., and Berns, K.I., 1980, Adenovirus early region Ib gene function required for rescue of latent adeno-associated virus, *Virology* **104:**502.

Pritchard, C., Stout, E.R., and Bates, R.C., 1981, Replication of parvoviral DNA. I. Characterization of a nuclear lysate system, *J. Virol.* **37:**352.

Rekosh, D.M.K., Russell, W.C., Bellett, A.J.D., and Robinson, A.J., 1977, Identification of a protein linked to the ends of adenovirus DNA, *Cell* **11:**183.

Revie, D., Tseng, B.Y., Grafstrom, R.H., and Goulian, M., 1979, Covalent association of protein with replicative form DNA of parvovirus H-1, *Proc. Natl. Acad. Sci. USA* **76:**5539.

Rhode, S.L., 1973, Replication process of the parvovirus H-1. I. Kinetics in a parasynchronous cell system, *J. Virol.* **11:**856.

Rhode, S.L., 1974, Replicative process of the parvovirus H-1. II. Isolation and characterization of H-1 replicative form DNA, *J. Virol.* **13:**400.

Rhode, S.L., 1976, Replication process of the parvovirus H-1. V. Isolation and characterization of temperature-sensitive H-1 mutants defective in progeny DNA synthesis, *J. Virol.* **17**:659.

Rhode, S.L., 1977, Replication process of the parvovirus H-1. VI. Characterization of a replication terminus of H-1 replicative-form DNA, *J. Virol.* **21**:694.

Rhode, S.L., 1978a, Defective interfering particles of parvovirus H-1, *J. Virol.* **27**:347.

Rhode, S.L., 1978b, Replication process of the parvovirus H-1. X. Isolation of a mutant defective in replicative form DNA replication, *J. Virol.* **25**:215.

Richardson, W.D., Carter, B.J., and Westphal, H., 1980, Vero cells injected with adenovirus type 2 mRNA produce authentic viral polypeptide patterns: Early mRNA promotes growth of adenovirus-associated virus, *Proc. Natl. Acad. Sci. USA* **77**:931.

Roizman, B., 1980, Herpes simplex viruses. in: *Molecular Biology of Tumor Viruses* (J. Tooze, ed.), 2nd edition, Vol. 2, p. 615, Cold Spring Harbor, New York.

Rose, J.A., 1974, Parvovirus reproduction, *Comprehensive Virol.* **4**:1.

Rose, J.A., and Koczot, F., 1971, Adenovirus-associated virus multiplication. VI. Base composition of the deoxyribonucleic acid strand species and strand-specific *in vivo* transcription, *J. Virol.* **8**:771.

Rose, J.A., and Koczot, F., 1972, Adenovirus-associated virus multiplication. VII. Helper requirement for viral deoxyribonucleic acid and ribonucleic acid synthesis, *J. Virol.* **10**:1.

Rose, J.A., Hoggan, M.D., and Shatkin, A.J., 1966, Nucleic acid from an adeno-associated virus: Chemical and physical studies, *Proc. Natl. Acad. Sci. USA* **56**:86.

Rose, J.A., Berns, K.I., Hoggan, M.D., and Koczot, F.J., 1969, Evidence for a single-stranded adenovirus-associated virus genome: Formation of a DNA density hybrid on release of viral DNA, *Proc. Natl. Acad. Sci. USA* **64**:863.

Salzman, L.A., 1977, Evidence for terminal S_1-nuclease-resistant regions on single-stranded linear DNA, *Virology* **76**:454.

Salzman, L.A., and Fabisch, P., 1979, Nucleotide sequence of the self-priming 3′ terminus of the single-stranded DNA extracted from the parvovirus KRV, *J. Virol.* **30**:946.

Salzman, L., White, W.L., and Kakefuda, T., 1971, Linear, single-stranded DNA isolated from Kilham rat virus, *J. Virol.* **7**:830.

Salzman, L.A., Fabisch, P., Parr, R., Garon, C., and Wali, T., 1978, *In vitro* synthesis of double-stranded DNA from the Kilham rat virus single-stranded DNA genome, *J. Virol.* **27**:784.

Samulski, R.J., Berns, K.I. Tan, M., and Muzyczka, N., 1982, Cloning of adeno-associated virus into pBR 322: Rescue of intact virus from the recombinant plasmid in human cells, *Proc. Natl. Acad. Sci. USA* **79**:2077.

Samulski, R.J., Srivastava, A., Berns, K.I., and Muzyczka, N., 1983, Rescue of adeno-associated virus from recombinant plasmids: Gene correction within the terminal repeats of AAV, *Cell* **33**:135.

Siegl, G., 1976, The parvoviruses, *Virol. Monogr.* **15**:1.

Siegl, G., and Gautschi, M., 1978, Purification and properties of replicative-form and replicative-intermediate DNA molecules of parvovirus LuIII. in: *Replication of Mammalian Parvoviruses* (D.C. Ward and P. Tattersall, eds.), p. 315, Cold Spring Harbor, New York.

Singer, I.I., and Rhode, S.L., 1977, Replication process of the parvovirus H-1. VIII. Partial denaturation mapping and localization of the replication origin of H-1 replicative-form DNA with electron microscopy, *J. Virol.* **21**:724.

Singer, I.I., and Rhode, S.L., 1978, Ultrastructural studies of H-1 parvovirus replication. VI. Simultaneous autoradiographic and immunochemical intranuclear localization of viral DNA synthesis and protein accumulation, *J. Virol.* **25**:349.

Smith, H.O., and Birnstiel, M.L., 1976, A simple method for DNA restriction site mapping, *Nucleic Acids Res.* **3**:2387.

Spear, I.S., Fife, K.H., Hauswirth, W.W., Jones, C.J., and Berns, K.I., 1977, Evidence for two nucleotide sequence orientations within the terminal repetition of adeno-associated virus DNA, *J. Virol.* **24**:627.

Straus, S.E., Ginsberg, H.S., and Rose, J.A., 1976a, DNA-minus temperature-sensitive mutants of adenovirus type 5 help adenovirus-associated virus replication, *J. Virol.* **17**:140.

Straus, S.E., Sebring, E., and Rose, J.A., 1976b, Concatemers of alternating plus and minus strands are intermediates in adenovirus-associated virus DNA synthesis, *Proc. Natl. Acad. Sci. USA* **73**:742.

Straus, S.E., Sebring, E.D., and Rose, J.A., 1978, Self-primed replication of adeno-associated virus DNA. in: *Replication of Mammalian Parvoviruses* (D.C. Ward and P. Tattersall, eds.), p. 243, Cold Spring Harbor, New York.

Sumper, M., and Luce, R., 1975, Evidence for *de novo* production of a self-replicating and environmentally adapted RNA structure by bacteriophage QB replicase, *Proc. Natl. Acad. Sci. USA* **72**:162.

Tattersall, P., and Ward, D.C., 1976, Rolling hairpin model for replication of parvovirus and linear chromosomal DNA, *Nature* (*London*) **263**:106.

Torikai, K., Ito, M., Jordan, L.E., and Mayor, H.D., 1970, Properties of light particles produced during growth of type 4 adeno-associated satellite virus, *J. Virol.* **6**:363.

Van der Vliet, P.E., Levine, A.J., Ensinger, M.J., and Ginsberg, H.S., 1975, Thermolabile DNA binding proteins from cells infected with a temperature sensitive mutant of adenovirus defective in viral DNA synthesis, *J. Virol.* **15**:348.

Wang, E.C., Furth, J.J., and Rose, J.A., 1978, Purification and characterization of a DNA single strand specific endonuclease from human cells, *Biochemistry* **17**:544.

Ward, D.C., and Dadachanji, D., 1978, Replication of minute-virus-of-mice DNA. in: *Replication of Mammalian Parvoviruses* (D.C. Ward and P. Tattersall, eds.), p. 197, Cold Spring Harbor, New York.

CHAPTER 2

Biology of Adeno-Associated Virus

GEORGE CUKOR, NEIL R. BLACKLOW, M. DAVID HOGGAN, AND KENNETH I. BERNS

I. INTRODUCTION

The helper-dependent adeno-associated viruses (AAV) are widely distributed throughout nature and have been isolated as infecting agents in almost every animal species where a serious attempt has been made to find them. These natural hosts include man (Blacklow *et al.*, 1967; Hoggan, 1970; 1971; Hoggan *et al.*, 1973; Parks *et al.*, 1970), various species of monkeys (Rapoza and Atchison, 1967; Hoggan *et al.*, 1973), cattle (Luchsinger *et al.*, 1970; 1971), dogs (Domoto and Yanagawa, 1967), quail and chickens (Dutta and Pomeroy, 1967; Yates *et al.*, 1976), and horses (Dutta, 1975).

Although a number of epidemiological and experimental animal studies have been carried out with these agents, our biological knowledge of them remains in many ways an enigma: we know more about their physical, chemical, and immunological characteristics at the molecular level than we know about where they come from or what they do to or for their respective hosts.

GEORGE CUKOR AND NEIL R. BLACKLOW • Division of Infectious Diseases, University of Massachusetts Medical Center, Worcester, Massachussetts 01605. M. DAVID HOGGAN • Laboratory of Viral Diseases, National Institute of Allergy and Infectious Diseases, National Institutes of Health, Bethesda, Maryland 20205. KENNETH I. BERNS • Department of Immunology and Medical Microbiology, College of Medicine, University of Florida, Gainesville, Florida 32610.

II. ASSOCIATION WITH ADENOVIRUS

In man, epidemiological studies of adeno-associated virus infection in both children and adults (Blacklow *et al.*, 1967; 1968a,b; 1969; Parks *et al.*, 1970) show a strong association of AAV with an active adenovirus (Ad) infection. In fact, serologic evidence for AAV infection is found significantly more often in adenovirus-infected children ill with respiratory disease than in asymptomatic Ad-infected control subjects (Blacklow *et al.*, 1971). These results are assumed to be a manifestation of the greater Ad replication seen in the diseased patients when compared to asymptomatic ones. In these studies of natural human disease, no amelioration of symptoms of Ad infection or correlation with specific disease syndromes could be noted. This was a little surprising since a number of *in vitro* studies showed that AAV has a strong inhibitory effect on Ad replication (Hoggan *et al.*, 1966; Casto *et al.*, 1967a; Parks *et al.*, 1968; and Carter *et al.*, 1979; see Chapters 8,9). AAV has also been shown to reduce the oncogenic potential of Ad in hamsters (Kirschstein *et al.*, 1968) as well as that of herpes simplex virus-transformed cells (Blacklow *et al.*, 1978; see below).

It is interesting to point out that on at least two occasions AAV-2 contamination was found to exhibit such strong inhibitory effects on Ad type 12 replication that no cytopathic effect or increase in Ad production could be measured, and in fact these preparations could not be serially passaged until the concentration of AAV was reduced with specific antibody (M.D. Hoggan, unpublished data). In the first instance, a pool of Ad type 12 strain 97830 supplied by R. Rafajko was subjected to direct electronmicroscopic examination because it appeared to be nonviable. It was found to be heavily contaminated with AAV-like particles which did not react to antibody prepared against the then known AAV-type 1. An aliquot of the virus pool was used to prepared purified AAV particles and an antiserum produced in guinea pigs. This antiserum did not react with AAV-type 1; it did, however, react by both complement fixation and serum neutralization with the AAV-like particles found in the 97830 strain of Ad type 12 (Hoggan *et al.*, 1966a) and was subsequently used to define AAV-type 2 (Hoggan *et al.*, 1966b).

The second instance of complete Ad suppression occurred during studies of coinfection with AAV-1 and Ad type 7 carried out on a specific lot of primary human embryonic kidney (HEK) cells. During the course of these experiments AAV-2 antigen was produced in the Ad control tubes (Hoggan, unpublished observations). It was further found that after adding fresh media containing anti-AAV-2 serum to previously unopened companion HEK tubes they developed Ad CPE in 2–3 days while other tubes from the same lot renewed with fresh media without antibody remained unchanged. The CPE was subsequently shown to be the result of an Ad 12 infection, a virus that had not been used in the laboratory for several years. Although many lots of perinatal as well as embryonic HEK cells are occultly infected with AAV-2 (Hoggan *et al.*, 1972), this was the only instance known in which a lot of HEK cells was covertly infected with both an Ad and AAV.

Of interest is the fact that of the three AAV serotypes either known (AAV-2, AAV-3) or suspected (AAV-1) of infecting humans, AAV-2 not only appears to have the greatest inhibitory effect on helper Ad—it also appears to have a greater inhibitory effect on AAV-1 and AAV-3 replication than on itself (see below and Chapters 8,9 for detailed discussion of AAV inhibitory effects). In a series of unpublished experiments various mixtures of AAV (AAV-1 + AAV-2, AAV-2 + AAV-3; and AAV-1 + AAV-3 equilibrated to contain equal numbers of physical particles as well as infectious virions) were inoculated onto HEK cells preinfected with adenovirus 7. When AAV-2 was mixed with either AAV-1 or AAV-3 it was always produced in the greater quantity. AAV-1 predominated when mixed with AAV-3. It is of further interest to point out that AAV-3 was originally isolated from a pool of AAV-1 treated with anti-AAV-1 sera (Hoggan *et al.*, 1966b). When AAV-3 was isolated from humans it was usually isolated from children who had previously been infected with AAV-2 and had existing antibody to AAV-2 (Blacklow *et al.*, 1969b). We also noted that most of our human AAV-3 isolates contained small levels of AAV-2 upon isolation which could be detected only after 2–4 serial cell culture passages in the absence of AAV-2 antibody. After 1–3 additional passages AAV-2 became the dominant population.

III. HOST RANGE

Although AAV-1 has not been directly isolated from man, a few AAV-1 monospecific antisera have been encountered (Blacklow *et al.*, 1968a). It is of further interest that even though AAV-1 appears to be associated with several different species of captive monkeys (Rapoza and Atchison, 1967) Rhesus monkeys from India bled in the wild appear free of AAV antibody while sera from Rhesus monkeys housed in U.S. and foreign zoos as well as from various commercial holding colonies had high levels of antibody to AAV types 1,2,3 and to a lesser extent AAV-4 (Blacklow *et al.*, 1968b).

In general the AAV appear less host specific than their nondefective cousins. To cite a few examples AAV types, 1,2, and 3 replicate well in canine kidney cells with infectious canine hepatitis virus (ICHV) as a helper (Hoggan *et al.*, 1966). Similarly AAV, 1,2, and 3 can be replicated in primary bovine kidney cells in the presence of bovine Ad (Hoggan and Laine, unpublished data) as well as in various mouse cell cultures with Ad (Blacklow and Hoggan, unpublished data; Lipps and Mayor, 1980a,b). AAV-1,2, and 3 will also replicate in chicken embryo fibroblasts with CELO virus as helper (Blacklow *et al.*, 1968b). By contrast, the other well-studied primate AAV (AAV-4) grows very poorly in cells other than primary African green monkey cells. This observation correlates with the finding that while AAV-4 antibody is common in African green monkeys (59–67%) it is found less frequently in other primates and only in animals which also have high levels of antibody against one or more of other primate AAV (Blacklow *et al.*, 1968b).

Although antibody to bovine AAV (AAVX7) has been reported in man (Luchsinger and Wellemans, 1971), our attempts to grow this virus in human cells with human Ad failed. We have also been unsuccessful in our attempts to demonstrate significant replication of the avian AAV (AAAV) in *Hep*II or HEK cells with Ad types 2,7, or 12. Yates *et al.* (1981), however, reported that AAAV can replicate in *Hep*II or HeLa cells with human Ad and presented supporting epidemiological data that AAAV infects man with concomitant human Ad infection. When using serum neutralization as an indicator they found a 67% AAAV infection rate among poultry workers compared to a 16% infection rate in an unselected population.

A small virus particle found in association with the Matsuda strain of ICH virus was first described by Sugimura and Yanagawa (1968) and characterized as an AAV the following year (Domoto and Yanagawa, 1969). Later Onuma and Yanagawa (1972a) reported that the canine AAV (CAAV) was serologically closely related to the primate AAV-3 and that antibody against this virus was common in both the human and canine population of Japan (Onuma and Yanagawa, 1972b). Our results do not agree with those of Onuma and Yanagawa. After receiving a preparation of ICHV, strain Matsuda, containing CAAV (kindly supplied by Dr. Yanagawa), we were able to grow the CAAV in both primary canine kidney and the continuous MDCK cell line. Attempts to transfer the CAAV to human cells with Ad types 2, 7, and 12 failed. We also found only limited replication of CAAV even in canine cells with three separate freshly plaqued strains of ICHV; these included the Ultricht and Cornell strains as well as a field isolate. On the other hand plaque-purified ICHV, matusda strain, proved to be an adequate helper. After purifying the CAAV and producing hyperimmune sera in guinea pigs the CAAV was distinct from AAV 1,2,3, and AAV 4 and AAVX7 by complement fixation, virus neutralization, immunofluorescence (FA), and immune-electron microscopy (IEM) (Thomas and Hoggan, 1974). In a survey of canine sera collected from around the world and provided by Leonard Binn, we found CAAV antibody was present in most dogs, but none was found in over 100 human sera (Hoggan and Thomas, unpublished data). On the other hand, no AAV-3 antibody was present in any of the canine sera.

In an attempt to clarify these conflicting results we requested new reagents from Dr. Yanagawa. All tests were repeated using a separate glass sealed vial of CAAV which had not been previously opened in our laboratory. Again, no cross-reactivity was found between AAV-3 or CAAV when using either hyperimmune sera prepared against purified virions or positive infection sera. However, when CAAV-infected canine cells were tested by FA using sera prepared against SDS acrylamide gel purified polypeptides (Johnson *et al.*, 1972) positive fluorescence was seen. It should be pointed out that AAV polypeptide sera reacts in FA against all AAV (including avian and bovine strains) but is negative for all Ad helpers and all nondefective parvoviruses for which it has been tested (Hoggan, unpublished data).

Replication of AAV-1 in mouse cells with mouse Ad (MAV) as helper has been reported by Lipps and Mayor (1980a). These workers further described the transplacental infection of mice with AAV-1 (Lipps and Mayor, 1980b) by demonstrating the presence of AAV-1 in the lungs and kidneys of fetuses and newborns from mothers who had been infected subcutaneously with AAV-1 plus MAV. They also reported that the development of fetuses was retarded, the litter number was reduced, and the average weight of surviving mice was smaller in those mice whose mothers had been injected with AAV + MAV compared to those from mothers who were injected with MAV alone. Such results are important considering the fact that the nondefective parvoviruses have been incriminated in a number of *in utero* and neonatal disease syndromes. These include cerebellar hypoplasia in kittens caused by feline panleukopenia virus (Johnson *et al.*, 1967); infertility, abortion, and fetal mummification *in utero* caused by porcine parvovirus (Cartwright and Hutch, 1967, Mengling *et al.*, 1975); and the osteolytic lesions reported in hamsters after infection with H-1 and RV (Toolan *et al.*, 1960; Kilham, 1961).

Because of these results and the findings that 1–2% of human embryos are covertly infected with AAV-2 and 10–20% of African green monkey cells covertly infected with AAV-4 (Hoggan *et al.*, 1973), we were prompted to see if any correlation could be found between AAV infection and the occurrence of reproductive failure or teratogenic effects in man. Therefore, a series of about 80 paired sera from mothers with histories of reproduction failure, stillbirth, or spontaneous abortion were tested. No significant difference in the frequency or level of antibody to AAV 1,2, or 3 was found (Blacklow and Hoggan, unpublished results).

IV. EXPERIMENTAL INFECTION *IN UTERO*

Because of the high incidence of AAV-1 antibody in captive monkeys (Raponyza and Atchison, 1967; Blacklow *et al.*, 1968b) and the good growth of AAV-1 in Rhesus monkey cells with the simian Ad SV-15 (Atchison *et al.*, 1965) we felt that the Rhesus monkey might provide an ideal model for investigating the results of an *in utero* AAV infection under controlled laboratory conditions. With the expertise of Dr. W.A. London in *in utero* infection of monkeys, it was possible to test this model (Hoggan, Blacklow, and London, unpublished results). After testing a group of pregnant Rhesus monkeys for the presence of SV-15 and AAV-1 antibody a group of eight seronegative animals was selected and isolated to protect from intercurrent infection. At 70–90 days after conception the fetuses were injected *in utero* with different combinations of AAV-1 and helper (SV-15). This was accomplished by performing a laparotomy on each mother, exposing the fetus to view, and injecting it through the uterine wall. The clinical status of each mother was followed daily with throat and anal swabs as well as blood samples taken about every two weeks. Except in the two instances where spontaneous abortion occurred, a cesarean section was performed 4–5 days prior to term. The results of these experiments can be summarized as follows:

1. The two fetuses injected with AAV-1 alone showed no clinical or serological evidence of AAV-1 infection and both were delivered by cesarean section 4–5 days prior to expected term. It should be pointed out, however, that one of these animals had a low (1:20) CF titer against adenovirus group antigen at birth which disappeared by $2\frac{1}{2}$ weeks after delivery. Its mother had developed a similar titer about 3 weeks prior to delivery but neither the infant or its mother had neutralizing antibody against SV-15. When surviving monkeys were challenged intravenously with SV-15 4–8 weeks after delivery they developed SV-15 antibody but not AAV-1 antibody; thus we were unable to demonstrate any latency in these animals.
2. No fetus survived that was injected with SV-15 alone (2 animals) or SV-15 + AAV-1 (3 animals). They were either aborted (1 of 2 animals injected with SV-15 aborted at 15 days) or found as mummified or partially resorbed fetuses at cesarian section. AAV-1 was isolated from virtually all fetal tissues as well as placenta along with SV-15 in those animals injected with AAV-1 and helper. No AAV was isolated from animals injected with helper alone.
3. Although injection of SV-15 resulted in the death of all fetuses, none of the mothers showed clinical signs of respiratory disease but SV-15 was isolated regularly from anal and throat swabs. Both CF and neutralizing antibody were found in all mothers.
4. AAV-1 was isolated from both anal and throat swabs from the mothers whose fetuses were injected with AAV-1 and helper. In one case AAV-1 was isolated from sera taken 2 weeks postinjection.
5. As an additional control one fetus was injected with a combination of AAV-1, SV-15, and AAV-1 antibody. This fetus was aborted 6 weeks after injection and AAV-1 was isolated from both fetal and placental tissue as well as from a throat swab taken from the mother soon after she aborted. SV-15 was also isolated from an anal swab taken 2 weeks after the abortion by which time she had developed a low level of CF antibody (1:8 vs. AAV-1, 1:4 vs. SV-15). Two weeks later low levels of neutralizing antibody were found.

Although of necessity the number of experimental animals used was low and it was not possible to demonstrate any AAV latency, it was concluded that in these experiments adenovirus was necessary for AAV replication and although SV-15 caused no obvious respiratory symptoms in the mothers all fetuses were killed by SV-15 with or without AAV.

V. LATENT INFECTION

The outstanding biological feature of the AAV is their absolute defectiveness. In the absence of coinfection with a suitable helper virus, either Ad or one of the herpesviruses, the AAV virion can attach to a cell and penetrate to the nucleus where the viral genome is uncoated. However, in the absence of a helper virus, no AAV-specific DNA, RNA, or protein synthesis can be detected by traditional methods (Rose and Koczot, 1972). In a biological sense, therefore, the defectiveness of AAV poses an interesting question: namely, how can AAV assure the biological continuity of its genome? Clearly, a suitable helper virus will not always be present to assure a productive coinfection. An increasing accumulation of evidence would suggest that AAV surmounts this problem by being able to efficiently establish a latent state in the infected cell in the absence of helper virus coinfection. The ability to maintain a latent state is a property AAV shares with other types of nuclear DNA viruses. The unique feature of AAV latent infection is that the AAV genome is efficiently rescued and replicated upon coinfection of the latently infected cell by an appropriate helper virus. As with other nuclear DNA viruses the process of AAV latent infection is potentially a useful way to escape destruction by host defense mechanisms, but unlike autonomous viruses, the defectiveness of AAV would seem to make the process an integral step in the replicative cycle of the virus.

Although the existence of latent infection by AAV has only been recognized in the past ten years, in many respects the molecular mechanisms underlying the process are better understood than for any other DNA virus. This is primarily a consequence of two special features: (1) the small size of the AAV genome (mol. wt. = 1.5×10^6; Gerry *et al.*, 1973) facilitates detailed analysis, and (2) because of the defectiveness of AAV it is possible to readily establish AAV latent infection in cultures of cells from the normal host. The latently infected cell cultures seem to behave in a manner entirely consistent with the properties of AAV latent infection observed *in vivo.* In succeeding sections the properties of AAV latent infection both *in vivo* and *in vitro* are spelled out in detail.

A. Latent Infection *in Vivo*

AAV latent infection *in vivo* was discovered as a consequence of a government program in the United States to screen primary cell cultures intended for use in production of viral vaccines for the presence of endogenous viruses (Hoggan *et al.*, 1972). None of the primary human or monkey cell lots tested were positive for AAV when assayed for infectivity or for the presence of AAV structural antigens. However, when the cell lots were challenged by infection with either human or simian adenoviruses, 1–2% of the human cell lots and up to 20% of the monkey cell lots produced infectious AAV. It was thus apparent that in monkeys, at least, AAV latent infection was a fairly common phenomenon. The relatively low rates observed in human cell lots may be attributable to the facts that the cells were of embryonic origin and that even under institutional conditions most people do not convert to seropositivity for AAV before the age of 7 or 8 years (Blacklow *et al.*, 1968b).

Although not as clearly documented as above, it has been suggested that avian AAV may also be able to latently infect chickens (El Mishad *et al.*, 1975; Yates *et al.*, 1976, 1979). This hypothesis arises from the observation of the ubiquity of avian AAV contamination of stocks of chicken embryo lethal orphan (CELO) virus, an avian adenovirus.

B. Latent Infection *in Vitro*

Because of a desire to study the events underlying AAV latent infection at the cellular and subcellular levels, attempts were made to establish model systems in cell culture very shortly after the discovery of latent infection *in vivo*. This proved quite easy to do. Infection of Detroit 6 cells (a continuous line of human cells) with a high multiplicity of AAV2 (250 tissue culture infectious doses, $TCID_{50}$/cell) led to a cell culture which continued to release AAV for more than 100 passages upon challenge with adenovirus (Hoggan *et al.*, 1972; Berns *et al.*, 1975). In the absence of adenovirus infection the cells were negative for AAV infectivity and structural antigens. The latent infection was not inhibited by the presence of AAV antiserum in the maintenance medium. At lower multiplicities of infection latent AAV infection was apparently established but was not maintained indefinitely by blind passage. Cells infected with 2.5 $TCID_{50}$/cell remained positive after challenge with adenovirus for 34 passages while those infected with 0.025 $TCID_{50}$/cell remained positive for only 8 passages. After 39 passages individual cells were cloned from the culture originally infected with 250 $TCID_{50}$/cell. Of 63 such clones tested, 18 produced AAV after infection with Ad. One of the positive clones was recloned. All of the subclones obtained by this procedure were positive for AAV production upon Ad challenge.

Very similar results were obtained upon infection of KB cells (a second continuous line of human cells) with AAV 1 (Handa *et al.*, 1977). Cultures were infected at a multiplicity of 500 fluorescent infectious units/cell in microwells. Upon challenge with Ad 12, 1 in 50–60 microwell clones was positive for AAV 1 antigens. Positive clones were subcloned and 2 clones were obtained which remained positive for AAV for 40 and 35 passages, respectively.

In both cases the latently infected cell cultures seemed to be biologically equivalent to *in vivo* latent infection. The latent infection in cell culture yielded cells of human origin which were negative for AAV antigens and infectivity until challenged by infection with Ad, exactly as observed with primary cultures of latently infected human cells.

1. The Proviral State

With the existence of clones of latently infected human cells it was possible to begin investigations on the state of the viral DNA in latently infected cells. Berns *et al.* (1975) were able to estimate the number of AAV genome copies per cell by measuring the ability of total cellular DNA to increase the rate of reannealing of a very small amount of highly radiolabeled AAV DNA. Because the annealing process is a bimolecular reaction, addition of cellular DNA containing AAV sequences increased the total amount of AAV DNA in the reaction, leading to an increase in the reannealing rate. From the observed enhancement it was possible to calculate that the two clones of latently infected Detroit 6 cells examined contained 3–5 copies of the AAV genome per diploid amount of cell DNA.

Whether the AAV genome present in latently infected cells is integrated into host DNA or exists as a plasmid was first addressed by Handa *et al.* (1977). They isolated total cell DNA from latently infected KB cells and separated high-molecular-weight chromosomal DNA from smaller DNA (which would have presumably included plasmid DNA) by the "network" technique. In this method the cell DNA is denatured and reannealed for a very brief period sufficient to allow annealing only of highly reiterated sequences. Such sequences are only present in chromosomal DNA, but sequences of the same family are present throughout the chromosomal DNA so that the brief period of reannealing leads to formation of a high-molecular-weight network of partially reannealed chromosomal DNA which may be selectively salt precipitated to separate it from low-molecular-weight DNA that does not contain highly repetitive sequences. The two fractions were assayed for the presence of AAV sequences by their ability to enhance the rate of reannealing of highly radiolabeled purified AAV DNA as described above. By these means these investigators found that 80% of the AAV sequences were associated with the high-molecular-weight chromosomal DNA which had precipitated. The association with chromosomal DNA was presumed to be covalent because it resisted severe denaturing conditions. The authors concluded that at least 80% of the AAV DNA sequences in the two clones they examined were integrated into cellular DNA. Furthermore, because only 80% of the chromosomal DNA was precipitated by the "network association" procedure, it was possible that all of the AAV DNA present represented integrated sequences.

The arrangement of AAV DNA sequences in several clones of Detroit 6 cells has been investigated in detail; using the Southern blotting procedure (1975). In one study (Cheung *et al.*, 1980), the state of the AAV DNA was studied in cells from both early (passages 8 and 9) and late (passage 118) passages of a clone. When total cell DNA was analyzed by electrophoresis through agarose gels there was an interesting discrepancy observed between the DNAs from early and late passage cells. In early passage cells all of the AAV sequences were associated with high-molecular-weight DNA ($>20 \times 10^6$ daltons); however, in the DNA preparations from late passage cells a small but definite amount of the AAV DNA migrated at a rate identical to that of the linear annealed form of DNA purified from virions. The majority of the AAV DNA in late passage cells was again found in a high-molecular-weight form. Further analysis using restriction endonucleases of the low-molecular-weight AAV DNA present in late passage cells demonstrated that this DNA was identical in terms of size and physical map to AAV virion DNA at the level of resolution afforded by such studies. Thus, it was concluded that the late passage cells contained some AAV DNA in a free form. The origin of the free DNA is not clear. Because it was not observed in early passage cells the free form was presumably derived from the high-molecular-weight form. The process by which this rescue or "popping out" occurred is unknown, although the fact that the high-molecular-weight forms observed contain a tandem repeat of the AAV genome could have contributed to recombination out of single copies (see below). Also undetermined at this time is whether the "free form" can replicate or is simply diluted out with continued passage. If the latter were true then the rescue process would presumably operate at a slightly faster rate than the dilution to account for the ready detection of the free form in late passage cells.

Further analysis of the high-molecular-weight DNA forms after digestion with restriction enzymes gave essentially identical results for both early and late passage cells. The high-molecular-weight species could have represented large plasmids containing many repeats of the AAV genome or they could have represented integration of the AAV genome into cellular DNA. To resolve this question total cell DNA was digested with several enzymes that do not cleave AAV DNA (*Pvu*I, *Bgl*II, *Hpa*I). A molecule containing only AAV sequences would not be cleaved and thus retain its original mobility. However, in at least 2 of the 3 cases the molecular weight of the high-molecular-weight species was significantly decreased, presumably by cleavage of host sequences flanking an integrated AAV genome(s). Only one band could be resolved after digestion; the molecular weight was about 20×10^6 daltons. Because the reannealing rate studies cited above had indicated several copies of AAV DNA per diploid amount of cell DNA in the clone under investigation, the possibility arose that the several copies of the AAV sequence were integrated at a single site as a linear tandem repeat. However, the difficulty of determining exact molecular weight in this size range on the agarose gels employed renders such a conclusion hypothetical at best.

Purified AAV virion DNA is linear; however, potential intermediates in the integration process might be either linear or circular. In the former case the joint between viral and cellular DNA would be most likely to involve the termini of the viral genome. In the latter case junctions between viral and cellular DNA could involve any region of the viral genome, as has been observed in the case of the papovaviruses (Ketner and Kelly, 1976; Botchan *et al.*, 1976). To characterize the portions of the AAV genome directly linked to cellular DNA, total cell DNA was digested with *Hind*II which cleaves AAV at three sites. All sequences normally contained in the two internal fragments from virion DNA were recovered in species exactly corresponding to the two internal fragments, with one exception. Some of the internal sequences were found in a species corresponding to a fusion product of the two internal fragments. Apparently the middle *Hind*II site had been altered in some of the integrated sequences. Three types of species containing terminal sequences were seen. None corresponded to terminal fragments from virion DNA. The first contained sequences from both termini of AAV virion DNA and corresponded in size to a direct fusion product of the two terminal fragments. The second also contained sequences from both ends but was larger than a simple fusion product. Whether the additional sequences were of viral or cellular origin was not determined. The last type of species contained sequences from only either the left or the right end of the viral genome and migrated more slowly than the corresponding terminal fragments from virion DNA. These appeared to represent the fragments containing the junctions between viral and cellular DNAs. Similar results were obtained with a second 3 cut enzyme (*Pst*I). These data indicated that the AAV terminal DNA sequences were specifically present at the site of integration into cellular DNA and were consistent with integration via a linear intermediate. However, on the basis of these data, it is not possible to rule out a circular intermediate which recombines into cellular DNA via a specific sequence in the circle. This hypothesis is similar to the Campbell model for bacteriophage integration (1962) except that the sequences involved in the integration process would be the original terminal sequences of the linear virion DNA.

The presence of species representing fusions between both ends of the viral genome in the experiments cited above was also in accord with the previously mentioned possibility that the integrated AAV sequences exist as a linear tandem repeat. The tandem repeat would have to be of the head-to-tail variety to correspond to the bringing together of sequences from both ends of the viral genome. To further test this point, total cell DNA from early passage cells (no free forms of AAV DNA) was digested with either of two enzymes (*BamH*I and *Hind*III) which cleave virion AAV DNA once. If the linear head-to-tail tandem repeat hypothesis were correct, such digestions should have yielded some species which corresponded in mobility to unit length AAV DNA. These species would have corresponded to the species generated by *Hind*III digestion that contained sequences from both termini with a molecular weight equal to the sum of the two normal terminal fragments. In fact digestion with either *BamH*I or *Hind*III produced unit length species in accord with the tandem repeat model.

Digestion of the integrated AAV sequences with *Hind*II or *Pst*I had indicated that the site of integration was specific with regard to viral sequences. Studies with an independently derived clone of Detroit 6 cells latently infected with AAV2 gave comparable results (K. Berns, and M. Lewis, unpublished experiments). However, junction fragments produced by digestion of the two clones with the same restriction enzymes differed in mobility. From these data it was concluded that the site of the integration of AAV DNA into cellular DNA was not specific with regard to the host sequences involved at the level of resolution provided by restriction enzyme analysis.

In summary, in studies of several clones of human cells latently infected by AAV the data were all consistent with the integration of viral DNA as a linear tandem repeat into cellular DNA. The integration process seemed to be specific for the terminal sequences of the viral genome but random with regard to cellular sequences.

2. Rescue

AAV is absolutely defective for replication. The defect is at the level of macromolecular synthesis and may be overcome by coinfection by any adenovirus or herpesvirus for which the system is permissive. Rescue of AAV from a latent infection has exactly the same requirement for a helper virus coinfection. A salient question given these facts is whether there is an additional helper virus function required for AAV rescue.

Existing evidence would indicate there is indeed an additional helper function required. Ad host range (*hr*) mutants have been described which can grow in Ad-transformed human cells (H31) but cannot multiply in HeLa cells (Harrison *et al.*, 1977). H31 cells have the left 12% of the Ad genome integrated and express the left 11% which contains Ad early regions 1a and 1b (Graham *et al.*, 1977). Early region 1a expression seems to be required for all subsequent Ad transcription (Berk *et al.*, 1979; Jones and Shenk, 1979) while early region 1b seems to be required for, among other things, either stable or abortive cellular transformation by Ad. The role of early region 1b in lytic infection is unclear. Ad *hr*3, a mutant mapped to early region 1a, is DNA negative (Lassam *et al.*, 1978) and supports neither AAV replication nor rescue in a coinfection (Ostrove and Berns, 1980). Ad *hr*6 which has been mapped in early region 1b is a DNA positive mutant which is quite leaky for replication but which appears to be absolutely unable to transform cells (Graham *et al.*, 1978). In an exogenous coinfection with AAV, *hr*6 is able to support AAV DNA replication as well as wild-type virus. However, *hr*6 appears to be unable to rescue AAV in the sense that in cells latently infected by AAV, challenge with *hr*6 infection does not lead to detectable AAV DNA synthesis. Thus, the *hr*6 mutation seems to define one Ad function which is specifically required for rescue of AAV DNA from its integrated form in the latent state. The function that *hr*6 is known to be deficient in is the ability to synthesize detectable amounts of the 58K Ad T-antigen. One obvious possibility, therefore, is that it is the Ad T-antigen itself that is required for AAV rescue. This interesting phenomenon might be related to the ability of AAV infection to inhibit Ad oncogenicity (see below). Ad-transformed hamster cells in culture have their oncogenicity significantly reduced when they are infected with AAV prior to inoculation into newborn hamsters. At the molecular level the AAV infected cells contain a greatly reduced quantity of immunoprecipitable 58K T-antigen (Ostrove *et al.*, 1981).

C. Rescue of AAV DNA Cloned in pBR322

The intact AAV2 genome in its double-stranded form has been cloned into pBR322 in *Eschericia coli* (Samulski *et al.*, 1982). Cloning was achieved by addition of short poly dC tails onto the 3′ ends of AAV DNA and annealing the product to pBR322 DNA which had been converted to the linear form by digestion with *Pst*I and tailed at the 3′ termini with poly dG. The recombinant DNA molecule produced after transfection into *E. coli* and cloning has the exciting property of being biologically active in human cells. Upon transfection of either H31 cells or HeLa cells in the presence of a concomitant Ad infection the AAV DNA is rescued from the recombinant plasmid, replicated, and packaged into infectious virions. From the point of view of analyzing the rescue process in AAV latent infection, the distinct possibility exists that the recombinant AAV-pBR322 plasmid represents an accurate model of AAV latent infection with a completely defined nucleotide sequence [i.e., both pBR322 and AAV DNA have been completely sequenced (Sutcliffe, 1978; Lusby and Berns, 1981; Srivastava, *et al.*, 1983)]. With such a system it is possible to begin to ask specific quantitative questions with regard to the nucleotide sequence and helper function requirements of the rescue process. For instance, if the helper functions apparently lacking in Ad *hr*6 have been correctly assessed (Ostrove and Berns, 1980), it might be possible to detect differences in the fate of the recombinant plasmid after transfection in the absence of Ad coinfection into H31 cells which express the Ad T-antigen as opposed to transfecting the plasmid into HeLa cells lacking the Ad T-antigen. Indeed, in preliminary experiments of this type the AAV sequences can be detected by blotting of agarose gels, but no new forms have been detected. This suggests that additional Ad functions are required for rescue. Similarly, early experiments have begun to demonstrate nucleotide sequence requirements for complete rescue. Recombinant plasmids in which one of the two AAV inverted terminal repetitions has been completely deleted cannot be rescued, although those with smaller deletions within the terminal repetition may be if the deletions are confined to only one end of the genome. This system promises to greatly enhance our knowledge in this area.

D. Outstanding Questions

Our rapidly increasing knowledge of the structure and function of the AAV genome now places us in a position to begin asking much more fundamental questions about the mechanisms underlying the phenomenon of AAV latent infection. One important aspect of this problem involves the mechanism of viral DNA integration. Current data suggest that this process is likely to occur via a linear intermediate and that the inverted terminal nucleotide sequence repetitions of the AAV genome are important. However, there is no information about which enzymes may be important in the recombination event and whether any may be coded for by the viral genome. Because of its terminal repetitions and ability to recombine in and out of the host cell DNA AAV would seem to share some attributes with bacterial transposons. Might it also code for its own transposase as an essential element in integration?

Another critical question concerns how a linear single-stranded virion DNA is incorporated in a double-stranded form. Several explanations are possible: (1) Although AAV DNA replication cannot be detected by traditional techniques of DNA hybridization, it is possible that the incoming DNA can use cellular enzymes to synthesize a complementary strand and that the critical dependence of AAV DNA replication on the helper virus has to do with synthesis of a site-specific nuclease to cleave the hairpin presumably formed at the initiation site for DNA replication (Straus *et al.*, 1976; Berns *et al.*, 1978; see Chapter 1). Preliminary evidence to support the notion that the incoming AAV viral strand can direct synthesis of a complement in the absence of helper virus has been obtained from both *in vitro* and *in vivo* experiments (K. Berns, W.W. Hauswirth, and R. Bohenzky, unpublished experiments). (2) Alternatively, AAV DNA may begin to insert as a single strand and in the process of recombination be converted to the duplex form. This notion would be in accord with the model proposed by Galas and Chandler (1981) for transposition, which features single-strand insertion into a double helix as a first step.

Finally, in the two clones of human cells latently infected by AAV that have been intensively studied, the inserted viral DNA sequences exist as a linear head-to-tail tandem repeat. Again this observation might be explained by two possible mechanisms: (1) Recombination of viral DNA molecules prior to insertion into cell DNA could create a suitable tandem. DNA transfected into mammalian cells is known to recombine into larger units before integration (Perucho *et al.*, 1980). (2) The initial insertion of a viral DNA into cellular DNA might be the rate-limiting step in the process. Because the clones so far investigated have been the results of high multiplicities of infection, presumably other copies of viral DNA in the nucleus could more readily integrate into the first integrated copies of the viral genome because of the increased sequence homology.

The molecular mechanism of AAV rescue is still uncertain. Two major possibilities could include excision by recombination or by a special mode of selective replication of the integrated AAV sequences, as has been suggested for the rescue of SV40 (Botchan *et al.*, 1979). The relationship of the early region 1b both to the rescue process and to synthesis of the Ad T-antigen is also of interest. A current hypothesis (Cairns, 1981) is that transformation of a normal cell to a tumor cell is much more likely to result from a recombinational event than from a simple mutation. If, indeed, the way that the Ad T-antigen works is to promote recombinational events, then its relationship to the rescue of the AAV genome from the integrated state would be readily comprehensible. Indeed the existence of the recombinant AAV-pBR322 plasmid could potentially represent a useful *in vitro* substrate to test this hypothesis. Why, on the other hand, AAV infection of an Ad-transformed cell inhibits the synthesis of, sequesters, or causes the degradation of the Ad T-antigen is much more difficult to grasp.

Although data on the subject are not exhaustive, it seems probable that AAV latent infection is common and widespread among several species, including man. Probably the most significant biological question in this area concerns the consequences of AAV latent infection. With regard to the virus, the advantages of an efficient mechanism of latent infection are evident and have already been suggested above. The consequences for the host, however, are much more uncertain. AAV has never been demonstrated to be associated with disease in a causative role in any species so that any deleterious consequences of the latent infection are problematic. On the other hand, in many ways AAV is akin to the defective interfering particles of other viruses, insofar as its helper viruses are concerned. AAV is known to inhibit the lytic replication of Ad (Atchison *et al.*, 1965; Hoggan *et al.*, 1966; Casto *et al.*, 1967a,b) and the oncogenicity of both Ad and herpes simplex virus (Kirchstein *et al.*, 1965; Mayor *et al.*, 1973; Blacklow *et al.*, 1978; see below). Thus one possible consequence of AAV latent infection may be that it can exert a protective effect upon the host. According to this hypothesis upon active infection of the host by a helper virus the latent AAV would be rescued and replicated. During the process the helper virus infection would be inhibited and the potential harm to the host reduced accordingly. This hypothesis has not been directly tested although several epidemiological studies have been carried out in the human population. In one such study patients at a major cancer center who suffered from carcinoma of either the cervix or the prostate, both epidemiologically associated with HSV type II, were compared to matched normal controls in terms of serum antibody titers against both HSV type II and AAV (Mayor *et al.*, 1976). In the controls >85% had positive titers against both viruses. Thus a large fraction of the controls had the possibility of latent AAV. Among the patients studied, however, although >85% had positive serum titers against HSV type II, only 14% had measurable antibodies against AAV. This type of data suggests that the question of a possible protective effect of latent AAV infection probably deserves more extensive study.

VI. INHIBITION OF ADENOVIRUS ONCOGENICITY

The absolute defectiveness of AAV forms the basis for an intimate relationship between AAV and its helper viruses, Ad and the herpesviruses. This is particularly true in the case of Ad, which is probably the natural helper most of the time. Although completely unrelated to Ad structurally, AAV behaves biologically as the equivalent of a defective interfering particle. In a lytic coinfection AAV may significantly inhibit the replication of Ad (Hoggan *et al.*, 1966; Casto *et al.*, 1967a,b; Parks *et al.*, 1968; Carter *et al.*, 1979; see Chapter 9). Additionally, some human Ad are oncogenic *in vivo* (Trentin *et al.*, 1962; Huebner *et al.*, 1965) and all Ad have the ability to transform cells in culture to an oncogenic phenotype (Freeman *et al.*, 1967; Williams, 1973; Graham *et al.*, 1977). AAV has been reported to inhibit both of these functions. The ability of AAV to inhibit these important biological activities of Ad emphasizes the interrelatedness of the two viruses which probably involves at least several Ad gene functions. Thus, it is possible to use AAV as a probe of Ad gene function. The ability of AAV to inhibit Ad lytic infection is reviewed in Chapter 9. In this chapter the inhibition of Ad oncogenicity by AAV is explored in detail. In order to place this inhibition in context the known properties of Ad oncogenicity are summarized below.

A. Adenovirus Oncogenicity

Studies of Ad oncogenicity have been focused for the most part on human Ad. In terms of *in vivo* oncogenicity no human Ad has been associated with induction of tumors in people. Tumor induction has been most carefully studied in neonatal Syrian hamsters which are a semipermissive host for human Ad replication. The tumors which arise have been characterized as undifferentiated sarcomas (Williams, 1973). In this system the human Ad have been divided into three groups, A, B, and C, based on the frequency of tumor induction after viral infection (Trentin *et al.*, 1962; Huebner *et al.*, 1967). Those in group A (represented by serotypes 12, 18, and 21) cause tumors at a high frequency and with a relatively short induction period, while those in group B (represented by serotypes 3, 7, 11, and 16) cause tumors less frequently with a longer induction period. Group C Ad (serotypes 1, 2, and 5) do not cause tumors after infection. The Ad in each of the three groups are closely related in terms of DNA sequence homology and G + C content. Group A viruses share 85–90% homology, group B viruses 70–100%, and group C viruses 85–90%, while the intergroup homology is only 10–25% (Lacy and Green, 1965; Fujinaga *et al.*, 1969, Green, 1970). Group A viruses have DNA with a G + C content of 48–49%, group B viruses 49–52% G + C, and group C viruses 57–59% G + C (Pina and Green, 1965). Interestingly the G + C content of the highly oncogenic group A viruses most closely approximates the 42–44% G + C content of mammalian cell DNAs. In contrast to the results observed after infection of intact animals, all human Ad have the ability to transform hamster cells in culture to an oncogenic phenotype (Freeman *et al.*, 1967; Williams, 1973; Graham *et al.*, 1977). Why the intermediate step of first infecting cells in culture allows all Ad to express an oncogenic potential is uncertain, but a reasonable hypothesis would be that the rate of cell transformation for the different groups of Ad differs sufficiently so that after group C Ad infection animals have sufficient time to mount a successful host immune response.

Detailed studies have been performed at the molecular level to identity those portions of the Ad genome responsible or required for transformation. The Ad genome is a linear, double-stranded DNA molecule approximately 30 kb long (Pina and Green, 1965). Two special features of the genome are an inverted terminal nucleotide sequence repetition of approximately 100 nucleotides (Garon *et al.*, 1972; Wolfson and Dressler, 1972; Steenbergh *et al.*, 1977) and a 55-Kd protein covalently linked to each 5′ terminus by a phosphate diester (Robinson *et al.*, 1973; Rekosh *et al.*, 1977; Desiderio and Kelly, 1981). Five regions of the genome have been identified as coding for early viral functions expressed prior to viral DNA synthesis (Berk and Sharp, 1977,1978; Galos *et al.*, 1979; Philipson *et al.*, 1979; Wilson *et al.*, 1979). Early regions 1 and 3 are on the *r* (rightward 3′-5′) strand; 2, 4, and 5 on the 1 strand. Early region 2 (75–62 map units) codes for a single-stranded DNA binding protein (72K protein) while early region 5 (29–11 map units) may code for the 55K protein found at the termini of virion DNA. Two small nonmessenger RNAs (VA) are also coded for in or near early region 2. It has been suggested that these may be involved in mRNA splicing. No specific Ad antigens or Ad functions have yet been ascribed to early regions 3 (76–86 map units) or 4 (99–91 map units). Interestingly, though, early region 4 RNA apparently codes for the specific helper function required for AAV DNA replication (Richardson and Westphal, 1981, see Chapter 3). Early region 1 maps from 0–11 map units and appears to encode the functions required for Ad transformation. This region has been divided into two parts, early regions (e) 1a and 1b. Promoters have been mapped at 1.5 and 4.5 map units, respectively (Wilson *et al.*, 1979). All transformed cells in culture contain the left end of the Ad genome integrated into cellular DNA. Physiological transformation has been observed after transfection with the left 4.5% of the genome which corresponds to early region 1a and codes for the Ad-specific 49K, 33–55K, 25K, and 11K proteins (Levinson and Levine, 1977; Ross *et al.*, 1980). However, the major biological function of early region 1a seems to be to turn on all subsequent Ad transcription. Cells transformed with the left 8% of the Ad genome produce a 19K T-antigen in addition to the early region 1a T-antigens (Schrier *et al.*, 1979). However, the major 58K Ad T-antigen is not observed until the left 15% of the genome is used to transform cells in which case all the Ad T-antigens are synthesized (Schrier *et al.*, 1979). Finer mapping has placed the right limit of early region 1b at 11 map units. Although several RNAs have been mapped in this region the number of polypeptides observed is greater, creating a dilemma, still unresolved, with regard to the specific origins of some of the antigens.

Experiments with Ad mutants have generally supported the results obtained by tranfection with specific DNA fragments. Host range (*hr*) mutants of Ad have been obtained which do not replicate in HeLa cells but do grow in 293 cells (Harrison *et al.*, 1977). The latter are human cells which have been transformed by Ad 5. They contain the left terminal 12% and the right terminal 9% of the genome. Only the left 11% of the genome is transcribed and translated (Graham *et al.*, 1977; Aiello *et al.*, 1979). The Ad *hr* mutants are divided into two groups. The group I mutants map between 0 and 4.4 in early region 1a and the group II mutants between 4.4 and 9 in early region 1b. The group I mutants are DNA negative in lytic infections but can abortively transform cells while the group II mutants synthesize DNA, are deficient in late protein synthesis, and most importantly are absolutely defective in transformation (Harrison *et al.*, 1977; Graham *et al.*, 1978; Lassam *et al.*, 1978; Frost and Williams, 1978; Berk *et al.*, 1979; Lassam *et al.*, 1979). Thus where early regions 1a and 1b are present together an intact early region 1b is absolutely required for transformation. This is somewhat paradoxical with regard to the transfection data where the left 4.4% of the genome was adequate to give some evidence of abortive transformation. Similarly the ability of group I (e1a) mutants to abortively transform cells suggests that some 1b transcription can occur even with a serious defect in 1a.

In spite of these somewhat paradoxical results and an almost total lack of information about the functions of the various Ad T-antigens it is clear that a significant amount of insight has been gained at the molecular level with regard to Ad transformation. It has been established that in Ad transformation specific parts of the Ad genome are integrated into host cell DNA (Sambrook *et al.*, 1974) and that early region 1b or its equivalent is absolutely required for transformation after infection with an intact Ad genome (Lassam *et al.*, 1978). Thus it is now possible to study the inhibition of Ad oncogenicity by AAV at the molecular level as well as at the levels of the cell and whole animal.

B. Inhibition *in Vivo*

Inhibition of Ad oncogenicity by AAV was first reported by Kirchstein *et al.* (1968). Human Ad 12 (group A) were injected into neonatal Syrian hamsters in either the absence or presence of AAV 1. In the two experiments reported 26/43 animals inoculated with Ad 12 alone developed tumors. The mean latent period until tumors were detectable was 45 days with a range from 33 to 81 days. Of those animals inoculated with both Ad 12 and AAV 1 only 12/58 animals developed tumors. The mean latent period was lengthened to >60 days with a range from 47 to 115 days. Under both conditions the tumors induced were undifferentiated sarcomas that were histologically indistinguishable. The authors concluded that AAV could partially inhibit the oncogenicity of Ad without altering the morphology of the transformed cells.

Mayor *et al.* (1973) were also able to demonstrate an inhibition of Ad oncogenicity. Two sets of experiments were performed. In the first the incidence of tumors induced by Ad 31 inoculation of neonatal Syrian hamsters was reduced from 50% to 0 when Ad inoculation was preceded by AAV 1 inoculation 24 hr earlier. One inconsistency in this experiment is that the control group were inoculated with Ad 31 on the first day of life whereas the experimental group were not inoculated with Ad for a further 24 hr. That this might have been an important difference is suggested by the results obtained in the second set of experiments reported. In these there was no difference in tumor incidence noted when animals were inoculated during the second day after birth, regardless of whether they had been inoculated with AAV 1 on day 1. Similarly there was no decrease in Ad oncogenicity when 2-day-old animals were simultaneously inoculated with AAV and Ad. Thus, in detail, these results differ from those of Kirchstein *et al.* (1968). In an additional experiment Mayor *et al.* (1973) found that no protection was conferred upon newborn animals if the mother had been inoculated with AAV 1, 8 days prior to birth. In fact, Ad tumor induction in female offspring of mothers inoculated with AAV increased to 100% from 27% and was associated with a decreased average latent period. Male offspring were not affected. However, this result should be viewed with some caution because it represents only a single experiment which was apparently not repeated. In the first set of experiments, where inhibition of Ad oncogenicity by AAV was noted, there was no inhibition of SV40 oncogenicity measured in a parallel experiment. Thus the AAV inhibition of Ad oncogenicity seemed to have a degree of specificity.

Gilden *et al.* (1968a,b), on the other hand, were unable to find any inhibition of Ad 12 oncogenicity when newborn Syrian hamsters were inoculated with Ad 12 alone or with mixtures of Ad 12 (10^7 $TCID_{50}$/ml) and either added AAV 1 or AAV 2 ($TCID_{50}$ ranging from 10^6 to 10^{10}/ml). The inability to demonstrate an inhibition was attributed to the very low level of infectivity in the AAV 1 and AAV 2 stocks used (Gilden *et al.*, 1968b). However, this explanation for the paradoxical results observed in these experiments may be questioned in the light of subsequent experiments reported by de la Maza and Carter (1981) that both AAV defective interfering (DI) particles and specific fragments of AAV DNA could inhibit Ad oncogenicity.

AAV infection leads to the production of defective interfering (DI) particles. Such particles contain viral genomes with deletions of varying sizes. The intact AAV genome is a linear single-stranded DNA of molecular weight 1.5×10^6 (Gerry *et al.*, 1973) with an inverted terminal nucleotide sequence repetition of 145 nucleotides. The deletions in the DNA molecules isolated from AAV DI particles always involve internal sequences—the smaller the DNA the greater the internal deletion. To date all DI DNAs which have been characterized have contained the inverted terminal repetition sequence. De la Maza and Carter (1981) were able to confirm the observations of Kirchstein *et al.* (1968) and Mayor *et al.* (1973). Two-day-old hamsters were inoculated with 5×10^6 $TCID_{50}$ Ad 12. In those which had been inoculated on day 1 with AAV 2 (7.7×10^{11} particles) the incidence of tumor formation was reduced from 49 to 12% with an increase in the mean latent period of tumor formation from 39 to 97 days. No tumors were observed after inoculation of animals with Ad 12 after inoculation on day 1 with AAV DI particles. This observation bears on the suggestion by Gilden *et al.* (1968b) that infectious AAV might be required to inhibit Ad oncogenicity. AAV DI particles cannot inhibit Ad replication and apparently do not replicate in the absence of coinfection with intact AAV (Carter *et al.*, 1979).

The conclusion that AAV DNA replication is not required to inhibit Ad oncogenicity is supported by additional experiments reported by de la Maza and Carter (1981). Inoculation of 1-day-old hamsters with purified AAV DNA also inhibited Ad oncogenicity. Interestingly, intact DNA had no protective effect, but sonicated DNA reduced the frequency of tumor formation from 49 to 30%. Even more dramatic results were obtained with DNA from DI particles. As noted above the DI DNA is enriched for terminal sequences suggesting some specifity for these sequences in the phenomenon of inhibition observed. DNA from DI particles missing over half the internal sequences reduced tumor formation from 49 to 18% while DNA isolated from DI particles containing little more than the terminal repetition completely abolished tumor formation. Amazingly, to the level of quantitation permitted by the nature of the experiments, AAV DNA was as effective in inhibiting tumor formation as intact particles. This is in contrast to the infectivity observed with transfection which is usually several orders of magnitude less than that observed with intact virions. This again suggests that AAV replication is not required for the inhibition of Ad oncogenicity. An additional observation by de la Maza and Carter was that poly(I): poly(C) did not have a protective effect, suggesting that the inhibition by AAV was not mediated by interferon.

C. Inhibition of Cellular Transformation

In order to be able to study the molecular mechanism by which AAV can inhibit the oncogenicity in Ad it is essential to be able to develop a model in cell culture. The first experiments at this level were reported by Casto and Goodheart (1972). They found that AAV 1 was able to inhibit the transformation of hamster embryo cells by Ad 12 or by two simian adenoviruses SA7 and SV11. The amount of AAV added to the culture was measured in terms of plaque-inhibiting units (PIU, the ability to inhibit plaque formation by Ad 1 in the hamster embryo cell monolayers). The inhibition of transformation was AAV dose dependent. At 5.6 AAV PIU/cell transformation (number of foci) by SA7 and SV11 was reduced 41% and 64%, respectively. At the maximum dose of 3.0 PIU/cell used, transformation by Ad 12 was reduced 66%. The authors concluded that AAV 1 could inhibit cellular transformation by these three adenoviruses and that inhibition of transformation was less efficient than the ability to inhibit Ad 1 replication in hmaster embryo cells (1 PIU/cell would have completely inhibited Ad 1 plaque formation).

The above experiments demonstrated that prior infection with AAV could inhibit Ad transformation of hamster embryo cells. Two sets of experiments have also been reported which demonstrate that AAV infection can decrease the oncogenicity of continuous lines of hamster cells already transformed by Ad. De la Maza and Carter (1981) obtained hamster cells transformed by Ad 12. If these cells were infected with AAV (10 $TCID_{50}$/cell) 24 hr prior to inoculation into 1-day-old hamsters the incidence of tumor formation was reduced from 53 to 25%.

Ostrove *et al.* (1981) reported a more extensive study on the decrease in oncogenicity of Ad-transformed hamster cells after AAV infection. H14b cells are a continuous line of hamster cells transformed by Ad 5 (Williams, 1973). The cells contain 5.5 copies of the left 40% of the Ad genome, but only the left 11% of the Ad genome, representing early regions 1a and 1b, is transcribed. Although Ad 5 is a group C virus H14b cells are oncogenic when inoculated into 1- to 2-day-old hamsters. When the hamsters were inoculated with 2×10^5 cells 75%–100% of the animals formed palpable tumors at the inoculation site 21–28 days later. When H14b cells were infected with AAV (2×10^3 fluorescent focus forming units/cell) prior to inoculation only 17–56% of the animals developed palpable tumors. The tumors that did develop had an increased latent period (up to 45 days) and in one experiment were 1,000-fold smaller (volume) than the tumors in control animals. The oncogenicity of the H14b cells was reduced both when the AAV infection preceded inoculation into hamsters by 2 and 72 hr, as well as at intermediate times of 6 and 24 hr. The statistical significance for five such experiments was $P<0.001$. Similar attempts to reduce the oncogenicity by AAV infection of chemically transformed hamster cells [dimethyl benzanthracene induced (Lausch and Rapp, 1971)] were not successful, again suggesting specificity in the inhibition of oncogenicity of the Ad-transformed H14b cells. The inhibition observed was not an all-or-none effect. When cells which had been infected with AAV 24 hr prior to inoculation were tested, only 16% of animals inoculated with 10^5 cells developed tumors. However, 50% of animals inoculated with 2×10^5 cells and 100% of the animals inoculated with 10^6 cells developed tumors. Thus the oncogenicity of the H14b cells was inhibited but not abolished by AAV infection. Additionally the tumors which developed were undifferentiated sarcomas regardless of whether or not cells had been preinfected with AAV.

The cells were also assayed in terms of several biological properties to see if these had also been altered as a consequence of AAV infection. For three days after infection there was no change in the growth rate, but by 5 days both control and infected cells had stopped growing and the density of the infected cells was only $\frac{1}{3}$–$\frac{1}{2}$ that of the controls. This difference was not relieved by addition of fresh medium. However, when the infected cells were replated at 6 days the subsequent growth rate was indistinguishable from that of the control H14b cells. Apparently the inhibitory effect of AAV infection on saturation density was temporary. Another mark of biological cell transformation is the ability of H14b cells to form clones in methyl cellulose (i.e., loss of anchorage dependence). In five experiments AAV infection of H14b cells reduced the number of colonies formed. The percent inhibition ranged from 38 to 92%. Thus by two measurements of transformation *in vitro*, AAV infection was able to reduce the transformed phenotype without having an apparent effect on the rate of cell division.

Neither AAV virion production nor DNA replication has ever been observed in a transformed cell, but in light of the conflicting suggestions of Gilden *et al.* (1968b) and de la Maza and Carter (1981) the ability of AAV to replicate its DNA in H14b cells was tested. Electrophoresis of Hirt (1967) supernatant fluid containing the small DNA synthesized in H14b cells after AAV infection failed to reveal any species characteristic of AAV. Thus, these data would be in agreement with the conclusion by de la Maza and Carter (1981) that AAV DNA replication is not required to inhibit Ad oncogenicity. However, when the synthesis of AAV RNA after infection of H14b cells was measured by Northern blotting (Alwine *et al.*, 1977) a significant level of synthesis was detected. This was the first instance in which AAV RNA synthesis had ever been observed in the absence of AAV DNA synthesis (Rose and Koczot, 1972). Among the species detected was one corresponding in mobility to the major AAV 2.3 kb transcript seen in productively infected cells (Laughlin *et al.*, 1979; Green and Roeder, 1980; Green *et al.*, 1980). The working hypothesis developed as a consequence of these data was that the Ad helper function required for AAV transcription was supplied by early region 1, which is all that is expressed in H14b cells (Williams, 1973). This is of interest in light of the observation that early region 1a RNA synthesis seems to be required for transcription of all subsequent Ad genes.

An attempt was made to characterize alteration(s) induced in H14b cells subsequent to AAV infection in terms of the Ad-specific macromolecules present. Ad DNA sequences integrated into cell DNA (5.5 copies of the left 40% of the Ad genome) were checked by Southern blotting (1975) and restriction enzyme analysis. No significant alterations in the DNA were seen. Similarly, no changes in the amount of the major 2.2-kb Ad RNA species thought to code for the 58K major T-antigen were detected in Northern blots. However, there was a consistent alteration in the pattern of minor species corresponding to approximately 5.9 kb in length. Control H14b cells had two bands in this region of the gel while the RNA from AAV-infected cells contained three bands, none of which corresponded to the two observed in the control cells. The significance of this apparent alteration in the minor high-molecular-weight RNA species is not understood nor indeed is the function of the minor RNA species in the control cells. Such high-molecular-weight RNA species have also been observed in other lines of Ad-transformed cells (Sawada and Fujinaga, 1980).

The major difference noted in the H14b cells after AAV infection was in the amount of the major Ad 58K T-antigen which could be observed on electropherograms of ^{35}S-labeled Ad proteins after immunoprecipitation. The amount of the Ad-specific 58K T-antigen which could be detected was reduced by 80% using an anti-T antiserum raised specifically against H14b cells in hamsters. No data are available which suggest the mechanism by which the decrease in the Ad 58K T-antigen is achieved. Possibilities which have to be considered would include inhibition of the processing or translation of Ad mRNA(s) and binding of the T-antigen by AAV DNA plus a possibly concomitant degradation of T-antigen. The latter possibilities are particularly interesting in light of the possible involvement of the 58K T-antigen in rescue of the AAV genome from the latent state (Ostrove and Berns, 1980; Samulski, Berns, and Muzyczka, unpublished experiments, see above).

D. Significance

The biological significance of the inhibition of Ad oncogenicity is difficult to assess at this time. As suggested above, in several of its interactions with Ad, AAV behaves as the equivalent of a defective interfering virus. Thus, the ability of AAV to inhibit Ad multiplication would suggest the possibility that a concomitant AAV infection or activation of a latent AAV infection might serve to moderate the virulence of Ad infection *in vivo*. In terms of Ad oncogenicity, however, the question of the significance of inhibition *in vivo* is difficult to evaluate because Ad has not been convincingly demonstrated to be associated with any human cancers nor have Ad which infect other species been demonstrated to induce tumors in the normally permissive host. In spite of this lack of evidence that Ad is oncogenic other than in model systems it is well accepted that a study of Ad transformation *in vitro* is a useful way to study the manner in which DNA viruses can induce cellular transformation at the molecular level. In this sense the ability of AAV to inhibit induction of the transformed state as well as the maintenance of the transformed state in cells transformed by Ad provides a potentially useful probe of the molecular mechanisms of Ad transformation. The available data indicate that the inhibition is at the level of reduction of T-antigens, especially the major 58K T-antigen. Still to be determined is the manner in which the observed reduction occurs. Additionally, the Ad 58K T-antigen has been suggested to be required for the rescue of the AAV genome from the latent integrated state (Ostrove and Berns, 1980). The existence of a recombinant AAV-pBR322 plasmid which responds to Ad early region 1 gene expression in a way which is indistinguishable from latent infection in cell culture or *in vivo* (Samulski *et al.*, 1982) suggests that the recombinant plasmid may prove to be a useful *in vitro* substrate with which to characterize a mechanism of T-antigen activity at the molecular level.

Although Ad has not been shown to be oncogenic in natural hosts, one AAV helper virus, herpes simplex virus (HSV), has been epidemiologically associated with some human tumors (Rawls *et al.*, 1968, 1970, 1973; Nahmias *et al.*, 1970; Rotkin, 1973) and the ability of AAV to inhibit HSV oncogenicity has also been reported (Blacklow *et al.*, 1978; see below).

VII. INTERACTIONS WITH HERPESVIRUSES

In contrast to the complete helper functions provided by adenoviruses, several studies have indicated that herpesviruses only partially assist in the replication of AAV. Coinfection of cells with herpesvirus and AAV results in the appearance of AAV-specific immunofluorescent-stainable antigens in the cell but without a detectable increase in the titer of infectious AAV. This observation, which was first made in 1970 (Atchison), has been confirmed and extended in a number of laboratories over the past decade (Blacklow *et al.*, 1970; Boucher *et al.*, 1971; Henry *et al.*, 1972; Rose and Koczot, 1972; Dolin and Rabson, 1973; Young and Mayor, 1979b). However, a recent report (Buller *et al.*, 1981) presents contradictory findings which indicate that herpes simplex virus (HSV) can indeed provide complete helper functions with the resulting efficient production of infectious progeny AAV. The reason for the discrepancy between this report and past studies is not resolved at the present time. In the earlier studies, all of the herpesviruses of humans, i.e., HSV types 1 and 2, cytomegalovirus, varicella-zoster virus, and Epstein–Barr virus (Atchison, 1970; Blacklow *et al.*, 1970) as well as two herpesviruses of animals, bovine rhinotracheitis virus and herpesvirus saimiri (Atchison, 1970; Dolin and Rabson, 1973) were shown to provide a partial helper function for AAV. Except for adenovirus and herpesvirus, no other viruses have been found to possess AAV helper activity (Atchison, 1970). Herpesvirus-transformed cells have also been observed to provide a restricted partial helper function for AAV (Dolin and Rabson, 1973; Blacklow, 1975).

Investigations of the interrelationship of herpesvirus and AAV may be divided into two categories: (1) the use of herpesvirus to elucidate the nature of AAV defectiveness, and (2) studies of the influence of AAV on the oncogenicity of herpesviruses.

A. Complementation of AAV by Herpesvirus

The nature of AAV defectiveness and the mechanism of helper activity are incompletely understood at present. AAV requires no helper function for normal cellular adsorption or penetration (Rose and Koczot, 1972). Several studies have indicated that with herpes simplex virus (HSV) as a helper, AAV-specific proteins, RNA, and infectious nonencapsidated DNA are produced (Atchison, 1970; Blacklow, *et al.*, 1970; Rose and Koczot, 1972; Boucher *et al.*, 1971). These observations have led to the hypothesis that in the absence of a helper virus, AAV manifests at least two groups of defects. One set of defects occurs in replication, specifically in DNA and RNA synthesis (Carter and Rose, 1972; Carter *et al.*, 1973) and another defect or group of defects involves viral maturation (Boucher *et al.*, 1971; Salo and Mayor, 1979). It thus appears that HSV can complement the defects in replication but not in maturation, whereas adenovirus provides both types of helper functions. Of course, it is alternately possible that while HSV complements AAV replication, one of the products of HSV infection inhibits AAV virion assembly (Berns and Hauswirth, 1979).

A recent report contradicts the previous findings that HSV can only act as a partial helper for AAV. Buller *et al.* (1981) have found that both HSV-1 and HSV-2 can be complete AAV helpers with AAV infectivity yields in some cell types being similar to the yields obtained with an adenovirus helper. The only difference noted by these investigators between adenovirus and herpesvirus helper functions was the time course of AAV macromolecular synthesis. AAV synthesis occurred earlier in coinfection with HSV than with adenovirus, apparently corresponding to the more rapid replicative cycle of HSV. No obvious explanation can be offered at this time to reconcile this study with previous reports. Buller *et al.* (1981) state that it is unlikely that either HSV or AAV multiplicities of infection were critical factors to explain their findings because the multiplicities used in earlier studies were within the range employed by these authors.

Temperature-sensitive (*ts*) mutants of HSV, which are defective in various virus-specific functions, have been useful in the analysis of AAV defectiveness. Several DNA-negative *ts* mutants of HSV are able to complement AAV protein and DNA synthesis at the nonpermissive temperature, indicating that AAV functions are not dependent on DNA synthesis by the helper virus (Drake *et al.*, 1974; Mayor and Young, 1978). Two distinct DNA-negative complementation groups of HSV *ts* mutants failed to support AAV synthetic activity. This observation implies that at least two HSV cistrons may be involved in the potentiation of AAV. However, the inability to complement AAV appears to be independent of the HSV mutant's failure either to replicate its own genome or synthesize the HSV-specific DNA polymerase (Mayor and Young, 1978). These findings are in agreement with the results of experiments employing phosphonoacetic acid (PAA), an inhibitor of the activity of the HSV-coded DNA polymerase (Young and Mayor, 1979a). The addition of PAA inhibited the AAV helper function of HSV. That the inhibitory effect was not exerted directly on AAV was indicated by the demonstration that AAV synthesis proceeded normally in the presence of PAA when either adenovirus or a PAA-resistant herpesvirus was used as a helper. However, restoration of HSV DNA polymerase activity by release of the PAA block did not result in initiation of AAV replication unless further *de novo* protein synthesis had also taken place. No data are presently available to indicate which HSV proteins, produced following the reversal of a PAA block, are important for AAV replication, but the absence of the direct requirement for HSV coded DNA polymerase activity is further supported by these data. Another inhibitor of HSV, 2-deoxy-D-glucose (Young and Mayor, 1979a), also promises to be useful in future investigations dissecting the complex, multilevel interactions of HSV and AAV.

B. Herpesvirus Oncogenicity and AAV

Evidence associating HSV-2 with neoplastic disease of the female genital tract continues to accumulate (Schwartz and Naftolin, 1981). Studies with HSV-transformed animal cells and seroepidemiological surveys of human populations have provided some indication that AAV has a suppressive effect on the expression of HSV oncogenicity. It is already known that AAV partially inhibits the tumorigenicity of adenovirus in hamsters if both viruses are administered together (Kirschstein *et al.*, 1968; Mayor *et al.*, 1973).

1. Studies with HSV-Transformed Cells

The observation (Blacklow, 1975) that AAV replication can be partially potentiated by an oncogenic hamster embryo cell line which was originally transformed by UV-irradiated HSV-2 (Duff and Rapp, 1971) has provided an appropriate system for studying the effect of AAV on the expression of HSV-2 oncogenicity. The HSV-2 transformed cell line, 333-8-9, is tumorigenic in young hamsters (Duff and Rapp, 1971) and is demonstrably free of infectious herpesvirus and adenovirus. When this cell line was infected with AAV-3 alone, fluorescent antibody-stainable antigen was produced which was detectable by AAV whole virion antiserum as well as by specific antisera to two of the three AAV viral proteins, VP-1 and VP-3 (Blacklow, 1975). The staining patterns of the 333-8-9 cells were characteristic of the results obtained in the presence of helper virus, with VP-1 antiserum showing cytoplasmic staining and VP-3 antiserum treatment resulting in predominantly intranuclear fluorescence. However, no antigen reacting with VP-2 antiserum was noted. If AAV-3 inoculated 333-8-9 cells were coinfected with either an adenovirus or herpesvirus helper, all three virion proteins appeared. Another transformed hamster cell line, the virus-free BHK-21, did not potentiate AAV. Inoculation of 333-8-9 cells with AAV-1 yielded the same results as with AAV-3 (Blacklow *et al.*, 1978) but AAV-4 was not potentiated by these cells (Mayor and Drake, 1974). It would appear that 333-8-9 cells, at the passage levels used in these experiments, contained sufficient HSV-2 genetic material to partially complement AAV. This observation opens the intriguing possibility of employing AAV potentiation as a sensitive marker for HSV transformation in naturally occurring tumors.

AAV infection of HSV-2 transformed cells also resulted in a striking inhibition of their tumorigenicity. When 333-8-9 cells were treated with AAV-1 prior to inoculation into syngenec LSH hamsters, the appearance of palpable primary tumors was delayed by 50–150% as compared with the development of tumors in animals receiving untreated cells (Cukor *et al.*, 1975). In contrast, AAV treatment of non-herpes-transformed hamster cell lines caused no delay in tumor latency. In addition, the metastasizing of the primary tumor to the lung was observed to be specifically restricted in animals inoculated with AAV-treated 333-8-9 cells (Blacklow *et al.*, 1978). AAV was not directly toxic to cultures of 333-8-9 cells or to any of 15 clones derived from that cell line. The exponential growth rate continued unaltered after AAV treatment. One hypothesis to account for the inhibition of 333-8-9 cell oncogenicity is that the AAV-specified antigens appearing on the surface of the tumor cells served as a target for an enhanced antitumor immune response by the syngeneic animal. The added inhibition of metastases is consistent with this hypothesis since the occurrence of 333-8-9 cell metastases was shown to be a sensitive indicator of the immunological status of the host (Duff *et al.*, 1973). An alternate hypothesis to account for the inhibitory effects of AAV could involve the direct interaction of AAV with HSV-2 genetic material in the tumor cell, which would render the 333-8-9 cell less oncogenic. No data have been presented to date indicating the effect of AAV on the transforming efficiency of HSV in cell culture.

2. Seroepidemiologic Studies

AAV infection of man commonly occurs, although it is not known to be associated with any specific symptoms. The virus has been recovered from the human gastrointestinal and respiratory tracts in association with adenovirus infection (Blacklow *et al.*, 1967), and most adult Americans show serologic evidence of prior AAV infection (Blacklow *et al.*, 1968b). However, in certain groups of cancer patients, the incidence of AAV serum antibody was lower than expected (Sprecher-Goldberger *et al.*, 1970, 1971). Extending these observations, Mayor *et al.* (1976) reported that the lowered incidence of AAV antibody is particularly marked in cervical cancer patients. They found that only 14% of this group possessed AAV antibody whereas a matched control group exhibited the expected 80% antibody prevalence. Antibody to HSV was equally prevalent in both groups. Thus, it seems that HSV-infected individuals, who are not also infected with AAV, have an increased risk of developing cervical carcinoma. These observations taken together with the known ability of AAV to establish latent infection (Berns *et al.*, 1975; Berns and Hauswirth, 1979; Cheung *et al.*, 1980) suggest the hypothesis that the oncogenicity of HSV is inhibited by the concomitant presence of AAV. In addition, the report by Buller *et al.* (1981) raises the possibility that HSV, in the absence of adenovirus, could serve to perpetuate AAV in nature.

REFERENCES

Aiello, L., Guifoyle, R., Heubner, K., and Weinmann, R., 1979, Adenovirus 5 DNA sequences present and RNA sequences transcribed in transformed human embryo kidney cells (HEK-Ad-5 or 293), *Virology* **94:**460.

Alwine, J.C., Kemp, D.J., and Stark, G.R., 1974, Method for detection of specific RNAs in agarose gels by transfer to diazobenzlyoxymethyl-paper and hybridization with DNA probes, *Proc. Natl. Acad. Sci. USA* **74:**5350.

Atchison, R.W., 1970, The role of herpesvirus in adenovirus-associated virus replication *in vitro*, *Virology* **42:**155.

Atchison, R.W., Casto, B.C., and Hammon, W.McD., 1965, Adenovirus-associated defective virus particles, *Science* **194:**754.

Berk, A.J., and Sharp, P.A., 1977, Sizing and mapping of early adenovirus mRNAs by gel electrophoresis of Sl endonuclease-digested hybrids, *Cell* **12:**721.

Berk, A.J., and Sharp, P.A., 1978, Structure of the adenovirus 2 early mRNAs, *Cell* **14:**695.

Berk, A.J., Lee, F., Harrison, T., Williams, J., and Sharp, P.A., 1979, Pre-early adenovirus 5 gene product regulates synthesis of early viral messenger RNAs, *Cell* **17:**935.

Berns, K.I., and Hauswirth, W.W., 1979, Adeno-associated viruses, *Adv. Virus Res.* **25:**407.

Berns, K.I., Pinkerton, T.C., Thomas, G.F., and Hoggan, M.D., 1975, Detection of adeno-associated virus (AAV)-specific nucleotide sequences in DNA isolated from latently infected Detroit 6 cells, *Virology* **68:**556.

Blacklow, N.R., 1975, Potentiation of an adenovirus-associated virus by herpes simplex virus type-2-transformed cells, *J. Nat. Cancer Inst.* **54:**241.

Blacklow, N.R., Hoggan, M.D., and Rowe, W.P., 1967, Isolation of adenovirus-associated viruses from man, *Proc. Natl. Acad. Sci. USA* **58:**1410.

Blacklow, N.R., Hoggan, M.D., Kapikian, A.Z., Austin, J.B., and Rowe, W.P., 1968a, Epidemiology of adenovirus-associated virus infection in a nursery population, *Am. J. Epidemiol.* **8**:368.

Blacklow, N.R., Hoggan, M.D., and Rowe, W.P., 1968b, Serologic evidence for human infection with adenovirus-associated viruses, *J. Nat. Cancer Inst.* **40**:319.

Blacklow, N.R., Hoggan, M.D., and McClanahan, M.S., 1970, Adenovirus-associated viruses: Enhancement by human herpesviruses, *Proc. Soc. Exp. Biol. Med.* **134**:952.

Blacklow, N.R., Hoggan, M.D., Sereno, M.S., Brandt, C.P., Kin, H.W., Parrot, R.H., and Chanock, R.M., 1971, Seroepidemiologic study of adenovirus-associated virus infection in infants and children, *Am. J. Epidemiol.* **94**:359.

Blacklow, N.R., Cukor, G., Kibrick, S., and Quinnan, G., 1978, Interactions of adeno-associated viruses with cells transformed by herpes simplex virus. in: *Replication of Mammalian Parvoviruses* (D.C. Ward, and P. Tattersall, eds), pp. 87–97, Cold Spring Harbor Laboratory, Cold Spring Harbor, New York.

Botchan, M., Topp, W., and Sambrook, J., 1976, The arrangement of simian virus 40 sequences in transformed cell DNA, *Cell* **9**:269.

Botchan, M., Topp, W., and Sambrook, J., 1979, Studies on SV40 excision from cellular chromosomes, in: *Cold Spring Harbor Symposia on Quantitative Biology*, Vol. 43, pp. 709–719, Cold Spring Harbor Laboratory, Cold Spring Harbor, New York.

Boucher, D.W., Melnick, J.L., and Mayor, H.D., 1971, Nonencapsidated infectious DNA of adeno-satellite virus in cells coinfected with herpesvirus, *Science* **173**:1243.

Buller, R.M.L., Janik, J.E., Sebring, E.D., and Rose, J.A., 1981, Herpes simplex virus type 1 and type 2 completely help adenovirus-associated virus replication, *J. Virol.* **40**:241.

Cairns, J., 1981, The origin of human cancer, *Nature* **289**:353.

Campbell, A., 1962, Episomes, *Adv. Genet.* **11**:101.

Carter, B.J., and Rose, J.A., 1972, Adenovirus-associated virus multiplication. VIII. Analysis of *in vivo* transcription induced by complete or partial helper viruses, *J. Virol.* **10**:9.

Carter, B.J., Koczot, F., Garrison, J., Rose, J.A., and Dolin, R., 1973, Separate helper functions provided by adenovirus for adenovirus-associated virus multiplication, *Nature New Biol.* **244**:71.

Carter, B.J., Laughlin, C.A., De La Maza, L.M., and Myers, M., 1979, Adeno-associated virus autointerference, *Virology* **92**:449.

Cartwright, S.F., and Huch, R.A., 1967, Viruses isolated in association with herd infertility, abortions and stillbirths in pigs, *Vet. Rec.* **81**:196.

Casto, B.C., and Goodheart, C.R., 1972, Inhibition of adenovirus transformation *in vitro* by AAV-1, *Proc. Soc. Exp. Biol. Med.* (*N.Y.*) **140**:72.

Casto, B.C., Atchison, R.W., and Hammon, W.McD., 1967b, Studies on the relationship between adeno-associated virus type 1 (AAV-1) and adenovirus. I. Replication of AAV-1 in certain cell cultures and its effect on helper adenovirus, *Virology* **32**:52.

Cheung, A.K-M., Hoggan, M.D., Hauswirth, W.W., and Berns, K.I., 1980, Integration of the adeno-associated virus genome into cellular DNA in latently infected human Detroit 6 cells, *J. Virol.* **33**:739.

Cukor, G., Blacklow, N.R., Kibrick, S., and Swan, I.C., 1975, Effect of adeno-associated virus on cancer expression by herpesvirus-transformed hamster cells, *J. Nat. Cancer Inst.* **55**:957.

de la Maza, L.M., and Carter, B.J. 1981, Inhibition of adenovirus oncogenicity by adeno-associated virus DNA, *J. Nat. Cancer Inst.* **67**:1323.

Desiderio, S.V., and Kelly, T.J., Jr. 1981, Structure of the linkage between adenovirus DNA and 55,000 molecular weight terminal protein, *J. Mol. Biol.* **145**:319.

Dolin, R., and Rabson, A.S., 1973, Herpesvirus saimiri: Enhancement of adenovirus-associated virus, *J. Nat. Cancer Inst.* **50**:205.

Domoto, K., and Yanagawa, R., 1969, Properties of a small virus associated with infectious canine hepatitis virus, *Jpn. J. Vet. Res.* **17**:32.

Drake, S., Schaffer, P.A., Esparaza, J., and Mayor, H.D., 1974, Complementation of adeno-associated satellite viral antigens and infectious DNA by temperature-sensitive mutants of herpes simplex virus, *Virology* **60**:230.

Duff, R., and Rapp, F., 1971, Properties of hamster embryo fibroblasts transformed *in vitro* after exposure to ultraviolet irradiated herpes simplex virus type 2, *J. Virol.* **8**:469.
Duff, R., Doller, E., and Rapp, R., 1973, Immunologic manipulation of metastases due to herpesvirus tranformed cells, *Science* **180**:79.
Dutta, S.K., 1975, Isolation and characterization of an adenovirus and isolation of its adenovirus-associated virus in cell culture from foals with respiratory tract infection, *Am. J. Vet. Res.* **36**:247.
Dutta, S.K., and Pomeroy, B.S., 1967, Electron microscopic studies of quail bronchitis virus, *Am. J. Vet. Res.* **28**:296.
El Mishad, A.M., McCormick, K.I., Yates, V.J., *et al.*, 1975, Detection and serological identification of adeno-associated virus in avian adenovirus stocks, *Infect. Immun.* **11**:287.
Freeman, A.E., Black, P.H., Van der Pool, E.A., Henry, P.H., Austin, J.B., and Heubner, R.J., 1961, Transformation of primary rat embryo cells by adenovirus type 2, *Proc. Natl. Acad. Sci. USA* **58**:1205.
Frost, E., and Williams, J., 1978, Mapping temperature-sensitive and host-range mutants of adenovirus type 5 by marker rescue, *Virology* **91**:39.
Fujinaga, K., Pina, M., and Green, M., 1969, The Mechanism of viral carcinogenesis by DNA mammalian viruses: VI. A new class of virus-specific RNA molecules in cells transformed by group C human adenoviruses, *Proc. Natl. Acad. Sci. USA* **64**:225.
Galas, D.J., and Chandler, M., 1981, On the molecular mechanisms of transposition, *Proc. Natl. Acad. Sci. USA* **78**:4458.
Galos, R.S., Williams, J., Biger, M.H., and Flint, S.J., 1979, Location of additional early gene sequences in the adenoviral chromosome, *Cell* **17**:945.
Garon, C.G., Berry, K., and Rose, J., 1972, A unique terminal redundancy in adenovirus DNA molecules, *Proc. Natl. Acad. Sci. USA* **69**:2391.
Gerry, H.W., Kelly, T.J., Jr., and Berns, K.I., 1973, Arrangement of nucleotide sequences in adeno-associated virus DNA, *J. Mol. Biol.* **79**:207.
Gilden, R.V., Kern, J., Beddow, T.G., and Heubner, R.J., 1968a, Oncogenicity of mixtures of adeno-associated virus and adenovirus type 12, *Nature* **219**:80.
Gilden, R.V., Kern, J., Beddow, T.G., and Heubner, R.J., 1968b, Oncogenicity of mixtures of adeno-associated virus and adenovirus type 12, *Nature* **220**:1139.
Graham, F.L., Smiley, J., Russell, W.C., and Nairn, R., 1977, Characteristics of a human cell line transformed by DNA from human adenovirus 5, *J. Gen. Virol.* **36**:59.
Graham, F.L., Harrison, T., and Williams, J., 1978, Defective transforming capacity of adenovirus type 5 host-range mutants, *J. Virol.* **86**:10.
Green, M., 1970, Oncogenic viruses, *Ann. Rev. Biochem.* **39**:701.
Green, M.R., and Roeder, R.G., 1980, Transcripts of the adenovirus-associated virus genome: Mapping of the major RNAs, *J. Virol.* **36**:79.
Green, M.R., Straus, S.E., and Roeder, R.G., 1980, Transcripts of the adenovirus-associated virus genome: Multiple polyadenylated RNAs including a potential primary transcript, *J. Virol.* **35**:560.
Handa, H., Shiroki, K., and Shimojo, H., 1977, Establishment and characterization of KB cell lines latently infected with adeno-associated virus type 1, *Virology* **82**:84.
Harrison, T., Graham, F.L., and Williams, J., 1977, Host range mutants of adenovirus type 5 defective for growth in HeLa cells, *J. Virol.* **77**:319.
Hauswirth, W.W., and Berns, K.I., 1977, Origin and termination of adeno-associated virus DNA replication, *Virology* **78**:488.
Hauswirth, W.W., and Berns, K.I., 1979, Adeno-associated virus DNA replication: Nonunit-length molecules, *Virology* **93**:57.
Henry, C.J., Merkow, L.P., Pardo, M., and McCabe, C., 1972, Electron microscope study on the replication of AAV-1 in herpes-infected cells, *Virology* **49**:618.
Hoggan, M.D., 1970, Adenovirus-associated viruses, in: *Progress in Medical Virology* (J.L. Melnick, ed.), pp. 211–239, Vol. 12, Karger, Basel.
Hoggan, M.D., 1971, Small DNA viruses, in: *Comparative Virology* (K. Maramorosch and E. Kurstak, eds.), pp. 43–74, Academic Press, New York.

Hoggan, M.D., Blacklow, N.R., and Rowe, W.P., 1966, Studies of small DNA viruses found in various adenovirus preparations: Physical, biological and immunological characteristics, *Proc. Natl. Acad. Sci. USA* **55**:1467.

Hoggan, M.D., Thomas, G.F., Thomas, F.B., and Johnson, F.B., 1972, Continuous "carriage" of adenovirus associated virus genome in cell cultures in the absence of helper adenoviruses, in: *Proceedings of the Fourth Lepetit Colloquium, Cocoyac, Mexico,* pp. 243–249, North-Holland Publishing Company, Amsterdam.

Huebner, R.J., Casey, M.J., Chanock, R.M., and Scheel, K., 1965, Tumors induced in hamsters by a strain of adenovirus type 3: Sharing of tumor antigens and "neoantigens" with those produced by adenovirus type 7 tumors, *Proc. Natl. Acad. Sci. USA* **54**:381.

Johnson, R.H., Margolis, G., and Kilham, L., 1967, Identity of feline ataxia virus with feline panleucopenia virus, *Nature (London)* **214**:175.

Johnson, F.B., Blacklow, N.R., and Hoggan, M.D., 1972, Immunological reactivity of antisera prepared against the sodium dodecyl sulfate-treated structural polypeptides of adenovirus-associated virus, *J. Virol.* **9**:1017.

Jones, N., and Shenk, T., 1979, An adenovirus type 5 early gene function regulates expression of other early viral genes, *Proc. Natl. Acad. Sci. USA* **76**:3665.

Ketner, G., and Kelly, T.J., Jr., 1976, Integrated simian virus 40 sequences in transformed cell DNA: Analysis using restriction endonucleases, *Proc. Natl. Acad. Sci. USA* **73**:1102.

Kilham, L., 1961, Mongolism associated with rat virus (RV) infection in hamsters, *Virology* **13**:141.

Kirschstein, R.L., Smith, K.O., and Peters, E.A., 1968, Inhibition of adenovirus 12 oncogenicity by adeno-associated virus, *Proc. Soc. Exp. Biol. Med.* **128**:670.

Lacy, S., and Green, M., 1965, Adenovirus multiplication: Genetic relatedness of tumorigenic human adenovirus types 7, 12, and 18, *Science* **150**:1296.

Lassam, N.J., Bayley S.T., and Graham, F.L., 1978, Synthesis of DNA, late polypeptides, and infectious virus by host-range mutants of adenovirus 5 in nonpermissive cells, *Virology* **87**:463.

Lassam, N.J., Bayley, S.T., and Graham, F.L., 1979, Tumor antigens of human Ad5 in transformed cells and in cells infected with transformation-defective host-range mutants, *Cell* **18**:781.

Laughlin, C.A., Westphal, H., and Carter, B.J., 1979, Spliced adeno-associated virus RNA, *Proc. Natl. Acad. Sci. USA* **76**:5567.

Lausch, R.N., and Rapp, F., 1971, Concomitant immunity in hamsters bearing syngeneic transplants of tumors induced by paraadenovirus 7, simian adenovirus 7 or 9, 10-dimethylbenzanthracene, *Int. J. Cancer* **7**:322.

Levinson, A., and Levine, A.J., 1977, The isolation and identification of the adenovirus group C tumor antigens, *Virology* **76**:1.

Lipps, B.V., and Mayor, H.D., 1980a, Properties of adeno-associated virus (type 1) replicated in rodent cells by murine adenovirus, *J. Gen. Virol.* **51**:223.

Lipps, B.V., and Mayor, H.D., 1980b, Transplacental infection with adeno-associated virus type 1 in mice, *Intervirology* **14**:118.

Luchsinger, E., and Wellemans, G., 1971, A study on the presence of antibodies against adenovirus and associated virus X, in the sera of cattle and calves with respiratory disease, *Arch. Ges. Virusforsch.* **35**:203.

Luchsinger, E., Strobbe, R., Wellemans, G., Kekegel, D., and Goldberger, S., 1970, Hemagglutinating adeno-associated virus (AAV) in association with bovine adenovirus type 1, *Arch. Ges. Virusforsch* **31**:390.

Lusby, E., and Berns, K.I., 1982, Mapping the 5′ termini of two adeno-associated virus 2 RNAs in the left half of the genome, *J. Virol.* **41**:518.

Mayor, H.D., and Drake, S., 1974, Complementation of adeno-associated satellite virus antigens in cells transformed by human herpesvirus, *Microbios.* **11A**:37.

Mayor, H.D., and Young, J.F., 1978, Complementation of adeno-associated virus by temperature-sensitive mutants of human adenovirus and herpesvirus, in: *Replication of*

Mammalian Parvoviruses (D.C. Ward and P. Tattersall, eds.), pp. 109–118, Cold Spring Harbor Laboratory, New York.

Mayor, H.D., Houlditch, G.S., and Mumford, D.M., 1973, Influence of adeno-associated satellite virus on adenovirus-induced tumours in hamsters, *Nature (London), New Biol.* **241**:44.

Mayor, H.D., Drake, S., Stahmann, J., and Mumford, D.M., 1976, Antibodies to adeno-associated satellite virus and herpes simplex in sera from cancer patients and normal adults, *Am. J. Obstet. Gynecol.* **126**:100.

Mengling, W.L., 1972, Porcine parvovirus properties and prevalence of strains isolated in the United States, *Am. J. Vet. Res.* **33**:2239.

Nahmias, A.J., Josey, W.E., Naib, Z.M., Lucez, C.F., and Guest, B.A., 1970, Antibodies to herpesvirus hominis, types 1 and 2 in humans. II. Women with cervical cancer, *Am. J. Epidemiol.* **91**:547.

Onuma, M., 1971, Distribution of antibodies in dogs versus adenovirus-associated satellite virus associated with infectious canine hepatitis virus and serological typing of satellite viruses, *Jpn. J. Vet. Res.* **19**:40.

Onuma, M., and Yanagawa, R., 1972a, Additional characteristics of adeno satellite virus associated with infectious canine hepatitis virus, *Jpn. J. Vet. Res.* **12**:13.

Onuma, M., and Yanagawa, R., 1972b, Distribution of antibodies in dogs versus adeno satellite viruses, *Jpn. J. Microbiol.* **16**:438.

Ostrove, J.M., and Berns, K.I., 1980, Adenovirus early region Ib gene function required for rescue of latent adeno-associated virus, *Virology* **104**:502.

Ostrove, J.M., Duckworth, D.H. and Berns, K.I., 1981, Inhibition of adenovirus-transformed cell oncogenicity by adeno-associated virus, *Virology* **113**:521.

Park, W.P., Green, M., Pina, M., and Melnick, J.L., 1967, Physiochemical characterization of adeno-associated satellite virus type 4 and its nucleic acid, *J. Virol.* **1**:980.

Park, W.P., Boucher, D.W., Melnick, J.L., Taber, H., and Yow, M.D., 1970, Seroepidemiological and ecological studies of the adenovirus-associated satellite viruses, *Inf. Imm.* **2**:716.

Perucho, M., Hanahan, D., and Wigler, M., 1980, Genetic and physical linkage of exogenous sequences in transformed cells, *Cell* **22**:309.

Philipson, L., 1979, Adenovirus proteins and their messenger RNAs, *Adv. Virus Res.* **25**:357.

Pina, M., and Green, M., 1965, Biochemical studies on adenovirus multiplication. IX. Chemical and base composition analysis of 28 human adenoviruses, *Proc. Natl. Acad. Sci. USA* **54**:547.

Rapoza, N.P., and Atchison, R.W., 1967, Association of AAV 1 with simian adenoviruses, *Nature (London)* **215**:1186.

Rawls, W.E., Gardner, H.L., and Kaufman, R.H., 1970, Antibodies to genital herpesvirus in patients with carcinoma of the cervix, *Am. J. Obstet. Gynecol.* **107**:710.

Rawls, W.E., Adam, E., and Melnick, J.L., 1973, An analysis of seroepidemiological studies of herpesvirus type 2 and carcinoma of the cervix, *Cancer Res.* **33**:1477.

Rawls, W.E., Bacchetti, S., and Graham, F.L., 1977, Relation of herpes simplex viruses to human malignancies, *Curr. Top. Microbiol. Immunol.* 77:71.

Rekosh, D.M.K., Russell, W.C., Bellet, A.J.D., and Robinson, A.J., 1977, Identification of a protein linked to the ends of adenovirus DNA, *Cell* **11**:283.

Richardson, W.D., and Westphal, H., 1981, A cascade of adenovirus early function is required for expression of adeno-associated virus, *Cell* **27**:133.

Robinson, A.J., Younghusband, H.B., and Bellett, A.J.D., 1973, A circular DNA-protein complex from adenoviruses, *Virology* **56**:54.

Rose, J.A., and Koczot, F., 1972, Adenovirus-associated virus multiplication. VII. Helper requirement for viral deoxyribonucleic acid and ribonucleic acid synthesis, *J. Virol.* **10**:1.

Ross, S.R., Flint, S.J., and Levine, A.J., 1980, Identification of the adenovirus early proteins and their genomic map positions, *Virology* **100**:419.

Rotkin, I.D., 1973, A comparison review of key epidemiological studies in cervical cancer related to current searches for transmissible agents, *Cancer Res.* **33**:1353.

Salo, R.J., and Mayor, H.D., 1979, Adenovirus-associated virus polypeptides synthesized in cells coinfected with either adenovirus or herpesvirus, *Virology* **92**:237.

Sambrook, J., Botchan, M., Gallimore, P., Ozanne, B., Pettersson, U., Williams, J., and Sharp, P.A., 1974, Viral DNA sequences in cells transformed by simian virus 40, adenovirus type 2, and adenovirus type 5, *Cold Spring Harbor Symp. Quant. Biol.* **39:**615.

Samulski, R.J., Berns, K.I., Tan, M., and Muzyczka, N., 1982, Cloning of AAV into pBR322: Rescue of intact virus from the recombinant plasmid in human cells. *Proc. Natl. Acad. Sci. USA* **79:**2077.

Sawada, Y., and Fujinaga, K., 1980, Mapping of adenovirus 12 mRNA's transcribed from the transforming region, *J. Virol.* **36:**639.

Schrier, P.I., van den Elsen, P.J., Hertoghs, J.J.L., and Van der Eb, A.J., 1979, Characterization of tumor antigens in cells transformed by fragments of adenovirus type 5 DNA, *Virology* **99:**372.

Schwartz, P.E., and Naftolin, F., 1981, Type 2 herpes simplex virus and vulvar carcinoma in situ, *New Engl. J. Med.* **305:**517.

Southern, E., 1975, Detection of specific sequences among DNA fragments separated by gel electrophoresis, *J. Mol. Biol.* **98:**503.

Sprecher-Goldberger, S., Dekegel, D., Otten, J., and Thiry, L., 1970, Incidence of antibodies to adenovirus-associated viruses in patients with tumors or other diseases, *Arch. gesamte Virusforsch.* **30:**16.

Sprecher-Goldberger, S., Thiry, L., Lefebvere, N., Dekegel, D., and DeHalleux, F., 1971, Complement-fixation antibodies to adenovirus-associated viruses, adenoviruses, cytomegaloviruses and herpes simplex viruses in patients with tumors and in control individuals, *Am. J. Epidemiol.* **94:**351.

Srivastava, A., Lusby, E.W. and Berns, K.I., 1983, Nucleotide sequence and organization of the adeno-associated virus 2 genome, *J. Virol.* **45:**555.

Steenberg, P.H., Maat, J., Van Ormondt, H., and Sussenbach, J.S., 1977, The nucleotide sequence at the terminus of adenovirus type 5 DNA, *Nucleic Acids Res.* **4:**4371.

Straus, S.E., Sebring, E.D., and Rose, J.A., 1976, Concatemers of alternating plus and minus strands are intermediates in adenovirus-associated virus DNA synthesis, *Proc. Natl. Acad. Sci. USA* **73:**742.

Sugimura, T., and Yanagawa, R., 1968, Effect of sodium deoxycholate on infectious canine hepatitis virus, an additional description of small associated viruses, *Jpn. J. Vet. Res.* **16:**1.

Sutcliffe, J.G., 1978, Complete nucleotide sequence of the *Eschericia coli* plasmid pBR322, *Cold Spring Harbor Symp. Quant. Biol.* **43:**77.

Thomas, G.F., and Hoggan, M.D., 1974, Immune electron microscopy (IEM) of canine AAV, in: *32nd Annual Proceedings of the Electron Microscopy Society of America* (C.J. Arcenaux, ed.), pp. 256, St. Louis, Missouri.

Toolan, H.W., 1960, Experimental production of mongoloid hamsters, *Science,* **131:**1446.

Trentin, J.J., Yabe, Y., and Taylor, G., 1962, The quest for human cancer viruses, *Science* **137:**835.

Williams, J.F., 1973, Oncogenic transformation of hamster embryo cells *in vitro* by adenovirus type 5, *Nature (London)* **243:**162.

Wilson, M.C., Fraser, N.W., and Darnell, J.E., Jr., 1979, Mapping of RNA initiation sites by high doses of UV irradiation: Evidence for three independent promoters within the left 11% of the Ad-2 genome, *Virology* **94:**175.

Wolfson, J., and Dressler, D., 1972, Adenovirus type 2 contains an inverted terminal repetition, *Proc. Natl. Acad. Sci. USA* **69:**3054.

Yates, V.J., Rhee, Y.O., Fry, D.E., *et al.*, 1976, The presence of adenoviruses and adeno-associated viruses in healthy chickens, *Avian Dis.* **20:**146.

Yates, V.J., Pronovost, A.D., Rhee, Y.O., *et al.*, 1979, Adenovirus-associated virus-precipitating antibody as an early indicator of an adenovirus infection, *Am. J. Vet. Res.* **40:**277.

Yates, V.J., Dawson, G.J., and Pronovost, A.D., 1981, Serologic evidence of AAAV infection in an unselected human population and among poultry workers, *Am. J. Epidemiology* **113:**542.

Young, J.F., and Mayor, H.D., 1979a, Studies on the defectiveness of adeno-associated virus (AAV). I. Effects of phosphonoacetic acid and 2 deoxy-D-glucose on the replication of AAV, *Virology* **94:**323.

Young, J.F., and Mayor, H.D., 1979b, Studies on the defectiveness of adeno-associated virus (AAV). II. Effect of growth in CV-1 cells on the replication of adeno-associated virus, *Virology* **94:**342.

CHAPTER 3

Adeno-Associated Virus Defectiveness and the Nature of the Adenovirus Helper Function

BARRIE J. CARTER AND CATHERINE A. LAUGHLIN

I. INTRODUCTION

Adeno-associated virus (AAV) was first observed as a contaminant in preparations of adenovirus (Brandon and Maclean, 1962; Archetti and Bocciarelli, 1963, 1964, 1965; Archetti *et al.*, 1965; Hoggan, 1965; Hull *et al.*, 1965; Melnick *et al.*, 1965) and subsequently recognized as a defective virus which was unconditionally dependent for its replication upon coinfection of the host cell with a helper adenovirus (Atchison *et al.*, 1965; Hoggan *et al.*, 1966; Parks *et al.*, 1966). It was also realized that one other group of viruses, namely, herpes viruses could provide helper functions for the production of AAV DNA, RNA, and proteins (Atchison, 1970; Blacklow *et al.*, 1970; 1971a; Boucher *et al.*, 1971; Carter and Rose, 1972; Carter *et al.*, 1972; Dolin and Rabson, 1973; Rose and Koczot, 1972). For some time, it was generally believed that the herpes virus helper function for AAV was incomplete and that either formation of AAV capsids or packaging of AAV DNA into capsids failed to occur (Atchison, 1970; Henry *et al.*, 1972). However, more recent work has demonstrated that

---

BARRIE J. CARTER AND CATHERINE A. LAUGHLIN • Laboratory of Cell Biology and Genetics, National Institute for Arthritis, Diabetes, and Digestive and Kidney Diseases, National Institutes of Health, Bethesda, Maryland 20205.

herpes viruses can provide a complete AAV helper function and allow production of fully infectious AAV particles (Buller *et al.*, 1981).

In this chapter, we concentrate on to the adenovirus helper functions for AAV. Most of the studies on AAV helper functions have employed adenovirus and there now exists a fairly extensive body of evidence which implicates certain specific adenovirus genes as having a significant role in the helper function. Furthermore the biochemical role of some of these adenovirus genes is becoming apparent. For AAV helper functions provided by herpes viruses much less information is available. This is largely a reflection of the slower development of the molecular biology of herpes viruses occasioned by the genome complexity of these viruses. The interaction between AAV and herpes viruses has been reviewed recently (Young and Mayor, 1979) and is discussed in a separate chapter in this volume (see Chapter 2).

We will discuss only briefly the earlier biological studies on AAV defectiveness and the requirement for a helper virus. This work has been reviewed in detail several times previously (Hoggan, 1970; Henry, 1973; Rose, 1974). We will emphasize discussion of more recent biochemical and genetic studies of AAV helper functions. These studies have been conducted over the last 8 to 10 years and are the direct result of a confluence of two separate lines of work. During this time the AAV genome has been extensively mapped and the location of DNA replication origins, RNA promoter sites and splice junctions has been determined. Also, the general details of AAV DNA replication, RNA synthesis, capsid protein synthesis and particle assembly have been described. During the same period similar studies have been performed on the structure, expression and replication of the adenovirus genome. Most of the AAV genome has been sequenced and, perhaps more remarkably, at least 70% (i.e., 25,000 nucleotides) of the genome sequence is known for human adenovirus serotype 2. This includes all the regions of Ad2 known to be required by AAV. In addition to these studies, many adenoviruses having accurately mapped and well characterized conditional-lethal mutations are becoming available. Thus, AAV and adenovirus will soon represent a completely defined set of interacting eukaryotic genes. Together, with the current developments in recombinant DNA technology and *in vitro* mutagenesis, it is likely that AAV and adenovirus will represent an important system for analysis of eukaryotic gene expression. Furthermore, both AAV and adenovirus genomes can integrate into cell chromosomes and thus both may be extremely useful as potential eukaryotic cloning vectors. In this respect also, it is important to define the molecular basis of the AAV helper function.

Before proceeding with a discussion of AAV defectiveness it is worth summarizing the differences between AAV and the autonomous parvoviruses. This is important because the overall growth cycle of these two genuses of parvoviruses may share more common features than differences. Furthermore, some of the differences between these two groups have recently become less distinct. Indeed, for some purposes, AAV can be usefully viewed as a parvovirus whose host requirement is for a cell in which certain adenovirus or herpes virus genes are being expressed rather than for a cell in S-phase.

It has been noted elsewhere in this volume (Chapter 6) that for certain reasons it might be more useful to refer to AAV and autonomous parvoviruses as helper-dependent and helper-independent viruses, respectively, since this is the most obvious difference between the two groups. However, some helper-independent parvoviruses may require a helper virus under certain circumstances. For instance, in human embryonic lung cells the hamster virus H-1 was reported to grow poorly unless augmented by coinfection with human adenovirus 12 (Ledinko and Toolan, 1968). Also a helper-independent goose parvovirus strain B, grew well in goose embryo fibroblasts which were in S-phase, but when DNA synthesis in the same cells was blocked by excess thymidine the parvovirus only grew well when a duck plague herpes virus was added as the helper (Kisary, 1979).

A second important difference is that both termini of AAV DNA have the same terminal palindrome whereas the left- and right-hand terminal palindromes of the autonomous viruses are different. This was thought perhaps to provide the basis for the separate encapsidation of either *plus* or *minus* strands of AAV DNA but only the *minus* (or viral) strand of autonomous parvovirus DNA. However it has recently been shown that at least one autonomous parvovirus, LuIII, can in certain cell lines separately encapsidate *minus* or *plus* strands with equal efficiency (G. Siegl and J. D. Tratschin, personal communication).

Autonomous parvoviruses have a requirement for a cell product present in late S-phase of the cell cycle in order to initiate or maintain replication of their genome (Tennant *et al.*, 1969; Rhode, 1973, 1974; Tattersall, 1972; Siegl and Gautschi, 1973a,b). Recent evidence suggests that this cell S-phase function may be needed to convert the infecting virus parental single-strand DNA to a parental RF duplex genome (Wolter *et al.*, 1980). Whether some similar cell function is required for AAV, in addition to the adenovirus helper function, is not known, but there is suggestive evidence that such a situation may pertain. AAV generally grows well only in cells which are in logarithmic growth, and it was recently noted in studies on an ovine AAV (Clarke *et al.*, 1979) that only rapidly dividing cells appear to produce AAV. It is possible that this might provide an explanation for the general observation that never more than about 50% of cells show accumulation of AAV antigens when adenovirus is the helper.

In this review we will first describe the biological background and then present a definition of the helper function. We will then discuss the helper function in terms of the AAV genome structure and its replication cycle. For this purpose it is necessary to summarize briefly the current state of studies of the adenovirus genome, its replication cycle and genetics. Finally, we will describe recent and more definitive experiments that have identified specific adenovirus genes involved in the helper function and in some cases yielded insights into the possible biochemical functions of these genes. Most of the information on helper functions as related to specific adenovirus genes has been obtained using the human AAV-2 and human adenoviruses 2 and 5 in human and monkey cells. This fortunate circumstance allows ready comparison of work in several laboratories. Some generally less specific studies have been reported using AAV-1 and AAV-4 and less well characterized mutants of human Ad12 and Ad31 or simian adenoviruses.

II. BIOLOGICAL BACKGROUND

Over the last 15 years a number of laboratories have searched without success for cell lines in which AAV would grow in the absence of any helper virus function. The only examples of cell lines which allow any AAV functions in the absence of added helper are some transformed cells which express certain adenovirus or herpes virus genes (see Section VII). Thus, AAV appears to be unconditionally defective for growth in laboratory cell cultures. Although there is no direct proof, there is circumstantial evidence from epidemiological studies that growth of AAV probably is defective in the normal human host. In these studies on infant children experiencing documented adenovirus infections, both AAV-2 and AAV-3 were isolated from children who subsequently sero-converted to AAV (Blacklow *et al.*, 1967a). High-titered mono-specific antibodies to AAV-1, AAV-2, and AAV-3 have been found in humans or to AAV-4 in most African green monkeys and AAV-4 is frequently isolated from cultured African green monkey kidney cells (Blacklow *et al.*, 1967a).

Only adenoviruses or herpes viruses appear to be helpers for AAV. Extensive testing of many other DNA and RNA viruses has failed to find any other helpers for AAV (Atchison, 1970). There is one, and perhaps a second, exception to this rule. Human adenoviruses do not grow in most monkey cells apparently because of blocks to transcriptional processing and translation of adenovirus late mRNAs (see below). However, monkey cells are permissive for AAV when simian adenoviruses are used as the helper (Atchison *et al.*, 1965; Casto *et al.*, 1967). AAV growth in monkey cells with a human adenovirus as a helper is apparently subject to a similar inhibition (Buller *et al.*, 1979). This host-range restriction to Ad and AAV is overcome by coinfection with AAV, adenovirus and SV40 (Rabson *et al.*, 1964; Buller *et al.*, 1979). Thus under these conditions SV40 can provide a helper function for AAV by complementing the adenovirus. SV40 cannot provide a helper function for AAV in the absence of adenovirus. Nevertheless, this complementation of Ad by SV40 indicates interesting functional similarities in the properties of SV40 T-antigen and the adenovirus early region 2 DNA binding protein (see below).

A recent study by Dreizin *et al.* (1979), suggested that helper-independent parvoviruses may be able to provide helper functions for AAV. These authors reported accumulation of AAV-4 antigens in the nucleus after superinfection of a pig embryo kidney cell line that was chronically infected with Kilham rat virus, strain RV-13. This report, if substantiated, would be particularly important for studies of parvovirus replication.

Evidence accumulated to date indicates that most adenoviruses and many herpes viruses which have been tested will help AAV. There are some host-range restrictions, such as that noted above for human adenoviruses in monkey cells, which may inhibit AAV growth. However, this host range defect in monkey cells apparently is a restriction exerted upon a specific gene product of human adenoviruses which is required by AAV (Myers *et al.*, 1980) and is not a restriction for AAV growth *per se* (see below). There is indeed no good evidence so far that the host range of AAV is restricted. Most serotypes of AAV appear to grow in any cell in which the appropriate adenovirus is used. For instance, it was suggested by Handa *et al.* (1978), using a human adenovirus helper that rodent cells might be nonpermissive for AAV but Lipps and Mayor (1980) recently reported that AAV-1 grew in primary mouse kidney cells or mouse L929 cells when a murine adenovirus was used as the helper. Thus AAV differs from the adenoviruses in its extended host-range. Most adenoviruses grow well only in cells derived from a naturally susceptible host (Pereira *et al.*, 1963). In some cases AAV will grow in spite of a restriction for adenovirus. For instance, AAV-1 was reported to grow efficiently in human amnion cells when helped by infectious canine hepatitus virus (ICHV) even though growth of ICHV is extremely poor or abortive in human cells (Smith and Gehle, 1967). This situation is reflected in the growth of AAV when supported by certain adenovirus mutants under nonpermissive conditions (see below).

AAV has not yet been reported to grow efficiently in adenovirus transformed cells. This appears to be accounted for largely by the observations that several adenovirus early gene regions are involved in the AAV helper function. Most of the adenovirus-transformed cell lines tested to date do not express several of these early regions. However, Handa *et al.* (1977a), did report "rescue" of an AAV helper function by fusion of adenovirus transformed rat cells with human KB cells.

It is of interest that growth of AAV in adenovirus transformed cells may decrease their oncogenicity (de la Maza and Carter, 1981; Ostrove *et al.*, 1981). This suggests an interaction between AAV and gene products of the adenovirus early region 1 transforming genes. Indeed, inhibition by AAV of either oncogenic or lytic replication functions of adenovirus is one useful approach to analysis of the helper function.

Interference by AAV with adenovirus is of practical importance in view of the frequent contamination of adenovirus stocks with AAV or the possibility of inadvertant rescue of AAV from latently infected cells. In some cases such contaminations have led to serendipitous discoveries.

For instance, contamination of stocks of Ad12 with AAV resulted in the original observations that AAV can decrease or inhibit oncogenicity of adenovirus in newborn hamsters (Kirschstein *et al.*, 1968). More recently, contamination of adenovirus type-2 stocks with AAV (perhaps as a result of rescue from latently infected HeLa cells) led to the demonstration that under some conditions during replication a protein may be attached near the AAV DNA termini (B. Stillman, personal communication).

III. DEFINITION OF THE HELPER FUNCTION

Although we shall use the term "helper function" in a rather broad fashion we intend it to apply to a precise concept in the following way. Cells which are not infected with any adenovirus are nonpermissive for AAV growth. Such cells become permissive after infection with adenovirus and expression of certain adenovirus genes. There is good evidence that some adenovirus genes must be expressed to allow AAV growth. AAV growth does not result simply from adsorption of adenovirus to cells or by the direct action of any component of the adenovirus particle. We will therefore use the term "AAV helper function" to encompass all those adenovirus genes which must be expressed in order to convert the host cell to a permissive state for AAV growth. According to this general concept, alterations in the expression of any adenovirus gene whether by mutation, deletion, or metabolic effectors which result in specific inhibition of AAV growth would tentatively include such a gene as part of the AAV helper function.

Significant progress, particularly with respect to analysis of defined adenovirus mutations, has now been made in identifying adenovirus genes involved in the AAV helper function. Analysis of specific adenovirus mutations or deletions has clearly implicated adenovirus early gene regions 1a, 1b, 2a, and 4 in the helper function (Laughlin *et al.*, 1982; Ostrove and Berns, 1980; Janik *et al.*, 1981; Myers *et al.*, 1980; Myers and Carter, 1981; Jay *et al.*, 1981; Richardson and Westphal, 1981; Carter *et al.*, 1983). Analysis of other mutations has suggested that adenovirus early regions 2b and 3 are not required (Straus *et al.*, 1976a, 1976b; Myers *et al.*, 1980; Carter *et al.*, 1983—see below). However, the failure of a mutation in adenovirus to affect growth of AAV (Straus *et al.*, 1976a,b; Ostrove and Berns, 1980) must be interpreted with caution since certain adenovirus proteins (e.g., the region 2 DNA binding protein) are pleiotropic effectors and many adenovirus regions express multiple overlapping mRNAs coding for different proteins. Studies using metabolic effectors or analogs to identify adenovirus helper functions have been attempted (Mayor and Gorman, 1975; Young and Mayor, 1979). These types of studies frequently are much more difficult to interpret because of the likelihood of more indiscriminate effects of metabolic inhibitors.

According to our definition, various adenovirus genes may be part of the AAV helper function in different ways. Firstly, as described below, adenovirus early gene expression apparently involves a strongly phased cascade control mechanism. Thus, some adenovirus gene products may be intimately required for AAV replication by direct interaction with the AAV genome or with an AAV product. Other adenovirus genes may be required indirectly to control the expression of or to modify the action of other more directly acting adenovirus genes or gene products. A third type of helper function may occur by modification, activation, or inhibition of cellular gene products or of other adenovirus gene products. Finally, it must also be noted that some AAV helper functions may act simply to increase the frequency of an event which otherwise can occur, albeit rarely. Such genes (e.g., region 1a) appear to exist for adenovirus and are not absolutely required at high multiplicity.

Whether the helper function can be resolved into a single, ultimate directly acting function is not yet known. However, a series of concurrent studies in several laboratories using cells variously infected with mutant adenoviruses (Myers *et al.*, 1980; Jay *et al.*, 1981; Laughlin *et al.*, 1981), transfected with purified adenovirus DNA fragments (Janik *et al.*, 1981) or microinjected with purified adenovirus mRNA (Richardson *et al.*, 1980; Richardson and Westphal, 1981) may converge to distinguish direct from indirectly acting functions. For instance transfection of DNA fragments or microinjection of mRNA might bypass the requirement for certain indirect functions. One of the most salient features of the AAV helper function studies may be the development of AAV as an independent probe to help analyze the complex function and control of adenovirus genes.

IV. GENOME STRUCTURE AND GROWTH CYCLE OF AAV

A. AAV Genome Structure

The genome structure and growth cycle of AAV are described in detail elsewhere in this volume but are reviewed briefly here. This provides a basis for consideration of the likely points of action of the AAV helper function.

The structure and map of the AAV genome are summarized in Fig. 1. The orientation of the AAV DNA strands was determined by restriction endonuclease cleavage of 5′–^{32}P-labeled DNA and subsequent strand separation of the DNA fragments (Carter *et al.*, 1975). The AAV RNA species have been analyzed by a variety of procedures. The genome location and spliced structures of these RNAs were determined using both electrophoretic analysis and electron microscopy of AAV DNA–RNA hybrid molecules (Laughlin *et al.*, 1979b; Green and Roeder 1980a; Marcus *et al.*, 1981). The mapping has been refined and the precise locations of the AAV promoter sites at map positions 5, 19, and 40 (designated P5, P19,

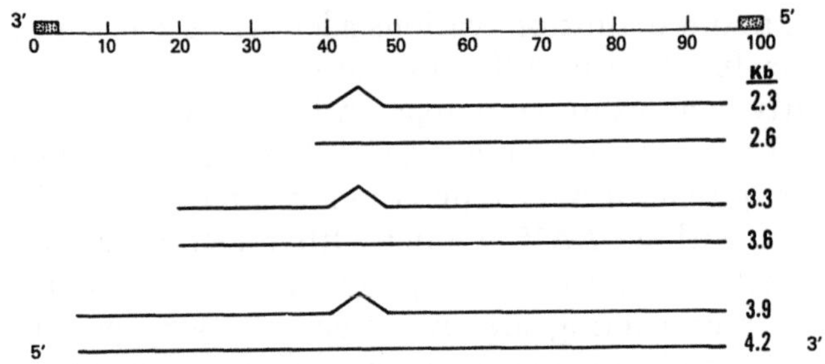

FIGURE 1. AAV genome map and *in vivo* RNA. The upper lines show the minus strand of AAV DNA with polarity as shown (Carter *et al.*, 1975) in accordance with the general nomenclature scheme for parvovirus genomes (Armentraut *et al.*, 1978). The genome is divided into 100 map units. One map unit is equivalent to 47 nucleotides. The stippled boxes indicate the extent of the AAV DNA terminally repetitious sequence. The lower lines indicate the major AAV RNA species transcribed in infected cells (Laughlin *et al.*, 1979; Marcus *et al.*, 1981). Some of these RNA species were also mapped by Green and Roeder (1979a,b). Caret marks indicate the intron which is removed by RNA splicing. The 2.3-kb mRNA is by far the most abundant species and codes for the three AAV capsid proteins (Jay *et al.*, 1981).

and P40, respectively) have been determined by direct sequencing, and by *in vitro* synthesis (Green and Roeder, 1980a,b; Lusby and Berns, 1981). These studies confirmed evidence from earlier work that all the *in vivo* AAV RNAs are complementary to the AAV DNA minus strand (Carter *et al.*, 1972, 1976). It has recently been demonstrated that the most abundant spliced 2.3-kb AAV RNA directs the synthesis of the three AAV capsid proteins VP1, VP2, and VP3 (Jay *et al.*, 1981). These three proteins apparently are derived by cleavage of a precursor of the size of VP1 as suggested by Buller and Rose (1978). It is not yet known if the larger AAV RNAs code for any protein product, although there are open reading frames that could specify 5 or 6 proteins present in the left-hand half of the genome as discussed elsewhere in this volume (Chapter 5).

B. Kinetics of Macromolecular Synthesis

The kinetics of macromolecular synthesis during the virus one-step growth cycle were studied in human KB cells infected with AAV2 and using adenovirus 2 as the helper (Carter *et al.*, 1973). Viral DNA and RNA were detected by pulse-labeling and hybridization to viral DNA immobilized on nitrocellulose filters. After simultaneous infection with AAV and adenovirus, both AAV and adenovirus DNA synthesis were first detected at 6 to 7 hr whereas AAV RNA synthesis was first detected at 9 to 10 hr after infection. This lag between the onset of AAV DNA and RNA synthesis was eliminated if the cells were infected with the helper adenovirus for 10 hr prior to infection with AAV. Furthermore, in the adenovirus preinfection, both AAV DNA and RNA synthesis were detected much earlier (within 3 to 4 hr of AAV infection) than in the simultaneous infection. Analogous results were obtained in other more indirect experiments in which AAV antigen production was measured using a fluorescent antibody assay (Blacklow *et al.*, 1967b). Similar kinetics of AAV DNA and RNA synthesis were observed in simultaneous infection experiments using a DNA-negative adenovirus mutant as the helper at the nonpermissive temperature (Straus *et al.*, 1976a). The presence of 5-fluoro-deoxyuridine from 3 hr prior to infection prevented synthesis of AAV RNA and adenovirus "late" RNA but not adenovirus "early" RNA. This observation, together with the fact that DNA negative mutants of adenovirus (Ito and Suzuki, 1970; Ishibashi and Ito, 1971; Mayor and Ratner, 1972; Handa *et al.*, 1975; Straus *et al.*, 1976a) or herpes viruses (Drake *et al.*, 1974) will help AAV RNA synthesis, indicates that AAV RNA synthesis is directly dependent upon synthesis of AAV DNA. However, once begun, AAV RNA synthesis is not dependent upon continued AAV DNA synthesis (see Chapter 5). Thus, AAV apparently has no "early" RNA species analogous to those of most viruses. Similar conclusions have also been reached by Handa *et al.* (1976) from experiments using cycloheximide to inhibit any potential early AAV protein synthesis. Most (nondefective) viruses synthesize early RNA prior to genome replication. The absence of an early function provides a teleological rationale for the defectiveness of AAV. Also, since the infecting genome is single-stranded and infection with AAV shows single hit kinetics (Blacklow *et al.*, 1967b), transcription of an early RNA would have implied synthesis from a single-stranded template.

The reason for the lag between detection of AAV DNA and RNA synthesis in cells infected simultaneously with the helper is not clear. We previously inferred from this observation that at least two independent adenovirus functions acted respectively at the level of AAV DNA replication and RNA synthesis (Carter *et al.*, 1973). It was further implied that these helper functions must be adenovirus early functions (Carter, 1978). It is now known that at least three adenovirus early functions are required for AAV: a region 1b and a region 4 function appear necessary to allow AAV RF DNA to accumulate and a region 2 function is required for accumulation of certain AAV RNAs, AAV capsid proteins and progeny SS DNA (see below). However, the precise correlation between these biochemical analyses and the physiology of the growth cycle remains to be resolved. This phenomenon should be reinvestigated using the much more sensitive technology now available for study of DNA and RNA synthesis.

C. Biochemistry of AAV Replication

Our current knowledge of the biochemistry of AAV growth allows the following general scheme to be envisaged as outlined in Fig. 2. Fol-

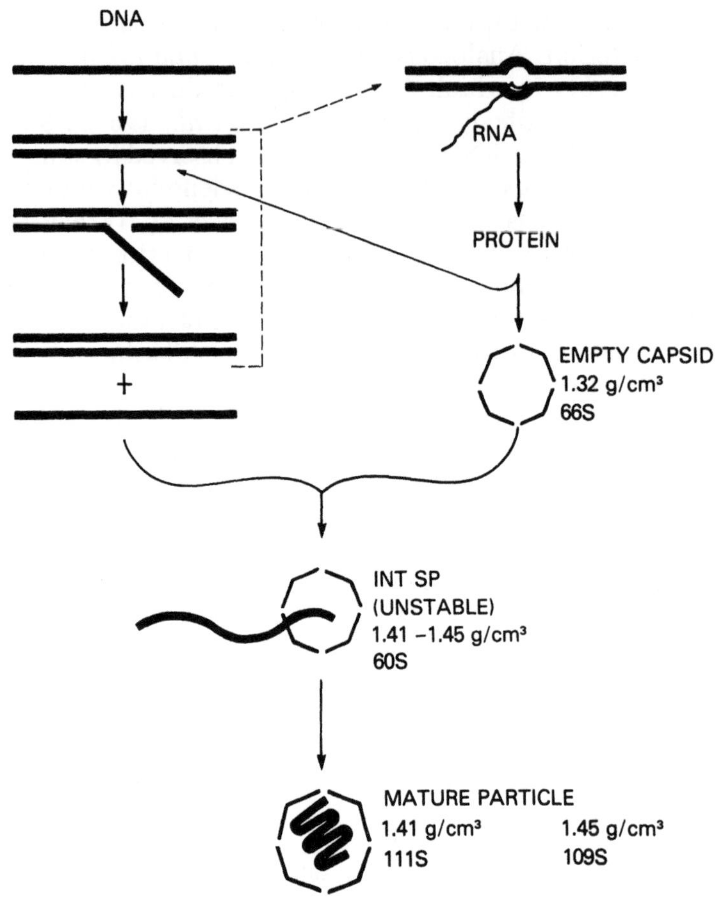

FIGURE 2. General scheme of AAV replicative cycle (see text for details).

lowing adsorption of the virus and uncoating, the parental single-strand genome must be converted to a duplex, parental RF structure, although this process has not been directly analyzed for AAV. We describe this first step as *parental RF replication*. It was pointed out above that this step possibly might be dependent in part on some host-cell function. Following this, it is likely that additional progeny RF molecules are produced. This amplification of the RF pool is described as *progeny RF replication* and may occur by initiation at the terminal replication origin using the palindromic properties of the 3′ terminal sequence to provide a base-paired, self-priming structure (Cavalier-Smith, 1974). Progeny RF replication also results in oligomeric RF structures and reinitiation at the replication origins may require site-specific nicking, terminal hairpin transfer (Tattersall and Ward, 1976), and consequent inversion ("flip-flop") of the terminal sequence. RF molecules having structures consistent with this scheme were observed to be precursors to progeny SS DNA

(Straus *et al.*, 1967b). The replication origin of AAV DNA is located in the terminus (Hauswirth and Berns, 1978b) and the terminal palindrome exhibits the expected "flip-flop" as shown by direct sequencing (Berns *et al.*, 1978b). It must be noted that not all initiation events on AAV DNA have been proven to occur by self-priming at a terminal hairpin structure.

Progeny RF molecules are converted to progeny single-stranded DNA (*progeny SS DNA replication*) as shown by isotopic pulse-chase experiments (Straus *et al.*, 1976b). This is believed to occur by a strand displacement mechanism. The displaced single strand apparently is immediately associated with a viral capsid for packaging and there appear to be few if any free progeny SS DNA molecules because in the absence of viral capsids, AAV single strands do not accumulate (Myers and Carter, 1981). It is not clear whether viral capsid proteins or assembled capsids are needed to drive SS DNA displacement, as was suggested for parvoviruses (Tattersall and Ward, 1976) or simply to sequester SS DNA and prevent either its degradation or reentry into the replication pool.

Studies on assembly of AAV particles (Myers and Carter, 1980) suggested that AAV SS DNA is probably encapsidated by packaging into a performed capsid rather than by condensation of capsid subunits around the DNA. These studies showed that the progeny SS DNA is rapidly associated with a viral capsid concomitantly with, or immediately after, synthesis to form an intermediate structure (Myers and Carter, 1980). This assembly intermediate forms a band in CsCl at a density of 1.41 to 1.45 g cm^{-3} but sediments in sucrose gradients at 60S as compared to 66 S empty capsids and 110 S mature paticles. In this intermediate structure the DNA is sensitive to DNase digestion whereas that in mature particles is not (Myers and Carter, 1980; de la Maza and Carter, 1980b). The 60 S assembly intermediate matures to a stable 110 S particle over a period of several hours and this maturation can occur in the absence of additional AAV DNA synthesis.

Studies of the structure and biological properties of AAV DI particles suggested that there may be an AAV gene product required for progeny RF replication and perhaps for progeny SS DNA replication also (Carter *et al.*, 1979; Laughlin *et al.*, 1979a; de la Maza and Carter, 1980a). This putative AAV *rep* function is discussed in more detail elsewhere in this volume (Chapter 6). This putative *rep* product may be the known AAV capsid protein or more likely a product coded by one of the RNAs from the left hand half of the genome.

A similar *rep* protein has also been postulated for the autonomous parvovirus, H1 (Rhode, 1978). Very recent evidence from studies on another autonomous parvovirus, MVM, show the existence of a *rep* protein. This *rep* protein is coded by the left-hand half of the MVM genome and is covalently attached to the 5′-terminus of RF DNA molecules (D. Ward, personal communication). A similar protein is present on the 5′ terminus of H-1 RF DNA (Revie *et al.*, 1979).

D. Possible Sites of Action of Helper Functions in the AAV Growth Cycle

The most revealing studies on AAV helper functions have come from analysis of well-characterized adenovirus mutants or from transfection or microinjection of specific adenovirus DNA fragments or mRNAs. Before describing these studies in detail it is worth discussing the helper function in the context of the general model of the AAV growth outlined above (IV.C).

1. Adsorption, Penetration, and Uncoating

It is clear that no adenovirus function is required for adsorption and penetration of AAV into the host cell nor apparently for uncoating of the AAV genome. Rose and Koczot (1972) used AAV particles containing ^{32}P-labeled DNA to show that the rate of adsorption of the virus to KB cells and the rate of appearance, and proportion, of ^{32}P-DNA that appeared in the cell nucleus was identical in the presence or absence of adenovirus. Generally at least half of the added AAV particles adsorb to the cell (Rose and Koczot, 1972; Laughlin *et al.*, 1979a) and at least half of the genomes from the adsorbed virus particles reach the cell nucleus (Rose and Koczot, 1972; Laughlin *et al.*, 1982). In this context, it is worth nothing that for purified AAV (1.41) particles, the particle to infectivity ratio is about 1:50 to 1:100 (Laughlin *et al.*, 1979a; de la Maza and Carter, 1980b). Thus, when assayed using an indirect fluorescent antibody assay for nuclear accumulation of AAV antigens (Carter *et al.*, 1979) one "infectious hit" (i.e., one infectious unit) per cell requires the addition of 50 to 100 added virus particles or the penetration to the nucleus of about 10 to 20 AAV DNA strands. These observations are important for consideration of certain aspects of the AAV helper function and adenovirus early region 1b.

2. Integration and Excision of AAV DNA into the Host-Cell Genome

When KB cells are infected with AAV in the absence of any helper virus the intact, unreplicated single-stranded genomes can readily be observed in the cell nucleus 20 hr after infection (Laughlin *et al.*, 1982; C. Marcus, unpublished). When host cells (human Detroit 6 or KB cells) infected only with AAV are passaged they retain the ability to produce infectious AAV particles upon superinfection with adenovirus (Hoggan *et al.*, 1973; Handa *et al.*, 1977b). This rescue of AAV following adenovirus superinfection is enhanced by halogenated pyrimidines. Berns *et al.*, (1975) used DNA–DNA hybridization reassociation kinetics to show that these latently infected or "carrier" cells contained three to five AAV genome copies per diploid cell. Handa *et al.* (1977b) suggested on the basis of hybridization "network" experiments that most of the AAV genomes in these cells are covalently integrated into the host genome. More definitive evidence on this point was provided by Cheung *et al.* (1980), who performed restriction endonuclease cleavage and DNA blotting experiments. Experiments of both of these groups (Handa *et al.*, 1977b; Cheung *et al.*, 1980) indicated that the latently infected cells may carry also some noncovalently integrated AAV genomes which appeared during continued passage of the cells.

These observations show that integration of AAV DNA into host-cell DNA does not require an adenovirus helper function. In addition, the spontaneous appearance of free AAV genomes in these cells appears to argue that adenovirus is not absolutely required for excision of integrated AAV genomes (Cheung *et al.*, 1980). This is in contrast to the recent suggestion by Ostrove and Berns (1980) that an adenovirus early region 1b function may be required for rescue of integrated AAV genomes. As discussed in more detail below (Section VI.D) this latter observation may reflect the requirement for an adenovirus region 1b function for AAV DNA replication rather than for the actual excision process.

3. AAV DNA Replication

Experiments from many laboratories have established that in the absence of helper virus function there is no appreciable replication of AAV DNA. Clearly an adenovirus helper function is required for AAV DNA replication. Analysis of adenovirus mutants has recently provided evidence for at least two adenovirus genes which specifically affect AAV DNA synthesis. Mutations in adenovirus type 5 early region 2a prevent accumulation of AAV progeny SS DNA (Myers *et al.*, 1980; Myers and Carter, 1981). Deletion of adenovirus type 5 early region 1b results in failure to replicate any AAV DNA in KB cells (Laughlin *et al.*, 1982). A critical point which has not previously been addressed in studies of AAV DNA replication is to analyze the initial stages of AAV DNA replication and particularly the production of the parental RF. For instance it is conceivable that synthesis of the first (parental) RF duplex might not require any adenovirus function. Recent studies (Laughlin *et al.*, 1982) have analyzed this possibility by using a DNA blotting analysis and *in vitro* labeled AAV DNA probes under carefully calibrated conditions such that a level of 0.1 AAV genomes per diploid cell nucleus could be detected. These studies provide no evidence for conversion of parental SS DNA to duplex RF in the absence of adenovirus functions and suggest that the requirement for an adenovirus function for AAV DNA replication may indeed be absolute.

The enzymology of DNA replication in eukaryotic cells is not yet well understood and the functions of the three classes of cellular DNA polymerases α, β, γ (Weissbach *et al.*, 1975) are not defined. For cell DNA replication, as well as for papovaviruses such as SV40 or polyoma, the polymerase α appears to be the most critical (Waqar *et al.*, 1978; Edenberg *et al.*, 1978; Vander Viret and Kevant, 1978; Huberman, 1981). In contrast, mitochondrial DNA is apparently replicated by DNA polymerase γ (Bolden *et al.*, 1977); Bertazzoni *et al.*, 1977; Hubscher *et al.*, 1977). The polymerase γ has also been implicated as the major polymerase activity in elongation of adenovirus DNA replication (Van der Vliet and Kwant, 1978) although some other evidence has also suggested the involvement of polymerase α in replicating adenovirus DNA (Ito *et al.*, 1975; Arens *et al.*, 1977; Brison *et al.*, 1977; Krokar, *et al.*, 1979; Ikeda *et al.*, 1981; Abboud and Horwitz, 1979; Longian *et al.*, 1979).

Handa and Shimojo (1977a) isolated an AAV DNA replication complex from human KB cells infected at 40.5°C with AAV-1 and the temperature-sensitive mutant adenovirus, Ad5*ts*125. These complexes incubated *in vitro* incorporated radioactive precursors into duplex DNA molecules which were mainly 14.5S monomers but included some longer duplex molecules which resembled *in vivo* RF structures. In alkaline conditions, the *in vitro* labeled AAV DNA was shorter than unit length. These studies were extended (Handa and Carter, 1979) to similar AAV DNA replication complexes isolated from a line of African green monkey kidney (Vero) cells infected with AAV-2 and using as helper either Adt*ts*125 (at the nonpermissive temperature) or herpes simplex virus (HSV) type I. In this latter study it was clearly shown that the major DNA polymerase activity present in the AAV replication complex isolated from *ts*125 helped cells was the cellular DNA polymerase γ. In the AAV complexes from the HSV helped cells the major polymerase activity was the HSV-coded DNA polymerase.

It has been pointed out (Van der Vliet and Kwant, 1978; Handa and Carter, 1979; Weissbach, 1979) that the putative association of cellular DNA polymerase γ with mitochondrial DNA, adenovirus DNA, and AAV DNA is of some interest since all of these genomes apparently replicate by a strand-displacement (Lechner and Kelly, 1977; Straus *et al.*, 1976b) or D-loop mechanism (Robberson *et al.*, 1972) rather than the more usually proposed Y-fork replication mechanism (Kornberg, 1974). This might suggest that there are two types of replication mechanisms in eukaryotic cells for strand displacement replication using mainly polymerase γ or Y-fork replication using mainly polymerase α.

It is not clear if the apparent involvement of polymerase γ in AAV DNA replication should be considered part of the helper function. This might reflect either an intrinsic property of AAV DNA replication as implied above or might indeed result from some more direct involvement such as activation or stimulation of this polymerase by adenovirus. In this respect, it is interesting that for the helper-independent bovine parvovirus (BPV), DNA polymerase α has been implicated as the main activity required for BPV DNA replication (Pritchard *et al.*, 1978, 1981). This would suggest some significant differences between DNA replication of AAV and helper-independent parvoviruses. However, recent experiments have provided evidence for the involvement of DNA polymerase γ as well as polymerase α in the replication of H-1 DNA, both for parental RF replication and for progeny RF replication (Kollek and Goulian, 1981).

The result of analysis of the AAV replication complex from the HSV-helped Vero cells (Handa and Carter, 1979) does suggest that the helper virus may influence the choice of enzymes involved in AAV DNA replication and that some properties of the HSV-coded polymerase may be utilized.

4. AAV RNA Synthesis

The studies on the growth cycle kinetics of AAV (Carter *et al.*, 1973) were interpreted as suggesting that there may be an adenovirus helper function involved in AAV RNA transcription. There is no compelling evidence for any adenovirus function that is directly required for AAV RNA synthesis other than those functions mentioned above which are required to allow prior AAV DNA replication. This point is discussed in more detail below (VII). There is evidence that the adenovirus early region 2 DNA binding M_r 72,000 protein may be required for either the efficient splicing or stability of certain AAV RNAs (Myers *et al.*, 1980) as described below (Section VI.D). It is also known that adenovirus region 1a is required for efficient transcription of most of the other adenovirus early region genes (Jones and Shenk, 1979; Shenk *et al.*, 1980; Berk *et al.*, 1979, 1980) and deletion or mutation of region 1a results in failure to help AAV (Ostrove and Berns, 1980; Janki *et al.*, 1981, Laughlin *et al.*, 1982). Whether the adenovirus region 1a has any similar function in enhancing AAV transcription is not clear. It is not yet possible to discriminate any direct effect of adenovirus region 1a products on AAV from those effects exerted by control of other adenovirus early regions such as 1a or 2a (see below).

5. AAV Protein Synthesis and Particle Assembly

It is not yet known if all the AAV coded proteins have been identified (see below), but synthesis of the AAV capsid proteins is apparently under a translational control influenced at least in part by the helper adenovirus (Jay *et al.*, 1981). The three AAV capsid proteins VP1 (M_r, 85,000), VP2 (M_r, 72,000), and VP3 (M_r, 61,000) are specified by AAV mRNA (Buller and Rose, 1978). Recently Jay *et al.* (1981) reported evidence that all three capsid proteins are specified by the spliced, 2.3-kb AAV mRNA as indicated by *in vitro* translation in a reticulocyte lysate. These results are also consistent with the suggestion of Buller and Rose (1978) that VP2 and VP3 are derived by posttranslational cleavage of VP1. Buller and Rose (1978) also identified two nonstructural proteins, M_r, 24,000 and M_r, 16,000, respectively, which were suggested, but not proven, to be AAV-specific products resulting from the posttranslational cleavage of VP1.

Mayor and associates (1979b) have suggested in several studies that adenovirus may provide a late function necessary for assembly of AAV proteins and DNA to yield mature particles. It was then further suggested that failure to provide this assembly function would explain the partial helper function of herpes viruses. However, it is difficult to reconcile these suggestions with other studies using well-characterized adenovirus mutants or employing microinjection of adenovirus RNA (see Section VI). Furthermore, it is now known that HSV is a complete helper of AAV (Buller *et al.*, 1981).

Mayor and associates have studied AAV replication in metabolically altered cells. Depletion of arginine is known to inhibit assembly of adenovirus (Rouse and Schlesinger, 1972). Although arginine depletion did allow production of viable AAV, albeit in reduced amount (Mayor and Ratner, 1972), some of the particles appeared to be less stable (Mayor and Gorman, 1975).

Although AAV capsids are not glycosylated, accumulation of AAV proteins in adenovirus infected human epidermoid carcinoma (Hep2) cells was reduced, but not eliminated, by the metabolic inhibitor 2-deoxy-D-glucose (Young and Mayor, 1979a). AAV DNA synthesis was only reduced to about one-third of the amount in the absence of the inhibitor. Accumulation of the adenovirus early region 3 glycoprotein and the late gene fiber protein was inhibited by 2DG. However, other studies provide strong evidence that neither of these glycosylated adenovirus proteins is necessary for the AAV helper function (see Section VI. D).

It is more likely that the effects of these metabolic aberrations are due to generalized affects upon the infected cell rather than direct inhibition of some specific AAV helper functions. One possible explanation may be in the slow maturation of the unstable assembly intermediate described by Myers and Carter (1980). Generalized metabolic aberrations might interfere with this maturation in a nonspecific fashion. It is also possible that failure of this step in cells infected with herpes virus at high multiplicity due to rapid cell death may account for the previously observed partial or incomplete helper function of these viruses.

In another study, Mayor *et al.* (1977) assayed several adenovirus type-31 temperature-sensitive mutants as a helpers for AAV in human HEK and KB cells. Unfortunately although all these mutants behave phenotypically as late mutants, the temperature sensitive lesions in them were unmapped. Thus the significance of these results for AAV helper function studies is unclear. For instance one mutant, *ts*94, synthesizes adenovirus DNA and late proteins but apparently is defective at some stage in assembly. When the mutant, *ts*94, was used as the helper at the nonpermissive temperature AAV DNA and antigen synthesis occurred although no quantitation was reported and AAV infectivity, in either HEK or KB cells, was decreased 100-fold (Mayor *et al.*, 1977). It is not clear if this represented complete or partial inhibition of AAV infectivity because the yield of AAV per infected cell was not reported. In the same study another adenovirus type-31 temperature-sensitive mutant, *ts*38, which was defective for adenovirus fiber protein production, remarkably, appeared to be *cold* sensitive for AAV infectivity in HEK cells but neither cold sensitive nor temperature sensitive for AAV in KB cells. Two other temperature-sensitive mutants, *ts*7 and *ts*9, both of which were defective for adenovirus hexon and fiber protein but not penton protein production, appeared to help AAV as efficiently as did wild type, at either permissive or nonpermissive temperature in KB cells but were deficient for AAV and HEK cells at both temperature. In view of the apparent nonrequirement of any adenovirus type 2 or 5 late genes (see below) and the incomplete genome mapping or characterization of the type-31 mutants, a clear interpretation of these data is difficult to envisage. In addition, it is worth noting that a mutant of adenovirus type 2, *ts*215, which synthesizes DNA and all the known late proteins but which fails to assemble mature adenovirus particles, was found to help AAV as efficiently as wild-type adenovirus 2 (Myers and Carter, 1980).

V. GENOME STRUCTURE AND GROWTH CYCLE OF ADENOVIRUS

Advances in our understanding of the adenovirus helper functions required by AAV have necessarily been dependent upon concurrent analysis of the structure and expression of adenovirus genes and the supply of adenovirus mutants. The molecular biology and genetics of adenoviruses have been reviewed in detail recently (Flint and Broker, 1980; Flint, 1980a,b; Ginsberg and Young, 1977; Philipson, 1979). However, an understanding of the current state of adenovirus molecular biology, particularly with respect to the early gene regions, is important for discussion of recent studies on AAV helper functions. We will review briefly the adenovirus genome map and salient features of the control of expression of adenovirus early genes.

The adenovirus growth cycle is conventionally divided into an *early* phase (before the onset of DNA replication) and a *late* phase (after the onset of DNA replication). The *early* functions are concerned with subversion of cellular metabolism for lytic replication in permissive cells or with transformation of nonpermissive cells. The late functions are mostly concerned with synthesis of structural proteins and the processing and assembly of these proteins into mature virions. AAV shares no known structural proteins in common with adenoviruses. Thus, in retrospect, it is hardly surprising that the only AAV helper functions provided by adenovirus involve certain early gene functions. Current evidence strongly implies that there are no late adenovirus genes required by AAV. This is reflected in the overall growth cycles of the two viruses (Carter *et al.*, 1973). AAV DNA synthesis begins between 6 and 8 hr after infec-

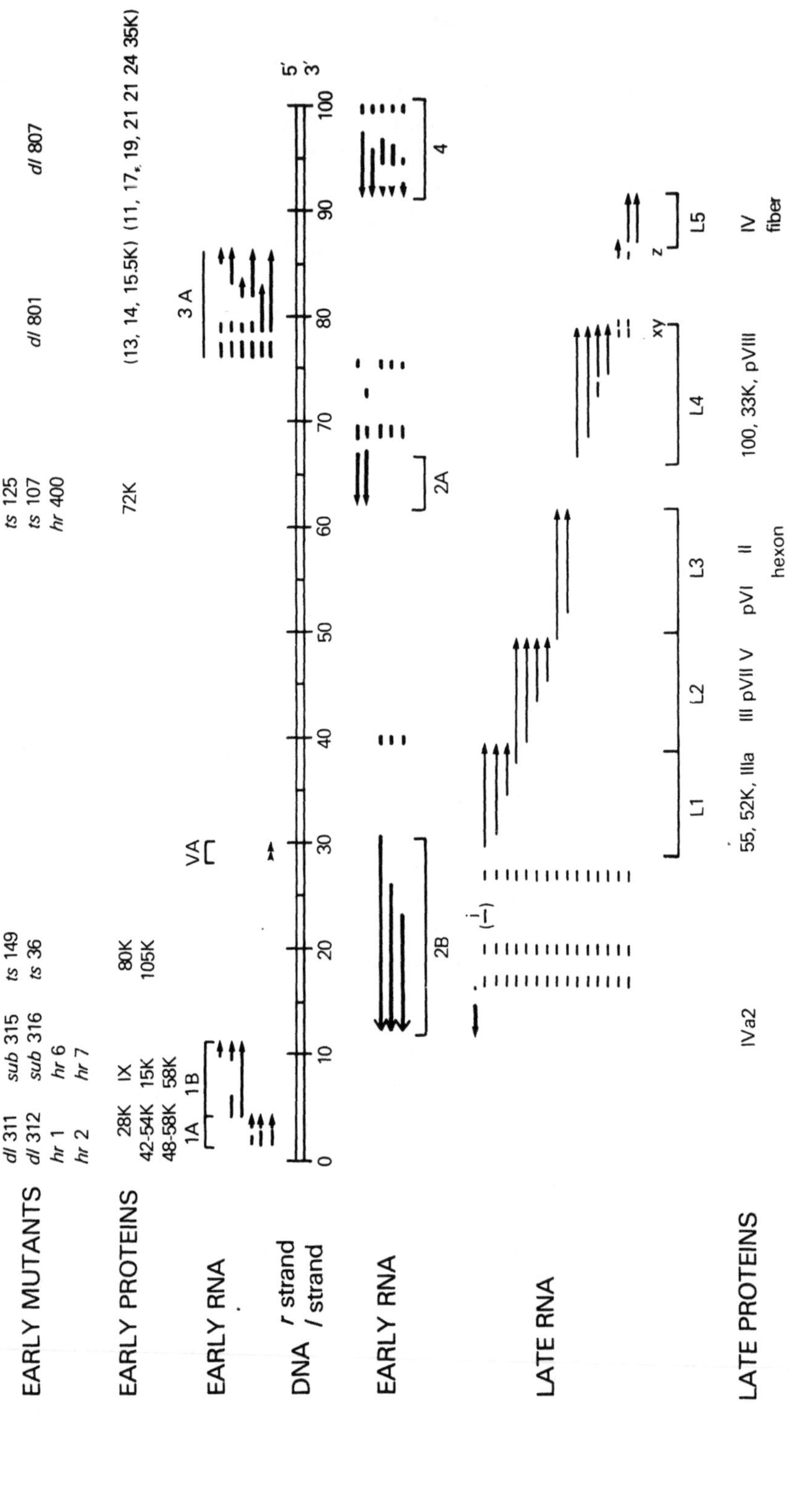
EARLY MUTANTS
dl 311
dl 312
hr 1
hr 2
dl 313
sub 315
sub 316
hr 6
hr 7
ts 149
ts 36
ts 125
ts 107
hr 400
dl 801
dl 807
EARLY PROTEINS
28K
42-54K
48-58K
IX
15K
58K
80K
105K
72K
(13, 14, 15.5K)
(11, 17, 19, 21 21 24 35K)
EARLY RNA
1A
1B
VA
3 A
DNA
r strand
l strand
5'
3'
0
10
20
30
40
50
60
70
80
90
100
EARLY RNA
2B
2A
4
LATE RNA
i
L1
L2
L3
L4
L5
xy
z
LATE PROTEINS
IVa2
55, 52K, IIIa
III pVII V
pVI
II
hexon
100, 33K, pVIII
IV
fiber

FIGURE 3. Genome map of adenovirus types 2 or 5. The location of the mRNA species has been determined by electron microscopy and nuclease digestion of RNA-DNA heteroduplexes. The 36.5-kb genome is divided into 100 map units (one map unit is equivalent to 365 base pairs). The cytoplasmic RNA species are indicated by arrows. The 5′ ends of the arrows (indicated by vertical brackets) correspond to the location of promoters. The 3′ ends (indicated by the arrow heads) correspond to the polyadenylation sites. The known adenovirus proteins and the approximate location of genes coding for them are indicated by their size in *K* (1000 molecular weight units) or by Roman numerals for virion structural proteins. The map locations of specific adenovirus early mutants are indicated. The data are redrawn from Broker and Chow (1980) and Shaw and Ziff (1980) and are a composite of evidence derived from two C-type adenoviruses, types 2 and 5 which have a very similar genome organization. Note that early gene region 2B (Stillman *et al.*, 1981) has also been called early region 5 (Galos *et al.*, 1979).

tion at the same time that adenovirus DNA replication and the transition from early to late adenovirus gene expression begins.

The general arrangement of genes in most adenovirus serotypes appears to be similar, but the most extensively mapped genomes are those for adenovirus 2 and 5, which are very similar. In Fig. 3 the genome map of adenovirus 2 is shown. The location and structure of the known mRNA species and the proteins for which they code are indicated. Adenovirus genes are expressed by transcription from nine promoters. Transcripts from each promoter are then processed by differential termination, polyadenylation or splicing to yield in most cases overlapping families of mRNAs which are coterminal at both 3′ or 5′ or both ends. Translation of individual mRNAs of each family may then yield different protein products. Nearly all the late genes are represented by the five families of single-strand mRNAs, L1, L2, L3, L4, and L5, derived from transcripts initiated at the late promoter near map position 16.5 (Broker and Chow, 1980; Shaw and Ziff, 1980). Two other RNAs synthesized mainly late are that for protein IX from within region 1B near map position 10 and that for protein IVa2.

Adenovirus early genes are transcribed from six different promoter sites. Initially five separately promoted early regions 1A, 1B, 2A, 3, and 4 were recognized (Kitchingman *et al.*, 1977; Berk and Sharp, 1977a,b; Chow *et al.*, 1979; Evans *et al.*, 1977; Sehgal *et al.*, 1979; Wilson *et al.*, 1980). More recently two additional early regions were discovered. The first of these contains single-strand mRNAs in the region between 27.0 and 12.0 map units which apparently are promoted from one of the same promoter sites used for region 2A and mRNA (Galos *et al.*, 1979; Chow *et al.*, 1979). This region was originally called region 5 (Galos *et al.*, 1979; Broker and Chow, 1980) but more recently was designated region 2B (Stillman *et al.*, 1981). It is not yet clearly determined if region 5 and region 2B are indeed synonymous. Finally, it was recently shown that the major late promoter at map position 16.5 is active at early times. However, the only RNA from this region which accumulates early is one of the L1 species, apparently because early transcription is terminated or attenuated between map positions 40 and 60 (Lewis and Mathews, 1980; Thomas and Mathews, 1980; Shaw and Ziff, 1980; Nevins and Wilson, 1981). Two additional RNAs, for protein IX and protein IVa2, appear at intermediate times but prior to or in the absence of DNA synthesis.

All of the adenovirus mRNAs are synthesized by RNA polymerase II. Two additional RNAs, the 5.5 S VA_I RNA and 5.2 S VA_{II} RNA are synthesized by RNA polymerase III from two genes located close to map position 29.0 and 30.0 (Akusjarvi *et al.*, 1980).

The proteins coded by many of the adenovirus mRNA families have now been identified (Fig. 3). Since the adenovirus late genes do not appear to be important for AAV we will discuss only the early gene proteins. The proteins from regions 1A, 1B, and 2A are described in more detail below. Early after infection two region 1A mRNAs code for at least four proteins with molecular weights between 42,000 and 58,000. Late after infection an additional unspliced 1A mRNA codes for a 28,000 molecular weight protein. Two early 1B mRNAs specify two related proteins M_r 58,000 M_r 15,000, and at intermediate times an additional, separately promoted 1B mRNA specifies the structural protein IX. Region 2 mRNAs code for a M_r 72,000 DNA-binding protein which has some remarkable pleiotropic properties. This protein is described in more detail below. Region 3 mRNAs specify several proteins of which the most well characterized is an M_r 14,000 protein which when glycosylated has an apparent size of 19,000 (Persson *et al.*, 1978, 1980; Ross and Levine, 1979; Ross *et al.*, 1980b). This glycoprotein is associated with the cell membrane and apparently complexed with cellular transplantation antigens. Region 4 mRNAs code for several proteins for which there are as yet no clear indications regarding possible functions.

One protein coded by a region 2B mRNA appears to be the M_r 87,000 (87K) protein that is covalently linked to the 5′ terminus of replicating Ad DNA (Challberg *et al.*, 1980; Stillman *et al.*, 1980; Stillman, 1981; Desiderio and Kelly, 1980). This protein is cleaved after covalent binding and during virus maturation by an adenovirus-coded protease to yield a covalently bound M_r 55,000 (55K) protein (Robinson and Bellett, 1975); Rekosh *et al.*, 1977). In addition, a 13.5K protein is specified by an mRNA transcribed from map position 17.0 to 21.5. The region L1 mRNA which occurs early codes for two related proteins, 55K and 52K (Miller *et al.*, 1980; Lewis and Mathews, 1980; Thomas and Mathews, 1980), but the IIIa protein is coded by another L1 mRNA present late but not early (Miller *et al.*, 1980). An early L1 mRNA also apparently specified a 13.6K protein about map position 21.0 to 22.0, and this may be located in the i-leader present in early but not late L1 mRNA (Lewis and Mathews, 1980).

It has become clear from a great deal of recently accumulated evidence that the division of adenovirus genes into early and late is a gross oversimplification. Early gene expression may be rather strongly phased by a complex cascade control mechanism. There is good evidence for an ordered appearance of mRNAs and gene products and both negative and positive controls appear to operate. However, the mechanism of these controls is quite unclear, and even the precise interrelationships between different early regions is not well understood. This cascade control is extremely important for consideration of AAV helper functions. For instance, a mutation in an early adenovirus gene resulting in a defect in AAV growth might indicate either a direct involvement of that gene product with AAV or an indirect effect on the control of another adenovirus gene.

Analysis of the time of appearance of adenovirus early RNAs and early proteins provided evidence for a cascade control. Nevins *et al.* (1979) analyzed the time course of appearance of adenovirus type-2 early RNAs. Region 1A mRNA synthesis, which was detected first within 45 min of infection, was maximal at about 3 hr and maintained this level for several hours. Synthesis of RNA from early regions 3 and 4 was maximal by 3 hr and thereafter declined. Region 2A RNA synthesis was the last detected, beginning at about 2 hr and reaching a maximum at 7 hr. Analysis of early protein synthesis (reviewed by Philipson, 1979) also showed results consistent with the RNA analysis of Nevins *et al.* (1979). Thus, early region 3 and 4 proteins clearly accumulated before the 72K early region 2 protein.

Additional evidence for an adenovirus cascade control comes from studies with specific inhibitors of protein synthesis and the use of adenovirus early gene mutants. Study with the region 2A temperature-sensitive mutant Ad5*ts*125 suggests that the 72K protein is required for the decline in mRNA accumulation of regions 3 and 4, as well as its own region 2A (Carter and Blanton, 1978; Blanton and Carter, 1979). Whether this down modulation of mRNA by the 72K protein is due to a decrease in synthesis or stability of mRNA remains unclear (Blanton and Carter, 1979; Nevins and Winkler, 1980).

Studies of adenovirus type-5 deletion or host-range mutants in region 1A suggest that expression of this region is required for transcription of the other early regions 1B, 2A, 2B, 3, and 4 (Jones and Shenk, 1979a,b; Shenk *et al.*, 1979; Berk *et al.*, 1979, 1980). However, the defect in transcription of other early regions can be overcome by increased multiplicity of the region 1A mutants (Shenk *et al.*, 1980; Berk *et al.*, 1980).

These effects of region 1A and region 2A mutations on transcription of various early regions may account for a variety of studies with protein synthesis inhibitors which suggested that incomplete protein synthesis inhibition with cycleheximide added 1 to 2 hr after the infection could result in increased levels of some mRNA (Craig and Raskas, 1974; Harter *et al.*, 1976; Eggerding and Raskas, 1978), whereas more stringent inhibition with emetine resulted in synthesis of region 1A mRNA but not region 1B, 2, 3, or 4 RNA (Lewis *et al.*, 1980).

More recently Lewis and Mathews (1980) reported that when cell protein synthesis was stringently inhibited with anisomycin prior to virus infection only the mRNA coding for a 13.5K protein (map position 17 and 21.5 map units) and the L1 RNA coding for the 52,55K proteins were not depressed in amount. RNAs for all the other early regions, including 1A, 1B, 2A, 3, 4, and the 87K (2B) protein and the 13.6K protein (encoded to the right of 21.5 map units) were decreased 20- to 200-fold in amount. The 13.5K and 52,55K proteins appear to be immediate early products.

Lewis and Mathews (1980) interpreted their experiments and the previous work of others to define several stages in adenovirus gene expression:

1. In the *immediate early* phase two genes (13.5K and 52,55K) are expressed in the absence of protein synthesis. One or both of these genes may be a positive effector for region 1A expression.
2. In the *pre-early* phase region 1A is expressed which is a prerequisite for expression of the other early regions.

3. In the *delayed early* phase all the other early regions are expressed but even within this phase there may be a cascade control.
4. After all the early genes are expressed DNA replication and expression of late genes occur.

As noted by Sharp (1980), studies based on inhibition of protein synthesis must be interpreted with caution. Such stringent inhibition of protein synthesis may alter the normal physiological controls. For instance inhibition of protein synthesis may drastically alter the stability of some mRNAs. In addition, the model of Lewis and Mathews (1980) is difficult to reconcile with synthesis of 1A and 1B mRNAs in transformed cells which do not contain any adenovirus sequences between 11.4 and 39.0 map units (Sharp, 1980). Stages 2, 3, and 4 in the above scheme are well established. It is also clear that the 13.5K and 52,55K proteins are expressed early from the L1 promoter. However, the role of these latter proteins in effecting region 1A expression is clearly not proven.

A very recent study by Persson *et al.* (1981) suggests that, in addition to the cascade controls apparently operating to affect the levels of mRNA accumulation, there may also be a translational control exerted on adenovirus early mRNA by an early region 1 product. Persson *et al.* (1981) used inhibition with anisomycin under conditions similar to Lewis and Mathews (1980) but concluded that while the translation of early mRNAs was inhibited, any decrease in the level of these mRNAs may be a secondary phenomenon due to destablization of mRNA in the absence of protein synthesis.

The observations of Persson *et al.* (1981) are particularly interesting in view of recent experiments by Jay *et al.* (1981) which suggest that translational control of the AAV 2.3-kb mRNA encoding the capsid proteins is exerted by the adenovirus E2A 72K DNA binding protein. In this case, destablization or incorrect processing of the AAV mRNA in the absence of translation apparently is not a factor (Myers *et al.*, 1980; Jay *et al.*, 1981).

In summary, although cascade control of adenovirus early gene expression appears to be a real phenomenon, it may be very complex and no underlying mechanisms can be discerned at present. This current situation must necessarily color the interpretation of experiments analyzing the effect of early gene mutations on the replication of AAV. However, analysis of some of these adenovirus control functions may be approachable by using AAV as an independent probe.

VI. ROLE OF SPECIFIC ADENOVIRUS GENES IN AAV HELPER FUNCTION

Recent work provides clear evidence that AAV growth requires expression of only a subset of adenovirus early genes. Studies with specific adenovirus mutants or with transfection or microinjection of Ad DNA fragments or Ad mRNA have provided good evidence for the involvement in the AAV helper function of adenovirus type-2 or type-5 early regions 1A, 1B, 2A, and 4 and the apparent nonrequirement for early regions 2B or 3 and all the late genes. In addition some DNA transfection experiments provided evidence for the involvement of the adenovirus viral-associated (VA) RNA genes.

A. AAV Helper Function of Adenovirus Mutants

The most interesting adenovirus mutants for analysis of AAV helper functions are obviously those in which the adenovirus mutation results in a defect in AAV growth. Such mutants will most likely allow identification and biochemical analysis of adenovirus genes providing AAV helper functions. In this respect the earliest experiments with adenovirus mutants and AAV were disappointing since all the mutants tested, including several that were DNA-negative for adenovirus, apparently helped AAV growth efficiently. Nevertheless, these early experiments did provide the first evidence that AAV DNA replication and growth can occur in the absence of adenovirus DNA synthesis and by implication in the absence of adenovirus late gene functions.

Ito and Suzuki (1970) reported that the mutant Ad31*ts*13, a DNA-negative, temperature-sensitive mutant of adenovirus type 31, grown at the nonpermissive temperature in human embryonic kidney (HEK) cells efficiently helped the production of infectious AAV-1. Studies with this mutant were extended to an analysis of AAV DNA replication which appeared to be similar with the mutant or wild type virus as helper (Mayor and Ratner, 1973; Handa *et al.*, 1976; Handa and Shimojo, 1977b).

Ishibashi and Ito (1971) analyzed a series of temperature-sensitive mutants of an avian adenovirus, chick embryo lethal orphan (CELO) virus. These mutants were not mapped but were phenotypically divided into several groups and at least some appeared to be DNA-negative ("early") mutants and others were likely late mutants. In chick kidney cells all these CELO mutants helped AAV-1 as efficiently as did wild-type CELO virus at both permissive and nonpermissive cultures.

Two groups analyzed several temperature-sensitive, DNA-negative, early mutants of adenovirus types 5 and 12. Handa *et al.* (1975) reported that in HEK cells at the nonpermissive temperature three adenovirus 12 temperature-sensitive mutants, Ad12*ts*A275, Ad12*ts*B22, and Ad12*ts*C295 all helped AAV-1 as efficiently as wild-type adenovirus 12. These three adenovirus 12 mutants represent different complementation groups (Shiroki and Shimjo, 1974) but have not been clearly mapped. Handa *et al.* (1975) and Straus *et al.* 1976a) examined adenovirus type-5 mutants in regions 2A (Ad 5*ts*125) and 2B (Ad5*ts*149, Ad5*ts*36) and reported that both types of mutants helped AAV. Handa *et al.* (1975) used HEK cells and AAV-1 and Straus *et al.*, (1976) used KB cells in spinner culture and AAV-2. The Ad12*ts*A275 mutation results in a thermolabile DNA binding protein and probably, like Ad5*ts*125, resides in region 2A (Van der Vliet *et al.*, 1975; Rosenwirth *et al.*, 1975). Also heterotypic complementation indicates that the Ad31*ts*13 mutation may be located in the same gene (the region 2A 72K DNA-binding protein) as Ad12*ts*A275 (Suzuki and Shimojo, 1974).

Myers *et al.* (1980) subsequently reexamined growth of AAV-2 in KB cells using the two mutants Ad5*ts*125 and Ad5*ts*149. In agreement with the other two laboratories, the *ts*149 mutant at the nonpermissive temperature helped growth of AAV as efficiently as wild-type adenovirus. In contrast to the previous reports, Myers *et al.* (1980) found that at the nonpermissive temperature of 40°C the Ad5*ts*125 mutant was an inefficient helper and the yield of infectious AAV or AAV particles was decreased by up to 100-fold. This was the first report of an adenovirus early gene mutant that failed to help AAV and provided the first direct evidence that some of the pleiotropic properties of the adenovirus early region 2A DNA binding protein were important for AAV growth. Myers *et al.* (1980) also showed that a closely related adenovirus 5 early region 2A mutant, Ad5*ts*107, also was defective as a helper for AAV-2 in KB cells at 40°C. It is not yet clear why the observation of Myers *et al.*, (1980) differs from the earlier reports, but this is discussed more extensively below. In retrospect, two unrelated earlier studies (Buller *et al.*, 1979; Klessig and Grodzicker, 1979) when considered together provide indirect circumstantial evidence that adenovirus region 2A is involved in AAV growth. Buller *et al.* (1979) reported that growth of AAV in certain monkey cells exhibited the same host-range defect as do the human adenoviruses. Klessig and Grodzicker (1979) reported that mutations mapping in early region 2A of human adenovirus 2 or 5 overcome the host restriction in monkey cells for adenovirus. The involvement of the adenovirus region 2A in host restrictions for virus growth may explain in part the difference between the results of Myers *et al.* (1980) and the previous work (Handa *et al.*, 1975; Straus *et al.*, 1976a). In addition recent experiments by Janik *et al.* (1981) provide good evidence that AAV growth in human cells requires adenovirus region 2A.

In summary, the results with the adenovirus DNA-negative mutants provide suggestive evidence that adenovirus late genes are not required by AAV since none of the early mutants synthesized significant levels of late gene products. This conclusion is strongly supported by other experiments involving transfection or microinjection of adenovirus DNA or RNA into host cells. The only current anomaly in this general scheme is the report by Mayor *et al.* (1977) that an uncharacterized but phenotypically late adenovirus type-31 mutant, Ad31*ts*94 may be deficient in helping assembly of stable, mature AAV particles. This was discussed above.

More recent studies with adenovirus type-5 host-range mutants having mutations in early region 1A or 1B have demonstrated the importance of these regions in the AAV helper function (Laughlin *et al.*, 1979c; Carter *et al.*, 1980; Ostrove and Berns, 1980; Janik *et al.*, 1981; Laughlin *et al.*, 1982). These studies provide support for an earlier contention by Handa *et al.* (1977a) that the transforming region of adenovirus 12 may under certain conditions supply AAV helper functions.

B. Analysis of Helper Function by Microinjection or Transfection of Adenovirus DNA or RNA

There are now a variety of procedures available for physically introducing genes or gene products into eukaryotic cells in culture. Two of these procedures, microinjection of RNA through glass capillaries (Graessman *et al.*, 1980) or transfection of DNA using the calcium phosphate precipitation technique (Graham and van der Eb, 1973), have recently been applied to studies of the AAV helper function (Richardson *et al.*, 1980; Janik *et al.*, 1981). Both procedures provide strong evidence that only a subset of adenovirus type 2 or 5 early gene functions are required for AAV growth.

Richardson *et al.* (1980) employed capillary microinjection to introduce adenovirus mRNA into the cytoplasm of a line of African green monkey kidney (Vero) cells which are permissive for growth of both adenovirus type 2 and AAV-2. Following injection of the mRNA into the cytoplasm the cells were infected with AAV-2 particles. Preparations of cytoplasmic, poly-A containing RNA extracted from KB cells early or late after infection with adenovirus 2 were used. The early mRNA preparations were prepared from infected cells grown for 10 hr in the presence of a low level of cycloheximide to enhance production of adenovirus early mRNA (see above). The early mRNA preparation was extensively characterized by electron microscopy and contained no late mRNA but did contain the L1 transcripts known to be present early (see above). Both early and late mRNA preparations were shown to synthesize authentic adenovirus early proteins both by *in vitro* translation and after microinjection into Vero cells. When early mRNA preparations were injected into cells that were subsequently infected with AAV-2 particles, synthesis of the AAV capsid proteins VP1, VP2, and VP3 could be demonstrated by labeling with ^{35}S-methionine followed by immunoprecipitation with specific anti-AAV-2 IgG and electrophoresis. Late adenovirus mRNA preparations were inactive in potentiating synthesis of AAV proteins. Also mixtures of the late and early mRNA preparations were inactive. This may mean either that the late mRNAs present in high copy numbers compete efficiently with the early mRNAs for the cellular translation apparatus or that some late mRNA products may be inhibitory of expression of early mRNAs.

Richardson *et al.* (1980) also performed additional experiments to demonstrate that, in Vero cells injected with early adenovirus mRNA

TABLE I. Production of Infectious AAV2 in Vero Cells Infected with AAV 2 and Injected with Adenovirus Early mRNA[a]

Innoculum in Vero cells	AAV yield after passage of Vero cell lysate in KB cells (infectious units)
AAV2 particles	10^2
AAV2 particles + Ad2 particles	11.0×10^7
AAV2 particles + Ad2 early mRNA	6.0×10^7

[a] Approximately 300 Vero cells were infected with AAV 2 (5 infectious units/cell) and either Ad2 virions (100 particles/cell) or microinjected with early mRNA from adenovirus infected KB cells (about several hundred to several thousand viral mRNA molecules per cell). After virus adsorption cells were treated with anti-AAV2 IgG for 30 min. At 40 hr after infection cells were lysed, heated at 60°C for 15 min and passaged on 2×10^5 KB cells in the presence of added adenovirus 2 (200 particles/cell). After 48 hr the cells were lysed and the total AAV infectivity was determined (Carter *et al.*, 1979). Based on a calibration by passage of known quantities of AAV2 in control KB cell cultures it was estimated that the titer of AAV2 in the original Vero cells lysates was amplified about 10^4-fold in the passage on KB cells (from Richardson *et al.*, 1980).

and infected with AAV-2, infectious AAV particles as well as AAV proteins were produced at levels comparable to a normal infection. In these experiments lysates of Vero cells were heated to inactivate any adenovirus and passaged once on human KB cells to amplify the expected AAV yield. The amount of AAV infectivity from this KB cell passage was then determined by assay on KB cells. As shown in Table I, the yield of AAV originating from AAV-infected Vero cells was similar whether the Vero cells were injected with adenovirus early mRNA or infected with adenovirus particles. The RNA microinjection experiments provided strong evidence that only adenovirus early mRNAs (and presumably expression of these RNAs) were required for a normal yield of fully infectious AAV. More recent experiments suggest that only a subset of these early Ad RNAs are required for AAV growth.

Richardson and Westphal (1981) microinjected purified adenovirus mRNAs from separate early regions into the cytoplasm of Vero cells or injected into the nucleus of the same cells, purified DNA fragments comprising individual Ad early regions. The microinjected cells were then infected with AAV particles. Injection of DNA comprising only early regions 1, 2A, and 4 was sufficient to allow expression of the AAV capsid protein. Microinjection of purified mRNAs required only region 4 mRNA for the same result. Injection of Ad DNA fragments containing early region 1 and 2A could be replaced by region 2A mRNA. Thus, injection of Vero cells with region 2A mRNA and region 4 DNA helped AAV capsid protein synthesis. This may reflect a cascade function in which early region 1 turns on early region 2A which then activates early region 4 transcription. Region 4 products may then allow AAV growth. This interpretation (Richardson and Westphal, 1981) is difficult to reconcile with the evidence that adenovirus early region 4 transcription precedes that from region 2 and that a region 2A product may actually turn off or des-

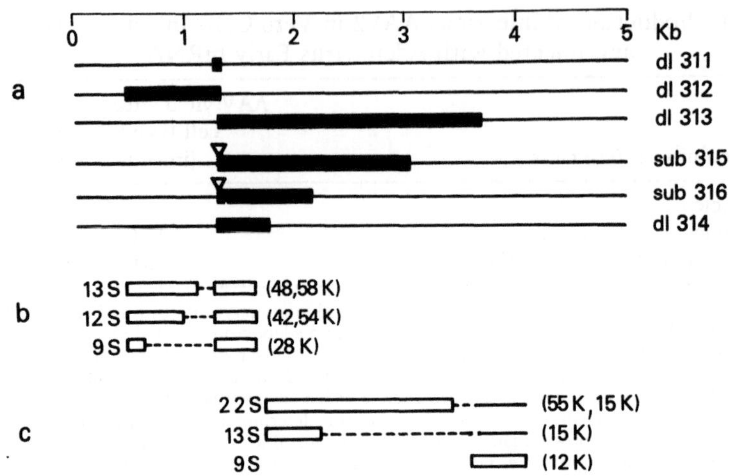

FIGURE 4. The left-hand region of the adenovirus type-5 genome. (a) The regions deleted in individual mutants are indicated by a shaded box. Mutants *sub*315 and *sub*316 have additional insertions of 1.5 and 1.35 kb, respectively, as indicated by the symbol ▽. (b) and (c) show RNAs transcribed from the region E1a or E1b respectively in permissive cells. ▭, exon; ---, intron; —— untranslated portion of exon. Proteins coded by individual transcripts are indicated in parenthesis. The maps were derived from the data of Jones and Shenk (1979a,b), Chow *et al.* (1979), Perricaudet *et al.* (1980), and Galos *et al.* (1980).

tabilize region 4 transcription (see Section V above). It should also be noted that in the latter experiments of Richardson and Westphal (1981) only accumulation of AAV DNA and AAV capsid proteins were assayed. Production of infectious AAV virions was not measured.

It is worth noting that cells microinjected with Ad early region 4 mRNA alone accumulated only 10^3 to 10^4 AAV genome copies whereas a normal infection with AAV and adenovirus results in synthesis of more than 10^6 AAV genomes per cell. This may indicate that other adenovirus genes augment region 4 in promoting AAV DNA replication. For instance, Laughlin *et al.* (1982) suggested that Ad region 1B is required to prevent AAV DNA degradation. Indeed in the experiments reported by Richardson and Westphal (1981) in which AAV is grown in Vero cells microinjected with Ad early region 4 mRNA alone, there is clear evidence for a great deal of AAV DNA degradation.

Janik *et al.* (1981) employed the calcium phosphate transfection procedure to introduce adenovirus DNA fragments into cells. In these experiments human 293 cells were used. These cells are human embryonic kidney cells that were transformed with sheared adenovirus 5 DNA (Graham *et al.*, 1975, 1977). Adenovirus 5 DNA sequences from map positions 0/12.0 and 91.0/100 are present in these 293 cells, but only the early region 1A and 1B genes between 0 and 12.0 map units are expressed (Aiello *et al.*, 1979). Thus transfection experiments on these cells can only determine if regions other than 1A and 1B are necessary for AAV.

Janik *et al.* (1981) infected 293 cells with AAV-2 and then transfected them with various combinations of purified restriction fragments of adenovirus types 2 or 5. The cells were then assayed for production of AAV capsid proteins using the immunofluorescent assay. One series of experiments showed that three regions of the adenovirus genome 0/59.5, 59.5/75.9, and 89.7/100 were required in the transfection to obtain an efficient yield of AAV. The fragments with coordinates 59.5/75.9 and 89.7/100 contain adenovirus early regions 2A and 4, respectively, and Janik *et al.* concluded that expression of both regions was required for AAV. These results are in good agreement with those of others (Richardson *et al.*, 1980; W. Richardson, unpublished; Myers *et al.*, 1980; Jay *et al.*, 1981).

Janik *et al.* (1981) then further fractionated the adenovirus DNA in the map region 0/59.5 and assayed smaller fragments for AAV helper activity in the DNA transfection assay. The only portion of this region which appeared to be required in the transfection (in addition to the region 0/12.0 supplied by the 293 cells) mapped in a small 28.5/29.4 fragment. This region is known to contain the gene for the adenovirus 5.5 S VA_I RNA, and Janik *et al.* concluded that this RNA is required for AAV growth.

C. Helper Function Involvement of Specific Adenovirus Early Gene Regions

1. Early Regions 1A and 1B

Two sets of adenovirus type-5 mutants in regions 1A and 1B are available. These are the *hr* mutants (Harrison *et al.*, 1977) and and the *dl* and *sub* mutants (Jones and Shenk, 1979a,b; Shenk *et al.*, 1980). All of these mutants are defective for transformation of hamster embryo cells and each set falls into two complementation groups mapping in region 1A and region 1B, respectively. All of the region 1A mutants are adenovirus DNA-negative as are the region 1B *dl* mutants. In contrast, the region 1B *hr* mutants are DNA-positive for adenovirus. All of these 1A and 1B mutants are host-range mutants. Growth is restricted in KB or HeLa cells but occurs readily in human 293 cells that contain and express the left-hand 11% (early regions 1A and 1B) of the adenovirus type 5 genome. The structure of these mutants is shown in Fig. 4.

Jones and Shenk (1979a,b) isolated and described a series of mutants of adenovirus 5 that had deleted the endonuclease *Xba*-1 cleavage site located at about 4.0 units on the genetic map. Ad5*dl*309 has a 2 base deletion at map position 29.0 but grows as well as Ad*wt* on either 239 or KB cells. *dl*311 is missing 58 base pairs at the right-hand end of region 1A, *dl*312 lacks 902 base pairs in region 1A, and *dl*313 has deleted 2350 base pairs including the right-hand end of 1A and most of 1B. Figure 4 illustrates the mRNA species that are transcribed from Ad region 1A and

TABLE II. AAV Antigen Synthesis with Helper Adenoviruses Containing Deletions in Early Gene Region 1[a]

		KB-3 Cells Percent fluorescent nuclei		293 Cells Percent fluorescent nuclei		
Adenovirus	AAV-2	Ad[b]	AAV[c]	Ad	AAV	Ratio of AAV yield in KB-3:293
None	−	<0.5	<0.5	<0.5	<0.5	
	+	<0.5	<0.5	<0.5	<0.5	
Ad5wt	−	70	<0.5	80	<0.5	
	+	ND[d]	26.0	ND	39	0.67
*dl*309	−	95	<0.5	95	<0.5	
	+	ND	61	ND	58	1.05
*dl*311	−	<0.5	<0.5	35	<0.5	
	+	ND	7	ND	55	0.13
*dl*312	−	<0.5	<0.5	95	<0.5	
	+	ND	<0.5	ND	39	0.01
*dl*313	−	<0.5	<0.5	75	<0.5	
	+	ND	<0.5	ND	53	0.01

[a] Cells were grown in monolayer cultures and infected with AAV-2 and the indicated adenovirus each at a multiplicity of 10. Cultures were fixed 30 hr after infection and stained for virus antigens using indirect immunofluorescence. The data are taken from Laughlin *et al.* (1981).
[b] Cells were stained with anti-adenovirus 5 antiserum.
[c] Cells were stained with anti-AAV 2 IgG.
[d] Not done.

1B and the proteins derived from them. *dl*311 synthesizes normal mRNAs from all four early regions (regions 1, 2A, 3, and 4) except that the 1A species are all shortened after the splice by 58 nucleotides. This mutant grows, albeit poorly, on KB cells and retains transforming ability. Human cells infected with *dl*312 at low multiplicity of infection do not synthesise virus specific RNA or proteins (Jones and Shenk, 1979b; Shenk *et al.*, 1980). *dl*313 infected cells synthesize 1A RNA species which contain also the 3′ end of region 1B. However, these hybrid RNAs do permit the expression of the other early regions of adenovirus and therefore seem to function as wild type early region 1A mRNAs.

Laughlin *et al.* (1982) analyzed the *dl* and *sub* mutants for their AAV helper function. The *dl* mutants were all efficient helpers of AAV capsid antigen synthesis in 293 cells (Table II). Adenovirus 5 wild type and *dl*309 were equally efficient helpers of AAV in both KB and 293 cells. In contrast AAV capsid protein synthesis was not detected in KB cells when either *dl*312 or *dl*313 was the helper. A small proportion of KB cells coinfected with AAV and *dl*311 did accumulate AAV antigens, which presumably reflects the fact that *dl*311 grows to a limited extent in KB cells. In KB cells infected with *dl*312 or *dl*313, AAV RNA was not detected by liquid hybridization or by the endonuclease S1 mapping procedure whereas *dl*311-infected KB cells supported a low level of AAV RNA synthesis. These results were consistent with the assays for fluorescent antigen.

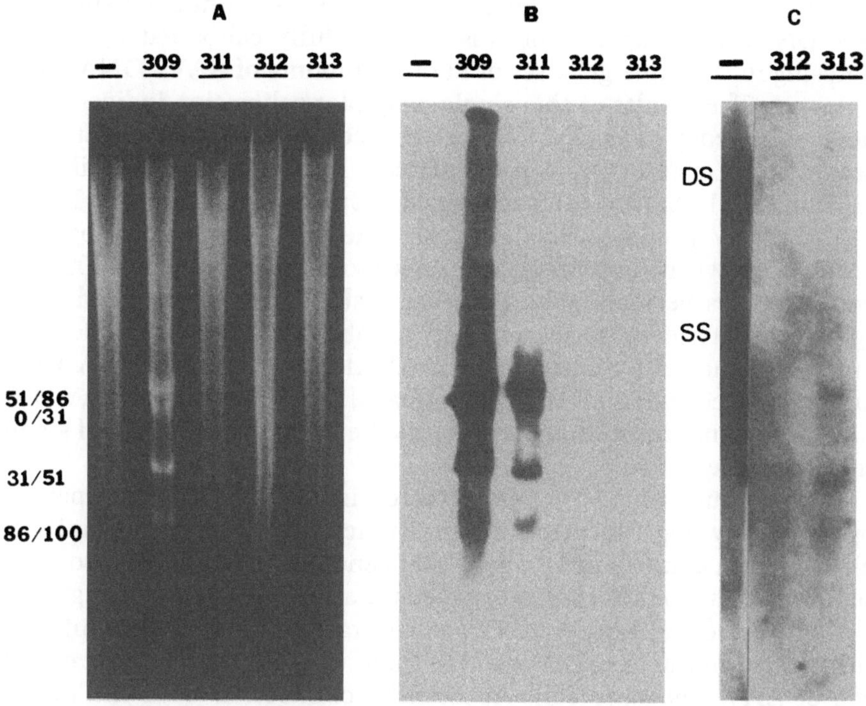

FIGURE 5. Replication of AAV DNA in mutant helped cells. KB monolayer cells were infected with AAV2 and either no added helper (—), or *dl*309 (309), *dl*311 (311), *dl*312 (312), or *dl*313 (313) as helper. At 20 hr after infection total cell DNA was extracted, digested with *Hinc*II and electrophoresed in a 1.4% agarose gel. Panel A: The DNA in the gel was viewed by UV illumination after staining with ethidium bromide. The *Hinc*II fragments of AAV DNA are indicated by their map coordinates at the left. Fragments 0/31 and 86/100 are resolved into two species because they are terminal fragments. Panel B: The gel in panel A was blotted into AMB paper, annealed with nick-translated AAV ^{32}P-DNA and autoradiographed. Each track contains 4 μg of DNA equivalent to approximately 6 × 10^4 cells. Panel C: Same tracks as in panel B but exposed for a longer time to sensitivity of 0.1 genome copies per cell. The position of single-stranded (SS) and reannealed double-stranded (DS) parental genomes are indicated. The material migrating faster than the SS DNA in the left track is an artifact due to the very large amounts of AAV *Hinc*II 51/86 and 0/31 fragments in the neighboring track (track 309 of panel B). From Laughlin *et al.* (1982).

Laughlin *et al.* (1982) also analyzed AAV DNA synthesis with the *dl* mutants as helpers in KB cells. This was performed by extracting total cell DNA, cleaving with endodeoxyribonuclease *Hinc*II, followed by electrophoresis in agarose gels. The AAV DNA was then detected by "Northern" blotting onto derivatized nitrocellulose paper and hybridizing with a nick-translated probe. This experiment is illustrated in Fig. 5, which shows that AAV DNA extracted from *dl*309 helped helped cells contained the expected *Hinc*II fragments. *dl*311 helped cells contained lesser amounts of AAV DNA and the *dl*312 or *dl*313 cells contained almost no AAV DNA. This experiment was also carefully calibrated by electrophoresing in the same gel a set of known amounts of AAV2 DNA *Hinc*II fragments. The results of this single-copy blot calibration indicated that one genome copy of AAV DNA per cell could be readily detected and that the limit of detection was 0.1 per cell. According to this calibration *dl*309 and *dl*311 helped cells contained, respectively, about 5×10^5 and 2×10^4 AAV genomes. It should be noted that the autoradiographic exposure shown in Fig. 5B is at a sensitivity of only about 1000 AAV genome copies per cell. When the same slab gel was examined at the higher sensitity level, as seen in Fig. 5C, about 100 AAV genomes were detected in the cells infected with AAV alone. This AAV DNA which was not cleaved by *Hinc*II and that migrated more slowly than AAV DNA *Hinc*II fragments most likely represents the unreplicated parental single-strand genomes.

The amount of AAV DNA detected in the absence of helper was approximately that expected for a multiplicity of 10 infectious units per cell, which is equivalent to about 500 genomes per cell and about one quarter of this reaches the cell nucleus (Carter *et al.*, 1979; Laughlin *et al.*, 1979; Rose and Koczot, 1972). In contrast, at 20 hr after coinfection with the Ad mutants *dl*312 or *dl*313 there was less than one intact genome of AAV DNA per cell. Apparently, coinfection with the mutant adenoviruses resulted in degradation of most of the infecting AAV genomes.

The host-range dependence of the Ad region 1 *dl* mutants is not absolute (Shenk *et al.*, 1980). If the multiplicity of *dl*311 or *dl*312 is increased sufficiently, the mutants will grow in KB cells. This apparently reflects a requirement for a 1A product to increase the efficiency of transcription of other Ad early regions. This requirement for 1A functions may be overcome by increased gene dosage. To a much lesser extent, this is also true for *dl*313. Both mutants *dl*311 and *dl*312 showed extensive multiplicity-dependent leakiness for AAV helper function whereas *dl*313 showed little or no leakiness for AAV growth (Laughlin *et al.*, 1982). Also, neither *dl*312 nor *dl*313 at low multiplicity (10 p.f.u/ml) helped AAV in KB cells even when the AAV multiplicity was increased to 500 infectious units (i.e., 25,000–50,000 particles) per cell. This result indicates that Ad region 1A does not act on AAV DNA simply to increase transcription as it does for other Ad early regions.

The mutants *dl*311 and *dl*312 can complement *dl*313 for adenovirus lytic growth (Ross *et al.*, 1980a; Shenk *et al.*, 1980) in KB cells. All these deletion mutants are complemented at 40°C by the early region 2A temperature-sensitive mutant Ad5*ts*125. However, Ad5*ts*125 is itself a defective helper of AAV at the nonpermissive temperature (see below).

As summarized in Table III, Laughlin *et al.* (1982), showed that the two 1A mutants *dl*311 and *dl*312 did not complement each other for AAV helper function in KB cells but were complemented by the region 1B

TABLE III. Complementation for AAV Helper Function[a]

Coinfecting adenoviruses	Percent cells fluorescent for AAV: KB	Percent cells fluorescent for AAV: 293
*dl*309	56.3	38.8
*dl*311	0.5	20.4
*dl*312	0.5	44.8
*dl*313	0.5	ND[b]
*dl*312 × *dl*312	1.0	55.2
*dl*311 × *dl*313	15.0	ND
*dl*312 × *dl*313	60.0	ND
*ts*125[c]	2.0	0.5
*ts*125 × *dl*311	34.0	6.8
*ts*125 × *dl*312	50.0	40.7
*ts*125 × *dl*313	23.0	ND

[a] KB and 293 cells were grown in monolayer cultures and infected ith AAV (10 i.u./cell) and either one adenovirus mutant (10 p.f.u./cell) or two adenovirus mutants at 5 p.f.u./cell each. The cells were fixed 30 hr after infection and stained for AAV capsid antigens as described in Methods.
[b] Not done. Data were taken from Laughlin *et al.* (1981).
[c] For all infections with mutant *ts*125 the cells were maintained at 40.5°C from the beginning of the adsorption period until fixation.

mutant *dl*313. All three deletion mutants were able to help AAV when complemented with Ad5*ts*125 at the nonpermissive temperature. These results indicate that at least three adenovirus early genes (1A, 1B, and 2A) must be functional to provide helper activity for AAV replication and that these genes are *trans*-acting.

The apparent requirement by AAV for an Ad early region 1 function was examined further using three additional mutants, Ad5*dl*314, Ad5*sub*315, and Ad5*sub*316 (Laughlin *et al.*, 1982). These mutants (detailed in Fig. 4) have smaller deletions than does *dl*313. In addition, *sub*315 and *sub*316 contain substitutions of 1.5 and 1.35 kb, respectively. *Sub*315 and *sub*316 are phenotypically 1B mutants and complement the 1A mutant *dl*312. The mutant *dl*314 behaves phenotypically as a 1A mutant and does not complement *dl*312. Apparently *dl*314 produces a large fusion protein consisting of 1A and most of 1B which is inactive for 1A functions. As shown in Table IV, in KB cells, at low multiplicity, *dl*314, *sub*315, and *sub*316 were all very inefficient helpers for accumulation of AAV antigens. Both *sub*315 and *dl*314 showed a low degree of multiplicity-dependent leakiness for AAV helper activity. *Sub*316, which contains more of early region 1B than *sub*315, showed a greater degree of multiplicity-dependent leakiness. Those observations were reflected when AAV DNA synthesis was examined (Laughlin *et al.*, 1982).

Ostrove and Berns (1980) and Janik *et al.* (1981) analyzed growth of AAV employing the Ad5 *hr* mutants isolated by Harrison *et al.* (1977). The mutants *hr*1 and *hr*3 map in region 1A and are phenotypically similar

TABLE IV. AAV Antigen Synthesis in Cells Infected with Ad Early Region 1B Mutants[a]

Helper adenovirus	Multiplicity p.f.u./cell	Proportion of nuclei fluorescent for AAV antigen in 293 cells (%)	Proportion of nuclei fluorescent for AAV antigen in KB cells (%)
*sub*316	8	60	1
*sub*315	10	57	0.5
*dl*314	8	50	0.5

[a] Cells were infected with AAV-2 at 10 infectious units per cell and adenovirus at the indicated multiplicity and assayed for nuclear accumulation of AAV antigen as described for Table I. Taken from Laughlin *et al.* (1981).

to *dl*312. The mutants *hr*6 and *hr*7 map in region 1B and are complemented by *dl*312 but not *dl*313 (Galos *et al.*, 1980). The *hr*1 and *hr*3 mutants, like *dl*312, did not help AAV DNA replication. In contrast to the results with *dl*313, both *hr*6 and *hr*7 helped AAV efficiently. However, these results are not inconsistent since *hr*6 and *hr*7 are phenotypically DNA-positive whereas *dl*313 is DNA-negative for adenovirus DNA replication in HeLa or KB cells (Lassam *et al.*, 1978; Shenk *et al.*, 1980). The *hr*6 and *hr*7 mutations are point mutants or much smaller deletions than the *dl*313 deletion and map between 6.1 and 8.5 map units on the adenovirus genome, most probably within the intervening sequence of the 13S mRNA (Galos *et al.*, 1980). This is the region bounded by the right-hand ends of the deletions in *sub*315 and *sub*316. However, neither of these mutants complements *hr*6 or *hr*7 (Ross *et al.*, 1980) although *sub*316, but not *sub*315, yields recombinants with both *hr* mutants (Galos *et al.*, 1980). Thus AAV helper function may be a useful assay for distinguishing phenotypically between some Ad region 1 mutants.

Because adenovirus region 1A products appear to control the other Ad early gene regions, any AAV requirement for region 1A may be rather indirect. This idea is supported by the multiplicity-dependent leakiness of region 1A mutants and by the fact that the *dl*312 mutation completely eliminates all 1A gene products even at high multiplicity.

Several lines of evidence now indicate that the control on adenovirus early gene expression exerted by region 1A requires only the 1A 13S mRNA and its products. First the region 1A 9S mRNA is synthesized only late after infection. Second, Montell *et al.* (1982) introduced into adenovirus 2 DNA a single base transversion (TG) at nucleotide 975 which prevents splicing of the 12S mRNA but does not alter the proteins encoded by the 13 S mRNA because of the degeneracy of the genetic code (see Fig. 4). This mutant Ad2/5pm975 is completely viable for lytic growth. Another Ad5 mutant, *hr*440, which has alternations at both nucleotides 978 and 979, fails to splice 12S mRNA but also causes premature termination during translation of the 13S mRNA. This mutant is not viable (DNA-negative) and grows only on 293 cells (Solnik, 1981).

A third class of region 1A mutants have lesions which occur within the intron of the 12S RNA and thus affect only expression of the 13S mRNA. Two such mutants are Ad5*hr*1 and Ad5*in*500 (Carlock and Jones, 1981). Both mutants are DNA-negative and grow only on 293 cells. *hr*1 has single base deletion at nucleotide 1055 which causes a reading frame shift and thus a stop codon (TGA) at position 1085–1088 (Ricciardi *et al.*, 1981). This results in premature termination. Ad5*in*500 contains an octanucleotide insertion at nucleotide 1010. This mutant also causes a frameshift and premature termination as is seen for *hr*1.

As noted above, at low multiplicity in HeLa or KB cells, *hr*1 fails to express other Ad early regions and fails to help AAV (Janik *et al.*, 1981). In contrast, *in*500 at low multiplicity remains DNA-negative for adenovirus but does express other Ad early regions (Carlock and Jones, 1981) and does help AAV (C. Laughlin and B. Carter, unpublished). These results support the conclusion that the requirement for adenovirus region 1A in the AAV helper function is mediated via control of other adenovirus early genes. This conclusion is also supported by the microinjection experiments of Richardson and Westphal (1981) as described above.

The studies with the region 1B *dl* and *sub* mutations (Laughlin *et al.*, 1982) suggest that a 1B function is required at least when AAV is helped by infecting with adenovirus particles. Laughlin *et al.* (1982) suggested that some function in the Ad-infected cell may otherwise interfere with AAV DNA replication but become blocked or inactivated following expression of a region 1B product. Apparently, a region 1B product protects AAV DNA against this degradation. It is not clear if this DNA degradation function is normally present in the cell, is induced or activated by Ad infection, or is contained in the infecting Ad adenovirus virion, but the results of Laughlin *et al.* (1982) argue against the first of these hypotheses. It is also unclear whether this putative helper function might involve the 15K or 58K proteins of region 1B. The simplest model is that the 15K protein is the required AAV helper function. In summary, early region 1A is probably required only to turn on other adenovirus early regions whereas region 1B is required by AAV to prevent DNA degradation.

Support for an adenovirus-induced DNA degradation function also comes from two other studies. D'Halluin *et al.* (1979) described an early mutant adenovirus 2 apparently containing a temperature-sensitive lesion in a gene required for protection of viral and cellular DNA from degradation. This mutant, Ad2*ts*111, maps between coordinates 0 and 29 in a complementation group different from those of early regions 2A or 2B (D'Halluin *et al.*, 1982). Also Eron *et al.* (1981) have shown that *cyt* mutants of adenovirus 12 cause degradation of newly synthesized cellular and viral DNA. Further R. L. Fatt and S. Mak (personal communication) showed that Ad5*dl*313 exhibits similar properties and does not complement the *cyt* mutants for lytic growth. These latter results particularly implicate region 1B in prevention of DNA degradation. In another report (Nass and Frenkel, 1978), inhibition of an adenovirus-induced DNAse was ascribed to the Ad region 2A DNA binding protein.

With respect to these apparently unusual properties of some adenovirus region 1B mutants, the report of Ostrove and Berns (1980) must be mentioned. These workers reported that mutant *hr*6 was unable to rescue AAV from the carrier (integrated) state in human Detroit 6 cells, whereas when the same cells were superinfected with AAV the mutant was an effective helper. This result was interpreted to indicate that a region 1B function is required for AAV rescue but not for replication. However, this conclusion apparently is contradicted by the observation of Cheung *et al.* (1980) that continued passage of latently infected cells results in excision of free copies of AAV DNA in the absence of any adenovirus. Also the experiments of Ostrove and Berns (1980) were not quantitated for the level of AAV genome detection. It is possible that their results might be accounted for by the suggested model of Laughlin *et al.* (1982). Thus, if a mutant such as *hr*6 is biochemically leaky, the very low level of free AAV genomes expected initially in a rescue event might be much more sensitive to degradation than the much larger number (by several orders of magnitude) present in a superinfection with AAV particles.

2. Early Region 2A

As noted above, it was initially reported that the region 2A mutant of adenovirus 5, Ad5*ts*125, was able to help AAV (Handa *et al.*, 1975; Straus *et al.*, 1976a). However, this mutant was reinvestigated by Myers *et al.* (1980) and reported to be a defective helper for AAV. When Ad5*ts*125 was used at 40°C, the yields of AAV antigen and AAV infectivity were decreased by about two logs (Table V). This decrease paralleled the decrease in production of Ad structural antigens. At 32 or 37°C, the Ad5*ts*125 mutant behaved as wild type for both Ad and AAV growth. Also, the inability of Ad5*ts*125 to help AAV at 40°C could not be overcome by a high multiplicity of infection with the mutant. These results show clearly that the Ad5*ts*125 mutant is a deficient helper for AAV at the nonpermissive temperature. This is not, however, a general property of adenovirus temperature-sensitive mutants as indicated by the efficient helper function provided by the late (assembly) mutant Ad2*ts*215.

A second early region 2A mutant, Ad5*ts*107, was also deficient for AAV helper function at 40°C. In contrast, Ad5*ts*149, an early region 2B mutant, helped AAV efficiently at all temperatures. Thus, the inability to help AAV was a discrete property of the Ad5*ts*125 and Ad5*ts*107 mutations in the Ad E72 protein and was not a general property of adenovirus DNA negative mutants. Additional studies (Myers *et al.*, 1980; Myers and Carter, 1981; Jay *et al.*, 1981) analyzed several biochemical defects

TABLE V. Ad5*ts*125 is Defective for AAV Helper Function at the Nonpermissive Temperature[a]

Adenovirus helper	Temperature (°C)	Proportion (%) of cell nuclei fluorescent for: Ad	AAV	Yield of infectious AAV particles AAV/Cell ($\times 10^{-4}$)
Ad5*ts*125	32	100	55	1.5
	37	90	48	2.5
	40	2	0.5	0.01
Ad5*ts*107	32	73	52	1.7
	37	7	5	0.5
	40	1	0	0.01
Ad5*ts*149	32	83	53	3.0
	37	25	41	—
	40	0	49	2.0
Ad2*ts*215	32	100	45	1.5
	37	100	35	2.0
	40	100	51	1.25
Ad5wt	32	100	47	1.0
	37	100	41	1.0
	40	100	49	1.0
None	40	0	0	0.0001

[a] KB monolayer cells were infected with AAV2 (10 I.U./cell) and the indicated adenovirus (10 p.f.u./cell) and grown at the temperatures shown. Cells were either fixed and assayed by immunofluorescence for Ad or AAV capsid antigens or were lysed and assayed for AAV infectivity. The infectivity data show the number of infectious units obtained per cell. The data are summarized from Myers *et al.* (1980).

in AAV replication with the *ts*125 mutant helper at 40°C. These defects include decreased accumulation of AAV SS progeny DNA and certain spliced AAV mRNAs, inhibition of *in vivo* AAV capsid protein synthesis, and failure to assemble empty or full capsids. Thus a single mutation in the pleiotropic adenovirus DNA binding protein appears to have several independent effects on AAV RNA structure and protein synthesis. These effects are described more fully below.

The reason for the difference between the observations of Myers *et al.* (1980) and previous workers is not clear. Straus *et al.* (1976a) did not assay AAV antigen synthesis, protein synthesis, or particle production and their analysis would not have detected the type of differences in RNA species observed by Myers *et al.* (1980). They also used KB cell suspension cultures at 39.5°C and the yields of infectious AAV, even with an Adwt helper were at best only of the order of 30 to 100 infectious units per cell. Similarly Handa *et al.* (1975) used HEK cells and obtained similar yields of AAV. These yields are about two logs lower than observed by Myers *et al.*, with a wild type or Ad5*ts*149 helper at 40°C. This last point may be important since the defect with Ad5*ts*125 for AAV growth may be affecting the efficiency of AAV growth in an "all-or-none" fashion. The few cells exhibiting fluorescence for AAV at 40°C show a normal amount of staining. Whether this reflects the mechanism of the helper function or the innate degree of "leakiness" of the mutant is not clear. More recently it has been shown that the lack of helper function with Ad*ts*125 is not dependent on the AAV stock or multiplicity used. Both crude lysates of AAV and purified CsCl-banded AAV particles having either no AAV DI particles or a wide range of AAV DI particles present and used at multiplicities from 0.25 to 200 infectious units per cell yielded similar results with the mutant helper (B. Carter, unpublished). In addition, the mutant helper does not appear to induce or increase the amount of AAV DI genome production (Myers and Carter, 1981). It is important to note that the Ad5*ts*125 mutant has a very high shutoff temperature for some phenotypic properties and it is important to ensure that the temperature of the culture is actually maintained at least at 40°C (Myers *et al.*, unpublished; T.H. Carter, personal communication). Additional evidence that the mutation in Ad5*ts*125 lies in a protein required for AAV was obtained by use of a spontaneous, wild-type revertant, R7, of the Ad5*ts*125 mutant (Carter and Blanton, 1978). The revertant, Ad5R7, helps AAV at 40°C in KB cells as efficiently as wild-type virus (B. Carter, unpublished). Since Ad early region 2A is involved in adenovirus host range, the different results with Ad5*ts*125 for AAV helper ability may be due to subtle differences in the cell lines used in different laboratories.

Replication of AAV DNA using the Ad5*ts*125 mutant as a helper at the nonpermissive temperature was studied by Myers and Carter (1981). Intracellular viral DNA labeled with [^{3}H]thymidine was selectively extracted using the Hirt (1967) procedure under conditions where single-stranded (SS) DNA did not reanneal. The DNA was analyzed by sedimentation in neutral sucrose gradients (Fig. 6). With the mutant adenovirus helper there was a specific inhibition of accumulation of AAV 22S progeny SS DNA but only a very small effect on accumulation of the duplex 14.5S (monomer) or 18S (dimer) RF DNA (Fig. 6A). There was also little change in the amount of AAV 10S DI genomes. The temperature-sensitive mutation abolished all Ad DNA synthesis as shown by the absence of 31 S DNA (Fig. 6A or 6B). In addition, Fig. 6 shows the inhibition of adenovirus DNA synthesis by AAV coinfection.

These experiments suggested that the region 2A 72K DNA binding protein is not needed for AAV RF DNA replication in contrast to its clear stoichiometric requirement for adenovirus DNA replication. Additional evidence suggests that the failure to accumulate AAV SS DNA using the mutant helper may be due to the absence of AAV capsids or AAV proteins rather than a requirement for the 72K DNA binding protein itself. The *ts*125 mutant helper also results in a failure to synthesize al three AAV proteins VP1, VP2, and VP3. Furthermore, when production of VP3 is prevented with the amino acid analog L-canavanine (Buller and Rose, 1978) accumulation of AAV SS DNA but not RF DNA is prevented (Myers and Carter, 1981). These results are analogous to those obtained with

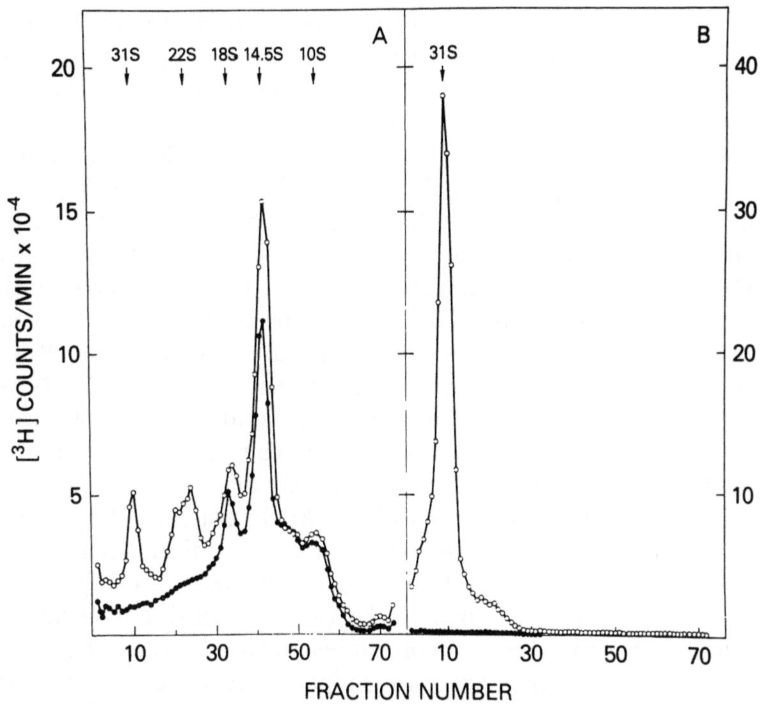

FIGURE 6. Effect on intracellular AAV DNA of a temperature-sensitive mutation in early region 2A of the helper virus. Cells were simultaneously infected with purified AAV2 (10 infectious units/cell) and adenovirus (10 plaque-forming units/cell) and grown at 40°C. Intracellular DNA was labeled with [^{3}H]thymidine from 12 to 20 hr after infection and sedimented in neutral sucrose gradients. (A) Cells were infected with AAV and either Ad5wt (○——○) or Ad5*ts*125 (●——●). (B) Cells infected with Ad5wt (○——○) or Ad5*ts*125 (●——●) alone. From Myers and Carter (1981).

mutants of the autonomous parvovirus H-1 having temperature-sensitive lesions in the viral capsid proteins (Rhode, 1976). These mutants, at the nonpermissive temperature, failed to accumulate H-1 capsid protein or progeny SS DNA, but RF DNA replication still occurred.

The simplest interpretation of the observations just described is that one or more AAV proteins or the assembled AAV capsid is required to accumulate SS DNA. AAV proteins or capsids might be required either directly to aid in displacement of SS DNA from a replicating structure (Tattersall and Ward, 1976) or merely to sequester SS DNA in particles and thus avoid recycling through the replicating pool. These two hypotheses are not mutually exclusive, and further studies are required to distinguish between them. However, the evidence is consistent with the idea that free SS DNA exists only when sequestered in a capsid. These experiments also provided two independent ways to dissociate AAV SS DNA accumulation from RF replication and should provide possible approaches to analyze the mechanism of AAV SS DNA production. It must be noted that an AAV-coded protein other than the three capsid proteins might be involved in SS DNA replication since there are additional reading frames in the left-hand half of the AAV genome (Lusby and Berns, 1982); whether the expression of such a putative protein would be affected by the Ad5*ts*125 mutant is not known.

The Ad5*ts*125 mutant also affected the accumulation of certain AAV mRNAs. Myers *et al.* (1980) analyzed cytoplasmic RNA using the procedure of Berk and Sharp (1977a, 1978) in which DNA–RNA hybrids are formed by annealing cytoplasmic, poly(A)-containing RNA with AAV ^{32}P-DNA. The hybrids are then digested with the single-strand specific endonuclease S1 and the products are analyzed by electrophoresis at neutral pH. When the Ad5wt helper was used in KB cells grown at 40°C, the digestion products observed corresponded to the normal AAV RNAs previously observed (Fig. 1). When the temperature-sensitive mutant, Ad5*ts*125, was the helper at the nonpermissive temperature, the relative amounts of some AAV RNAs were apparently decreased. This was most obvious for the spliced 3.3- and 3.9-kb RNAs. In contrast the amounts of the most abundant, spliced 2.3-kb mRNA were similar with either mutant or wild-type helper. The failure to accumulate the less abundant spliced AAV RNAs with the mutant helper was not due to inhibition of RNA synthesis because analysis of nuclear RNA (Marcus *et al.*, unpublished) showed that with the mutant helper there were normal amounts of the three unspliced AAV RNAs. It is not clear whether failure to accumulate these RNAs represents a defect in splicing or in stability of the RNA. It is interesting that in the experiments of Myers *et al.* (1980) the AAV 2.3-kb mRNA, which apparently contains the same splice as the other AAV RNAs (Laughlin *et al.*, 1979; Green and Roeder, 1980b), was spliced and accumulated to normal levels. This may reflect different intrinsic splicing efficiencies of the AAV RNAs (see Chapter 5).

In addition to the defects in AAV DNA and RNA synthesis with the Ad5*ts*125 mutant helper, synthesis of AAV capsids (both empty and full) and AAV capsid antigens was inhibited (Myers *et al.*, 1980; Myers and Carter, 1981). This defect was also reflected in a greatly decreased accumulation of AAV capsid proteins in infected cells (Myers *et al.*, 1980; Jay *et al.*, 1981). This is illustrated by the experiment shown in Fig. 7, in which the *in vivo* synthesis of ^{35}S-labeled AAV proteins at 40°C in KB cells using either Ad5*wt* or Ad5*ts*125 was analyzed by immunoprecipitation and gel electrophoresis. As shown in Fig. 7, the AAV VP1, VP2, VP3 proteins were immunoprecipitated from *wt* helped cells in proportions similar to those observed in purified particles. With the mutant helper (Fig. 7, tracks f and g) VP1, VP2, and VP3 were greatly reduced.

The failure to accumulate AAV protein in *ts*125 infected cells at 40°C was particularly remarkable in view of the demonstration that all three proteins are coded for by the AAV 2.3-kb mRNA (Jay *et al.*, 1981). This observation, together with the evidence that accumulation of this AAV 2.3-kb mRNA was relatively normal in the mutant-helped cells, indicated that there may be a translation control exerted upon the AAV mRNA

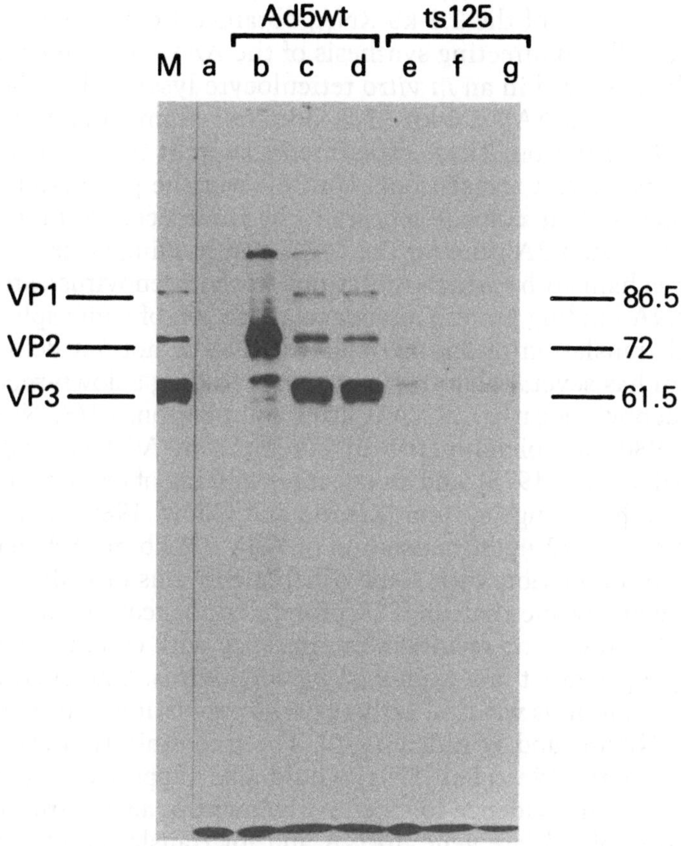

FIGURE 7. Analysis of AAV proteins synthesized *in vivo* at 40°C with a mutant Ad5*ts*125 or wild-type helper. ^{35}S-labeled AAV proteins from infected cells were detected by immunoprecipitation with AAV antiserum and electrophoresis in 9% acrylamide gels as described in Methods. Track M, ^{35}S-labeled AAV proteins from purified particles. Track a, uninfected cells. Track b, Ad5wt (10 p.f.u./cell). Track c, Ad5wt + AAV2 (25 i.u./cell). Track d, Ad5wt + AAV2 (100 i.u./cell). Track e, Ad5*ts*125 (10 p.f.u./cell). Track f, Ad5*ts*125 + AAV2 (25 i.u./cell). Track g, Ad5*ts* + AAV2 (100 i.u./cell). In the samples from Adwt infected cells (track b) there was some nonspecific precipitation of adenovirus proteins including the E72K protein (migrating very close to VP2) as well as lesser amounts of hexon protein (migrating slower than VP1) and Ad fiber protein (migrating slightly slower than VP3). Accumulation of these Ad proteins is greatly inhibited by an increasing multiplicity of AAV infection or by the mutation in Ad5*ts*125. These Ad proteins were also precipitated nonspecifically with normal rabbit serum whereas AAV proteins were not precipitated by normal serum. Furthermore, no AAV proteins were precipitated by AAV antiserum from cells uninfected or infected with Adwt or Ad5*ts*125. The Ad E72 protein precipitation is probably due to formation of aggregates under certain salt conditions, Schechter *et al.* (1980). Reprinted from Jay *et al.* (1981) with permission.

that is affected by the mutation in Ad5*ts*125. Further evidence in support of this idea was provided by Jay *et al.* (1981), who demonstrated that the biological activities of the 2.3-kb RNA, obtained from either mutant or *wt*-infected cells, in directing synthesis of the AAV capsid proteins were similar when assayed in an *in vitro* reticulocyte lysate. Thus the biological activity of the AAV 2.3-kb RNA does not seem to be impaired by the Ad5*ts*125 mutation. These experiments support the concept that adenovirus may exert a translational control over the production of AAV capsid proteins. This control appears to be influenced by mutations in the Ad early region 2A gene for the 72K DNA binding protein.

The mechanism by which mutations in the adenovirus early region 2A 72K DNA binding protein affect accumulation of some spliced AAV RNAs and translation of another spliced RNA is not yet defined. The 72K protein has several pleiotropic effects including a down-modulation of some adenovirus early mRNA (Carter and Blanton, 1978; Nevins and Winkler, 1980), a stoichiometric involvement in Ad DNA replication (Van der Vliet *et al.*, 1975), and an effect on splicing of certain adenovirus mRNAs in a host-range system (Klessig and Chow, 1980). The effect of the mutation in E72 upon translation of AAV 2.3-kb mRNA may be indirect via in interaction with some other adenovirus or cellular component. It is conceivable that the 72K protein is not required for AAV but that the *ts*125 mutation results in interference with certain other potential AAV helper functions expressed by adenovirus. One such example might be interactions with Ad early region 4 expression (Carter and Blanton, 1978; Nevins and Winkler, 1980). The microinjection experiments of Richardson and Westphal (1981) would also support this idea. There is also very recent evidence for the involvement of adenovirus VA RNA in translation of Ad late gene mRNA and for translation of AAV RNA (see below).

Whatever the mechanism of the region 2A mutations, it is clear that AAV now offers a useful experimental system to analyze an apparent translational control exerted on a eukaryotic mRNA as well as controls on either splicing or stability of RNA. An additional feature of this work is that AAV may provide a novel probe to dissect several pleiotropic functions of the 72K protein since the adenovirus DNA replication function of this protein is not required for AAV DNA replication. Thus, the properties of E72K manifested later in the Ad replication cycle (i.e., after Ad DNA replication begins) and which affect expression of adenovirus late genes can be assayed in human permissive cells using AAV as the probe. For adenovirus, these functions can only be directly assayed in a host-range system.

The growth of human adenovirus is restricted in most monkey cells but this restriction can be overcome by coinfection of the cells with SV40 virus (Rabson *et al.*, 1964). This host restriction results in failure to express late functions though early functions and DNA synthesis occur normally. This host restriction to adenovirus is accounted for, at least in part, at the level of post-transcriptional processing. Klessig and Chow (1980) demonstrated that when human adenovirus was grown in monkey cells most of the fiber protein mRNA was incorrectly spliced. This phenomenon may be analogous to the observations for AAV 3.3- and 3.9-kb mRNA with the *ts*125 mutant. This concept is supported by the demonstration that mutations in human adenovirus types 2 or 5 which overcome the host restriction in monkey cells map in the early region 2A 72K protein gene (Klessig, 1977; Klessig and Grodzicker, 1979).

There probably is, in addition, a defect in translation of human adenovirus late mRNAs in monkey cells although this is more controversial and definitive evidence has proven difficult to obtain (Klessig and Anderson, 1975; Eron *et al.*, 1975; Farber and Baum, 1978). Coinfection with SV40 in monkey cells allows adenovirus to grow efficiently. The fractional protein of SV40 T antigen induced by the nondefective Ad2-SV40 hybrid virus (Ad2 ND2) which allows adenovirus to overcome this restriction in monkey cells has been shown to bind specifically to the 40S ribosomal subunit *in vivo* and *in vitro* (Jay *et al.*, 1977). This supports the idea of a translational control.

The studies with AAV and the adenovirus region 2A mutants therefore provide strong suggestive evidence that the pleiotropic effects of region 2 which are exerted in the host-range system also occur, and can be assayed using AAV, in the permissive human cells.

There is also evidence that at least some of these same pleiotropic effects of adenovirus region 2A are also exerted upon AAV in the host range monkey cell system. As already noted, growth of AAV occurs efficiently in monkey cells when a simian adenovirus is the helper. In primary African green monkey kidney (AGMK) cells growth of AAV is restricted, when human adenovirus is the helper, but addition of SV40 virus to the infection (i.e., a triple infection) allows AAV as well as adenovirus to overcome the host restriction (Blacklow *et al.*, 1967b). Buller *et al.* (1979) analyzed growth of AAV in AGMK cells using adenovirus type 7 as the helper and demonstrated that AAV RF DNA was synthesized as well as AAV RNA but there was no accumulation of the three AAV capsid proteins. In addition, in these conditions AAV SS DNA did not accumulate (Buller and Rose, personal communication). Furthermore, although there was no intracellular accumulation of AAV proteins, AAV RNA extracted from the AGMK cells could be translated *in vitro* to yield the three AAV proteins (Buller and Rose, personal communication). These results appear to be analogous to those obtained with Ad5*ts*125 and AAV in permissive human cells at the nonpermissive temperature and agree well with the other described properties of adenovirus region 2A.

Ad2*hr*400 is a mutant human adenovirus type 2 obtained by Klessig (1977) which grows efficiently on monkey cells. *hr*400 also helps AAV in CVC cells, a subline of African green monkey kidney cells (C. Laughlin, unpublished). This again is consistent with the properties of the 72K protein.

SV40 overcomes the host-range defect for both adenovirus and AAV. Also SV40 can complement the mutation in Ad5*ts*125 when assayed in human cells. Furthermore, in monkey cells, the 72K protein is produced and is normally active for its stoichiometric requirement for adenovirus DNA replication. However, SV40 alone is not a helper for AAV in either human or monkey cells. These facts together also provide suggestive evidence that the effects of the 72K protein on post-transcriptional events and translation are mediated through interactions with, or control of, another cellular or adenovirus product or most likely both. The involvement of other adenovirus genes, such as early region 4 or VA RNA, would then provide a rationale for the lack of direct AAV helper ability of SV40.

It is likely that these interactions between three different viruses and the host cell will prove to be highly complex. AAV is clearly an important and interesting additional probe for these studies.

3. VA RNA

Before discussing the other early regions of adenovirus we will first describe recent results with adenovirus VA RNA since these shed light both on the function of VA RNA per se and its involvement in the AAV helper function. Furthermore, these results may be directly relevant to the involvement of adenovirus early region 2A in growth of AAV.

Until very recently, the function of the adenovirus VA RNAs remained obscure. The genes for 5.5S VA_I RNA and the 5.2 S VA_{II} RNA map closely together (see Fig. 3) and both RNAs are synthesized by RNA polymerase III (Weinmann *et al.*, 1974). Both RNAs accumulate in the cytoplasm late after infection but are present at lower levels at early times (Soderlund *et al.*, 1976). It has been suggested that VA_I RNA might be involved in splicing of some adenovirus mRNA by hybridizing around the splice junctions (Murray and Holliday, 1979; Mathews, 1980). VA_I RNA does exhibit homology with the splice junctions of some adenovirus late mRNAs (Akusjarvi *et al.*, 1980) and with the DNA sequence round the splice junction of the AAV 2.3-kb mRNA (R. McPherson, unpublished). A similar involvement of small cellular RNAs with splicing of mRNA has been proposed (Lerner and Steitz, 1979; Lerner *et al.*, 1980). However, homology of small RNAs with spliced mRNAs might indicate a function in postsplicing events such as transport, stability, or translation of mRNA rather than with the mechanism of splicing per se.

Very recent work by B. Thimmappaya, C. Weinberger, and T. Shenk (personal communication) has provided clear evidence that VA_I RNA is involved in translation of adenovirus late mRNA. These workers constructed an adenovirus containing a mutation in the VA_I gene. This mutant grows 20-fold less well than wild-type adenovirus. Biochemical analysis reveal that this mutant synthesizes relatively normal levels of late mRNA but accumulation of late protein is significantly reduced. The mutation in VA_I RNA results in a decreased rate of initiation of protein synthesis on late Ad mRNAs; these RNAs were on average associated with only one or two ribosomes rather than six to seven ribosomes as in wild-type infected cells. Thus adenovirus exerts a translational level control on its late mRNA via VA_I RNA. These results are reminiscent of the results with AAV mRNA obtained with the early region 2A mutants (Jay *et al.*, 1981) or in the host-range system employing primary African green monkey cells (Buller *et al.*, 1979).

It was noted above that the transfection experiments of Janik *et al.* (1981) suggested a requirement for VA_I RNA for growth of AAV. More recent transfection experiments by Janik *et al.* (1982) are in accord with the results of Thimmappaya *et al.* and indicate that VA_I RNA may be required for translation of AAV mRNA. The apparent requirement for VA_I RNA in the AAV helper function may not be absolute. Indeed, the data of Janik *et al.* (1981) suggest that the presence of VA_I-containing DNA fragments variously increased, by 4- to 10-fold, the number of 293 cells which became immunofluorescent for AAV. However, even in the absence of VA_I genes some cells were clearly positive for AAV capsid antigens. The absolute requirement for VA_I RNA is also not supported by the results of Richardson *et al.* (1980), who showed that microinjected adenovirus polyA containing early mRNA helped AAV. This RNA should not contain VA_I RNA which is not polyadenylated. More recently Richardson and Westphal (1981) reported that cleavage of adenovirus DNA with an enzyme that cleaves in the VA_I gene did not destroy the AAV helper activity in the microinjection assay. Also, in the experiments which showed that microinjection of Ad early region 4 mRNA alone elicited AAV capsid protein synthesis, the region 4 RNA was purified by polyA selection and hydridization to region 4 DNA which should eliminate VA RNA.

There is one significant difference between the experiments of Janik *et al.* (1981) and those of Richardson *et al.* (1980) and Richardson and Westphal (1981). The former workers used human 293 cells and the latter used monkey Vero cells. There might be important host-range differences involving functions such as VA RNA. Certainly, as described below, there are host-range differences regarding region 2A function and there might possibly be pleiotropic effects of this region on region 4 or VA RNA. Indeed, in contrast to the results with Vero cells, microinjection of Ad early region 4 mRNA into primary African green monkey kidney (AGMK) cells failed to help AAV (W. Richardson, personal communication).

The observations to date with adenovirus region 2A and VA RNA suggest that these genes apparently control translation of AAV mRNA. However, this control may not be absolute and might be affected by host range. If the underlying mechanism involves effects on the efficiency of initiation of translation on AAV mRNA it would not be surprising if this efficiency were subtly altered by variations in cell type or cell physiology. Whether the region 2A product and VA_I RNA interact with each other, as well as cellular components, remains to be determined. In this respect, it is interesting that Vero cells may represent a bypass of the requirement for interaction with VA_I RNA because the Ad5 early region 2A mutants, *ts*125 and *ts*107, were still mutant for AAV when assayed in Vero cells (B. Carter *et al.*, unpublished).

4. Early Region 4

Because of the absence of mutants in this region, the function of the adenovirus early region 4 genes was completely unknown. Indeed, the AAV helper function (Janik *et al.*, 1981; Richardson and Westphal, 1981; B. Carter *et al.*, 1983) remains the only easily measureable property of adenovirus early regions 4. This early region gives rise to 6 to 9 RNAs and perhaps a similar number of protein products (Fig. 3). T. Shenk *et al.* (personal communication) constructed a deletion mutant in Ad2 early region 4. Somewhat disappointingly, this mutant eliminated certain of the region 4 RNA transcripts but remained as viable as wild-type virus. Apparently some region 4 functions are non-essential for normal lytic growth.

A more useful mutant was constructed by Challberg and Ketner (1981). This mutant, Ad2*dl*807, has deleted a portion of adenovirus extending from the middle of early region 3 through the fiber gene and extending into region 4 to about map position 95.0. Thus, the promoter distal (carboxyl terminal) portion of early region 4 is deleted in this mutant. Because early region 3 is nonessential (see below) and the fiber gene is clearly not required for AAV (as discussed above), the mutant *dl*807 can be used to test the involvement of region 4 in AAV growth. When AAV is grown in KB cells using Ad2*dl*807 as the helper, the yield of infectious AAV is decreased at least 10-fold as compared to a wild-type helper (Carter *et al.*, in preparation). Also with the *dl*807 helper synthesis of AAV RF DNA, SS DNA and capsid proteins were all decreased approximately tenfold. These results are in contrast to the results obtained with the adenovirus early region 2A mutants which affected SS DNA and capsid protein synthesis but not RF DNA synthesis (see above). Similarly, Richardson and Westphal (1981) reported that the absence of early region 4 expression resulted in failure to synthesize AAV DNA. These results are most easily interpreted as indicating a requirement for an early region 4 product to allow AAV RF DNA synthesis, perhaps for conversion of the parental SS DNA to RF DNA or for subsequent replication of RF DNAs. The apparent leakiness of the *dl*807 mutant for AAV helper function may be intrinsic to the mutation since only part of early region 4 is deleted. More likely, the leakiness may result from contamination with residual wild-type adenovirus genomes. *dl*807 is not viable and must be complemented with wild-type virus, then mutant and wild-type particles are separted by several cycles of isopycnic centrifugation in CsCl (Challberg and Ketner, 1981).

It is of interest to note that *dl*807 is not at all defective for Ad DNA synthesis but appears to be defective in expression (perhaps translation) of some adenovirus late genes. Also stable early region 4 mRNAs are not accumulated in *dl*807 infected cells (G. Ketner, personal communication).

4. Early Regions 2B and 3

Neither of these regions appears to be important for AAV. This is supported by the microinjection experiments of Richardson and Westphal 1981 and the transfection experiments of Janik *et al.* (1981). Furthermore, region 2B temperature-sensitive mutants Ad5*ts*36 and Ad5*ts*149 are efficient helpers of AAV. Also, the nondefective hybrid adeno-SV40 viruses which contain varying deletions of portions of early region 3, including some, such as Ad2ND1, which do not synthesize the region 3 glycoprotein, help AAV readily (B. Carter unpublished). Finally, Challberg and Ketner (1981) isolated a mutant adenovirus Ad2*dl*801 in which all of early region 3 is deleted. This mutant, which is viable and grows as efficiently as Adwt, also helps AAV as efficiently as wt (B. J. Carter *et al.*, 1983).

5. Early Region L1

As noted above, Lewis and Mathews (1980) proposed that the L1 region products which map between position 30 and 40 on the adenovirus genome are required as immediate early functions to turn on region 1A which in turn activates the other early region. Current evidence from AAV studies is not consistent with this. Transfection (Janik *et al.*, 1981) or microinjection (Richardson and Westphal, 1981) of Ad DNA did not require any adenovirus DNA between map positions 30 and 40 in order to help AAV efficiently.

VII. GROWTH OF AAV IN TRANSFORMED CELLS AND INHIBITION OF ADENOVIRUS FUNCTIONS

One approach to analysis of potential helper functions for AAV is to study inhibition of adenovirus functions by AAV. Coinfection of cells with AAV and adenovirus can result in inhibition of adenovirus DNA replication (Carter *et al.*, 1979) or adenovirus *in vitro* transformation of hamster embryo cells (Casto and Goodheart, 1972). More interestingly, AAV was shown to inhibit tumor production by group A adenoviruses in newborn Syrian hamsters (Kirschstein *et al.*, 1968; Mayor *et al.*, 1973).

De la Maza and Carter (1981) studied further the inhibition of adenovirus 12 tumorigenesis in newborn hamsters. The inhibition of tumorigenesis was obtained either by standard AAV particles or DI particles which contained AAV genomes having only the terminal DNA sequences present (de la Maza and Carter, 1980a). It was further shown that purified standard or DI AAV DNA could inhibit the tumorigenesis. de la Maza and Carter also showed that prior infection of adenovirus 12 transformed hamster cells with AAV decreased their oncogenicity in newborn hamsters.

These results imply that it is the DNA component of AAV, and perhaps the terminal sequences, which mediate the inhibition possibly by decreasing or sequestering an essential adenovirus transformation function or T-antigen.

Evidence for this hypothesis was recently obtained by Ostrove *et al.* (1981), who studied AAV infection of the adenovirus-transformed hamster embryo fibroblast cell line (HT12B) originally isolated by Williams (1973). HT14B cells contain 5.5 copies of the left-hand 40% of adenovirus 5 DNA and express regions 1A and 1B (Flint *et al.*, 1976; Flint and Weintraub, 1977). Infection of these cells with a very large amount of AAV (2000 focus forming units, which is equivalent to about 200,000 particles per cell according to Carter *et al.*, 1979) resulted in decreased oncogenicity of these cells upon subsequent injection into 4- to 5-day-old hamsters. Interestingly, analysis of the AAV infected HT14B cells showed a decreased amount of the early region 1B 58K protein (see Fig. 3). This suggests a possible interaction between this protein and AAV and is interesting in view of the effect upon AAV of mutations in region 1A or B as was described above. Of course expression of other adenovirus region 1A or 1B proteins might also be altered by AAV infection but this was not determined.

Evidence that adenovirus early region 1 might be important for the AAV helper function was in fact first proposed by Handa *et al.* (1977a) based on their studies of growth of AAV in adenovirus transformed cells. Handa *et al.* (1977a) used a series of cells lines derived by adenovirus 12 transformation of a rat cell line, 3Y1, derived from a Fischer rat embryo. After infection of these cells with AAV in the absence of any added adenovirus no synthesis of AAV antigens or infectivity could be detected using the immunofluorescent assay. When these transformed rat cells were fused with human KB cells production of AAV antigens could be detected in about 1% of the heterokaryons. Also a low level of infectious AAV was detected in heterokaryons. Several of the transformed cell lines contained adenovirus 12 DNA only from early regions 1A and 1B. Thus Handa *et al.* (1977) suggested that the transforming region functions of adenovirus 12 (i.e., adenovirus T-antigens or early region 1 products) could, when introduced into KB cells, induce a low level of adenovirus helper function. Handa *et al.* also pointed out that AAV might indeed be a useful tool for analysis of the biological functions of adenovirus T-antigens. This prediction clearly is being fulfilled.

Handa *et al.* (1977a) suggested that growth of AAV required a host-range factor and an adenovirus-induced helper factor. They interpreted their experiments to indicate that the failure of AAV to grow in the transformed rat cells was due to absence of a host-range factor that was supplied by fusion with human KB cells. This immediately suggested that AAV should grow, in the absence of added adenovirus helper, in a human transformed cell line that expressed adenovirus region 1 functions. However, three laboratories demonstrated the absence of AAV DNA or protein synthesis in such a human cell line, the 293 cells described above (Janik *et al.*, 1981; Laughlin *et al.*, 1981; Ostrove and Berns, 1980).

There are several possible reasons for the failure of AAV to grow in 293 cells except in the absence of superinfection by adenovirus. These include quantitative (i.e., gene dosage) or qualitative changes in expression of region 1. Although the region 1 RNAs in 293 cells were reported to be similar to those of lytic infection (Berk *et al.*, 1979), some quantitative differences were recently reported (Spector *et al.*, 1980). The alternative possibility is that AAV requires adenovirus helper functions other than, or in addition to, those coded by region 1, as suggested by a variety of experiments described above.

In their study of the effect of AAV on the oncogenicity of Ad, Ostrove *et al.* (1981) also detected synthesis of AAV RNA in AAV infected HT14B cells in the absence of any added adenovirus. More recently AAV RNA synthesis has also been detected in AAV-infected Ad5-transformed human 293 cells in the absence of added adenovirus. These intriguing results post an interesting question as to the nature of the AAV template used. While this issue is not resolved, several comments can be made.

In the study of Ostrove *et al.* (1981) AAV DNA replication was only detected in the HT14B cells when a helper adenovirus was also present. However, AAV DNA synthesis was measured by a relatively insensitive method of *in vivo* labeling with ^{32}P followed by gel electrophoresis and the sensitivity of detection was only 5% of the control cells which were also infected with adenovirus. Since the normal yield of AAV DNA in coinfections with adenovirus is at least 10^6 genomes per cell (Laughlin *et al.*, 1981); a 5% detection level could have failed to detect up to 5×10^4 genomes per cell. Thus, it is not clear if the AAV RNA synthesis in the HT14B cells in the absence of added adenovirus resulted from a low level of synthesis of duplex AAV DNA or was directed from unreplicated single-stranded genomes.

If AAV RNA synthesis can occur in the absence of DNA synthesis it would indicate that *in vivo* RNA transcription can occur from a single-stranded DNA template. There is thus far no precedent for *in vivo* transcription from single-stranded DNA genomes (Cortese *et al.*, 1980). Recent experiments (Laughlin *et al.*, 1982) showed that in KB cells infected with AAV in the absence of adenovirus, even at multiplicities such that several thousand AAV genomes were present in the cell nucleus AAV DNA replication did not occur and AAV RNA synthesis was not detected. In the transformed cells it is possible that a low level of duplex AAV genomes could have arisen (1) by a very limited replication; (2) by integration of unreplicated genomes into cellular DNA; or (3) by annealing of plus and minus parental strands as a consequence of the high multiplicity of infection (2,000 infection units/cell).

In summary, the observation that AAV RNA synthesis can occur in an adenovirus-transformed cell that expresses only early region 1 functions (Ostrove *et al.*, 1981) is important but perplexing. The experiments of Laughlin *et al.* (1982) using adenovirus early region 1 mutants (e.g., *dl*312, 313) suggest that this region codes for a helper function for AAV DNA replication. Also, as described already (IVD3), there is strong, but not conclusive, evidence that an adenovirus helper function is required for conversion of the parental AAV genome to double-stranded structures. Thus, the observation of AAV RNA synthesis in transformed cells may mean that there is an additional helper function coded by early region 1 which can allow AAV transcription directly in the absence of AAV DNA synthesis. Alternatively, in these transformed cells, the replication helper function of region 1 simply may allow conversion of parental genomes to duplex structures thereby indirectly providing a transcription template for AAV RNA synthesis. These are not mutually exclusive possibilities and further experiments are needed to resolve the issue. Nevertheless, the observation of Ostrove and Berns that AAV RNA synthesis can occur in an adenovirus-transformed cell line which expresses only region 1 functions is important. In this respect, we note that synthesis of AAV proteins (and therefore AAV RNA) has been previously reported in a hamster cell line transformed by herpes simplex virus (see Chapter 2, this volume).

VIII. SUMMARY

Although we have not yet achieved a clear description of the biochemistry of the AAV helper function provided by adenovirus this area of molecular biology has reached an exciting stage. The recent rapid development of new technology for studying gene function together with the degree of sophistication obtained in dissecting the genome of adenovirus has allowed major advances in dissecting the AAV helper function. However, as with many areas of research, the reward for rapid advance is apparent paradox.

There is now strong collective evidence that the AAV helper function requires only adenovirus early genes. Furthermore only a subset of these early genes appears necessary. It seems fairly well established that only early regions 1A, 1B, 2A, and 4 and VA_I RNA are important for AAV growth.

The paradox arises when the experiments with some adenovirus-transformed cells are considered in relation to other studies using lytic infection with adenovirus. Some of the former studies (Handa *et al.*, 1977a) suggest that regions 1A and 1B alone are sufficient to provide a complete or partial AAV helper function. In contrast, other studies (Janik *et al.*, 1981; Myers *et al.*, 1980; W. Richardson, unpublished) indicate that region 2A and region 4 also play important and indispensable roles. Finally, some experiments (W. Richardson, unpublished) suggest that region 4 gene products alone may be sufficient to allow AAV DNA replication and protein synthesis. However, the transformed cell lines used by Handa *et al.* (1977a) and Ostrove *et al.* (1981) did not contain the DNA of regions 2A and 4.

There are a number of considerations which are important as a general basis for the resolution of the AAV helper function paradox. Several of these considerations arise as a result of the growing awareness of the complexity of the control mechanisms operating on adenovirus early gene expression. As discussed above, these early genes may be controlled by a cascade control mechanism but even some of the gross details of this cascade remain unclear.

The biochemical functions of many adenovirus early genes (e.g., region 4) remain completely unknown and for other regions only very general clues exist. The region 2A protein, in addition to its role in DNA replication, has a number of pleiotropic controlling functions, but so far there is very little evidence for the mechanism of these properties.

Some of the adenovirus early genes may function essentially to mimic cellular functions but might perform these functions either more efficiently than the analogous cell function of after this cell function has been inhibited or modified by adenovirus infection. Certain adenovirus early genes may also act simply to increase the efficiency of expression of other adenovirus genes. In addition there may be important host-range differences in regard to the expression or control of early genes. Thus, for AAV helper function studies, two considerations may be particularly important. Experiments reporting the presence or absence of AAV functions relative to specific adenovirus genes should be carefully quantitated. Second, the host range should be considered. For instance, the evidence to date suggests that AAV helper functions provided by adenovirus genes expressed in transformed cells may occur at low efficiency compared to lytic infection. This might indicate a gene dosage effect in transformed cells, but lytic infection with DNA-negative mutants such as Ad5*ts*149 at very low multiplicity (less than one pfu per cell) does elicit AAV helper function readily.

One useful approach to the solution of the helper function paradox may be to consider those adenovirus early genes that influence AAV as consisting of three sets of functions: those functions required solely for transformation and those required for lytic replication, with the lytic replication genes in turn being subdivided into those required directly for Ad DNA replication and those required for controlling late structural gene expression. These three sets of functions may overlap with respect to any given adenovirus early gene. Expression of these overlapping sets of functions may be controlled differently depending on the host cell. The further implication of this hypothesis is that there may be alternative mechanisms by which these adenovirus genes can control AAV functions.

An interesting point arises in consideration of this model of adenovirus. Adenovirus and AAV do not appear to use the same set of viral proteins for the actual mechanics of replicating their DNA although they may well use the same cellular DNA polymerases and other cellular proteins. The roles of the early region 2A 72K DNA binding protein and the early region 2B covalently bound protein are clearly established for Ad DNA replication. However, neither of these proteins appears to be involved in AAV DNA replication. AAV DNA replication likely requires AAV *rep* proteins (discussed elsewhere in this volume, e.g., Chapters 5 and 6), and perhaps the AAV capsid proteins as well. These differences presumably reflect, in part, two different approaches to the problem of replicating the termini of a linear DNA molecule employing different terminal structures and different proteins.

To follow this model of adenovirus early gene expression further, AAV may be considered as a set of genes whose *expression* is controlled by adenovirus early genes in a similar fashion to the expression of adenovirus late genes, but whose *replication* is dependent upon AAV specific genes as well as certain adenovirus early genes. This general model allows a rationalization of most of the available evidence.

Whatever the actual mechanism of the AAV helper function proves to be, it is very clear that AAV is an increasingly useful probe for studies of adenovirus. AAV may be useful both to study controlling functions of adenovirus genes and to identify particular biochemical functions. For instance AAV is clearly an excellent probe to study pleiotropic effects of adenovirus region 2A on post-transcriptional and translational events in a permissive cell system. In regard to translational control, AAV is also a useful probe to study the role of VA_I RNA. Also, AAV has provided good evidence for, and a useful discrete probe with which to analyze the role of, region 1B in protection of DNA against a DNA degradation function that is apparently induced upon adenovirus infection of KB cells. Also, AAV may be an important probe available to analyze the biochemical functions of adenovirus region 4. Thus AAV and adenovirus represent an important system for studying interaction and control of eukaryotic genes contained on two separately replicating genomes. These studies provide a model for similar analysis of the herpes virus helper functions for AAV and for the development of AAV as a similarly useful independent probe of some herpes virus gene functions. It seems reasonable to expect that the herpes virus helper function for AAV might also include functions to allow AAV RF replication and other functions to allow AAV gene expression.

ACKNOWLEDGMENTS. We thank Joan Mok for her painstaking preparation of this manuscript through many revisions. We thank also those people, including K. Berns, J. Janik, J. Rose, R. McPherson, H. Westphal, W. Richardson, and T. Shenk, who provided information in advance of publication.

REFERENCES

Abboud, M.M., and Horwitz, M.D., 1979, The DNA polymerases associated with the adenovirus type 2 replication complex: Effect of 2′-3′-dideoxythymidine-5′-triphosphate on viral DNA synthesis, *Nucleic Acids Res.* **6**:1025.

Aiello, L., Guifoyle, R., Huebner, K., and Weinmann, R., 1979, Adenovirus 5 DNA sequences present and RNA sequences transcribed in transformed human embryo kidney cells (HEK-Ad-5 or 293), *Virology* **94**:460.

Akusjarvi, G., Mathews, M.B., Andersson, P., Vennstrom, B., and Pettersson, U., 1980, Structure of genes for virus-associated RNA_I and RNA_{II} of adenovirus type 2, *Proc. Natl. Acad. Sci. U.S.A.* **77**:2424.

Archetti, I., and Bocciarelli, D.S., 1963, Struttura e caratteristiche biologiche di un piccolo virus non ancora tipizzato. Atti Del. IV Congresso Italiano di microscopia electindca Padova, Novembre 25, pp. 131–132.

Archetti, I., and Bocciarelli, D.S., 1964, On the structure of a small simian virus. Proc. 3rd. Europ. Reg. Conf. Electron Microscopy, Prague, p. 343.

Archetti, I., and Bocciarelli, D.S., 1965, Biological and structural characteristics of a still untyped small virus. *Ann. inst. Super. Sanita.* **1**:103.

Archetti, I., Bereczky, E., and Bocciarelli, D., 1965, A small virus associated with the simian adenovirus SV11, *Virology*, **29**:642.

Arens, M., Yamashita, T., Padmanabhan, R., Tsurno, T., and Green, M., 1977, Adenovirus DNA replication. Characterization of the enzyme component of a soluble replication system, *J. Biol. Chem.* **252**:7947.

Armentrout, R., Bates, R., Berns, K., Carter, B., Chow, M., Dressler, D., Fife, W., Hauswirth, W., Hayward, G., Lavelle, G., Rhode, S., Straus, S., Tattersall, P., and Ward, D., 1978, A standardized nomenclature for restriction endonuclease fragments, in: *Replication of Mammalian Parvoviruses* (D. Ward and P. Tattersall, eds.), pp. 523–526, Cold Spring Harbor Laboratory, New York.

Atchison, R.W., 1970, The role of herpes viruses in adenovirus-associated virus replication *in vitro*, *Virology* **42**:155.

Atchison, R.W., Casto, B.C., and Hammond, W.McD., 1965, Adenovirus-associated defective virus particles, *Science* **194**:754.

Berk, A.J., and Sharp, P.A., 1977a, Sizing and mapping of early adenovirus mRNAs by gel electrophoresis of S1 endonuclease digested hybrids, *Cell*, **12**:721.

Berk, A.J., and Sharp, P.A., 1977b, UV mapping of the adenovirus 2 early promoters, *Cell* **12**:45.

Berk, A.J., and Sharp, P.A., 1978, Structure of the adenovirus 2 early mRNAs, *Cell* **14**:695.

Berk, A.J., Lee, F., Harrison, T., Williams, J., and Sharp, P.A., 1979, Pre-early adenovirus 5 gene product regulates synthesis of early viral messenger RNAs, *Cell* **17**:935.

Berk, A.J., Lee, F., Harrison, T., Williams, J., and Sharp, P.A., 1980, Phenotype of adenovirus 5 host range mutants for early mRNA synthesis, *Cold Spring Harbor Symp. Quant. Biol.* **44**:367.

Berns, K.I., Pinkerton, T.C., Thomas, G.F., and Hoggan, M.D., 1975, Detection of adeno-associated virus (AAV)-specific nucleotide sequences in DNA isolated from latently infected Detroit 6 cells, *Virology* **68**:556.

Berns, K.I., Hauswirth, W.W., Fife, K.H., and Spear, I.S., 1978a, Terminal structure of Adeno-associated virus DNA, in: *Replication of Mammilian Parvoviruses* (D. Ward and P. Tattersall, eds.), pp. 179–192, Cold Spring Harbor Laboratory, New York.

Berns, K.I., Hauswirth, W.W., Fife, K.H., and Lusby, E., 1978b, Adeno-associated virus DNA replication, *Cold Spring Harbor Symp. Quant. Biol.* **43**:781.

Bertazzoni, U., Scovassi, A.I., and Brun, M.G., 1977 Chick-embryo DNA polymerase γ. Identity of γ-polymerases purified from nuclei and mitochondria, *Europ. J. Biochem.* **81**:241.

Blacklow, N.R., Hoggan, M.D., and Rowe, W.P., 1967a, Isolation of adenovirus-associated viruses from man. *Proc. Natl. Acad. Sci., U.S.A.* **58**:1410.

Blacklow, N.R., Hoggan, M.D., and Rowe, W.P., 1967b, Immunofluorescent studies of the potentiation of an adenovirus-associated virus by adenovirus 7, *J. Exptl. Med.* **125**:755.
Blacklow, N.R., Hoggan, M.D., and McClanahan, M.S., 1970, Adenovirus-associated viruses. Enhancement by human herpes viruses, *Proc. Soc. Exp. Biol. Med.* **134**:952.
Blacklow, N.R., Dolin, R., and Hoggan, M.D., 1971a, Studies of the enhancement of an adenovirus-associated virus by herpes simplex virus, *J. Gen. Virol.* **10**:29.
Blacklow, N.R., Hoggan, M.D., Sereno, M.S., Brandt, C.D., Kim, H.W., Parrott, R.H., and Chanock, R.M., 1971b, A seroepidemiologic study of adenovirus-associated virus infection in infants and children, *Am. J. Epidemiol.* **94**:359.
Blanton, R.A., and Carter, T.H., 1979, Auto-regulation of adenovirus type 5 early gene expression III Transcription studies in isolated nuclei, *J. Virol.* **29**:458.
Bolden, A., Noy, C.P., and Weissbach, A., 1977, DNA polymerase of mitochondria is γ-polymerase, *J. Biol. Chem.* **252**:3351.
Bos, J.L., Polder, L.J., Bernards, R., Schrier, P., van den Elsen, P.J., van der Eb. A.J., and van Ormondt, H., 1981, The 2.2 kb Elb mRNA of human Ad12 and Ad5 codes for two tumor antigens starting at different AUG triplets, *Cell* **27**:121.
Boucher, D.W., Melnick, J.L., and Mayor, H.D., 1971, Non-encapsidated infectious DNA of adeno-satellite virus in cells coinfected with herpes virus, *Science* **173**:1243.
Brandon, F.B., and McLean, I.W., Jr., 1962, Adenoviruses, *Adv. Virus Res.* **13**:157–193.
Brison, O., Kedinger, C., and Wilhelm, J., 1977, Enzymatic properties of viral replication complexes isolated from adenovirus type 2 infected Hela cell nuclei, *J. Virol.* **24**:423.
Broker, T.R., and Chow, L.T., 1980, Patterns and consequences of adenoviral RNA splicing. Ex pluribus unum-Ex unoplura, *Trends in Biochemical Sciences* **July**:174–178.
Buller, R.M.L., and Rose, J.A., 1978, Characterization of adenovirus associated virus-induced polypeptides in KB cells, *J. Virol.* **25**:331.
Buller, R.M.L., Straus, S.E., and Rose, J.A., 1979, Mechanism of host restriction of adenovirus-associated virus replication in African green monkey cells, *J. Gen. Virol.* **43**:663.
Buller, R.M.L., Janik, J., Sebring, E.D., and Rose, J.A., 1981, Herpes simplex virus types 1 and 2 completely help adenovirus-associated virus replication, *J. Virol.* **40**:241.
Carlock, L.R., and Jones, N.C., 1981, Transformation-defective mutant of adenovirus type 5 containing a single altered E1a mRNA species. *J. Virol.* **40**:657.
Carter, B.J., 1978, Parvovirus Transcription, in: *Replication of Mammalian Parvoviruses* (D. Ward and P. Tattersall, eds), pp. 33–52, Cold Spring Harbor Laboratory, New York.
Carter, B.J., and Rose, J.A., 1972, Adenovirus-associated virus multiplication. VIII. Analysis of *in vivo* transcription induced by complete or partial helper viruses, *J. Virol.* **10**:9.
Carter, B.J., Khoury, G., and Rose, J.A., 1972, Adenovirus-associated virus multiplication IX. Extent of transcription of the viral genome *in vitro*, *J. Virol.* **10**:1118.
Carter, B.J., Koczot, F.J., Garrison, J., Rose, J.A., and Dolin, R., 1973, Separate helper functions provided by adenovirus for adenovirus-associated virus multiplication, *Nature New Biol.* **244**:71.
Carter, B.J., Khoury, G., and Denhardt, D.T., 1975, Physical map and strand polarity of specific fragments of adenovirus-associated virus DNA produced by endonuclease, R. *Eco*R1, *J. Virol.* **16**:559.
Carter, B.J., Fife, K.H., de la Maza, L.M., and Berns, K.I., 1976, Genome localization of adeno-associated virus RNA, *J. Virol.* **19**:1044.
Carter, B.J., Laughlin, C.A., de la Maza, L.M., and Myers, M., 1979, Adeno-associated virus autointerference, *Virology* **92**:449.
Carter, B.J., Laughlin, C.A., Myers, M., and Jay, F., 1980, The involvement of adenovirus early regions in the AAV helper function, *J. Supramolecular Struct.* **Suppl 4**:695.
Carter, B.J., Marcus, C., Laughlin, C.A., and Ketner, G., 1983, *Virology*, **126**:505.
Carter, T.H., and Blanton, R.A., 1978, Possible role of the 72,000–dalton DNA binding protein in regulation of adenovirus type 5 early gene expression, *J. Virol.* **18**:156.
Casto, B.C., Atchison, R.W., and Hammon, W., McD., 1967, Studies on the relationship between adeno-associated virus type 1 (AAV-1) and adenoviruses. I. Replication of AAV-1 in certain cell cultures and its effect on helper adenovirus, *Virology* **32**:52.

Casto, B.C., and Goodheart, C.R., 1972, Inhibition of adenovirus transformation *in vitro* by AAV-1, *Proc. Soc. Exptl. Biol. Med.* **140:**72.

Cavalier-Smith, T., 1974, Palindromic base sequences and replication of eukaryotic chromosome ends, *Nature* **250:**467.

Challberg, S., and Ketner, G., 1981, Deletion mutants of adenovirus 2: Isolation and initial characterization of virus carrying mutations near the right end of the genome, *Virology* **114:**196.

Challberg, M.D., Desiderio, S.V., and Kelly, T.J., Jr., 1980, Adenovirus DNA replication in vitro: Characterization of a protein covalently linked to nascent DNA strands, *Proc. Natl. Acad. Sci. U.S.A.* **77:**5105.

Cheung, A.K.M., Hoggan, M.D., Hauswirth, W.W., and Berns, K.I., 1980, Integration of the adeno-associated virus genome into cellular DNA in latently infected human Detroit 6 cells, *J. Virol.* **33:**739.

Chow, L.T., Broker, R.R., and Lewis, J.B., 1979, Complex splicing patterns of adenovirus 2, *J. Mol. Biol.* **134:**265.

Clarke, J.K., McFerran, J.D., McKillop, E.R., and Curran, W.L., 1979, Isolation of an Adeno-associated virus from sheep, *Arch. Virol.* **60:**171.

Cortese, R., Harland, R., and Melton, D., 1980, Transcription of tRNA genes *in vivo*: Single-stranded compared to double-stranded templates, *Proc. Natl. Acad. Sci. U.S.A.* **77:**4147.

Craig, E.A., and Raskas, H.J., 1974, Effect of cycloheximide on RNA metabolism early in productive infection with adenovirus 2, *J. Virol.* **14:**26.

de la Maza, L.M., and Carter, B.J., 1980a, Molecular structure of adeno-associated virus variant DNA, *J. Biol Chem.* **255:**3194.

de la Maza, L.M., and Carter, B.J., 1980b, Heavy and light particles of adeno-associated virus, *J. Virol.* **33:**1129.

de la Maza, L.M., and Carter, B.J., 1981, Inhibition of adenovirus oncogenicity by adeno-associated virus DNA, *J. Nat. Cancer Inst.* **67:**1323.

Desiderio, S.V., and Kelly, T.J., Jr., 1981, Structure of the linkage between adenovirus DNA and the 55,000 molecular weight terminal protein *J. Mol. Biol.* **145:**319.

D'Halluin, J.-C., Allart, C., Cousin, C., Boulanger, P.A., and Martin, G.R., 1979, Adenovirus early function required for protection of viral and cellular DNA, *J. Virol.* **32:**61.

D'Halluin, J.-C., Cousin, C., and Boulanger, P.A., 1982, Physical mapping of adenovirus type 2 temperature-sensitive mutations by restriction endonuclease analysis of interserotypic recombinants, *J. Virol.* **41:**401.

Dolin, R., and Rabson, A.S., 1973, Herpes virus saimiri: Enhancement of adenovirus-associated virus, *J. Natl. Cancer Inst.* **50:**205.

Drake, S., Schaffer, P.A., Esparza, J., and Mayor, M.D., 1974, Complementation of adeno-associated satellite viral antigens and infectious DNA by temperature-sensitive mutants of herpes simplex virus, *Virology* **60:**230.

Dreizin, R.S., Zhuravel, T.F., Shalunova, N.V., Klenova, A.V., and Zolotarskaia, E.E., 1979, Helper viruses of adeno-associated virus type 4 replication, *Vopr. Virol.* **1:**43.

Edenberg, H.J., Anderson, S., and DePamphilrs, M.L., 1978, Involvement of DNA polymerase α in simian virus 40 DNA replication, *J. Biol. Chem.* **253:**3273.

Eggerding, F., and Raskas, H.J., 1978, Effect of protein synthesis inhibitors on viral mRNAs synthesized early in adenovirus type 2 infection, *J. Virol.* **25:**453.

Eron, L., Westphal, H., and Khoury, G., 1975, Post-transcriptional restriction of human adenovirus expression in monkey cells, *J. Virol.* **15:**1526.

Evans, R.M., Fraser, N., Ziff, E.B., Weber, J., Wilson, M., and Darnell, J.E., 1977, The initiation sites for RNA transcription in Ad2 DNA, *Cell,* **12:**733.

Everitt, E., Sundquist, B., Petterson, U., and Philipson, L., 1973, Structural proteins of adenoviruses. X. Isolation and topography of low molecular weight antigens from the virion of adenovirus type 2, *Virology, 62:*130.

Ezoe, H., Lai Fatt, R.B., and Mak, S., 1981, Degradation of intracellular DNA in KB cells infected with *cyt* mutants of human adenovirus type 12, *J. Virol.* **40:**20.

Farber, M.S., and Baum, S.C., 1978, Transcription of adenovirus RNA in permissive and non-permissive infections, *J. Virol.* **27:**136.

Flint, S.J., 1980a, Structure and genomic organization of adenoviruses, in: *DNA Tumor Viruses* (J. Tooze, ed.), pp. 383–442, Cold Spring Harbor Laboratory, New York.

Flint, S.J., 1980b, Molecular biology of adenoviruses, in: *Viral Oncology* (G. Klein, ed.), pp. 603–663, Raven Press, New York.

Flint, S.J., and Broker, T.R., 1980, Lytic infection by adenoviruses, in: *DNA Tumor Viruses,* (J. Tooze, ed.), pp. 443–546, Cold Spring Harbor Laboratory, New York.

Flint, S.J., and Weintraub, H.M., 1977, An altered subunit configuration associated with the actively transcribed DNA of integrating adenovirus genes, *Cell* **12:**783.

Flint, S.J., Sambrook, J., Williams, J.F., and Sharp, P.A., 1976, Viral nucleic acid sequences in transformed cells. IV. A study of the sequences of adenovirus 5 DNA and RNA in four lines of adenovirus 5 transformed rodent cells using specific fragments of the viral genome, *Virology* **72:**456.

Galos, R., Williams, J., Binger, M.H., and Flint, S.J., 1979, Location of additional early gene sequences in the adenoviral chromosome, *Cell* **17:**945.

Galos, R.S., Williams, J., Shenk T., and Jones, N., 1980, Physical location of host-range mutants of adenovirus type 5; deletion and marker rescue, *Virology* **104:**510.

Ginsberg, H.S., and Young, C.S.H., 1977, Genetics of adenoviruses, in: *Comprehensive Virology* (H. Fraenkal-Conrat and R. Wagner, eds.), Vol. 9, pp. 27–78, Plenum Press, New York.

Graessman, A., Graessman, M., and Mueller, C., 1980, Microinjection of cells, *Methods Enzymol.* **65:**816.

Graham, F.L., and Van der Eb, A.J., 1973, A new technique for the assay of infectivity in human adenovirus DNA, *Virology* **52:**456.

Graham, F.L., Abrahams, P.J., Mulder, C., Heijneker, H.L., Warnoor, S.O., de Vries, F.A.J., Fiers, W., and van der Eb, A.J., 1975, Studies on *in vitro* transformation by DNA and DNA fragments of human adenovirus and SV40, *Cold Spring Harbor Symp. Quant. Biol.* **39:**637.

Graham, F.L., Smiley, J., Russell, W.C., and Naiva, R., 1977, Characteristics of a human cell line transformed by DNA from human adenovirus type 5, *J. Gen. Virol.* **36:**59.

Green, M.R., and Roeder, R.G., 1980a, Transcripts of the adenovirus-associated virus genome: mapping of the major RNAs, *J. Virol.* **36:**79.

Green, M.R., and Roeder, R.G., 1980b, Definition of a novel promoter for the major adenovirus-associated virus mRNA, *Cell* **22:**231.

Handa, H., and Shimojo, H., 1977a, Isolation of the viral DNA replication complex from adeno-associated virus type 1 infected cells, *J. Virol.* **24:**444.

Handa, H., and Shimojo, H., 1977b, Viral DNA synthesis in vitro with nuclei isolated from adenovirus-associated virus type-1 infected cells, *Virology* **77:**247.

Handa, H., and Carter, B.J., 1979, Adeno-associated virus DNA replication complexes in Herpes simplex virus or adenovirus-infected cells, *J. Biol. Chem.* **254:**6603.

Handa, H., Shiroki, K., and Shimojo, H., 1975, Complementation of adeno-associated virus growth with temperature-sensitive mutants of human adenovirus types 12 and 5, *J. Gen. Virol.* **29:**239.

Handa, H., Shimojo, H., and Yamaguchi, 1976, Multiplication of adeno-associated virus type 1 in cells coinfected with a temperature-sensitive mutant of human adenovirus type 31, *Virology* **74:**1.

Handa, H., Shiroki, K., and Shimojo, H., 1977a, Helper factor(s) for growth of adeno-associated virus in cells transformed by adenovirus 12, *Proc. Natl. Acad. Sci. USA* **74:**4508.

Handa, H., Shiroki, K., and Shimojo, H., 1977b, Establishment and characterization of KB cell lines latently infected with adeno-associated virus type 1, *Virology* **82:**84.

Handa, H., Shiroki, K., and Shimojo, H., 1978, Expression of helper function for Adeno-associated virus in adenovirus transformed cells, in: *Replication of Mammalian Parvoviruses* (D. Ward and P. Tattersall, eds), pp. 99–107, Cold Spring Harbor Laboratory, New York.

Harrison, T.J., Graham, F.L., and Williams, J.F., 1977, Host range mutants of adenovirus type 5 defective for growth in Hela cells, *Virology* **77**:319.

Harter, M.L., Shanmugam, G., Wold, W.S., and Green, M., 1976, Detection of adenovirus type 2 induced early polypeptides using cyclohesimide pretreatment to enhance viral protein synthesis, *J. Virol.* **19**:232.

Hauswirth, W.W., and Berns, K.I., 1978, Initiation and termination of adeno-associated virus DNA replication, in: *Replication of Mammalian Parvoviruses* (D. Ward and P. Tattersall, eds.), pp. 257–267, Cold Spring Harbor Laboratory, New York.

Henry, C.J., 1973, Adenovirus-associated (Satellite) viruses, *Prog. Exp. Tumor Res.* **18**:273.

Henry, C.J., Merkow, L.P., Pardo, M., and McCabe, C., 1972, Electron microscope study on the replication of AAV-1 in Herpes-infected cells, *Virology* **49**:618.

Hirt, B., 1967, Selective extraction of polyoma DNA from infected mouse cell cultures, *J. Mol. Biol.* **26**:365.

Hoggan, M.D., 1965, Presence of small virus-like particles in various adenovirus type 2, 5, 7 and 12 preparations, *Fed. Proc.* **24**:248.

Hoggan, M.D., 1970, Adenovirus-associated viruses, *Progr. Med. Virol.* **12**:211.

Hoggan, M.D., Blacklow, N.R., and Rowe, W.P., 1966, Studies of small DNA viruses found in various adenovirus preparations: Physical, biological, and immunological characteristics, *Proc. Natl. Acad. Sci. U.S.A.* **55**:1467.

Hoggan, M.D., Thomas, G.F., and Johnson, F.B., 1973, Continuous carriage of adenovirus-associated virus genome in cell culture in the absence of helper adenovirus, in: *Possible Episomes in Eukaryotes* (L. Silvestri, ed.), pp. 243–249, North Holland, Amsterdam.

Huberman, J.A., 1981, New views of the biochemistry of eukaryotic DNA replication revealed by Aphidicolin, an unusual inhibitor of DNA polymerase α, *Cell* **23**:647.

Hubscher, U., Kuenzel, C.C., and Spadari, S., 1977, Identity of DNA polymerase γ from synaptosomal mituchondria and rat brain nuclei, *Eur. J. Biochem.* **81**:249.

Hull, R.N., Johnson, I.S., Culbertson, C.G., Reimer, C.B., and Wright, H.F., 1965, Oncogenicity of simian adenoviruses, *Nature* **150**:1044.

Ikeda, J.-E., Enomoto, T., and Hurwitz, J., 1981, Replication of adenovirus DNA-protein complex with purified proteins, *Proc. Natl. Acad. Sci. USA* **78**:884.

Ishibashi, M., and Ito, M., 1971, The potentiation of type 1 adeno-associated virus by temperature-sensitive conditional lethal mutants of CELO virus at the restrictive temperature, *Virology* **45**:317.

Ita, M., and Susuki, E., 1970, Adeno-associated satellite virus growth supported by a temperature-sensitive mutant of human adenovirus, *J. Gen. Virol.* **9**:243.

Ito, K., Arens, M., and Green, M., 1975, Isolation of DNA polymerase γ from an adenovirus replication complex, *J. Virol.* **15**:1507.

Janik, J.E., Huston, M.M., and Rose, J.A., 1981, Locations of adenovirus genes required for the replication of adenovirus-associated virus, *Proc. Natl. Acad. Sci. USA* **78**:1925.

Janik, J.E., Huston, M.M., Cho, K., and Rose, J.A., 1982, Requirement of adenovirus DNA-binding protein and VA-I RNA for production of adeno-associated virus polypeptides. *J. Cell Biochemistry* **Suppl 6**:209.

Jay, G., Jay, F.T., Friedman, R.M., and Levine, A.S., 1978, Biosynthesis, immunological specificity and intracellular distribution of the simian virus 40-specific protein induced by the non-defective hybrid Ad2ND1, *J. Virol.* **26**:411.

Jay, F.T., Laughlin, C.A., and Carter, B.J., 1981, Eukaryotic translational control: Adeno-associated virus protein synthesis is affected by a mutation in the adenovirus DNA binding protein, *Proc. Natl. Acad. Sci. USA* **78**:2927.

Jones, N., and Shenk, T., 1979a, Isolation of Ad5 host range deletion mutants defective for transformation of rat embryo cells, *Cell* **17**:683.

Jones, N., and Shenk, T., 1979b, An adenovirus type 5 early gene function regulates expression of other early viral genes, *Proc. Natl. Acad. Sci. USA* **76**:3665.

Kirchstein, R.L., Smith, K.O., and Peters, E.A., 1968, Inhibition of adenovirus 12 oncogenicity by adeno-associated virus, *Proc. Soc. Exp. Biol. Med.* **128**:670.

Kisary, J., 1979, Interaction in replication between the goose parvovirus strain B and duck plaque herpes virus, *Arch. Virol.* **59**:81.

Kitchingman, G.R., Lai, S.P., and Westphal, H., 1977, Loop structures in hybrids of early RNA and the separated strands of adenovirus DNA, *Proc. Natl. Acad. Sci. USA* **74**:4392.

Klessig, D.F., 1977, Isolation of a variant human adenovirus serotype 2 that multiplies efficiently on monkey cells, *J. Virol.* **21**:1243.

Klessig, D.F., and Anderson, C.W., 1975, Block to multiplication of adenovirus serotype 2 in monkey cells, *J. Virol.* **16**:1650.

Klessig, D.F., and Chow, L.T., 1980, Incomplete splicing of several late viral mRNAs in monkey cells infected by human adenovirus type 2, *J. Mol. Biol.* **139**:221.

Klessig, D.F., and Grodzicker, T., 1979, Mutations that allow human Ad2 and Ad5 to express late genes in monkey cells map in the viral gene encoding the 72K DNA binding protein, *Cell* **17**:957.

Kollek, R., and Goulian, M., 1981, Synthesis of parvovirus H-1 replicative form from viral DNA by DNA polymerase γ. *Proc. Natl. Acad. Sci. USA* **78**:6206.

Kornberg, A., 1974, *DNA Synthesis*, Freeman, San Francisco.

Krokan, H., Schaffer, P., and DePamphilis, M.L., 1979, Involvement of eukraryotic deoxyribonucleic acid polymerases α and γ in the replication of cellular and viral deoxyribonucleic acid, *Biochemistry* **18**:4431.

Lassam, N.J., Bayley, S.T., and Graham, F.L., 1978, Synthesis of DNA, late polypeptides, and infectious virus by host-range mutants of adenovirus 5 in non-permissive cells, *Virology* **87**:463.

Laughlin, C.A., Myers, M.W., Risin, D.L., and Carter, B.J., 1979a, Defective-interfering particles of the human parvovirus adeno-associated virus, *Virology* **94**:162.

Laughlin, C.A., Westphal, H., and Carter, B.J., 1979b, Spliced adenovirus-associated virus RNA, *Proc. Natl. Acad. Sci. USA* **76**:5566.

Laughlin, C.A., Jones, N., Shenk, T., and Carter, B.J., 1979c, AAV helper ability of adenovirus early deletion mutants, American Soc. Microbiol. Ann. Mtg. Abstr., p. 73.

Laughlin, C.A., Jones, N., and Carter, B.J., 1982, Effect of deletions in adenovirus early region 1 genes upon replication of adeno-virus, *J. Virol.* **41**:868.

Lechner, R.L., and Kelly, T.J., Jr., 1977, The structure of replicating adenovirus 2 DNA molecules, *Cell* **12**:1007.

Ledinko, N., and Toolan, H., 1968, Human adenovirus 12 as a "helper" for growth of H-1 virus, *J. Virol.* **2**:155.

Lerner, M.R., Boyle, J.A., Mount, S., Wolin, S., and Steitz, J.A., 1980, Are snRNPs involved in splicing? *Nature* **283**:220.

Lerner, M.R., and Steitz, J.A., 1979, Antibodies to small nuclear RNAs complexed with proteins are produced by patients with systemic lupus erythematosus, *Proc. Natl. Acad. Sci. USA* **76**:5495.

Lewis, J.B., and Mathews, M.B., 1980, Control of adenovirus early gene expression: a class of immediate early products, *Cell* **21**:303.

Lewis, J.B., Esche, H., Smart, J.E., Stillman, B.W., Harter, M.L., and Mathews, M.B., 1980, Organization and Expression of the left third of the genome of adenovirus, *Cold Spring Harbor Symp. Quant. Biol.* **44**:493.

Lipps, B.V., and Mayor, H.D., 1980, Properties of adeno-associated virus (type 1) replication in rodent cells by murine adenovirus, *J. Gen. Virol.* **51**:223.

Longiara, M., Ikeda, J.E., Jarkorsky, Z., Horwitz, S.B., and Horwitz, M.S., 1979, The effective of aphidicolon on adenovirus DNA synthesis, *Nucleic Acids Res.* **6**:3369.

Lusby, E., and Berns, K.I., 1981, Mapping the 5′ ends of adeno-associated virus RNAs, *J. Virol.* **41**:518.

Lusby, E., and Berns, K.I., 1982, Mapping the 5′ termini of two adeno-associated virus RNAs in the left-half of the genome. *J. Virol.* **41**:518.

Marcus, C.J., Laughlin, C.A., and Carter, B.J., 1981, Adeno-associated virus RNA transcription in vivo, *European J. Biochem.* **121**:147.

Mathews, M.B., 1980, Binding of adenovirus VA RNA to mRNAs: A possible role in splicing? *Nature* **285**:575.

Mayor, H.D., and Gorman, C., 1975, Growth of adeno-associated satellite virus under conditions of arginine deprivation, *Virology* **66**:25.

Mayor, H.D., and Ratner, J., 1972, Conditionally defective helper adenovirus and satellite virus replication, *Nature New Biol.* **239**:20.

Mayor, H.D., Houlditch, G.S., and Mumford, D.M., 1973, Influence of adeno-associated satellite virus on adenovirus-induced tumors in hamsters, *Nature New Biol.* **241**:44.

Mayor, H.D., Carrier, S., and Jordan, L., 1977, Complementation of adeno-associated satellite virus (AAV) by a temperature-sensitive mutants of adenovirus type 31, *J. Gen. Virol.* **35**:545.

Melnick, J.L., Mayor, H.D., Smith, K.O., and Rapp, F., 1965, Association of 20 millimicron particles with adenoviruses, *J. Bacteriol.* **90**:271.

Miller, J.S., Ricciard, R.P., Roberts, B.E., Paterson, B.M., and Mathews, M.B., 1980, Arrangement of messenger RNAs and protein coding sequences in the major late transcription unit of adenovirus 2, *J. Mol. Biol.* **142**:455.

Montell, C., Fisher, E.F., Caruthers, M., and Berk, A.J., 1982, Resolving the functions of overlapping viral genes by site-specific mutagenesis at a mRNA splice site, *Nature* **295**:380.

Murray, V., and Holliday, R., 1979, Mechanism for splicing of gene transcripts, *FEBS Lett.* **106**:5.

Myers, M.W., and Carter, B.J., 1980, Assembly of adeno-associated virus, *Virology* **102**:71.

Myers, M.W., and Carter, B.J., 1981, Adeno-associated virus replication. The effect of L-canavanine or a helper virus mutation on accumulation of viral capsids and progeny single-stranded DNA, *J. Biol. Chem.* **256**:567.

Myers, M.W., Laughlin, C.A., Jay, F.T., and Carter, B.J., 1980, Adenovirus helper function for growth of adeno-associated virus: Effect of temperature-sensitive mutations in adenovirus early gene region 2, *J. Virol.* **35**:65.

Nass, K., and Frenkel, G.D., 1978, Adenovirus-induced inhibition of cellular DNase. *J. Virol.* **26**:540.

Nevins, J.R., and Wilson, M.C., 1981, Regulation of adenovirus-2 gene expression at the level of transcriptional termination and RNA processing, *Nature* **290**:113.

Nevins, J.R., and Winkler, J., 1980, Regulation of early adenovirus transcription: A product of early region 2 specifically represses region 4 transcription, *Proc. Natl. Acad. Sci. USA* **77**:1893.

Nevins, J.R., Ginsberg, H.S., Blanchard, J.M., Wilson, M.C., and Darnell, J.E., 1979, Regulation of the primary expression of the early adenovirus transcription units, *J. Virol.* **32**:727.

Ostrove, J., and Berns, K.I., 1980, Adenovirus early region 1b gene function required for rescue of latent adeno-associated virus, *Virology* **104**:502.

Ostrove, J., Duckworth, D.H., and Berns, K.I., 1981, Inhibition of adenovirus transformed cell oncogenicity by adeno-associated virus, *Virology* **113**:521.

Parks, W.P., Melnick, J.L., Rongey, R., and Mayor, H.D., 1967, Physical assay and growth cycle studies of a defective adeno satellite virus, *J. Virol.* **1**:171.

Pereira, H.G., Huebner, R.J., Ginsberg, H.S., and van der Veen, J., 1963, A short description of the adenovirus group, *Virology* **20**:613.

Perricaudet, M., Akusjarvi, G., Virtanen, A., and Pettersson, U., 1980, Analysis of cloned mRNA sequences from the transforming region of adenovirus 2, *Cold Spring Harbor Symp. Quant. Biol.* **44**:471.

Persson, H., Oberg, B., and Philipson, L., 1978, Purification and characterization of an early protein (E14K) from adenovirus type 2 infected cells, *J. Virol.* **28**:119.

Persson, H., Jornvall, H., and Zabielski, 1980, Multiple mRNA species for the precursor to an adenovirus-encoded glycoprotein: Identification and structure of the single sequence, *Proc. Natl. Acad. Sci. USA* **77**:6349.

Persson, H., Monstein, H.J., Akusjarvi, G., and Philipson, L., 1981, Adenovirus early gene products may control viral mRNA accumulation and translation in vivo, *Cell* **23**:485.

Philipson, L., 1979, Adenovirus proteins and their messenger RNAs, *Adv. Virus Res.* **25**:357.

Pritchard, C.R., Bates, R.C., and Stout, E.R., 1978, Levels of cellular DNA polymerases in synchronized bovine parvovirus-infected cells, *J. Virol.* **27**:258.

Pritchard, C., Stout, E.R., and Bates, R.C., 1981, Replication of parvoviral DNA. I. Characterization of a nuclear lysate system, *J. Virol.* **37**:352.

Rabson, A.S., O'Conor, G.T., Berezesky, I.K., and Paul, F.J., 1964, Enhancement of adenovirus growth in African green monkey kidney cell cultures by SV40, *Proc. Soc. Exptl. Biol. Med.* **116**:187.

Rekosh, D.M., Russell, W.C., Bellett, A.J.D., and Robinson, A.J., 1977, Identification of a protein linked to the ends of adenovirus DNA, *Cell*, **11**:283.

Revie, D., Tseng, B.Y., Grafstrom, B.H., and Goulian, M., 1979, Covalent association of protein with replicative form DNA of parvovirus H-1. *Proc. Natl. Acad. Sci. USA* **76**:5539.

Rhode, S.L., 1973, Replication process of the parvovirus H-1.I. Kinetics in a parasynchromons cell system, *J. Virol.* **11**:856.

Rhode, S.L., 1974, Replication process of the parvovirus H-1, III, factors affecting H-1 RF DNA synthesis, *J. Virol.* **14**:791.

Rhode, S.L., III, 1976, Replication process of parvovirus H-1. V. Isolation of and characterization of temperature sensitive H-1 mutants defective in progeny DNA synthesis, *J. Virol.* **17**:659.

Rhode, S.L., III, 1978, Replication process of the parvovirus, H-1. X. Isolation of a mutant defective in replicative form DNA replication. *J. Virol.* **25**:215.

Ricciardi, R.P., Jones, R.L., Cepko, C.L., Sharp, P.A., and Roberts, B.E., 1981, Expression of early adenovirus genes requires a viral encoded acidic polypeptide, *Proc. Natl. Acad. Sci. USA* **78**:6121.

Richardson, W.D., and Westphal, H., 1981, A cascode of adenovirus early functions is required for expression of adeno-associated virus, *Cell* **27**:131.

Richardson, W.D., Carter, B.J., and Westphal, H., 1980, Vero cells injected with adenovirus type 2 mRNA produce authentic viral polypeptide patterns: Early mRNA promotes growth of adenovirus-associated virus, *Proc. Natl. Acad. Sci. USA* **77**:931.

Robberson, D.L., Kasamatsu, H., and Vinograd, J., 1972, Replication of mitochondrial DNA, *Proc. Natl. Acad. Sci. USA* **69**:737.

Robinson, A.J., and Bellett, A.J.D., 1975, A circular DNA-protein complex from adenovirus and its possible role in DNA replication, *Cold Spring Harbor Symp. Quant. Biol.* **39**:523.

Rose, J.A., 1974, Parvovirus reproduction, in: *Comprehensive Virology* (H. Fraenkal-Conrat and R. Wagner, eds.), Vol. 3, pp. 1–60, Plenum Press, New York.

Rose, J.A., and Koczot, F.J., 1972, Adenovirus-associated virus multiplication. VII. Helper requirement for viral deoxyribonucleic acid and ribonucleic acid synthesis, *J. Virol.* **10**:1.

Rosenwirth, B., Shiroki, K., Levine, A.J., and Shimojo, H., 1975, Isolation and characterization of adenovirus type 12 DNA binding proteins, *Virology.* **67**:14.

Ross, S.R., and Levine, A.J., 1980, The genomic map position of the adenovirus type 2 glycoprotein, *Virology* **99**:427.

Ross, S.R., Levine, A.J., Galos, R.S., Williams, J., and Shenk, T., 1980a, Early viral proteins in Hela cells infected with adenovirus type 5 host-range mutants, *Virology* **103**:475.

Ross, S., Flint, S.J., and Levine, A.J., 1980b, Identification of the adenovirus early proteins and their genomic map positions, *Virology* **100**:419.

Rouse, H.C., and Schlesinger, R.W., 1972, The effects of arginine starvation and macromolecular synthesis in infections with type 2 adenovirus, *Virology* **48**:463.

Salo, R.J., and Mayor, H.D., 1979, Adenovirus-associated virus polypeptides synthesized in cells coinfected with either adenovirus or herpes virus, *Virology* **93**:237.

Sehgal, P.B., Fraser, N.W., and Darnell, J.E., 1979, Early Ad2 transcription units: Only promoter proximal RNA continues to be made in the presence of DRB, *Virology* **94**:185.

Sharp, P.A., 1980, Workshop #3 Summary: Adenovirus/SV40/Polyoma II, in: *ICN-UCLA Symposium on Molecular and Cellular Biology Animal Virus Genetics* (B. Fields, R. Joenisch, and C. Fox, eds.), Vol. 38, pp. 799–807, Academic Press, New York.

Shaw, A.R., and Ziff, E.B., 1980, Transcripts from the adenovirus-2 major late promoter yield a single early family of 3′ coterminal mRNAs and five late families, *Cell* **22:**905.

Shenk, T., Jones, N., Colby, W., and Fowlkes, D., 1980, Functional analysis of adenovirus-5 host-range deletion mutants defective for transformation of rat embryo cells, *Cold Spring Harbor Symp. Quant. Biol.* **44:**367.

Shiroki, K., and Shimojo, H., 1974, Analysis of adenovirus 12 temperature-sensitive mutants defective in viral DNA replication, *Virology* **61:**474.

Siegl, G., and Gautschi, M., 1973a, The multiplication of parvovirus LuIII in a synchronized culture system I. Optimum conditions for virus replication, *Arch. Gesamte Virusforsch.* **40:**105.

Siegl, G., and Gautschi, M., 1973b, The multiplication of parvovirus LuIII in a synchronized culture system II. Biochemical characteristics of virus replication, *Arch. Gesamte Virusforsch.* **40:**119.

Smith, K.O., and Gehle, W.D., 1967, Replication of an adeno-associated virus in canine and human cells with infectious canine hepatitis virus as a helper, *J. Virol.* **1:**648.

Soderlund, H., Pettersen, U., Vennstrom, B., Philipson, L., and Mathews, M.B., 1976, A new species of virus-coded low molecular weight RNA for cells infected with adenovirus type 2, *Cell* **7:**585.

Solnik, D., 1981, An adenovirus mutant defective in splicing RNA from early region 1A, *Nature* **291:**508.

Spector, D.H., Halbert, D.N., and Raskas, H., 1980, Regulation of integrated adenovirus sequences during adenovirus infection of transformed cells, *J. Virol.* **36:**860.

Stillman, B., 1981, Adenovirus DNA replication in vitro: A protein linked to the 5′ end of nascent DNA strands. *J. Virol.* **37:**139.

Stillman, B.W., Lewis, J.B., Chow, L.T., Mathews, M.B., and Smart, J.E., 1981, Identification of the gene and mRNA for the adenovirus terminal protein precursor, *Cell* **23:**497.

Straus, S.E., Ginsberg, H.S., and Rose, J.A., 1976a, DNA-minus temperature-sensitive mutants of adeno-associated virus replication, *J. Virol.* **17:**140.

Straus, S.E., Sebring, E.D., and Rose, J.A., 1976b, Concatemers of alternating plus and minus strands are intermediates in adenovirus-associated virus DNA synthesis, *Proc. Natl. Acad. Sci. USA* **73:**742.

Suzuki, E., and Shimojo, H., 1974, Temperature-sensitive formation of the DNA replication complex in adenovirus 31-infected cells, *J. Virol.* **13:**538.

Tattersall, P., 1972, Replication of parvovirus MVM. I. Dependence of virus multiplication and plaque formation on cell growth, *J. Virol.* **10:**586.

Tattersall, P., and Ward, D.C., 1976, The rolling hairpin: A model for replication of parvovirus and linear chromosomal DNA, *Nature* **263:**106.

Tennant, R.W., Layman, K.R., and Hand, R.E., Jr., 1969, Effect of cell physiological state on infection by rat virus, *J. Virol.* **4:**872.

Thomas, G.P., and Mathews, M.B., 1980, DNA replication and the early to late transition in adenovirus infection, *Cell* **22:**523.

Van der Vliet, P.C., and Kwant, M.M., 1978, Role of DNA polymerase γ in adenovirus DNA replication, *Nature* **276:**532.

Van der Vliet, P.C., Levine, A.J., Ensinger, M.J., and Ginsberg, H.S., 1975, Thermolabile DNA binding proteins from cells infected with a temperature-sensitive mutant of adenovirus defective in viral DNA synthesis, *J. Virol.* **15:**348.

Wagar, M.A., Evans, M.J., and Huberman, J.A., 1978, Effect of 2′3′-dideoxy-thymidine-5′-triphosphate on Hela cell *in vitro* DNA synthesis: Evidence that DNA polymerase α is the only polymerase required for cellular DNA replication, *Nucleic Acids Res.* **5:**1933.

Weinmann, R., Raskas, H.J., and Roeder, R.G., 1974, Role of DNA-dependent RNA polymerases II and III in transcription of the adenovirus genome late in productive infection, *Proc. Natl. Acad. Sci. USA* **71:**3436.

Weissbach, A., 1979, The functional roles of mammalian DNA polymerase, *Arch. Biochem. Biophysics* **198:**386.

Weissbach, A., Baltimore, D., Bollum, F., Gallo, R., and Korn, D., 1975, Nomenclature of eukaryotic DNA polymerases, *Science* **190:**401.

Williams, J.F., 1973, Oncogenic transformation of hamster embryo cells *in vitro* by adenovirus type 5, *Nature* **243:**162.

Wilson, W.C., Nevins, J.R., Blanchard, J.M., Ginsberg, H.S., and Darnell, J.E., 1980, The metabolism of mRNA from the transforming region of ad2, *Cold Spring Harbor Symp. Quant. Biol.* **44:**447.

Wolter, S., Richards, R., and Armentrout, R.W., 1980, Cell cycle-dependent replication of the DNA of minute virus of mice, a parvovirus, *Biochimia et Biophysics Acta* **607:**420.

Young, J.F., and Mayor, H.D., 1979a, Studies on the defectiveness of adeno-associated virus (AAV). I. Effects of phosphonoacetic acid and 2-deoxy-D-glucose on the replication of AAV, *Virology* **94:**323.

Young, J.F., and Mayor, H.D., 1979b, Adeno-associated virus—An extreme state of viral defectiveness, *Prog. Med. Virol.* **25:**113.

CHAPTER 4

Autonomous Parvovirus DNA Structure and Replication

William W. Hauswirth

I. DNA STRUCTURE

A. Virion DNA

The genomes of the autonomous parvoviruses are linear, single-strand DNA molecules of about 5100 base pairs (bp) and of minus polarity (complementary to mRNA). There are only minor length differences between parvovirus genomes of various mammalian species including those from mouse, hamster, rat, pig, and man. As will be discussed, genome similarity also extends to secondary structure, replicative intermediates, and, in a more limited sense, to the primary nucleotide sequence. Closely related parvoviruses of rodents additionally exhibit the ability to complement each other for several replicative processes (McMaster *et al.*, 1981; Rhode, 1982), suggesting considerable functional conservation for both *cis* and *trans* acting elements. Bearing in mind that species variations do exist, it has been found thus far that all autonomous parvoviruses appear to share a common strategy for replicating their genomes. Therefore, although experimental data for a given structure or replicative process may exist for only one or two parvoviruses, they will be treated as characteristic of the entire virus subgroup.

WILLIAM W. HAUSWIRTH • Department of Immunology and Medical Microbiology, College of Medicine, University of Florida, Gainesville, Florida 32610.

Initial estimates for the genome molecular weight were made by sedimentation and electron microscopic analysis (Crawford *et al.*, 1969; McGeoch *et al.*, 1970). Subsequently, size estimates were refined by electrophoretic mobility in agarose and polyacrylamide gels, summation of fragments generated by restriction enzymes, and then made absolute by direct nucleotide sequencing of the entire genome. Presently, the exact sizc has been determined by nucleotide sequencing for two autonomous parvovirus genomes; 5081 bp for MVM DNA (Astell *et al.*, 1983) and 5176 bp for H-1 DNA (Rhode and Paradiso, 1983).

Genome strandedness was first suggested by its reactivity toward formaldehyde, its electron microscopic appearance, and its susceptability to single strand specific endonucleases (Crawford *et al.*, 1969). Most importantly, Crawford *et al.* (1969) reported the base composition was: A = 27%, T = 33%, G = 19%, C = 21%. The fact that this composition violated Chargaff's rules for duplex DNA coupled with nearest-neighbor frequencies which were also inconsistent with those expected for double stranded DNA, but in agreement with the base composition analysis, clearly demonstrated the single-stranded nature of the autonomous parvovirus genome. Finally, it was shown by lack of genome self-complementarity and annealing to viral RNA that MVM (and other parvoviruses) package >99% minus strands in their virions (Bourguignon *et al.*, 1976).

Genome linearity has been established by several techniques. Exonuclease I which digests only single-stranded DNA from its ends was initially reported to degrade KRV DNA (Salzman *et al.*, 1971). However, this may have been fortuitous because the very 3′-end of parvovirus DNA is usually resistant to such nucleases by virtue of its duplex nature (see below). Electron microscopic studies on several parvovirus genomes have found no evidence for circular forms of virion DNA (Crawford, 1966; Bourguignon *et al.*, 1976). The strongest evidence for genome linearity comes from more recent studies of nucleotide sequence arrangement using restriction enzyme digestion and DNA sequencing techniques. Restriction site mapping has demonstrated unique terminal fragments for MVM, KRV, H-1, and H-3 DNAs. Additionally, unique terminal nucleotide sequences have been determined for the 3′-ends of MVM, KRV, H-1, and H-3 DNAs (Astell *et al.*, 1979a,b; Salzman and Fabisch, 1979) and two populations of 5′-ends for H-1 DNA (Rhode and Klaassen, 1982) and MVM DNA (Astell *et al.*, 1983) which differ in primary sequence from the 3′-end. Thus, the autonomous parvovirus genome is a linear molecule containing no circular permutations.

Although generally single stranded, both the 3′ and 5′ termini of the parvovirus genome normally exist in a duplex configuration. The very 3′ terminal nucleotide is also part of the duplex because it is resistant to digestion by single-strand specific nuclease (Bourguignon *et al.*, 1976). Limit digestion with the single strand specific DNase S_1 and elution of the products from hydroxylapatite columns suggested that about 7% of the linear, single-stranded DNA existed in a double-stranded form (Bourguignon *et al.*, 1976; Salzman, 1977; Ward and Chow, 1978). Localization of these duplex regions to the genome ends was determined by radiolabeling both termini and showing that after S_1 DNase digestion all duplex segments contained label. Electrophoretic sizing of these fragments showed that the duplex region at the 3′ terminus of MVM DNA contains 115–120 nucleotides and that the one at the 5′ terminus contains approximately 160 nucleotides (Ward and Chow, 1978; Astell *et al.*, 1979a). These authors also showed that the terminal duplex regions would renature unimolecularly, i.e., "snap back," after heat denaturation. They therefore concluded that terminal palindromic sequences exist at each end of the DNA which can self-hybridize. The fact that the duplex structure found at each end differed in size suggested that, unlike AAV DNA, the terminal palindromes in the autonomous parvoviruses may not be identical.

Terminal palindromic regions have now been sequenced for several autonomous parvovirus genomes (Astell *et al.*, 1979a,b; Salzman and Fabisch, 1979; Rhode and Klaassen, 1982). The genomes of the rodent parvoviruses have very similar 3′ terminal sequences (Fig. 1). Between MVM, H-1, H-3, and KRV DNAs there are only two or three nucleotide replacements in the 115 or 116 bp palindromic sequence and very few other differences up to 150 bp from the genome end. As expected, when the divergence between species becomes greater the primary sequence of the genome ends becomes more unrelated (e.g., MVM vs. AAV). However, the palindromic nature of the 3′ termini is maintained, independent of the species of origin and primary sequence. Such evolutionary conservation of secondary structure and not primary sequence might mean that the structure and not the exact sequence is functionally important (see below).

In addition to the ability of all single-stranded 3′ termini to self anneal creating duplex structures, it may be significant that the most stable base pairing configuration in almost all cases is not a perfect hairpin but one having a T- or Y-shaped structure at the midpoint of the large palindrome (Fig. 1). This internal structure is stable because two small palindromic sequences exist within the overall palindrome (Fig. 1). These short self-complementary sequences generate the arms of the "T" (e.g., nucleotides 46–59 and 60–71 in MVM DNA). It may be significant that all the base pairs within these internal palindromic regions are G-C base pairs. This certainly increases the stability of the "T" and may be important if the structure plays a role in parvovirus biology. It has been recently suggested (Ward and Tattersall, 1982) that the alternating G-C sequence in one arm of the "T" in rodent parvoviruses (e.g., bp 45–51 and 52–58 in MVM DNA) may lead to a local z-DNA conformation in solution. This may serve as a topological marker for protein recognition during DNA replication. In addition to its position in the "T" at the end of viral DNA, the alternating G-C palindrome may also form a short z-DNA cruciform at the junction between viral and complementary strands

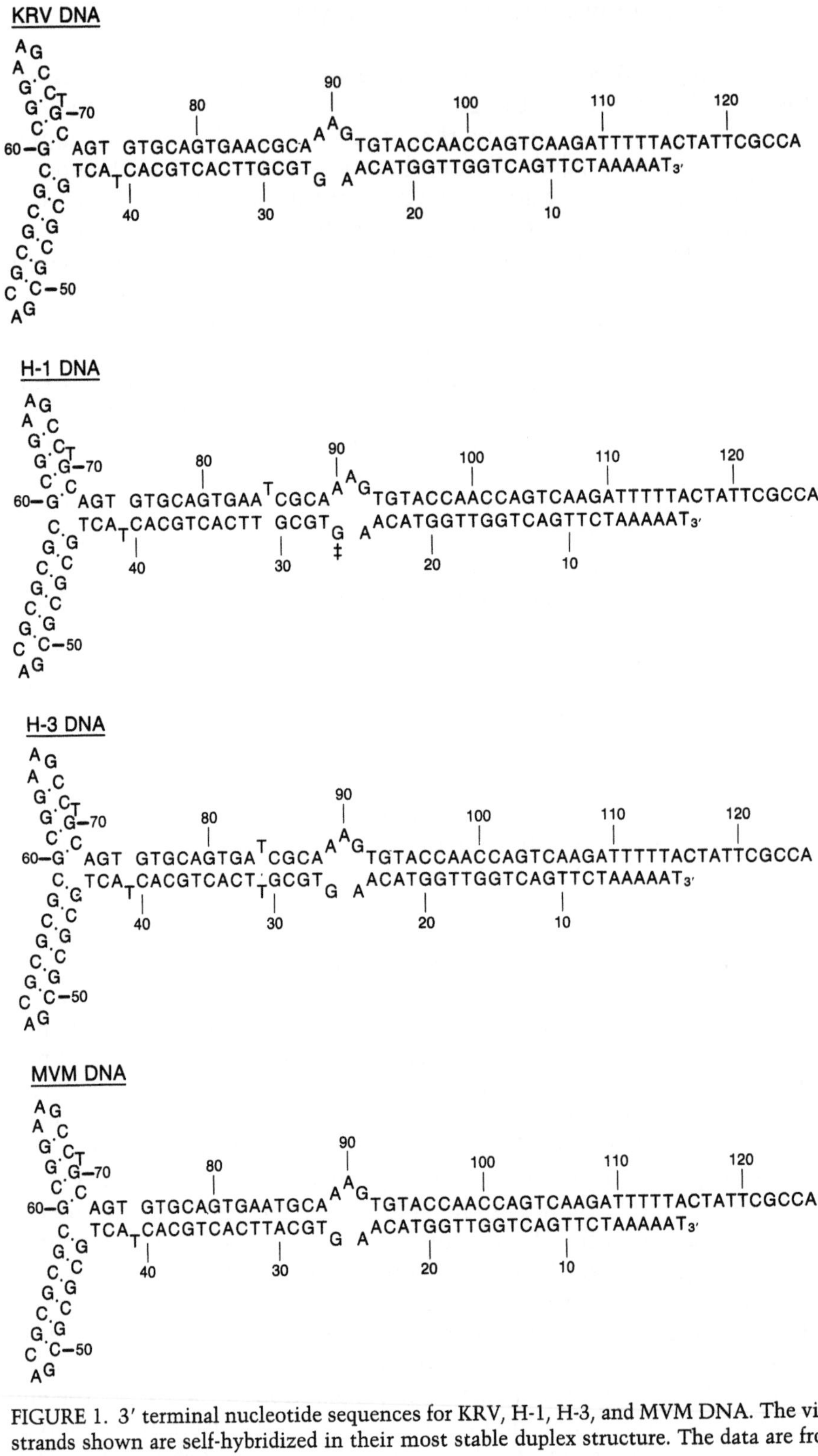

FIGURE 1. 3′ terminal nucleotide sequences for KRV, H-1, H-3, and MVM DNA. The viral strands shown are self-hybridized in their most stable duplex structure. The data are from Astell *et al.* (1979a).

in dimeric, duplex replicative intermediate molecules. In this position it may serve to signal the region to be processed (see Section III). Precedent for such structures and their recognition by proteins exist in bacterial (Wang *et al.*, 1979) and eukaryotic (Nordheim *et al.*, 1982) systems.

Regardless of the exact conformation of the "T" structure, it seems significant that rather divergent autonomous parvovirus, including those of rodent, bovine (BPV) and human (LuIII) origin (C. Astell and M. Chow, unpublished observations) as well as the defective human parvovirus AAV (see Chapter 1), have all retained this specific structure at the midpoint of their palindromic ends while the primary sequence has changed totally. Because this region contains no open reading frames (see Chapter 5) it must function in a structural or regulatory sense. It should be emphasized that in the above studies only one unique 3′ terminal fragment and nucleotide sequence was detected. No evidence could be found for the type of terminal sequence inversion found at both ends of the AAV genome. The 3′ terminus of the autonomous parvovirus genome is therefore a predominantly palindromic sequence of 115 or 116 bp existing in only one orientation.

The 5′ terminal sequence has proved more difficult to determine generally because 5′-end labeling of the genome with polynucleotide kinase has proven to be extremely inefficient. Recently the sequences of the 5′ terminal palindromes of MVM DNA (Ward and Tattersall, 1983; Astell *et al.*, 1983) and H-1 DNA (Rhode and Klaassen, 1982) have been determined and are 207 and 242 bp, respectively (Fig. 2). As a prerequisite to determining these sequences, detailed restriction fragment analysis at the 5′-end was carried out in both cases. This revealed the existence of two orientations of the 5′ terminal region which was then confirmed by the nucleotide sequence. As with AAV DNA, it proved possible to resolve two terminal restriction fragments for MVM and H-1 because the 5′ sequence is not a perfect palindrome (Fig. 2). For H-1 DNA, *Hha*I and *Taq*1 both cut within terminal sequences that are not part of the overall palindrome. Thus the size of the resultant fragments are dependent on the orientation of the overall palindrome and the fact that two sets of sites exist for each enzyme is evidence of two terminal sequence orientations. For example, the penultimate *Hha*I fragment is either 100 bp or 140 bp in length depending on the sequence orientation and both are found in approximately equal proportion (Fig. 2). As expected, neither *Taq*I nor *Hha*I cuts in this region if 5′-hairpinned DNA is the enzyme substrate because the recognition sequences are destroyed by base mismatches. For MVM DNA it was possible to show that, although the single *Rsa*I site is within the 5′ palindrome, two fragments are generated with approximately equal probability. They are separable, presumably, because they differ in sequence at the base mismatch and the hairpin turn-around positions.

Therefore, in contrast to the 3′ terminus where only one sequence orientation is found, the 5′ terminal palindrome exists in two mutually

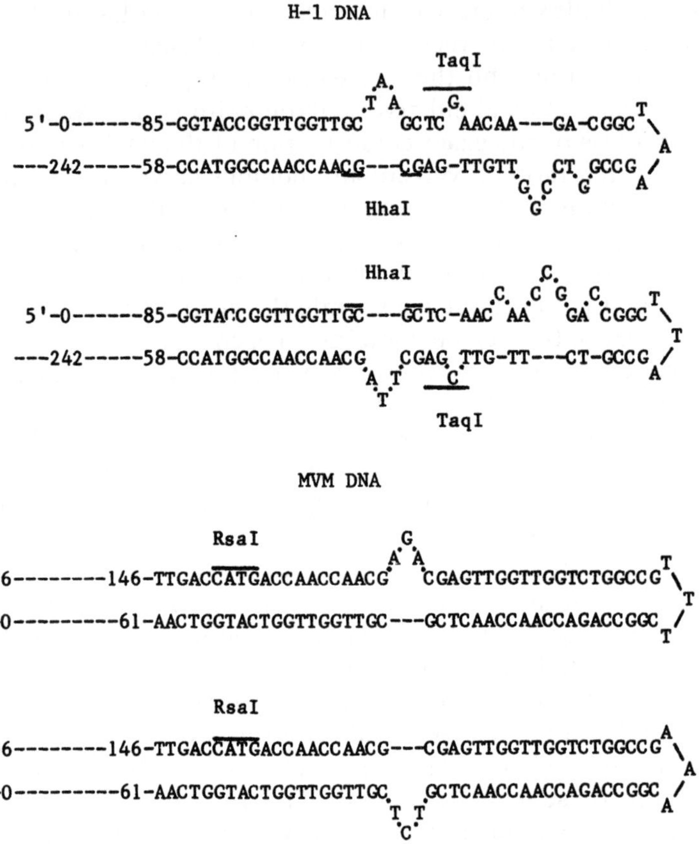

FIGURE 2. 5′ terminal nucleotide sequences of H-1 and MVM DNA. The viral strands shown are self-hybridized in their most stable duplex structure. H-1 DNA: the two mutually inverted 5′ terminal palindromic sequences are depicted. The overall palindrome is 242 bases long, but only bases 58–85 are shown because the sequence inversion only affects parts of this region. Also shown are the *Hha*I and *Taq*I sites with which the inversion was first detailed (see text). The data are redrawn from Rhode and Klaassen (1982). MVM DNA: the two mutually inverted 5′ terminal palindromic sequences are depicted. The overall palindrome in 206 bases long, but only the region affected by the inversion, bases 61 to 146 are shown. Also shown is the *Rsa*I site with which the inversion could be detected. The data are redrawn from Astell *et al.* (1983).

inverted orientations as do both ends of AAV. The sequencing studies confirm the earlier conclusion that the 3′ and 5′ terminal sequences are unrelated in size and additionally show them to be unrelated in primary sequence as well. This is again unlike AAV DNA which contains inverted terminal repetitions which are slightly larger and include the terminal palindromic sequence (see Chapter 1). These findings represent the major DNA structural differences between the autonomous and defective parvovirus genomes and will demand an essential difference in their mechanisms of replication.

B. Replicative Forms

1. DNA Structure

Isolation of intracellular parvovirus DNA routinely yields, in addition to unit length progeny genomes, a significant fraction of monomer and dimer length, double stranded forms of the genome. The kinetics of appearance of these forms relative to that of progeny genomes have suggested that both duplex monomer and multimeric forms are precursors to progeny strands for H-1, H-3, MVM, LuIII, and KRV DNA (Rhode, 1979; Ward and Dadachanji, 1978; Siegl and Gautschi, 1978; Hayward *et al.*, 1978; Li *et al.*, 1978). These structures have therefore been collectively termed replicative form (RF) molecules.

The initial phase in DNA replication appears to be formation of a monomer double-stranded RF from the input single-stranded virion DNA. Only at later times can dimeric and multimeric RF molecules be detected. The duplex nature of RF DNA has been generally deduced by its sedimentation properties on sucrose gradients and by chromatographic analysis on hydroxylapatite. Because virions contain only minus strand DNA, all characterization of the genome using restriction enzymes has relied on such duplex RF molecules or *in vitro* constructed versions of monomer RF using bacterial polymerases. Terminal hairpin structures at both genome ends were detected in RF DNAs by the ability of terminal restriction fragments to snap back reforming a duplex structure after heat denaturation and quick chilling. However, the hairpin structure which covalently links plus and minus strands must be short-lived because a large fraction of unit length strands can be isolated from the RF pool after denaturation and banding on CsCl density gradients using BUdR substituted H-1 RF DNA (Rhode, 1977a). This suggests the presence of an efficient endonucleolytic activity which unlinks viral (minus) and complementary (plus) strands by a site-specific nicking of the initial RF molecule.

In summary, the basic structural features of autonomous parvovirus RF DNA are (1) RF DNA contains the same 3′ and 5′ terminal sequences as virion DNA; (2) RF DNA contains both fully extended and hairpinned forms of these termini (Rhode, 1977a; Chow and Ward, 1978); (3) monomeric RF DNA has viral and complementary strands linked through a single hairpinned terminal sequence; (4) dimeric RF DNA has viral and complementary strands linked by terminal hairpins as well as by an internal copy of the 3′ terminal sequence (Rhode, 1977b); (5) the concentration of covalently linked viral and complementary strands is at a relatively low steady-state level.

Recent nucleotide sequencing studies of RF termini have added significant detail to this picture. The 3′-end of RF MVM DNA like the 3′-end of virion DNA contains a unique sequence orientation (Astell *et al.*, 1982). Thus, 3′ sequence inversion does not take place in either RF DNA or virion DNA, a critical feature of any proposed replication mechanism.

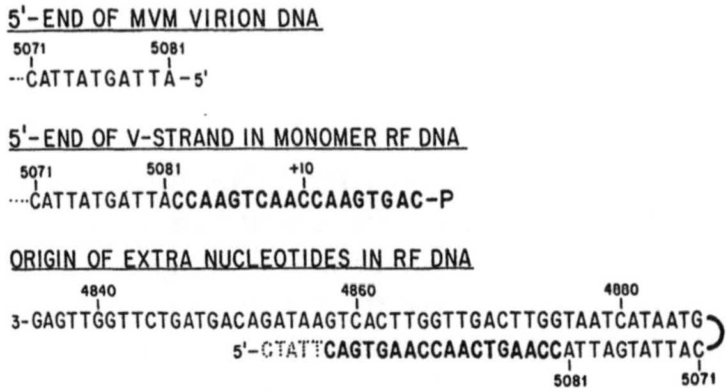

FIGURE 3. Relationship between the MVM viral DNA 5′ terminal sequence and the 5′ viral sequence found in RF molecules. "P" indicates the covalently attached 5′ terminal protein found in RF molecules. Also shown is the presumed origin of the difference between viral and RF 5′ ends, e.g., they arise by an extension of the hairpinned 5′ terminus. The data are from Astell *et al.* (1983). Solid line represents the remainder of the hairpin.

These authors also determined the nucleotide sequences for the mutually inverted viral strand 5′ termini of monomer RF DNA. Interestingly, the sequence was found to be 18 nucleotides longer than the 5′ end of virion DNA. These extra bases are complementary to a region of viral strand DNA immediately internal to the 5′ palindromic sequence. The relationship between viral and RF 5′ terminal sequences is shown in Fig. 3. It can be seen that the 18 extra nucleotides arise from an extension of the 5′ hairpin at its terminus. The authors offer an explanation for this by suggesting that, rather than nicking a viral 5′ hairpin at the normal 5′ terminus (bp 5081), the nick occurs on the complementary strand 18 bp from bp 5081 during formation of RF DNA. It should be noted that the sequence of the 5′ viral strand in the MVM RF was determined from its complement (the 3′ complementary strand) both because of the presence of a 5′ terminal protein (see below) which reduced labeling efficiency and because the 5′-end is slightly frayed and does not terminate at a specific base (Astell *et al.*, 1982).

The 5′ terminal sequence has also been reported for H-1 monomer RF DNA (Rhode and Klaassen, 1982). Again, both mutually inverted terminal 5′ sequences were determined. Unfortunately, the procedure used to clone the 5′ terminus utilized a restriction site 6 bp from the 5′-end and thus the terminal 6 base pairs were not sequenced. These workers therefore may have missed the potentially significant difference between virion and RF DNA at the 5′-end as reported for MVM.

2. 5′ Terminal Protein

The covalent association of a protein to the 5′ termini of RF molecules has been demonstrated for H-1 DNA (Revie *et al.*, 1979). These authors showed that viral RF banded in a CsCl gradient at a density less than that for pure DNA (viral or RF). Even after treatment with detergent at 60°C, 8 M urea at 25 or 100°C, or 1 M NaOH at 37°C (all of which disrupt capsids) a density difference remained. Only extensive protease treatment restored RF DNA to a viral DNA density. These results suggested that a protein, calculated to be 60,000–70,000 daltons, was covalently linked to RF molecules. Restriction fragment analysis and strand separation experiments demonstrated specific association of the protein with the 5′-ends of both viral and complementary strands of RF molecules but no evidence could be found for its existence at the 5′ end of virion DNA. Except for the commonly reported difficulty in enzymatically radiolabeling the 5′ terminus of virion DNA, no report has yet shown the association of such a protein with virion DNA. Removal of the 5′ protein apparently precedes or coincides with DNA packaging.

More recently Astell *et al.* (1982) have reported a similar protein associated to the 5′ termini of MVM RF DNA. These authors noted that RF DNA was refractory to 5′ end labeling using polynucleotide kinase and determined by phenyl-sepharose chromatography that a protein was complexed to the DNA. After restriction enzyme digestion only duplex DNA with fully extended terminal fragments was retained on phenyl-sepharose. Thus, protein was associated only with termini of RF molecules which were not hairpinned. Because the DNA–protein complex was a substrate for a 3′ → 5′ exonuclease (exonuclease III) but not for λ exonuclease (a 5′ → 3′ exonuclease) the protein–DNA linkage was deduced to be at the 5′-end of each RF DNA strand. Stability to 8 M urea, 6 M guanidine-HCl, and 2% SDS at 60°C suggested a covalent linkage. Stability to 0.3 M NaOH for 8 hr at 37°C suggested that is was not a phosphoserine or phosphothreonine bond and that it may be a phosphotyrosine linkage as described by Tse *et al.* (1980). If DNase digestion to release the protein is followed by radiodination and gel electrophoresis, an apparent molecular weight of 60,000 daltons is obtained for this protein, in general agreement with estimates for the H-1 DNA terminal protein.

The function and genetic origin of these proteins are clearly critical questions, particularly in view of their association only with RF DNA and only at the 5′-ends. There are presently no data to decide whether the protein is viral or host coded. However, parvovirus genomes have been recently shown to contain an open reading frame which is transcribed but for which no function is yet assigned (see Section II.C.2.). Therefore, because all viral structural proteins are coded elsewhere in the viral genome, the 5′ protein may well be virally coded. It is tempting to speculate that it may serve a replicative function, possibly in aiding hairpin formation and/or as the postulated site-specific nicking enzyme necessary to resolve progeny viral strands from RF molecules (see Section III).

C. Defective Genomes

Chapter 6 deals extensively with the structure and biological function of defective parvovirus particles. The structure of the viral DNA contained in such defective particles will only be briefly reviewed here because relationships between the genome structure of a defective virus and the ability to replicate DNA can serve to locate *cis*-acting determinants for DNA replication and define *trans*-acting viral genes if any are required for DNA replication.

The size and physical structure of genomes for defective interfering particles of autonomous parvovirus have been characterized to varying degrees (Muller *et al.*, 1978; Rhode, 1978a, 1978b; Faust and Ward, 1979). Some defective H-1 genomes contain both inserted and deleted sequences. The insertions were initially shown to constitute integral numbers (1–10) of an approximately 60 bp sequence near the 5′-end of the viral genome (Rhode, 1978a). Subsequently, sequence analysis of one such insertion mutant having one extra copy of the inserted sequence relative to the parental genome showed it to contain three tandem copies of a 55 bp sequence whereas the parental H-1 DNA contained two copies (Rhode and Klaassen, 1982). This sequence begins 348 bp from the 5′-end of the genome and is therefore not within the terminal palindrome. The relationship of this tandem repeat to DNA replication is unclear because it is outside the terminal palindrome and because the closely related H-3 virus contains only one 55 bp repeat and is equally viable. Perhaps this genome region is under minimal functional constraint. However, it appears to be within an open reading frame (Rhode and Paradiso, 1983) and so the reason for the plasticity of this genome region remains unclear.

Another class of H-1 defective interfering particles contain deleted genomes of variable sizes. Most deletions were located at map positions 32–44 or 80–90 on the genome (Rhode, 1978b). These genomes were capable of replication only in the presence of wild type virus but in this condition they could outgrow the wild-type helper and inhibit its replication. Such defective particles were shown to be quite sensitive to UV irradiation suggesting that they retain *cis*-acting nucleotide sequences presumably necessary for replication. Thus, deletion of significant internal genome regions does not destroy those sequences required presumably as recognition signals for the replication process. Additionally, the fact that wild-type helper virus is necessary for defective genome replication suggests that a virally coded protein is required for genome replication.

Two types of defective DNA molecules have been isolated and characterized from variant MVM virions (Faust and Ward, 1979). The first type is composed of variable length (10–70% unit length) single-stranded viral DNA molecules which retain the original 5′ and 3′ terminal palindromic sequences found in the intact genomes. These resemble the deleted H-1 DNA molecules described above. Although all the sequences found in the whole genome are present in this heterogeneous population, there is a severe underrepresentation of the central genome regions. These molecules have therefore lost significant internal regions. Deletion size varied widely with the largest leaving a molecule containing only 255 nucleotides from each end of the parental MVM DNA. It was noted that this general type of molecule, including the very shortest observed, was selectively amplified during continued passage in cell culture in the presence of wild type MVM. Therefore all *cis*-acting nucleotide sequences required for DNA replication must be contained within the terminal 255 nucleotides at the 5′- and 3′-ends of the genome.

The second type of defective molecules found upon high multiplicity passage of MVM are double-stranded hairpinned structures originating from the 5′-end of the genome. They contain a normal 5′-end and an internal hairpin at variable positions up to 15 map units from the 5′-end. Such structures suggest that during replication a complementary strand may have been displaced from the original template and then, with the replication complex intact, the nascent viral strand folded back to copy itself. These molecules were found to be diluted out with continued passage in cell culture. Thus, it was concluded that they do not contain *cis*-acting sequences required for reinitiating or continuing replication. However, in this case it cannot be ruled out that the internal covalent hairpin structure itself may inhibit continued replication. With this caveat, the overall implication from the replicative ability of various defective parvovirus genomes is that several hundred bases from *both* the 3′- and 5′-ends are the minimal *cis*-acting requirements for replication.

Recently, the replicative ability of a number of deleted MVM genomes created *in vitro* by cloning terminal fragments into the bacterial plasmid pBR322 was investigated by DNA transfection into mouse cells (D.C. Ward, personal communication). As before, it was found that both terminal palindromic sequences must be present on the same molecule for DNA replication to occur. Also suggested by the defective genome results, *in vitro* constructed deletions within the terminal 200 bp abolish replicative ability whereas substantial internal deletions do not effect replicative ability as long as intact helper is coinfected.

II. DNA REPLICATION

A. *In Vivo* Replication

1. Viral DNA Intermediates

The small size of autonomous parvovirus genomes and the fact that they replicate their DNA in the nucleus of the cell suggest that most of the functions required for viral DNA replication are supplied by the host cell. Histologic studies on H-1 infected cells suggest that DNA synthesis begins at specific euchromatic or nucleolar sites and then migrates to less distinct positions in the nucleus (Singer and Rhode, 1978). The early association of viral DNA with chromosomal DNA is consistent with the expectation that parvovirus DNA replication is heavily dependent on cellular functions and suggests that viral and cellular DNA replication may have many common features, particularly initiation events. Thus, a link between the efficiency of viral DNA replication and the infected cell cycle might be expected. It has indeed been generally noted that parvoviruses replicate best in actively dividing cells (Tennant *et al.*, 1969; Tattersall, 1972; Salzman *et al.*, 1972; Siegl and Gautschi, 1973). For several parvoviruses a link to S phase in the cell cycle has been clearly demonstrated. Using parasynchronous hamster cell cultures Rhode (1973) showed that H-1 DNA synthesis begins at a point late in S phase. Presumably, production of RF DNA from input viral DNA requires synthesis of host proteins specifically made during S phase. However, protein synthesis inhibition studies showed that once the replication process had started, host protein synthesis is not required for continued viral RF production. Thus, a host factor(s) may be necessary for initiation of DNA synthesis, but if so, its continued production is unnecessary for maintenance of RF formation. Ward and Dadachanji (1978) found, using parasynchonized A9 cells, that input ^{32}P-labeled MVM DNA began to appear intracellularly in a duplex form early in S phase. This is several hours earlier in the cell cycle than that reported for H-1. The data for MVM are probably more sensitive due to the availability of high specific activity radiolabeled viral DNA. Wolter *et al.* (1980) confirmed that MVM replication begins early in S phase by using the mitotic detachment technique to obtain highly synchronous RT cell cultures. If infected early in Gl, no RF MVM DNA could be detected until early S phase 10 hr later. Within the next 4 hr 50–100 RF molecules could be detected per infected cell. Thus, these authors and others (Tattersall, 1972; Rhode, 1973; Linser *et al.*, 1977, 1979) suggested that the S phase dependency of autonomous parvovirus replication may be due to a host cell replicative factor(s) necessary for the first step in viral DNA synthesis, e.g., the conversion of single-stranded viral DNA to a monomeric duplex RF. The possibility that this first step might be the integration of the MVM genome into chromosomal DNA was considered by Richards and Armentrout (1979). However, using the network technique in which viral specific sequences covalently linked to chromosomal DNA can be separated from free viral DNA, no integrated MVM DNA could be detected even after viral replication had proceeded to a point where 100 genome equivalents of free MVM DNA were produced per cell. Although clearly dependent on host DNA polymerases, the precise replication steps so intimately dependent on S phase proteins remain unclear.

One class of host protein synthesized in great abundance early in the S phase and closely associated with DNA synthesis are the histone proteins. Ben-Asher *et al.* (1982) investigated the possibility that cellular histones may associate with replicating MVM DNA. By gentle extraction, MVM DNA–histone complexes could be isolated and characterized. All cellular histones, including H-1, were present. Histone proteins were assembled into conventional nucleosome structures containing a nucleosome core with about 140 bp of viral DNA and a linker between cores containing another 50bp. Electron microscopic analysis showed that such minichromosomes contain either 26 (monomer genome length) or 52 (dimer genome length) beads on a viral DNA. These results suggest intimate histone association with all MVM monomer and dimer RF forms. Because these workers could find no free single-stranded viral DNA and no histone can be found in the MVM virion (Tattersall, 1978), it appears that progeny viral strands may be encapsidated as they are synthesized from such RF minichromosomes. The existence of autonomous parvovirus minichromosomes suggests an even closer resemblance between viral and host DNA replication.

The temporal appearance of intracellular forms of viral DNA is pivotal to understanding its replication. The fate of MVM DNA was followed in detail as a function of time after infection by Hirt (1967) extraction and analysis of single- and double-stranded viral DNA forms on hydroxylapatite (Ward and Dadachanji, 1978). Conversion of viral DNA to double-stranded RF DNA could be detected as early as 6 hr postinfection. Most of this DNA was able to snap back after denaturation, suggesting that parental viral strands were covalently linked to complementary strands, presumably through a 3′ terminal hairpin structure. By 16 hr postinfection, however, a much smaller fraction of duplex RF exhibited the snapback phenomenon, although hairpinned terminal fragments were still present. Thus, covalent hairpin linkage of strands appeared to be transitory, apparently due to single-strand nicking of the duplex structure soon after its formation.

To prove that such duplex forms of viral DNA were true replicative intermediates, ^{3}H-thymidine pulse-labeling experiments were done 14 hr postinfection. Gel electrophoretic sizing of the viral DNA which incorporated label showed that monomeric, dimeric, and higher oligomers were all being synthesized. Similar analysis after various chase periods showed that all classes of duplex RF molecules produced progeny viral DNA. It was not possible to clearly distinguish a precursor product relationship between oligomeric forms. Isolation of these RF structures and restriction enzyme analysis showed that only fragments containing terminal regions of the genome would snap back after denaturation. Thus, viral and complementary strands in dimeric and higher-order structures must be alternating duplex concatemers. As with monomer RF, denaturation of oligomeric forms after longer chase periods yielded mostly unit length strands, again demonstrating the transient nature of the covalent linkage between opposite polarity strands.

The origin and termination points for replication of viral DNA during RF DNA synthesis were determined for H1 (Rhode, 1977; Singer and

Rhode, 1977) by measuring the relative level of pulsed radiolabel as a function of genome position, analogous to the earlier experiments on SV40 by Danna and Nathans (1972). The results indicate unidirectional DNA replication beginning at only one end of the genome and terminating at the other. The initiation point for the replication of RF is at or near the 5′ terminus of the viral strand. Thus, there is a replicative asymmetry within the genome which is almost certainly related to the assymmetry in terminal nucleotide sequences. This is logically consistant with the situation for AAV DNA which contains symmetrical terminal sequences and which replicates its duplex RF unidirectionally for each strand starting from either end of the genome (Hauswirth and Berns, 1977).

The overall picture arising from work on the *in vivo* forms of autonomous parvovirus DNA is one in which progeny viral strands are produced after several stages of viral specific DNA synthesis (Ward and Tattersall, 1982). The first stage, clearly dependent on host cell DNA synthetic factor(s), is conversion of input viral strands to a monomer duplex molecule with a covalent hairpin structure linking viral and complementary strands. The second stage is replication of this RF to produce both more monomer RFs and oligomeric RFs containing concatemeric, alternating viral and complementary strands. It is unclear whether dimeric and higher-order RF molecules are an obligatory intermediate in parvovirus DNA replication. The third, and last, stage involves the production of progeny viral strands, presumably via an asymmetric viral strand specific displacement mechanism. Very little experimental data has accumulated on this final stage of DNA replication probably because very few unencapsidated viral strands can be found in infected cells. Progeny viral DNA is only detectable in close association with capsid proteins (Ward and Dadachanji, 1978) and it requires treatment with SDS and proteolytic enzymes to obtain pure intracellular viral DNA. Richards *et al.* (1977, 1978) also showed that the appearance of progeny DNA occurs simultaneously with encapsidation. This is consistent with the earlier suggestion that viral DNA packaging may drive strand displacement synthesis (Tattersall and Ward, 1976). It would be interesting to know whether the 5′ terminal protein present on RF molecules plays a role in this process.

As mentioned above, no evidence for encapsidated circular forms of viral genomes can be found. However, some evidence for intracellular circular forms has been reported. Electron microscopic examination of MVM infected cell lysates has demonstrated both circular forms and circular molecules with linear tails (Tattersall and Ward, 1976). By electron microscopic appearance both forms are duplex and of unit length. These molecules have been tentatively identified as parvovirus genomes because of their length and restriction fragment analysis. The origin of such structures has been considered more recently by Bratosin *et al.* (1979). They concluded that circular MVM forms are not replicative interme-

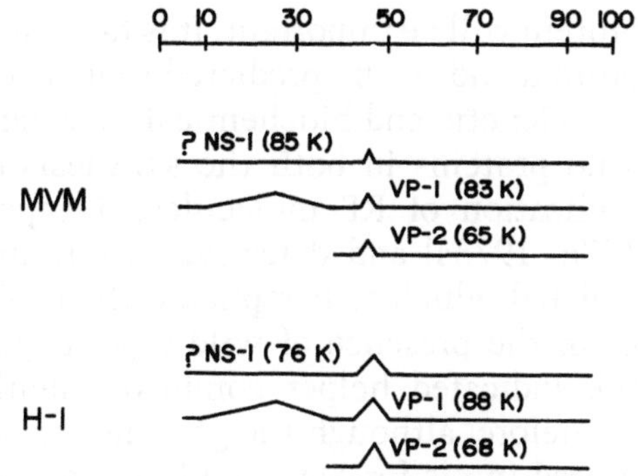

FIGURE 4. Transcriptional patterns for MVM and H-1 DNA. The approximate extent of viral transcripts on the genome is indicated as well as open reading frames (horizontal line), spliced regions (sloped lines), the encoded viron proteins (VP) and nonstructural viral proteins (NS), and their estimated molecular weights in parenthesis. The data for H-1 are redrawn from Rhode and Paradiso (1983) and for MVM it is unpublished material from D. C. Ward.

diates and are most likely products of replicative errors because no intermediate lengths were detectable and no single-stranded regions could be found. It still remains possible that such forms are intermediates important in transcription.

2. Involvement of Viral Proteins

Clearly, host proteins are involved in autonomous parvovirus DNA replication. The question of whether viral proteins play a role is unresolved. The genetic information necessary to code for autonomous parvovirus capsid protein (Chapter 7) would, in a classical manner, encumber almost the entire genome. However, the ability of DNA viruses to encode within the same DNA sequence information for multiple proteins by either splicing RNA or using alternate promoters in different reading frames opens the possibility that the autonomous parvoviruses may also code for nonstructural proteins. Very recently the nucleotide sequences of the H1 and MVM genomes have been determined (Rhode and Paradiso, 1983; Astell *et al.*, 1983) and the major viral mRNAs mapped to these sequences. The transcriptional pattern for both H1 and MVM are quite similar. There are essentially two overlapping open reading frames to which three major, polyadenylated, spliced RNAs map (Fig. 4). Relative to the question at hand, the relevant feature is that the messages for both capsid proteins map to the right-hand transcriptional unit. This leaves the identity of the protein(s) encoded within the left-hand open reading frame unknown. By *in vitro* translation of viral mRNA, two nonstructural viral proteins of 85,000 and 25,000 daltons are encoded in the left hand transcriptional unit of MVM DNA (D.C. Ward, Personal communication). Rhode and Paradiso (1983) used hybrid-arrest translation and calculated amino acid composition to assign one nonstructural viral protein of 76,000 daltons to this transcriptional unit in H-1 DNA. Interestingly, hamster anti-MVM serum precipitates this H-1 protein, suggesting that the 85,000 dalton MVM and 76,000 dalton H-1 proteins are antigenically related. Clearly, establishment of the function for this nonstructural viral protein will be important. It is tantalizing to hope that it is the 5′ terminal protein and/or the predicted sequence-specific DNA nicking enzyme.

Genetic and biochemical evidence also exist for the involvement of viral proteins in both the synthesis of viral DNA from RF and in the replication of RF molecules. Temperature sensitive mutants (Rhode, 1976, 1978a) and defective interfering particles (Chapter 6) have been isolated which will replicate their DNA at the permissive temperature or in the presence of wild-type helper, respectively. However, without the indicated helper condition, neither can undergo DNA synthesis. Therefore, although the genomes of both types contain all sequences required in *cis* for DNA replication, they cannot replicate apparently because they cannot synthesize viral proteins required for DNA synthesis. Additionally, Wolter *et al.* (1980) noted that 5-bromodeoxyuridine (BUdR) had a selective inhibitory effect on MVM DNA synthesis. If added before the onset of RF production (e.g., in G1 phase cells) incorporated BUdR severely restricted the amount of RF made, but if added after the beginning of RF production (early S1 phase) very little effect was seen. The authors suggested that the reason BUdR was effective in inhibiting viral DNA synthesis was because it inhibited transcription of monomer RF, the first conventional double-stranded transcription template made during the infection. If so, BUdR would be inhibitory only at this point in the infection because its incorporation may inhibit transcription of an early gene necessary for viral DNA replication. The results of all of these studies may now be understandable if the recently indentified nonstructural viral gene product discussed above proves to be involved in viral DNA synthesis, particularly in an initiation event.

B. *In Vitro* Replication

Development of subcellular systems capable of making or interconverting any of the intermediate replicative forms of the parvovirus DNAs *in vitro* would greatly enhance our understanding of viral DNA replication. As discussed above, all autonomous parvoviral DNA examined will serve as a primer template for bacterial DNA polymerases, whereas until recently it appeared that neither eukaryotic polymerase α, β, or γ would (Tattersall and Ward, 1982). Clearly, therefore, understanding the *in vivo* situation requires examination of a more complete set of host and viral components. Using a nuclear lysate system from KRV- or BPV-infected cells, Pritchard *et al.* (1981) detected virus specific DNA synthesis. The product was shown to be single-stranded viral DNA. The system was only able to elongate strands which had been initiated *in vivo* and could not initiate strand synthesis *de novo*. However, in these experiments the infected nuclei were isolated late in the infectious cycle when predominately progeny strand synthesis is occurring. Thus, the *in vitro* reaction at least appeared to be mimicking that part of the *in vivo* process which predominated late in the infected cell. Additionally, it should be noted that reinitiation of viral DNA synthesis has been generally the most elusive property of almost all *in vitro* viral systems. Pritchard *et al.* (1981) also reported that inhibitors of cellular DNA polymerase α were able to inhibit viral DNA synthesis. However, the presence of inhibitors of other DNA polymerases did not affect the production of viral DNA. The product in this case was identical in quantity, restriction enzyme analysis, and molecular weight to that made in the absence of inhibitors.

Recently, Kollek *et al.* (1982) have characterized an *in vitro* system using intact nuclei from H1 infected cells. Viral specific DNA systhesis was inhibited by both aphidicolin, an α polymerase inhibitor, and by 2′,3′-dideoxythymidine-5′-triphosphate (ddTTP), an inhibitor of eukaryotic DNA polymerases in the order $\gamma > \beta > \alpha$. Thus, both α polymerase, as before, and a ddTTP sensitive polymerase, probably polymerase γ are involved. A majority of the *in vitro* viral DNA product migrated on electrophoretic gels as duplex, monomer RF. In addition, smaller amounts of what may have been dimeric and oligomeric RF molecules as well as nascent RF molecules (e.g., partially duplex and partially single-stranded DNA) could be seen. No completely single-stranded viral DNA was detected. The kinetics of viral DNA synthesis in the presence of the inhibitors suggested that an early period of DNA synthesis carried out by polymerase α was followed by a period of DNA synthesis more sensitive to ddTTP. These results are consistent with a polymerase α involvement in strand displacement synthesis in RF molecules and with polymerase γ converting any displaced strands (and possibly also infecting viral DNA) back to duplex RF. It should be noted that the nuclei used in this study were infected at 18 hr postinfection, a time when progeny viral strand synthesis is minimal. Hence, as in the previous study, the *in vitro* products generally reflect the stage of the infection at which nuclei were isolated. Because such cell-free systems are making both accurate RF molecules and progeny strands further work offers promise of understanding the roles of specific proteins in DNA replication particularly in view of the fact that *in vitro* constructed deletion mutants are now available (see above).

III. MODEL FOR DNA REPLICATION

The existence of 3′ termini which could self-anneal, forming a hairpinned structure suitable for priming DNA synthesis, immediately suggested that the model of Cavalier-Smith (1976) for replicating the ends of eukaryotic chromosomal DNA might also apply to the replication of the linear, single-stranded DNA genomes of the parvoviruses (Ward and Tattersall, 1976; Berns and Hauswirth, 1978). For autonomous parvoviruses several recently appreciated structural features of the genome and intermediates in replication require some modification of the original model. However, the fundamental property that any DNA replication model must explain is the fact that only strands complementary to viral mRNA (i.e., the minus DNA strand) are encapsidated. This introduces a fundamental asymmetry to automonous DNA replication not present with defective parvoviruses which produce and encapsidate strands of both polarity with equal probability.

The parvovirus genome is sufficiently small that only one origin of replication should be necessary, and this has been confirmed by pulse-labeling experiments which place one origin at a position in duplex RF corresponding to the 5′ end of the viral strand (see Section II.A.1). This is consistent with the final step in the overall replication process being the displacement synthesis of viral strands from a duplex RF. However, the initial origin of DNA synthesis when the viral strand is first uncoated in the nucleus must be at the 3′-end of the viral strand so that a complementary plus strand can be synthesized to create the first monomer RF. This is consistent with the fact that the 3′ terminal sequence is palindromic allowing viral strands to fold over to serve as a primer for DNA synthesis. It is also analogous to the data obtained for AAV in which pulse labeling showed that *in vivo* DNA replication started nearly opposite the 3′ end of both template strands (Hauswirth and Berns, 1977). Thus, both the nucleotide sequence and the structure of the initial RF molecule agree with the Cavalier-Smith model which predicts that terminal sequences form hairpins and prime DNA synthesis, thus resulting in covalently linked viral and complementary strands. *In vitro* studies suggest that host DNA polymerases, most likely polymerase γ, function at this first stage in replication (Kollek *et al.*, 1982).

The next step, according to the original models, is binding of a sequence specific endonuclease to the hairpin structure and single strand nicking of the parental strand at a position opposite the hairpinned 3′ terminal nucleotide of the parental strand. This effects a transfer of parental 3′ nucleotides within the hairpin to the 5′ end of the complementary strand. Subsequent DNA synthesis from the new viral 3′ terminus using the hairpin sequence as the template will generate a new viral 3′ terminal sequence that is inverted related to the parental genome. This 3′ terminal sequence inversion is not supported by recent experimental data. Because autonomous parvovirus 3′ terminal sequences are not perfect palindromes, the indicated nicking and repair synthesis will lead to physically separable populations of 3′ terminal sequences. They will be distinguishable because of sequence differences in both the "T" structure at the hairpin turn around point and at the mismatched base positions (Fig. 2). Whereas experimental data clearly demonstrate the existence of two 3′ terminal sequences as a result of such an inversion for AAV (Chapter 1), both restriction enzyme analysis and direct nucleotide sequencing data show that the 3′ ends of autonomous parvovirus genomes are unique sequences (see Section I.A.). Therefore, the original models of autonomous parvovirus replication based on the Cavalier-Smith model proposal require modification. Finally, the existence and apparent involvement of a 5′ terminal protein in DNA replication also require its inclusion in an updated model.

Several recent proposals for autonomous parvovirus DNA replication account for these new features (Rhode and Klaassen, 1982; Astell *et al.*, 1983; Berns and Hauswirth, 1982). These models are basically similar and differ mainly in processes for which no experimental data yet exist. Common to all models is an initial conversion of the infecting parental DNA (V1 strand) (Fig. 5) to a monomer duplex RF by synthesis of a complementary strand (C1) initiated at the 3′-OH of the hairpinned viral strand (step 1, Fig. 5). Whether a 5′ terminal protein is added to this initial RF molecule is unknown, and a decision on this point may require a more precise understanding of its function. The next step is replication of the monomer, duplex RF by extension of the new complementary strand at its 3′-OH end (step 2). Priming may occur by 3′ hairpin formation from the extended end of the viral RF (Astell *et al.*, 1983; Berns and Hauswirth, 1982). However, Rhode and Klaassen (1982) argue that the stability of the terminal sequence in its fully extended conformation and their failure to find "rabbit-eared" H-1 RF molecules (i.e., complete, complementary terminal sequences in which both strands are hairpinned) demands an internal priming event and Okazaki fragment synthesis near the V1 strand 5′ end to initiate RF replication. However, evidence is lacking for Okazaki fragments in both H-1 RF molecules (Tseng *et al.*, 1979) as well as in AAV intermediates (W.W. Hauswirth unpublished observation). Additionally, terminal rabbit ear structures are expected to be extremely transient structures in an actively replicating system. Finally, overcoming the energy barrier to convert a fully duplex terminus into two hairpins may be one function of the 5′ terminal protein. It may act to destabilize a fully extended duplex terminal structure to facilitate hairpin formation. Therefore, 3′ hairpin initiation is presently the favored mechanism for initiating monomer RF replication (step 2, Fig. 5).

At this point, two possibilities exist; either DNA synthesis proceeds to the natural end of the template (5′ end of V1) creating a duplex, dimeric RF (Tattersall and Ward, 1976; Astell *et al.*, 1983) or, after the C1 strand has been completely copied, the 3′ end of the newly made V2 strand could hairpin and continue DNA synthesis using V2 as the template (Berns and Hauswirth, 1982). According to the latter proposal, two monomeric RF molecules would be generated, assuming the hairpin linking V2 to C1 were processed according to the Cavalier-Smith proposal. Contrary to this result, a significant amount of experimental data has identified dimeric RF molecules as replicative intermediates (although not necessarily obligatory intermediates) in the production of viral strands. Without further information, formation of a dimer RF will be assumed because this structure clearly exists in the intracellular replicating pool (step 3, Fig. 5).

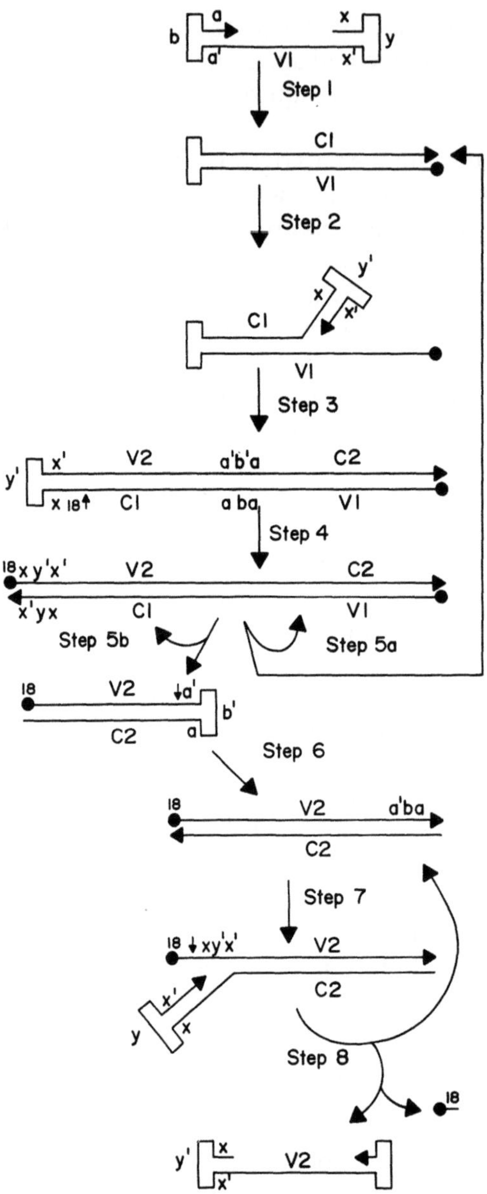

FIGURE 5. A model for replication of autonomous parvovirus DNA. The 3′ terminal palindrome on the infecting viral strand (V1) is shown as 3′aba′ and the 5′ terminal palindrome as 5′XYX′. Arrows on the DNA strands indicate the 3′ ends. Filled circles on DNA strands indicate covalently attached 5′ terminal protein. Details of the schematic model are discussed in the text.

Bearing in mind that the end result of the replication process must be a population of viral strands all having the same 3′ terminal sequence orientation and having one of two, mutually inverted 5′ terminal sequences, there are several alternative ways this may be achieved. Assuming that terminal sequences are preserved and replicated according to the Cavalier-Smith proposal, the critical intermediate structure from which both classes of viral strands can be directly made is very simply related to the infecting viral strand. It is a monomer, duplex RF containing a viral strand with an 3′ terminal sequence identical to the infecting viral strand and a complementary strand linked to it by a hairpin of the viral strand 5′ end. In all models previously proposed, this is the primary RF structure from which the two types of viral strands are made. A direct way to create this structure from the dimeric RF is to first repair the V2–C1 hairpin by nicking C1 at the 5′ end of the palindrome as indicated. Assuming a 3′-OH is produced at the nick site, replication here resynthesizes an extended viral 5′ terminus according to the Cavalier-Smith proposal (step 4). To account for the extra 18 nucleotides found at the 5′ end of viral strands in RF molecules (Astell *et al.*, 1983), this nick must occur 18 nucleotides outside of the 5′ end of V2 as indicated at step 4, Fig. 5. It may be that the extra 18 nucleotides are necessary for the terminal protein to attach to a 5′ end. If so, this dimer RF would be the first molecule to which a 5′ terminal protein would attach and it would contain only one protein at the V2 5′ end.

The dimeric RF now contains two complementary strand 3′ ends, either of which may hairpin to prime DNA synthesis precisely analogous to Step 2. Elongation of C2 (step 5a) recreates the dimer RF (with an inverted C2 3′ terminal sequence after Cavalier-Smith replication) and displaces a monomer RF identical to the original monomer RF. Elongation of C1 (step 5b) also recreates the dimer RF but now displaces a monomer RF having the critical structure, e.g., (V2)3′5′(C2). If the 5′ protein serves a function in hairpin formation to facilitate displacement synthesis and if it is only at the V2 5′ end because of the extra 18 nucleotides here, it is possible that displacement synthesis from the dimeric RF will be asymmetric, elongating only at C1 (step 5b). This may serve to eliminate the less directly productive elongation at C2 (step 5a).

Having created the pivotal intermediate, production of viral strands is now straightforward. Repair of the viral 3′ hairpin as indicated in step 6 and repeated rounds of synthesis using C2 as the template (steps 7 and 8) will displace viral strands all having a 3′ terminal sequence identical to the infecting strand (3′aba′) and alternately containing one of the two mutually inverted 5′ terminal sequences (5′XYX′ or 5′XY′X′).

The model requires two sequence-specific endonucleases. One recognizes the 5′ terminal hairpin of the viral strand and nicks only the complementary strand 18 nucleotides from the genomic 5′ end. The other is required so that the 3′ terminal hairpin of the viral strand is nicked only on the viral strand and only when the sequence is 3′ab′a′, not when it is 3′aba′. Therefore, a structural role in enzyme recognition is postulated for the 3′ palindromic "T" structure within the hairpin because it will present a slightly different secondary structure depending on which 3′ hairpin is present. The 5′ terminal protein is postulated to attach to the viral 5′ ends in RF molecules and aid in destabilizing the extended duplex terminal structure so that a complementary strand 3′ terminal hairpin can be created for priming DNA synthesis. Several other potential functions exist for the 5′ terminal protein. In its most efficient role, it may recognize the 5′ viral hairpin and simultaneously nick it appropriately and attach covalently to the 5′ viral end. In this way it may direct preferential displacement synthesis to make the key intermediate V2 (3′-5′) C1 as shown in step 5b, Fig. 5. It may also signal assembly of viral structural proteins to begin encapsidation of displaced viral strands (step 7). Finally, its attachment 18 nucleotides from the viral 5′ end provides a nucleotide sequence for its removal from nascent viral strands before complete encapsidation.

This model is derived from the Cavalier-Smith (1972) proposal and is a modification of the proposal of Tattersall and Ward (1976) specifically for autonomous parvovirus DNA replication. The major differences account for recent data concerning the orientation of the terminal nucleotide sequences and the existence of a 5′ terminal protein in RF molecules.

ACKNOWLEDGMENTS. The author would like to thank Dr. D.C. Ward and his co-workers for very generously providing a great deal of information concerning MVM DNA structure and replication prior to publication. He also wishes to thank Drs. N. Muzyczka and K.I. Berns for their critical reading of the manuscript.

REFERENCES

Astell, C.R., Smith, M., Chow, M.B., and Ward, D.C., 1979a, Structure of the 3′ hairpin termini of four rodent parvovirus genomes: Nucleotide sequence homology at origins of DNA replication, *Cell* **17**:691.

Astell, C.R., Smith, M., Chow, M.B., and Ward, D.C., 1979b, Sequence of the 3′-terminus of the genome from Kilham rat virus, a nondefective parvovirus, *Virology* **96**:669.

Astell, C.A., Thompson, M., Chow, M.B., and Ward, D.C., 1983, Structure and replication of minute virus of mice DNA, *Cold Spring Harbor Symp. Quant. Biol.* **47**, in press.

Ben-Asher, E., Bratosin, E.S., and Aloni, Y., 1982, Intracellular DNA of the parvovirus minute virus of mice is organized in a minichromosome structure, *J. Virol.* **41**:1044.

Berns, K.I., and Hauswirth, W.W., 1978, Parvovirus DNA structure and replication, in: *Replication of Mammalian Parvoviruses*, (D.C. Ward and P. Tattersall, eds.), pp. 13–32, Cold Spring Harbor Laboratory, Cold Spring Harbor, New York.

Bourguignon, G.J., Tattersall, P.J., and Ward, D.C., 1976. DNA of minute virus of mice: Self-priming, nonpermuted, single-stranded genome with a 5′-terminal hairpin duplex, *J. Virol.* **20**:290.

Bratosin, S., Laub, D., Tal, J., and Aloni, Y. 1979, Mechanism for circularization of linear DNAs: Circular parvovirus MVM DNA is formed by noose sliding in a lasso-like DNA structure, *Proc. Natl. Acad. Sci. USA* **76**:4289.

Chow, M.B., and Ward, D.C., 1978, Comparison of the terminal nucleotide structures in the DNA of nondefective parvoviruses, in: *Replication of Mammalian Parvoviruses*, (D.C. Ward and P. Tattersall, eds.), pp. 205–217, Cold Spring Harbor Laboratory, Cold Spring Harbor, New York.

Crawford, L.V., 1966, A minute virus of mice, *Virology* **29**:605.

Crawford, L.V., Follett, E.A.C., Burdon, M.G., and McGeoch, D.J., 1969, The DNA of a minute virus of mice, *J. Gen. Virol.* **4**:37.

Danna, K.J., and Nathans, D., 1972, Bidirectional replication of simian virus 40 DNA. *Proc. Natl. Acad. Sci. USA* **69**:309.

Faust, E.A., and Ward, D.C., 1979, Incomplete genomes of the parvovirus minute virus of mice: Selective conservation of genome termini, including the origin for DNA replication. *J. Virol.* **32**:276.

Hauswirth, W.W., and Berns, K.I., 1977, Origin and termination of adeno-associated virus DNA replication, *Virology* **79**:488.

Hayward, G.S., Bujard, H., and Gunther, M., 1978, Three distinct replicative forms of Kilham-rat virus DNA, in: *Replication of Mammalian Parvoviruses* (D.C. Ward and P. Tattersall, eds.), pp. 327–340, Cold Spring Harbor Laboratory, Cold Spring Harbor, New York.

Hirt, B., 1967, Selective extraction of polyoma DNA from mouse cell cultures, *J. Mol. Biol.* **26**:365.

Kollek, R., Tseng, B.T., and Goulian, M., 1982, DNA polymerase requirements for parvovirus H-1 DNA replication *in vitro*, *J. Virol.* **41**:982.

Li, A.T., Lavelle, G.C., and Tennant, R.W., 1978, DNA replication of Kilham rat virus: Characterization of intracellular forms of viral DNA extracted by guanidine hydrochloride, in *Replication of Mammalian Parvoviruses* (D.C. Ward and P. Tattersall, eds.), pp. 341–354, Cold Spring Harbor Laboratory, Cold Spring Harbor, New York.

Linser, P., Brunning, H., and Armentrout, R.W., 1977, Specific binding sites for a parvovirus, minute virus of mice, on cultured mouse cells, *J. Virol.* **24**:211.

Linser, P. Brunning, H., and Armentrout, R.W., 1979, Uptake of minute virus of mice into cultured rodent cells, *J. Virol.* **31**:537.

McGeoch, D.J., Crawford, L.V., and Follett, E.A.C., 1970. The DNAs of three parvoviruses. *J. Gen. Virol.* **6**:33.

McMaster, G.K., Beard, P., Engers, H.E., and Hirt, B., 1981. Characterization of an immunosuppressive parvovirus related to the minute virus of mice, *J. Virol.* **38**:317.

Muller, H.P., Gautschi, M., and Siegl, G., 1978. Defective particles of parvovirus LuIII, in: *Replication of Mammalian Parvoviruses* (D.C. Ward and P. Tattersall, eds.), pp. 231–240, Cold Spring Harbor Laboratory, Cold Spring Harbor, New York.

Nordheim, A., Tesser, P., Azorin, F., Kwon, Y.A., Moller, A., and Rich. A., 1982, Isolation of *Drosophila* proteins that bind selectively to left-handed z-DNA, *Proc. Natl. Acad. Sci. USA* **79**:7729.

Pritchard, C., Stout, E.R., and Bates, R.C., 1981, Replication of parvoviral DNA I. Characterization of a nuclear lysate system. *J. Virol.* **37**:352.

Revie, D., Tseng, B.Y., Grafstom, R.H., and Goulian, M., 1979, Covalent association of protein with replicative form DNA of parvovirus H-1, *Proc. Natl. Acad. Sci. USA* **76**:5539.

Rhode, S.L., 1973, Replication of the parvovirus H-1. Kinetics in a parasynchronous cell system, *J. Virol.* **11**:856.

Rhode, S.L., 1976, Replication process of the parvovirus H-1. V. Isolation and characterization of temperature-sensitive H-1 mutants defective in progeny DNA synthesis, *J. Virol.* **17**:659.

Rhode, S.L., 1977a, Replication process of the parvovirus H-1. VI. Characterization of a replication terminus of H-1 replicative-form DNA synthesis, *J. Virol.* **21**:694.

Rhode, S.L., 1977b, Replication process of the parvovirus H-1. IX. Physical mapping studies of the H-1 genome, *J. Virol.* **22**:446.

Rhode, S.L., 1978a, Replication process of the parvovirus H-1. X. Isolation of a mutant defective in replicative-form DNA replication, *J. Virol.* **25**:215.

Rhode, S.L., 1978b, Defective interfering particles of parvovirus H-1, *J. Virol.* **27**:347.

Rhode, S.L., and Klaassen, B., 1982, DNA sequence of the 5′-terminus containing the replication origin of parvovirus replicative form DNA, *J. Virol.* **41**:990.

Rhode, S.L., and Paradiso, P.R., 1983, Parvovirus genome: Nucleotide sequence of H-1 and mapping of its genes by hybrid arrest translation, *J. Virol.* **45**:173.

Richards, R.G., and Armentrout, R.W., 1979, Early events in parvovirus replication: Lack of integration by minute virus of mice into host cell DNA, *J. Virol.* **30**:397.

Richards, R., Linser, P., and Armentrout, R.W., 1977, Kinetics of assembly of a parvovirus, minute virus of mice, in synchronized rat brain cells, *J. Virol.* **22**:778.

Richards, R., Linser, P., and Armentrout, R.W., 1978, Maturation of minute virus of mice particles in synchronized rat brain cells, in: Replication of Mammalian Parvoviruses (D.C. Ward and P. Tattersall, eds.), pp. 447–458, Cold Spring Harbor Laboratory, Cold Spring Harbor, New York.

Salzman, L.A., 1977, Evidence for terminal S_1-nuclease-resistant regions on single stranded linear DNA, *Virology* **76**:454.

Salzman, L.A., and Fabisch, P., 1979, Nucleotide sequence of the self-priming 3′ terminus of the single stranded DNA extracted from the parvovirus KRV, *J. Virol.* **30**:946.

Salzman, L.A., White, W.L., and Kakefuda, T., 1971, Linear, single-stranded DNA isolated from Kilham rat virus, *J. Virol.* **7**:830.

Salzman, L.A., White, W.L., and McKerlie, L., 1972, Growth characteristics of Kilham rat virus and its effect on cellular macromolecular synthesis, *J. Virol.* **10**:573.

Salzman, L.A., Fabisch, P., Parr, R., Garon, C., and Wali, T., 1978, *In vitro* synthesis of double-stranded DNA from the Kilham rat virus single-stranded DNA genome, *J. Virol.* **27**:784.

Siegl, G., and Gautschi, M., 1973, The multiplication of parvovirus LuIII in a synchronized culture system. I. Optimum conditions for virus replication, *Arch. Gesamte Virusforsch.* **40**:119.

Siegl, G., and Gautschi, M., 1978, Purification and properties of replicative-form and replicative-intermediate DNA molecules of parvovirus LuIII, in: *Replication of Mammalian Parvoviruses* (D.C. Ward and P. Tattersall, eds.), pp. 315–325, Cold Spring Harbor Laboratory, Cold Spring Harbor, New York.

Singer, I.I., and Rhode, S.L., 1977, Replication process of the parvovirus H-1. VIII. Partial denaturation mapping and localization of the replication origin of H-1 replicative form DNA with electron microscopy, *J. Virol.* **21**:724.

Singer, I.I., and Rhode, S.L., 1978, Ultrastructural studies of H-1 parvovirus replication. VI. Simultaneous autoradiographic and immunochemical intranuclear localization of viral DNA synthesis and protein accumulation, *J. Virol.* **25**:349.

Tattersall, P., 1972, Replication of parvovirus MVM. I. Dependence of virus multiplication and plaque formation on cell growth, *J. Virol.* **10**:586.

Tattersall, P., 1978, Susceptibility to minute virus of mice as a function of host-cell differentiation, in: *Replication of Mammalian Parvoviruses* (D.C. Ward and P. Tattersall, eds.), pp. 131–149, Cold Spring Harbor Laboratory, Cold Spring Harbor, New York.

Tattersall, P., and Ward, D.C., 1976, Rolling hairpin model for replication of parvovirus and linear chromosomal DNA, *Nature (London)* **263**:106.

Tennant, R.W., and Hand, R.E., Jr., 1970, Requirement of cellular synthesis for Kilham rat virus replication, *Virology* **42**:1054.

Tennant, R.W., Layman, K.R., and Hand, R.E., Jr., 1969, Effect of cell physiological state on infection by rat virus, *J. Virol.* **4**:872.

Tse, Y.-C, Kirkegaard, K., and Wang, J.C., 1980, Covalent bonds between protein and DNA. Formation of phosphotyrosine linkage between certain DNA topoisomerases and DNA, *J. Biol. Chem.* **255**:5560.

Wang, A.H.J., Quigley, G.J., Kolpak, F.J., Crawford, J.L., van Boom, J.H., vander Marel, G., and Rich, A., 1979, Molecular structure of a lefthanded double helical DNA fragment at atomic resolution, *Nature (London)* **282**:680.

Ward, D.C., and Dadachanji, D.K., 1978. Replication of minute virus of mice DNA, in: *Replication of Mammalian Parvoviruses* (D.C. Ward and P. Tattersall, eds.), pp. 297–313, Cold Spring Harbor Laboratory, Cold Spring Harbor, New York.

Ward, P.C., and Tattersall, P.J., 1982, Minute viruses of mice, in: *The Mouse in Biomedical Research*, Vol. II, p. 313, Academic Press, Inc., New York.

Wolter, S., Richards, R., and Armentrout, R.W., 1980. Cell cycle-dependent replication of the DNA of minute virus of mice, a parvovirus, *Biochem. Biophys. Acta* **607**:420.

CHAPTER 5

Parvovirus Transcription

BARRIE J. CARTER, CATHERINE A. LAUGHLIN, AND CAROL J. MARCUS-SEKURA

I. INTRODUCTION

The last review devoted solely to parvovirus transcription (Carter, 1978) was written in December 1976. At that time most of the work on parvovirus transcription had been performed only on AAV. A single predominant 20 S AAV RNA species had been identified in infected cells. A similar KRV RNA had also been identified. On the basis of this knowledge and the limited coding potential of parvovirus genomes together with the known size of parvovirus capsid proteins it was thought that parvoviruses might be monogenic. Clearly, parvoviruses offered a useful system for studying eukaryotic transcription. This encouraged an increasing number of workers to study parvovirus transcription. Such increased interest, together with the application of recently developed and much more sophisticated techniques, rapidly cast doubt upon the apparent monogenicity of parvoviruses. It is now certain that parvoviruses have at least two genes. This change in the picture of parvovirus transcription occurred concurrently with dramatic changes in our view of eukaryotic transcription. This change resulted from two signal discoveries reported in 1977: the use of overlapping reading frames and the noncolinearity of many eukaryotic coding sequences. First, more than one protein may be coded from the same DNA sequence by using different reading frames (Sanger *et al.*, 1977). Second, RNA splicing in eukaryotic cells creates the potential for coding different proteins from single DNA

BARRIE J. CARTER, CATHERINE A. LAUGHLIN, AND CAROL J. MARCUS-SEKURA • Laboratory of Cell Biology and Genetics, National Institute for Arthritis, Diabetes and Digestive and Kidney Diseases, National Institutes of Health, Bethesda, Maryland 20205.

sequences as well as for proteins whose sequences are not colinear with the DNA. RNA splicing certainly does occur, and overlapping reading frames may occur, in parvoviruses.

It is now well established that *in vivo* transcription of most parvovirus genomes by RNA polymerase II produces a series of overlapping RNAs which have independent promoters. These RNAs are spliced. Furthermore, it is now clear that parvoviruses code for proteins other than their capsid protein. This idea was suggested by studies of AAV and H-1 DNA replication which indicated the existence of a *rep* protein (see Chapters 3 and 6). *In vitro* translation studies show that the AAV capsid proteins are coded by a single 2.3-kb mRNA transcribed from sequences in the right-hand half of the AAV DNA (Jay *et al.*, 1981). DNA sequence studies show that in the left-hand half of AAV DNA there exist additional open reading frames (Lusby and Berns, 1982). Recent studies also show that several parvoviruses including MVM, H-1, and PPV encode a nonstructural *rep* protein in the left-hand half of the genome (D. Ward, S. Rhode, and T. Molotor, personal communication).

Other recent progress has provided a good deal of information about the structure of eukaryotic promoters and enabled development of *in vitro* systems in which transcription is initiated faithfully at these promoters (Weil *et al.*, 1979; Manley *et al.*, 1980). These procedures are also being applied to studies of parvovirus transcription, since parvoviruses provide a useful system containing two to four transcription promoters of apparently differing strengths.

In this review we will focus on the work performed since 1977. Earlier work on parvovirus transcription is described in an earlier review (Carter, 1978). Before discussing particular studies on parvovirus transcription it is useful to summarize briefly the general procedures which have been developed in recent years for studying transcription of DNA genomes. We will then discuss the structure of parvovirus RNAs synthesized *in vivo* followed by a description of the state of *in vitro* transcription studies for several parvovirus genomes. We will then discuss post-transcriptional processing and modification of parvovirus RNAs. Finally we will discuss briefly the translation and coding potential of parvovirus RNAs.

II. METHODS FOR STUDY OF TRANSCRIPTION

Early studies to determine mRNA size employed sedimentation in sucrose gradients or electrophoresis in acrylamide or agarose gels. Accurate size estimates using these procedures were hampered because of effects of RNA conformation and aggregation. In attempts to overcome these problems various strongly denaturing solvents such as dimethyl sulfoxide, formamide, urea, or formaldehyde were introduced into the sedimentation or electrophoretic procedures. The problems of RNA conformation and aggregation in analysis of AAV RNA have been described

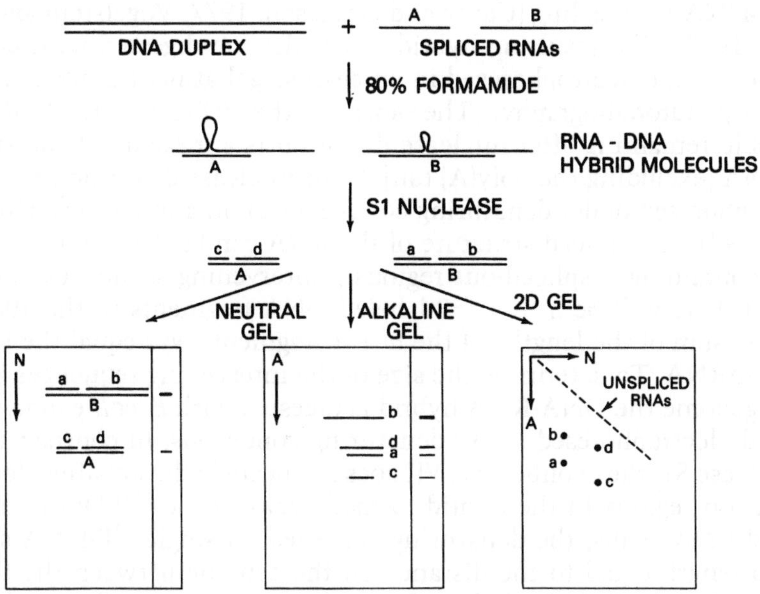

FIGURE 1. Schematic diagram of the Berk and Sharp nuclease-gel electrophoresis procedure for mapping the structure of DNA-RNA hybrids.

(Carter, 1974; Carter and Rose, 1974). More recently RNA aggregation and conformation have been much more effectively countered by performing gel electrophoresis either in the presence of millimolar concentrations of methyl mercury hydroxide (Bailey and Davidson, 1976) or after complete denaturation of the RNA by reaction with glyoxal (McMaster and Carmichael, 1977). The methyl mercury procedure has the advantage that biologically active mRNA can be recovered from the gel.

An alternative way to obtain fairly precise estimates of mRNA size (Khoury and Martin, 1972; Carter *et al.*, 1972) is to anneal the mRNA to its ^{32}P-labeled, complementary DNA template. The amount of DNA present in the resulting DNA–RNA hybrid is then measured by chromatography on hydroxylapatite (Gelb *et al.*, 1971) or by digestion with the single-strand specific endonuclease S1 (Sutton, 1971). This type of procedure is based on earlier procedures for analysis of DNA reassociation kinetics (Britten and Kohne, 1969) and can be used to calculate RNA abundance as well as size. In addition, annealing of RNA with specific DNA fragments or separated DNA strands allows mapping of the genome location of the RNA.

Based on these procedures Berk and Sharp (1977, 1978) developed a much more versatile system which can provide an accurate determination of the size and genome location of any given mRNA as well as determing the position of any intervening sequences. In this procedure, which is illustrated in Fig. 1, RNA is annealed with a denatured ^{32}P-DNA probe in conditions favoring RNA–DNA hybrid formation rather than

DNA–DNA reannealing (Casey and Davidson, 1977; Vogelstein and Gillespie, 1977). The resulting hybrid molecules are digested with endonuclease S1 and electrophoresed in an agarose gel at neutral pH and analyzed by autoradiography. The size of the ^{32}P-DNA–RNA duplex molecule remaining after nuclease digestion is a measure of the size of the RNA [excluding the poly(A) tail]. If the nuclease digestion product is electrophoresed under denaturing conditions at alkaline pH (McDonnell *et al.*, 1977) the spliced structure of the RNA can be determined. For an RNA containing n spliced-out regions ("intervening sequences" or "introns") there will be $n + 1$ ^{32}P-labeled DNA fragments in the alkaline gel. The sum of the lengths of the $n + 1$ fragments will equal the length of the mRNA. To determine the size of the intervening sequences in the DNA genome the DNA-RNA hybrid is digested with *E. coli* exonuclease VII and electrophoresed under denaturing conditions. In contrast to endonuclease S1, the exonuclease VII does not degrade the unpaired, looped-out intron regions in the hybrid formed between the ^{32}P-DNA and the spliced RNA. Thus, the denaturing gel reveals a single ^{32}P-DNA strand with a length equal to the distance on the genome between the 5′ and 3′ termini of the RNA including the intervening sequence [but excluding the poly(A) tail].

Several modifications of the Berk and Sharp procedure have been described. In some cases mung bean nuclease (a single-strand specific endonuclease) has proven to be more reliable than endonuclease S1 (D. Ward, personal communication). Favoloro *et al.* (1980), performed Berk and Sharp mapping using a 2-D gel system in which the S1 digestion products are first electrophoresed at neutral pH and then at alkaline pH in a direction perpendicular to the first dimension. In this procedure (see Fig. 1) unspliced RNAs yield digestion products that migrate to positions (depending on their size) along a diagonal whereas spliced RNAs yield two or more DNA fragments that migrate ahead of the diagonal in the alkaline dimension. Favoloro *et al.* (1980) further increased the sensitivity of detection by transferring (blotting) the digestion products from the 2-D gel onto nitrocellulose paper and annealing with very high specific activity nick-translated ^{32}P-DNA probes.

An additional refinement of the Berk and Sharp procedure employs a 5′-terminal ^{32}P-labeled DNA probe which partially overlaps the RNA (Weaver and Weissman, 1979). Nuclease digestion yields a ^{32}P-DNA fragment which is a precise measure of the distance between the 5′-terminus of the DNA probe and the 5′-terminus of the RNA. If this digestion product is analyzed on a sequencing gel the 5′ end of the RNA can be mapped on the DNA genome with an accuracy of 1 to 2 nucleotides. Similarly the 3′ end of RNA can be mapped using 3′ terminally labeled probes (Green and Roeder, 1980a).

Additional precise information on RNA map locations and structure can be obtained by a variety of sequencing procedures including direct sequencing of RNA or cloned cDNA copies of RNA. Alternatively, primer extension methods can be employed in which a small DNA fragment is annealed to RNA. The DNA primer is first extended using the RNA as a template for reverse transcriptase. The extended primer can then be readily sequenced.

Another important tool for the analysis of transcription has been the development of procedures for transfer ("blotting") of nucleic acids from gels onto nitrocellulose or derivatized paper sheets. The first such procedure, "Southern" blotting, was developed for transfer of DNA fragments from gels to nitrocellulose paper (Southern, 1975). The blotted DNA can then be analyzed by hybridizing with radioactive DNA or RNA probes. In a variation of this technique, "Northern" blotting, RNA or denatured DNA is transferred from gels to derivatized paper such that the nucleic acid is covalently bound (Alwine *et al.*, 1980). The advantage of Northern blotting is that the transferred RNA or DNA is covalently bound to the paper and can be hybridized successively with several different radioactive probes.

Together with these blotting procedures, the ease of preparation of cloned DNA probes and the labeling of these probes to very high specific activity using the *in vitro* nick translation reaction of *E. coli* DNA polymerase (Ribgy *et al.*, 1977) have increased the sensitivity of RNA detection and mapping by several orders of magnitude. For instance, the highest specific activity ^{32}P-DNA obtainable by *in vivo* labeling of viral genomes is about 10^6 counts/min per μg whereas *in vitro* nick-translation readily yields specific activities of $1\text{–}2 \times 10^8$ counts/min per μg.

The structure of mRNA can also be studied by electron microscopy, generally by direct observation of RNA–DNA heteroduplexes. Electron microscopy has some disadvantages. It is a very laborious technique when analysing RNAs with low abundance. Secondly, in heteroduplexes, regions of unpaired bases less than about 100 nucleotides long generally cannot be seen. An example of such complications occurred in an electron microscope study of MVM RNA (Tal *et al.*, 1979) in which only spliced MVM RNAs were scored. Thus, the most abundant MVM mRNA was not analyzed because this contains only one intron of less than 50 bases (D. Ward, personal communication). However, electron microscopy is especially useful to analyze mRNAs having many, or very large, introns, Indeed, RNA splicing was first discovered by electron microscopy (Berget *et al.*, 1977). Since a variety of artifacts can also occur in the Berk and Sharp procedure, it is generally best, whenever possible, to analyse mRNA both by electron microscopy and electrophoretic procedures.

III. RNA SYNTHESIS AND THE VIRUS GROWTH CYCLE

The collected evidence from many studies indicates that for all parvoviruses, synthesis (or accumulation) of RNA is detected only at the onset of, or after, virus DNA synthesis. Thus, unlike most viruses, parvoviruses appear to have no early RNA. This presumably reflects the single strandedness of the infecting parvovirus genome since it is generally believed that transcription *in vivo* requires a double strand template. Specific evidence for this hypothesis is sparse. In one case, for a tRNA transcribed by RNA polymerase III, it has been shown that single-stranded DNA, when injected into *Xenopus* oocytes, is not an *in vivo* template for transcription (Cortese *et al.*, 1980).

In KB cell suspension cultures which were infected with AAV and a helper adenovirus (Carter *et al.*, 1973), both AAV and Ad DNA synthesis were detected at 6 to 7 hr whereas AAV RNA synthesis was first detected at 9 to 10 hr after infection. This lag between the onset of AAV DNA and RNA synthesis was eliminated if the cells were infected with the helper adenovirus for 10 hr prior to infection with AAV. Similar kinetics of AAV DNA and RNA synthesis were observed when a DNA-negative adenovirus mutant was used as the helper at the nonpermissive temperature (Straus *et al.*, 1976). The presence of FUDR from 3 hr prior to infection prevented synthesis of AAV RNA and adenovirus "late" RNA but not adenovirus "early" RNA. This observation, together with the evidence that some DNA-negative mutants of adenovirus (Ito and Suzuki, 1970; Ishibashi and Ito, 1971; Mayor and Ratner, 1972; Handa *et al.*, 1975; Straus *et al.*, 1976; Myers *et al.*, 1980) or herpesviruses (Drake *et al.*, 1974) will help AAV RNA synthesis, suggests that AAV RNA synthesis is directly dependent upon synthesis of AAV DNA. However, continued AAV RNA synthesis does not require continuing AAV DNA synthesis (see Section VI below). Thus, AAV has no "early" RNA species analogous to those of most viruses. There has been one recent report (Ostrove *et al.*, 1981) that infection of HT14B cells (a line of hamster embryo fibroblasts transformed by human adenovirus serotype 5) with AAV resulted in synthesis of AAV-specific RNA in the apparent absence of AAV DNA synthesis. This is discussed elsewhere (see Chapter 3).

The explanation of the lag between the onset of AAV DNA and AAV RNA synthesis remains unclear. It was suggested that the lag reflected the action of at least two independent adenovirus functions acting at the level of AAV DNA synthesis and RNA synthesis, respectively (Carter *et al.*, 1973). It now appears that there are adenovirus helper functions for AAV which are required for AAV DNA synthesis or which may affect AAV RNA splicing. There is some suggestive evidence for a helper function that may under some conditions stimulate transcription (see Chapter 3). The lag may result simply from a delay in availability of transcribable templates for AAV RNA synthesis rather than intervention of another adenovirus helper function. In a preinfection with adenovirus, the initial rounds of AAV DNA synthesis might occur more rapidly. Other possibilities might include functional modifications of RNA polymerase molecules actively engaged in RNA synthesis following infection (Weinnman *et al.*, 1976).

Salzman and Redler (1974) studied Kilham rat virus (KRV) transcription in randomly growing rat nephroma cells using the KRV viral (v) strand DNA as a hybridization probe. They detected KRV-specific RNA complementary to the vDNA as early as 2 hr after infection. However, the temporal relationship between KRV RNA synthesis and DNA synthesis in rat nephroma cells was not clear because KRV progeny DNA synthesis, as measured by FUDR inhibition of infectious virus production, did not begin till 8 hr post infection (Salzman *et al.*, 1972). More recent experiments using synchronized RN cells indicate that KRV RNA accumulation begins at about the same time as KRV DNA synthesis at about 8 hr after infection (L. Salzman, personal communication). Similarly, Patton *et al.* (1979) reported that in synchronized primary bovine fetal spleen cells BPV RNA synthesis began by 8 hr after infection.

Synthesis of H-1 RNA has been examined in RL5E cells (rat cells transformed by murine sarcoma virus) growing in suspension culture (Green *et al.*, 1979). Similar H-1 RNA species were observed at both 10 hr and 24 hr after infection but the level of these RNAs at 10 hr was very low.

The absence of an early function, together with the apparent lack of dependence on a specific S-phase host cell function like the nondefective parvoviruses (Tennant and Hand, 1970; Tattersall, 1972; Siegl and Gautschi, 1973; Rhode, 1973) provides a teleological rationale for the defectiveness of AAV. Indeed it appears that early functions required by AAV are supplied by certain early functions of adenovirus or analogous functions of herpesviruses (see Chapter 3). Similar reasoning suggests that transcription of autonomous parvoviruses may be dependent only upon host-cell functions. Indeed, Green *et al.* (1979) reported that normal H-1 RNA synthesis and processing occurred even when protein synthesis was inhibited by cycloheximide added at a concentration of 25 μg/ml at 5 hr after infection (i.e., prior to the onset of viral RNA synthesis).

IV. DNA GENOME ORIENTATION AND RNA STRAND SPECIFICITY

Preparative separation of the complementary strands of AAV DNA was achieved (Berns and Rose, 1970) by substitution of AAV DNA thymidine residues with BUdR which resulted in a sufficient density difference in a CsCl density gradient at neutral pH. The strand which then banded at a heavier density in CsCl was defined as the "minus" strand because it is the strand that is transcribed into mRNA. The lighter strand was defined as the "plus" strand. By taking advantage of this finding, together with the ability to obtain specific fragments of DNA with bacterial restriction endonucleases, the AAV DNA genome was oriented with respect to DNA strand polarity (Carter and Khoury, 1975; Carter *et al.*, 1975). By convention (Armentrout *et al.*, 1978) the AAV map is drawn with the 3′ terminus of the minus strand at the left. Thus transcription using the minus strand as template proceeds from left to right. For the autonomous parvoviruses, the map is drawn in an analogous fashion with the 3′ terminus of the viral (v) strand at the left. This strand (the one packaged into virions) is also the template for *in vivo* RNA synthesis.

A variety of studies have demonstrated that parvovirus RNA in infected cells is complementary to the minus or viral strand only. Repeated attempts have failed to demonstrate any RNA that hybridizes to the plus strand. This was first demonstrated by Rose and Koczot (1971) who purified *in vivo* synthesized AAV ^{32}P-RNA by filter hybridization to AAV DNA and then annealed this RNA in solution with an excess of purified plus or minus strands of [^{3}H]-BUdR AAV DNA. The resulting DNA–RNA hybrid was centrifuged to equilibrium in a CsCl density gradient. This experiment showed that the AAV RNA annealed only to the DNA minus strand. Also, the base composition of the purified AAV RNA was complementary to that of the AAV DNA minus strand. Additional experiments employing annealing in solution of trace amounts of high specific activity ^{32}P-labeled AAV DNA with an excess or saturating amount of unlabeled AAV RNA confirmed this conclusion (Carter *et al.*, 1972; Carter, 1976; Jay *et al.*, 1979). In addition, viral transcription complexes isolated from infected cells using the sarkosyl lysis procedure of Gariglio and Mousset (1975) synthesized *in vitro* AAV-specific RNA that was complementary to the AAV DNA minus strand only (Jay *et al.*, 1978; C.A. Laughlin, unpublished).

More recently it was shown that all the AAV RNAs observed in gel electrophoresis analysis could be accounted for by RNAs observed by electron microscopy to anneal to the DNA minus strand (Laughlin *et al.*, 1979a). Similar or analogous observations have also been made for H-1 RNA (Green *et al.*, 1979), MVM (Tal *et al.*, 1979; D. Dadachanji and D. Ward, personal communication) and KRV (L. Salzman, personal communication).

V. *IN VIVO* TRANSCRIPTION. RNA STRUCTURE AND GENOME LOCATION

The earliest experiments to measure the size of parvovirus RNAs were performed by extracting ^{3}H-labeled RNA from the cytoplasm of infected cells and sedimenting it through sucrose gradients containing dimethylsulfoxide (Carter and Rose, 1972). The viral-specific ^{3}H-RNA was detected by annealing to virus DNA immobilized on nitorcellulose filters. These experiments revealed in the cytoplasmic RNA a single predominant peak of AAV specific 20 S RNA and a heterogenous population of smaller AAV RNAs ranging in size from 4 S to 18 S (Carter and Rose, 1972, 1974). Similar observations were made for RNA detected in the same way after fractionation on composite agarose-acrylamide gels

(Carter, 1974; Carter and Rose, 1974). Analysis of either KRV-specific or H-1 specific cytoplasmic RNA by sedimentation and filter hybridization showed results analogous to the AAV experiment (Salzman and Redler, 1974; Green *et al.*, 1979). In each case a major peak of 18 to 20S parvovirus RNA was detected.

Use of much more sophisticated techniques has now shown that the predominant cytoplasmic 18–20 S RNA detected on gradients represents the majority species of 2.3-kb RNA for AAV (Laughlin *et al.*, 1979a; Green and Roeder, 1980a) or approximately 2.8 to 3.0 kb for autonomous parvoviruses such as H-1 (Green *et al.*, 1979), MVM (Dadachanji and Ward, personal communication), or KRV (L. Salzman, personal communication). However, in each case there are additional different sized RNAs present in lower abundance. In the earlier studies, because of the limitation of the techniques available, these less abundant RNAs either could not be detected or could not be readily distinguished from residual aggregation artifacts.

We will first consider the structure and genome location of *in vivo* AAV RNAs and then discuss the *in vivo* RNAs transcribed from several autonomous parvovirus genomes. It is now becoming apparent that both the defective and autonomous parvovirus genomes generally have a very similar transcription pattern. In each case only the minus or viral DNA strand is transcribed. In each case the most abundant RNA is transcribed from a promoter located between 38.0 and 40.0 on the genome map while less abundant RNAs are transcribed from promoters near map position 5.0 at the left-hand end of the genome. All of the RNAs which have been accurately mapped terminate [presumably at a poly(A) addition site] at about map position 96.0. There are certain differences in detail such as the location of introns and the presence or absence of an additional promoter at map position 19.0.

In discussing map locations of RNAs or DNA restriction fragments we will use the conventions described by Armentrout *et al.* (1978). In addition, we will describe individual promoter sites for individual RNAs using the convention p_n where p indicates promoter and the subscript n indicates the map position of the 5′ end of RNA. Thus the promoter site for the 2.3-kb AAV RNA is described as $p_{40.0}$. Individual RNAs will be described according to their size in kilobases (kb) as measured by Berk and Sharp mapping or electron microscopy. This size designation does not include the size of the poly(A) tail which for AAV RNA is 0.15 to 0.25 kb (Carter, 1976). In some previous publications (e.g., Laughlin *et al.*, 1979a), various RNAs were designated according to their size calculated as a percentage of the viral genome.

A. Structure of AAV RNA Synthesized *in Vivo*

All of the studies on in vivo AAV RNA have employed AAV serotype 2 except for one report on AAV 4 by Muster *et al.* (1980). On the basis

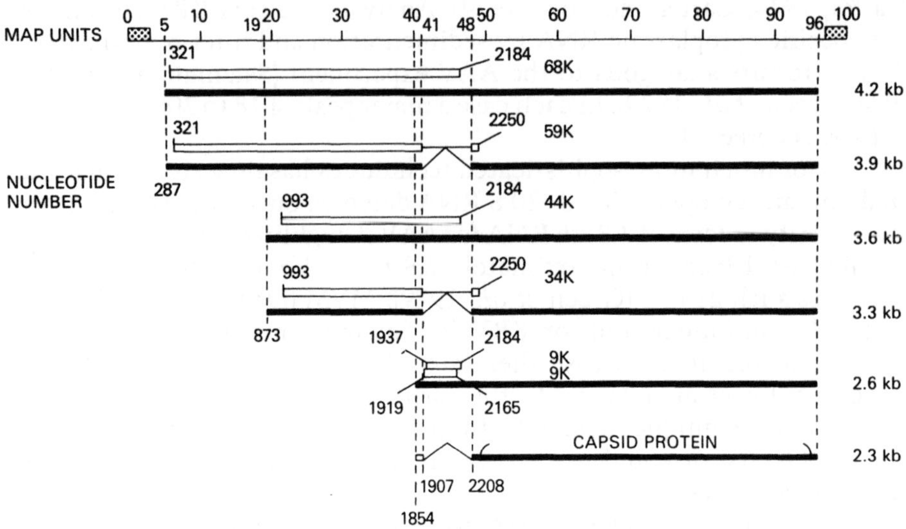

FIGURE 2. Diagram of poly(A)-containing AAV RNA species present in infected cells. The upper line represents the AAV DNA minus strand with polarity as shown and divided into 100 map units. One map unit is approximately 47 nucleotides. Stippled boxes indicate the terminal repetition in AAV DNA. The horizontal lines indicate AAV RNAs with the size in kilobases (kb) indicated at the right. All the RNAs are coterminal at position 96.0. The carets indicate the spliced out region (map position 41/48). The 5′ termini are at map positions 40.0, 19.0, and 5.0, respectively. All RNAs are complementary to the DNA minus strand. The location of 5′ start points and splice junctions is also indicated by the nucleotide number using the sequence data of Green and Roeder (1980b) and Lusby and Berns (1982) and the nomenclature of Lusby and Berns (1982) beginning from the left-hand end. Potential open reading frames on each RNA transcript are indicated by open boxes.

of inhibition by α-amanitin of synthesis of AAV RNA in *in vitro* transcription complexes (Jay *et al.*, 1978) or in isolated cell nuclei (Bloom and Rose, 1978) AAV RNA is synthesized by RNA polymerase II as in all eukaryotic mRNA (Roeder, 1976).

The structure and genome locations of the AAV2 RNAs synthesized in KB cells using human adenovirus types 2 or 5 as the helper have been precisely mapped by a variety of procedures including electron microscopy (Laughlin *et al.*, 1979a), Berk and Sharp analysis and Northern blotting procedures, or primer extension sequencing procedures (Laughlin *et al.*, 1979a; Green *et al.*, 1980; Green and Roeder, 1980a,b; Marcus *et al.*, 1981; Lusby and Berns, 1982). Most of these studies are in excellent agreement except for certain less abundant AAV RNAs which were either not detected or not analyzed in some studies.

The structures of the *in vivo* AAV RNAs which have been clearly described and characterized are summarized in Fig. 2. These RNAs comprise three overlapping families transcribed from promoters at map positions 5.0 (4.2- and 3.9-kb RNA), 19.0 (3.6- and 3.3-kb RNA), and 40.0 (2.6- and 2.3-kb RNA), respectively, but which are all coterminal at their

3′ ends at map position 96. For each of the three families both unspliced and spliced counterparts have been described. For each family the spliced RNA has a single detectable intron of about 350 kb between map positions 41 and 49. The evidence to date suggests, but does not rigorously prove, that the locations of the splice are the same for all RNAs from all three families.

1. Electron Microscopy of AAV DNA–RNA Hybrids

Laughlin *et al.* (1979a) fractionated in a neutral sucrose gradient poly(A)-containing, cytoplasmic RNA isolated from cells at 20 hr after infection. RNA from individual gradient fractions was annealed with the AAV minus strand DNA and examined by electron microscopy. When this DNA was annealed with RNA from the 18–20 S region of the gradient one predominant type of DNA–RNA duplex was seen (Fig. 3E). These structures are formed by annealing of the spliced 2.3-kb RNA containing a 2.25-kb main body, a 0.3-kb intron equivalent to the loop of intervening DNA sequences seen in the photograph, a small 0.05-kb 5′ leader and a presumptive poly(A) tail (indicated by the bifurcation near one end of the heteroduplex). Also, in the same region of the gradient, unspliced molecules (56% of genome length) equivalent to the 2.6-kb RNA were found (Fig. 3G). A similar analysis of the 22–24S RNA revealed analogous molecules (Fig. 3B) which contained a longer 5′ leader region equivalent to the spliced 3.3- or 3.9-kb RNAs. Also, in some RNA preparations the 20 S region contained unspliced molecules equivalent to the 3.6- and 4.2-kb RNAs such as the example shown in Fig. 3H. The term, "leader," is used merely for convenience and is not intended to imply whether or not this region is translated. Most probably for the 2.3-kb RNA it is not translated whereas for the 3.3- and 3.9-kb RNAs it may be translated if these RNAs are functional messages (see below).

The map orientation of the AAV RNA species was determined (Laughlin *et al.*, 1979a) by annealing the DNA–RNA hybrids containing the 2.3-kb RNA with the denatured DNA fragments, *Sal* 0/31 or *Pst/Sal* 11/31. Either of these fragments annealed to the sequences of the AAV minus strand no covered by RNA (Figs. 3C and 3D). This placed the main body of the 2.3-kb RNA (and by analogy that of the other RNA species also) between coordinates 49/96.

Measurement of the DNA–RNA duplexes of the type shown in Fig. 3E yielded map coordinates for the 2.3-kb RNA of 40.1 ± 2.7 and 41.5 ± 2.6 for the 5′ leader and 49.8 ± 2.5 and 95.9 ± 0.8 for the main body, respectively. Because of the lower abundance of the larger AAV RNA species, statistically precise measurements were more difficult to obtain in the electron microscope (Laughlin *et al.*, 1979a). However, all molecules had a 3′ end at map position 96 and measurements of individual molecules were consistent with the schematic diagram in Fig. 2. Analysis

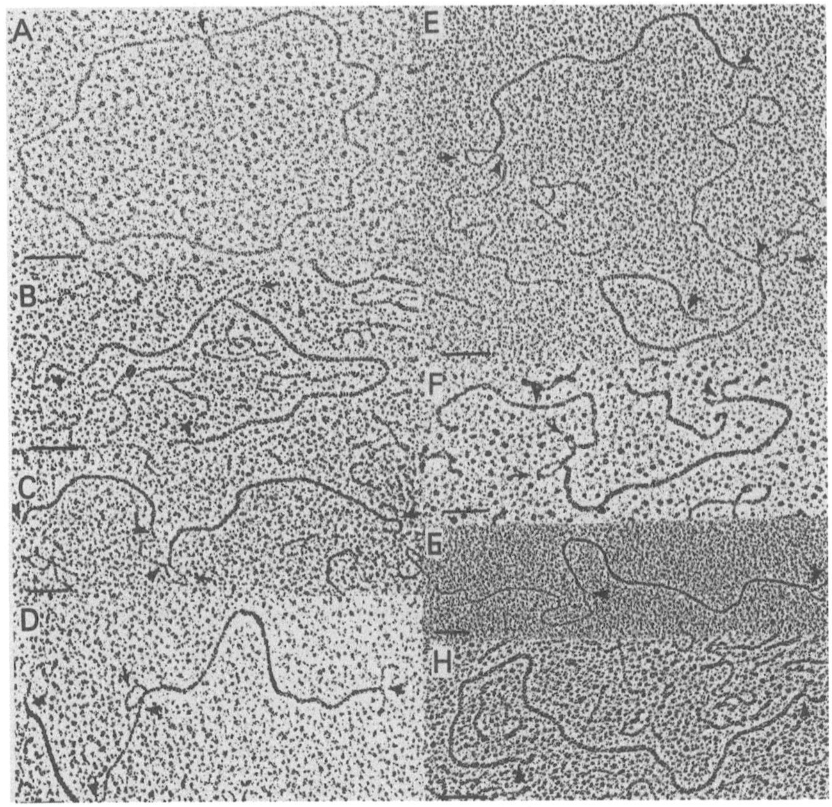

FIGURE 3. Electron micrographs of AAV RNA hybridized to template DNA. (A) The minus strand of AAV DNA. The arrow points to terminal inverted repeat sequences annealed to form a circle. The bar in this and subsequent panels indicates 0.1 μm. In panels (B) through (H) AAV RNA is seen hybridized to linear minus strand AAV DNA. Arrows point to loops of intervening DNA sequences. Double-stranded regions are between arrow heads. (B) 3.6-kb RNA. (C) 2.3-kb RNA. The molecule is oriented in rightward 5′ to 3′ polarity by a Sal 0/31 DNA fragment seen annealed at the left. At the right, the poly(A) tail of RNA and a short end of unhybridized template DNA may be discerned. (D) 2.3 kb RNA oriented by a *Pst/Sal* 11/31 fragment. (E) Two molecules of 2.3 kb RNA. (F) A rare RNA generating two loops of intervening DNA. The smaller of the two loops is the rare one. The larger loop represents the normal AAV intron. (G) 2.6-kb RNA. (H) 4.2-kb RNA. Reprinted from Laughlin *et al.* (1979a).

of total poly(A)-containing, cytoplasmic RNA Laughlin *et al.* (1979), revealed no other predominant RNA species.

The electron microscopy studies also provided a measure of the relative abundance of the AAV RNAs. The 2.3-kb spliced RNA was always the most abundant cytoplasmic species. In one particular preparation of 200 DNA–RNA hybrids from the total cytoplasmic poly(A)-containing RNA, 55% were derived from the 2.3-kb RNA, 32% from the 2.6-kb RNA, 7% from the spliced 3.3- and 3.9-kb RNAs, and 6% from the unspliced

3.6- and 4.2-kb RNAs. The relative abundance of the spliced RNAs was consistent in different RNA preparations but that of the unspliced RNAs varied and in some preparations was greatly diminished. This may indicate that some of the unspliced RNAs appear in the cytoplasm because of nuclear leakage (see below).

2. Electrophoretic Analysis of AAV *in Vivo* RNA

A variety of electrophoretic procedures have confirmed the electron microscope analysis and provided clear evidence for 6 *in vivo* synthesized AAV RNAs having sizes of 4.2, 3.9, 3.6, 3.3, 2.6, and 2.3 kb, respectively (Laughlin *et al.*, 1979a; Green and Roeder, 1980a; Marcus *et al.*, 1981).

Laughlin *et al.* (1979a) fractionated cytoplasmic poly(A)-containing AAV RNA in sucrose gradients and analyzed RNA from individual fractions by Berk and Sharp mapping with AAV ^{32}P-DNA as described in Fig. 4A. When the endonuclease digestion products were analyzed on a neutral gel the 18–20 S AAV RNA yielded a major species of 2.3 kb and a minor species of 2.6 kb. In addition to the two RNAs, the 20–24S fraction contained less abundant AAV RNAs of 3.3, 3.6, 3.9, and 4.2 kb. All six products were seen when whole cytoplasmic RNA was analyzed. Six AAV RNA species are also observed when cytoplasmic RNA is treated with glyoxal and electrophoresed in neutral gels and the AAV RNA detected by Northern blotting (Fig. 4B Marcus *et al.*, 1981). The size of the six AAV RNAs observed in the blotting analysis are the same as those observed in the Berk and Sharp analysis (Fig. 4A) when due allowance is made for the size of the poly(A) tail (which is removed in the latter procedure).

Analysis of the digestion product of a Berk and Sharp analysis either by electrophoresis in denaturing gels at alkaline pH (Laughlin *et al.*, 1979a) or by electrophoresis at neutral pH after denaturation with glyoxal (Marcus *et al.*, 1981) revealed that both the 3.3- and 3.9-kb RNAs are spliced (see Fig. 4B). In each case denaturation results in loss of the 3.3- and 3.9-kb ^{32}P-DNA bands and appearance of new bands at 1.0 and 1.6 kb, respectively (see Fig. 4B). In each case the bands corresponding to the main bodies of the 3.3- and 3.9-kb RNAs migrated at 2.25 kb and were not separated from the main body of the 2.3-kb RNA. This separation was achieved when the products of the Berk and Sharp analysis of poly (A)-containing cytoplasmic AAV RNA were analyzed by electropheresis in 2-D gels first at neutral, and then at alkaline pH (Marcus *et al.*, 1981). As shown in Fig. 5, the 3.9- and 3.3-kb spliced AAV RNAs each yield main body digestion products (spots 3.9′ and 3.3′, respectively) having sizes of about 2.3 kb. However, the 3.9- and 3.3-kb RNAs, yielded leader fragments of 1.6 kb (spot 3.9″) and 1.0 kb (spot 3.3″), respectively. It should be noted that in the experiments of Figs. 4B and 5 the Berk and Sharp analyses were blotted and hybridized with nick-translated AAV ^{32}P-DNA probes that have specific activities some 50- to 100-fold higher

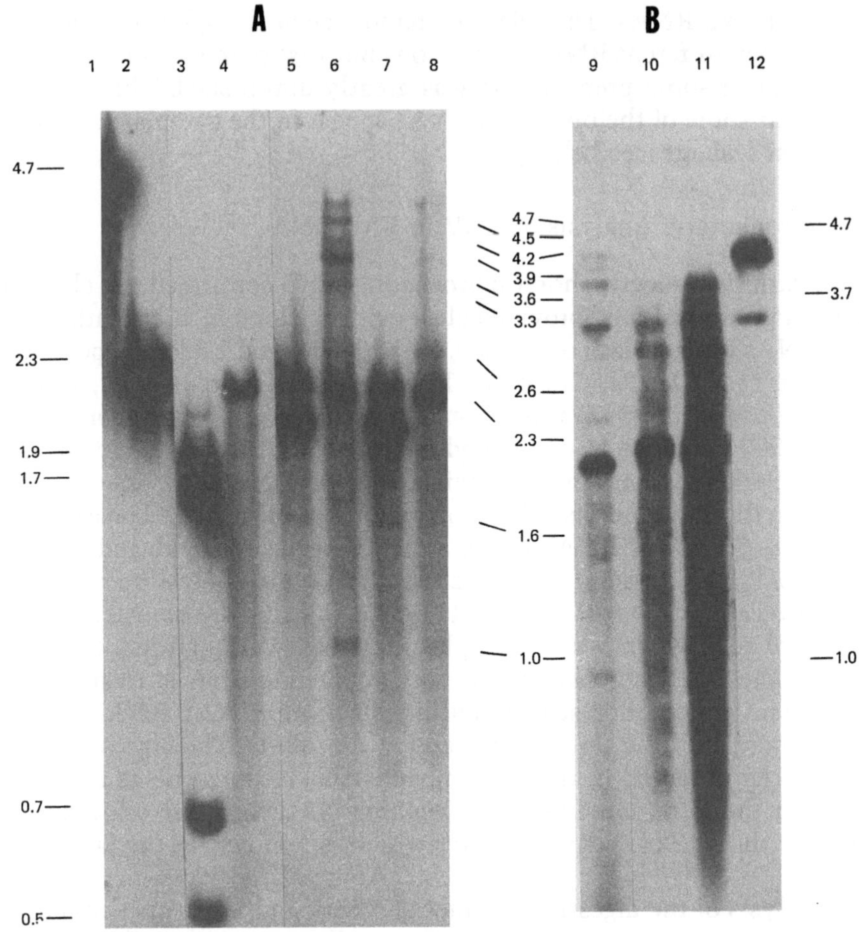

FIGURE 4. Electrophoretic analysis of AAV RNA. Cytoplasmic, poly(A)-containing AAV RNA was isolated 20 hr after coinfection of KB cells using adenovirus type 2 as a helper. RNA was electrophoresed in 1.4% agarose gels after subjecting it to either the Berk and Sharp procedure or glyoxal denaturation. (A) Berk and Sharp analysis of AAV RNA using *in vivo* labeled DNA probes. The cytoplasmic, poly(A)-containing RNA from infected cells was fractionated in a sucrose gradient. RNA from the 18S or 20S regions of the gradient was hybridized with heat-denatured intact AAV ^{32}P-DNA or the AAV *Pst* 42/92 ^{32}P-DNA fragment. Specific activity of the DNA was 0.5–1.0 × 10^6 counts/min/μg. The hybridization mixtures were then digested with endodeoxyribonuclease S1 and electrophoresed at neutral pH. The hybridization reactions contained 20S RNA annealed with intact AAV DNA (track 6) or the *Pst*-42/91 DNA (track 5) or 18S RNA annealed with intact AAV DNA (track 8) or the *Pst*-42/91 fragment (track 7). The ^{32} P-labeled DNA-RNA hybrid digestion products were revealed by autoradiography of the gel. The sizes of the digestion products in kilobases (kb) are indicated at the right of panel A. Molecular weight markers were provided from ^{32}P-AAV DNA as follows: intact AAV DNA (track 1), *Pst* 42/91 (track 2), Hae II digest (track 3). Fragment sizes are indicated in kilobases (kb). The control hybridization reaction (track 4) contained yeast RNA annealed with intact AAV ^{32}P-DNA. (B) Direct analysis of *in vivo* synthesized AAV RNA using glyoxal denaturation and Northern blotting with *in vitro*, nick-translated DNA probes (specific activity: 5 × 10^7 counts/min/μg). An RNA preparation similar to that used in panel A was denatured with glyoxal, electrophoresed on a 1.4% agarose gel, blotted onto DBM paper, and hybridized with nick-translated AAV ^{32}P-DNA

$(5 \times 10^7$ to 10^8 counts/min per mg) than the *in vivo* labeled ^{32}P-DNA probes used in the experiments of Laughlin *et al.* (1979a) such as that shown in Fig. 4A. However, the increased sensitivity did not result in detection of any additional AAV RNA species.

The spliced 3.3- and 3.9-kb AAV RNAs (Laughlin *et al.*, 1979a; Marcus *et al.*, 1981) were not described by Green and co-workers (Green *et al.*, 1980; Green and Roeder, 1980a). However, careful inspection of the data reported by these latter authors reveals good evidence at least for the presence of the 3.3-kb RNA in neutral pH, but not in alkaline pH gels of Berk and Sharp experiments or after glyoxal treatment of the RNA. It is probable that this discrepancy is explained by the low abundance of the spliced 3.3- and 3.9-kb RNAs and the fact that most of the experiments of Green *et al.*, employed whole cell RNA rather than cytoplasmic RNA which is enriched for the less abundant RNAs (see below).

The 2.3-kb RNA observed in the electrophoretic analyses corresponds to the spliced 2.3-kb RNA observed by electron microscopy. The spliced structure of this 2.3-kb RNA could be demonstrated when the Berk and Sharp analysis was performed using exonuclease VII digestion and analyzed in agarose gels (Laughlin *et al.*, 1979a; Green and Roeder, 1980a). For these experiments, Laughlin *et al.* (1979a) used AAV Bal 3/97 ^{32}P-DNA. The *Bal* restriction enzyme removes the terminal palindromic sequences that otherwise base-pair and inhibit the nuclease. This experiment, analysed on alkaline gels, yielded products of 2.6, 3.6, and 4.2 kb only. Thus, all the 2.3-kb RNA was represented by a 2.6-kb DNA fragment indicating that all of the 2.3-kb RNA was spliced. In an analogous way all of the 3.3- and 3.9-kb RNAs were shown to be spliced. The spliced structure of the 2.3-kb RNA was also shown by Green and Roeder (1980a) in similar experiments using exonuclease VII and the 5′-labeled *Hpa*II-36.8/51.0 AAV DNA fragment. This experiment yielded a 550 nucleotide digestion product which also accurately located the 5′ end of the spliced RNA at 39.7 map units (see below).

An interesting technical point arises from the exonuclease VII experiments. In an earlier experiment using the same enzyme to map AAV RNA, Green *et al.* (1980) used intact AAV ^{32}P-DNA as the probe. In this experiment only a small amount of the 2.3-kb AAV RNA appeared to be spliced as measured by replacement of the 2.3-kb band by a 2.6-kb band. This may have been because a higher annealing temperature was used which perhaps caused melting of the short leader in the hybrid. More interestingly, since intact AAV DNA was used, and the terminal hairpins

---

(track 2). Track 3 is a longer autoradiographic exposure of lane 2. Track 1, the AAV cytoplasmic poly A(+) was first annealed with denatured AAV DNA and digested with endonuclease S1 as in panel A track 6 and then treated with glyoxal before electrophoresis and blotting. Track 4, AAV ^{32}P-DNA and fragments *Bam*-A and *Bam*-B treated with glyoxal. Panel A is redrawn from Laughlin *et al.* (1979a). Panel B is reprinted from Marcus *et al.* (1981).

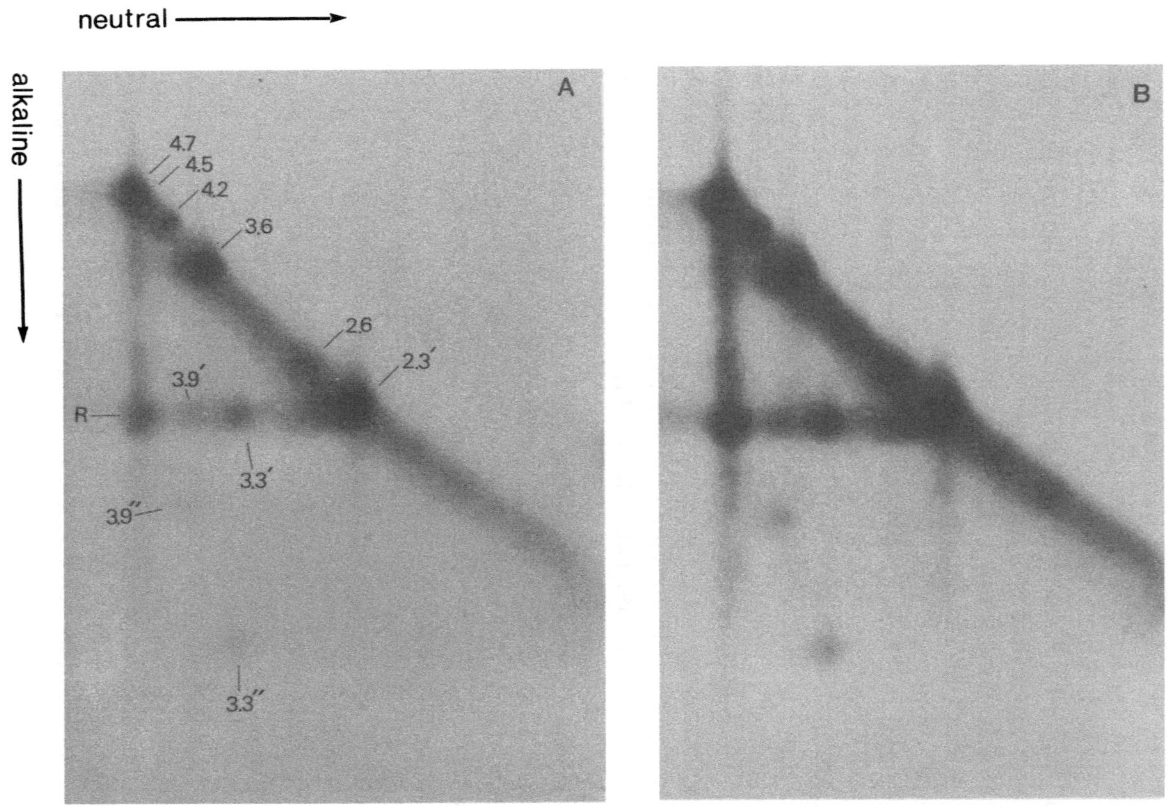

FIGURE 5. Two-dimensional agarose gel analysis of cytoplasmic polyA(+) AAV RNA–DNA hybrids. After hybridization of RNA to denatured AAV DNA and digestion with S1 nuclease, samples were run in neutral and then alkaline dimensions as described in the text. Gels were Northern blotted onto DBM paper then hybridized with nick-translated AAV ^{32}P-DNA and autoradiographed. Panel B shows a longer autoradiographic exposure of the same gel blot as in panel A. The figure is reprinted from Marcus *et al.* (1981).

which are 62 base pairs in length (Berns *et al.*, 1978) did not impede the enzyme (Green *et al.*, 1980) it is not surprising that the 50-base-pair leader was not detected. Apparently, under the conditions used the exonuclease was able to jump over 50 to 100 paired bases since the enzyme preparation used had no detectable endonuclease activity. These results illustrate the caution with which small RNA leaders must be analyzed and indicate the value of mapping RNAs both by electron microscopy and electrophoretic procedures. This latter point is also illustrated by the studies of MVM RNA structure described below.

It is not yet known whether the RNA transcription pattern observed for AAV serotype 2 is the same for all AAV serotypes though this seems likely. Muster *et al.* (1980) reported that AAV4 poly(A) containing RNA had a major peak at 18S in sucrose gradients. cDNA copies of this AAV4 mRNA were prepared by priming reverse transcriptase synthesis off oligo dT annealed to the poly(A) tail. This cDNA annealed to the minus strand component of the *Hinc*II-51/100 fragment of AAV4 DNA.

In the studies of AAV RNA it was important to ensure that none of the observed RNAs were transcripts from variant or DI genomes. This was especially important because AAV DI genomes arise extremely rapidly in AAV stocks and comprise molecules with internal deletions which could potentially give rise to apparently spliced RNA transcripts.

The proportion of AAV DI genomes in a productive infection can readily be altered by using differently passaged stocks of virus or by varying the multiplicity of infection with any given stock (for a discussion of AAV DI genomes see Chapter 6). This point was analyzed by Laughlin *et al.* (1979) and Marcus *et al.* (1981). None of the six major AAV RNAs can be accounted for by transcription of DI genomes. Most of the DI genomes are too small (less than 20–30% of genome length) to code for the discrete AAV RNAs observed. Laughlin *et al.* (1979a,b) annealed the DNA of DI particles with standard AAV DNA. In more than 200 heteroduplex molecules observed, none had a displacement loop resembling that of the DNA–RNA hybrids in size or position. Also, increasing the multiplicity of infection with AAV from 3 to 50 infectious units per cell raised the amount of DI genomes replicated (but not standard AAV genomes) by 10-fold but the ratio of the six AAV RNAs remained the same as measured by electron microscopy of AAV DNA–RNA duplexes. A few very rare RNA molecules that appeared to contain a second "splice" of varying map position (Fig. 3F) may represent transcripts from larger DI genomes that contained only small internal deletions or annealing of RNA with a minority population of DNA strands from such DI genomes. In similar experiments, in which increasing amounts of DI genomes actually inhibited the amount of standard AAV genome synthesis by at least 10- to 20-fold, there was no change in either the type or steady state amount of the AAV species as quantitated by Berk and Sharp analysis (Marcus *et al.*, 1981). This also indicates that the majority of the observed discrete AAV RNAs are not due to transcription of DI genomes. Two other implications of these observations are that (a) the pool of transcription templates is readily saturated and is not directly linked to the amount of DNA replication and (b) the primary point of interference by DI genomes is probably not at the level of transcription.

3. Genome Mapping of AAV RNA

The location of the AAV RNAs on the genome have been mapped quite precisely by three methods: (1) electron microscopy as described above, (2) Berk and Sharp mapping using various restriction fragments and (3) by sequencing procedures involving primer extension.

A series of Berk and Sharp experiments using a variety of AAV2 DNA restriction fragments as probes (Laughlin *et al.*, 1979a; Green and Roeder, 1980) allowed determination of the location of AAV RNA termini and splice junctions with a precision of about 10 to 20 nucleotides. An example of one such experiment using the AAV2 DNA *Pst* 42.9/91.0 fragment is shown in Fig. 4A. These experiments placed the location of AAV RNA 5′ termini at map positions 5.0, 19.0, and 40.0 and the location of the splice junctions at 41.0 and 48.5 and the 3′ terminus [i.e., location of the poly(A) addition site] at 96.0. These measurements were in excellent accord with the measurements made by electron microscopy. In addition, Green and Roeder (1980a) obtained further mapping data by performing Berk and Sharp mapping experiments using 5′ or 3′ terminal labeled DNA probes which again provided similar estimates for the location of the AAV RNA 5′ and 3′ termini.

All of the AAV RNA species identified by gel electrophoresis could be accounted for by RNAs that annealed to the minus strand as shown by electron microscopy (Laughlin *et al.*, 1979). This confirmed previous evidence that all the *in vivo* AAV RNAs were complementary to the DNA minus strand.

Laughlin *et al.* (1979a) proposed that there may also be a fourth family of RNAs of very minor abundance with 5′ termini at map position 13.0. This proposal was made primarily because Berk and Sharp mapping showed some evidence for a very low abundance digestion product consistent with a RNA transcript beginning at map position 13.0 and proceeding at least through the *Sal* and *Eco* R1 cleavage sites at positions 30.7 and 38.1, respectively. Assuming that such RNAs also terminated at position 96.0, an unspliced transcript would be 3.9 kb long and the spliced transcript would be 3.6 kb long. Because these putative RNAs would be of the same size as the spliced 3.9-kb (from the p_5 promoter) or the unspliced 3.6-kb (from the p19 promoter) RNAs, respectively, they could not be readily distinguished in the 1-D Berk and Sharp analysis of Laughlin *et al.* (1979a). However, 2-D gel analysis using more sensitive probes (Marcus *et al.*, 1981) as shown in Fig. 5 revealed no evidence for spliced or unspliced RNAs spanning map positions 13/96. Also, the experiments of Laughlin *et al.* (1979a) using exonuclease VII digestion of

DNA-RNA hybrids did not reveal the expected 3.9-kb digestion product (equivalent to 82% of genome length). In addition, as noted below, primer extension experiments by Lusby and Berns (1982) failed to find an RNA originating from position 13.0. Thus at present there is no strong evidence for the fourth family of RNA proposed by Laughlin *et al.* (1979a). Such RNAs, if they exist, must be present in very low abundance or only under some conditions of cell growth or virus infection.

4. Location of AAV RNA Termini and Introns on the DNA Sequence

Primer extension experiments have mapped the location of three AAV RNA 5′ termini very precisely (Green and Roeder, 1980b; Lusby and Berns, 1982). Lusby and Berns (1982) reported the sequence of the first 2118 nucleotides of AAV 2 DNA, i.e., map coordinates 0/45.0. In addition, Green and Roeder (1980b) determined the sequence of the AAV2 *Hpa*II-36.8/51.0 fragment (i.e., nucleotides 1734–2390 using the numbering scheme of Lusby and Berns). The agreement between the two sequences in the region of overlap (nucleotides 1734–2118) is excellent. Also the sequence of the *Eco*R1-38.1/42.6 fragment (nucleotides 1770–1986) determined by L. de la Maza (unpublished) agrees with the sequences determined by the other two groups. Thus the AAV2 DNA sequence determined thus far appears to have a high degree of reliability.

Green and Roeder (1980b) determined the location of the 5′ terminus and the intron of the AAV 2.3-kb RNA transcribed from p_{40}. In this experiment a small 41 nucleotide DNA primer (located between nucleotides 2236 and 2275) was labeled with ^{32}P at the 5′-end of its minus strand (nucleotide 2275) then annealed to RNA from AAV infected cells and extended with reverse transcriptase using the AAV RNA as template. The main 5′-terminally labeled primer extension product was about 105 nucleotides long. This showed that the first 41 nucleotides from the 5′ end were simply those of the primer and the next nine nucleotides were also colinear with the DNA genome immediately upstream from the primer (i.e., to nucleotide 2227). Following this the extension product omitted 320 nucleotides of genome sequence but then became colinear with the genome 320 nucleotides upstream for a stretch of 55 bases beginning at nucleotide 1908 and ending approximately at nucleotide 1855. This experiment located the intron as extending from position 1908 to 2227 and located the 5′ end of the 2.3-kb mRNA approximately at position 1855 (see Fig. 6).

Similar primer extension experiments (Lusby and Berns, 1982) located the 5′ ends of the less abundant RNAs transcribed from p_5 and p_{19} at nucleotide 286–288 and 871–874, respectively (Fig. 6). The 5′ termini of all three RNA families located by these primer extension studies correspond well with the positions determined by the electron microscopy and electrophoretic mapping experiments done previously. Lusby and Berns (1982) were also unable to obtain any evidence from primer exten-

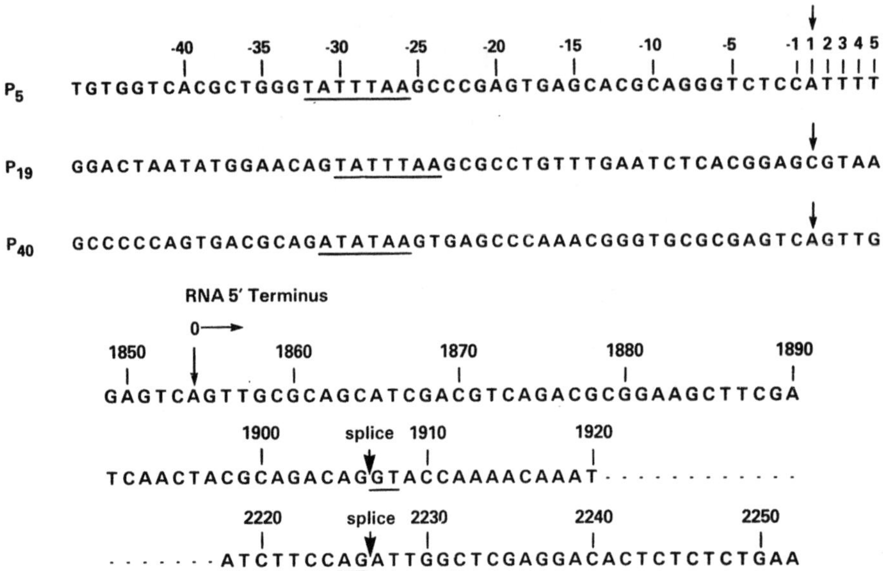

FIGURE 6. AAV2 DNA sequences surrounding RNA start points. The sequences surrounding the three AAV promoters p_5, p_{19}, and p_{40} are shown. The sequence shown is that of the AAV DNA *plus* strand and therefore has the same sense as the RNA transcripts. The approximate locations of the 5′ ends of the AAV RNA transcripts are shown by the arrows. The positions of sequences corresponding to the Goldberg–Hogness TATA box are underlined. Nucleotides are indicated by their position relative to the RNA 5′ start point (+1), by the number of the nucleotide according to the scheme of Lusby and Berns (1982), and by their approximate location in AAV map units (1 unit is equivalent to 1% of the genome or 47 nucleotides). The data were compiled from Green and Roeder (1980b) and Lusby and Berns (1982). The sequence and location of the AAV2 intron is also shown. The mapping data are from Laughlin *et al.* (1979a) and Green and Roeder (1980a). The location of the intron sequence is according to Green and Roeder (1980b).

sion studies for any RNA with a 5′ terminus located near map position 13.0.

The sequences at, and upstream from, the three AAV promoter sites p_5, p_{19}, and p_{40} have many features now thought to be common to eukaryotic promoters. This is discussed further below. One interesting feature of the three families of AAV RNAs is that, unlike those from some viral RNAs (Ghosh *et al.*, 1979; Legon *et al.*, 1979; Flavell *et al.*, 1979), they do not contain multiple distinct 5′ leader sequences nor exhibit significant 5′ terminal microheterogeneity.

5. Cellular Location of AAV RNAs

AAV RNA isolated from the nucleus of infected cells consists of the unspliced 4.2-, 3.6-, and 2.6-kb and the spliced 2.3-kb RNA (Green *et al.*, 1980; Marcus *et al.*, 1981) as shown by the two-dimensional Berk and Sharp analysis in Fig. 7. Although spliced 2.3-kb RNA is detected in the

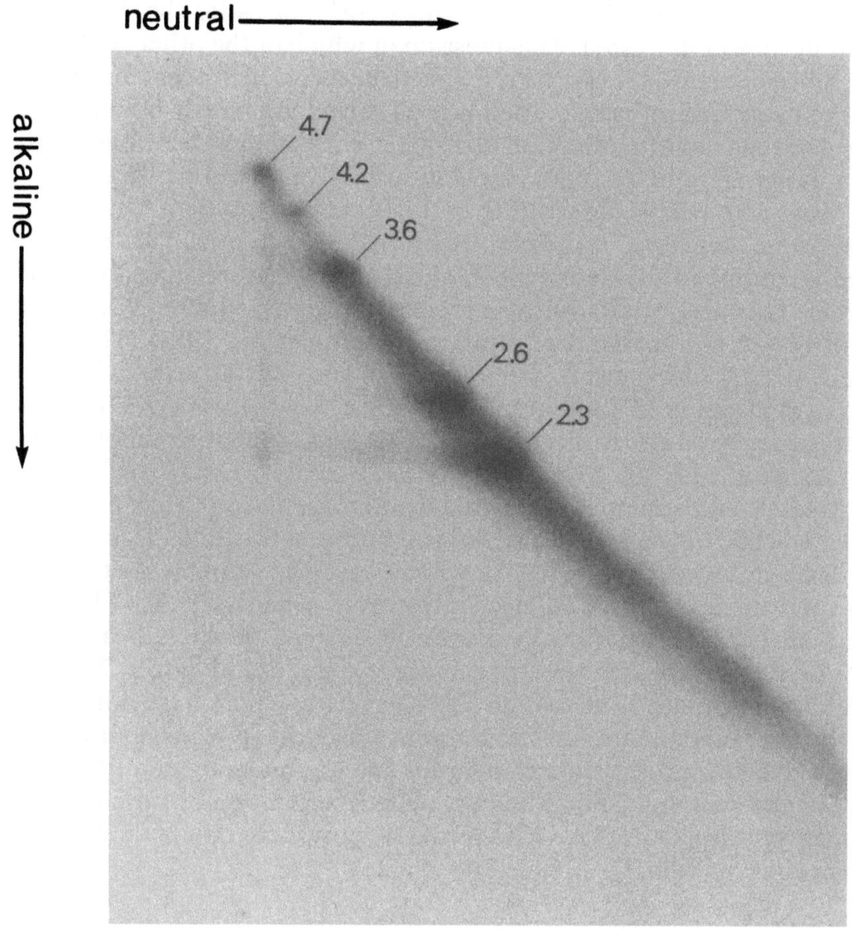

FIGURE 7. Two-dimensional agarose gel analysis of nuclear polyA(+) AAV RNA–DNA hybrids. The analysis was performed exactly as described for Fig. 5. Reprinted from Marcus *et al.* (1981).

cell nucleus the spliced 3.3- and 3.9-kb RNAs were absent (compare Fig. 7 and Fig. 5). The three unspliced AAV RNAs appear to represent the primary transcripts (see below) and it is likely that each is the precursor to the corresponding coterminal spliced RNA.

Laughlin *et al.* (1979a) reported that both the unspliced 4.2-, 3.6-, and 2.6-kb RNA as well as the spliced 3.9-, 3.3-, and 2.3-kb RNAs were present in the cytoplasm. Green *et al.* (1980) analyzed preparations of AAV RNA present on polysomes and detected only the most abundant 2.3-kb RNA. However, it is not clear if this latter experiment was sufficiently sensitive to detect the less abundant AAV RNAs. It is clear that the 2.3-kb AAV

RNA is a polysomal RNA since it apparently codes for the AAV capsid proteins (Jay *et al.*, 1981). The question of which of the other AAV RNAs is a functional mRNA species remains to be resolved. Appearance of RNA in the cytoplasm is not in itself a good criterion since it is known that late after adenovirus infection the nuclear membrane becomes particularly leaky and nuclear blebbing is induced (Lenk *et al.*, 1980).

Jay *et al.* (1979) observed that significant amounts of nuclear AAV RNA were recovered in the chromatin pellet when nuclei were fractionated according to the detergent–high salt precipitation procedure of Hirt (1967). This also occurs when a proteolytic digestion step was included prior to the chromatin precipitation (Marcus *et al.*, 1981). The significance of this observation is not clear. It does not appear to be due to physical trapping of RNA molecules since the bulk of AAV DNA and replicating AAV DNA molecules are not precipitated in this procedure (Straus *et al.*, 1976b). Marcus *et al.* (1981) analyzed nuclear RNA from Hirt pellet and supernatant fractions after separation of poly A(+) and poly A(−) RNA on oligo-dT cellulose. Both nuclear RNA fractions contained similar poly A(+) RNAs which were the same as that observed for total nuclear RNA (see Fig. 7). However, when poly A(−) RNA was examined the pellet fraction contained, instead of the 2.3- and 2.6-kb RNAs, two RNAs that were in each case about 0.2 kb larger, 2.5 kb and 2.8 kb, respectively. The two new RNA species, which were not observed in the Hirt supernatant, may represent transcripts in which synthesis has proceeded beyond the polyadenylation site (at map position 96.0) to the end of the genome. Thus, the Hirt pellet fraction may contain more recently synthesized RNA. Alternatively, polyadenylation may occur in this pellet fraction.

B. Structure of Autonomous Parvovirus RNA Synthesized *in Vivo*

The RNAs synthesized by several autonomous parvoviruses have been analysed. The most complete and detailed information available is that for MVM. Additional studies have been performed on H-1, KRV and BPV and PPV RNAs.

To assess the strand specificity of autonomous parvovirus RNA the complementary DNA strand (cDNA) must be synthesized *in vitro* using the vDNA strand as template. For MVM RNA, no *in vivo* RNA can be detected when the cDNA strand is used as a probe either for liquid hybridization or for analysis of Northern blots (Dadachanji and Ward, personal communication). For H-1 (Green *et al.*, 1979) or KRV (L. Salzman, personal communication) the same virus RNAs are detected using either the single strand viral DNA probe or the double strand *in vitro* synthesized probe.

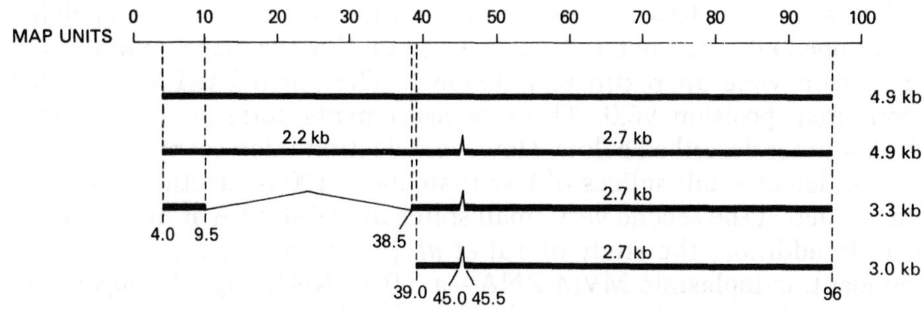

FIGURE 8. Transcription map of major stable species of MVM poly(A)-containing RNA in infected cells. The uppermost line represents the MVM viral (minus sense) DNA strand to which all the stable RNAs are complementary. The other lines represent the location of stable RNAs having the sizes (in kilobases) indicated at the right. The carets indicate the locations of introns (map coordinates 9.5/38.5 and 45.0/45.5) removed by splicing. The MVM genome is approximately 5150 nucleotides in length and is divided into 100 map units of approximately 51 nucleotides each. The splice at map position 38.5 in the 3.3-kb RNA removes the TATA box from the promoter site at this location. The map is derived from work by Dadachanji, Pintel, Tattersall, Merschlinski, Chow, Astell and Ward (personal communication). A very similar transcription map has been derived for the parvovirus H-1 (Lebowitz and Roeder, personal communication).

1. Structure of MVM RNA

The structure of the in vivo MVM RNA transcripts have been analyzed by electron microscopy (Tal *et al.*, 1979) and a combination of electrophoretic techniques (Dadachanji and Ward, personal communication). A summary map of the structure of the main MVM transcripts is shown in Fig. 8.

Preliminary experiments using liquid hybridization analysis established that total RNA or nuclear RNA from infected cells saturated 90% of the viral DNA strand and none of the complementary strand (Dadachanji and Ward, unpublished). Cytoplasmic RNA saturated only 70% of the viral DNA strand. About 90% of the nuclear or cytoplasmic MVM RNA contained poly(A). Low resolution sedimentation analyses in denaturing sucrose gradients containing dimethyl sulfoxide showed a main 16–17 S peak for cytoplasmic MVM RNA and a 20 S peak for nuclear RNA. These studies indicated that MVM nuclear RNA was larger than the cytoplasmic species.

Tal *et al.* (1979) analyzed, by electron microscopy, nuclear and cytoplasmic MVM poly(A)-containing RNA isolated from mouse A-9 cells at 20 hr after infection. Annealing of the nuclear RNA to viral DNA yielded heteroduplexes which appeared double stranded over 95% of their lengths. The cytoplasmic MVM RNA observed by Tal *et al.* (1979) yielded a single type of heteroduplex comprising an RNA with a main body of about 3.0 kb and a leader sequence of about 0.3 kb with a spliced-out region of about 1.6 kb. These RNAs appear to correspond to the 4.9- and 3.3-kb MVM RNAs shown in Fig. 8. The orientation of these molecules by Tal *et al.* was based upon the observation of a presumptive poly(A) tail at one end of the heteroduplex. Based on this the map coordinates of the intron were approximately 8.8/38.1. The poly(A) tail was located about map position 96.0. These measurements correspond well with other studies described below. However, electron microscopy procedures cannot detect small splices of less than about 100 nucleotides and thus did not detect the second very small splice in MVM RNA at map position 45.0. In addition, the study of Tal *et al.* (1979) did not detect the most abundant, cytoplasmic MVM RNA—a 3.0-kb RNA (Fig. 8)—apparently because only molecules observed to be spliced were scored in the microscope.

Dadachanji Ward (unpublished) have mapped MVM RNAs using a combination of Northern blotting and Berk and Sharp analysis. Total cell RNA from infected cells was electrophoresed in agarose gels containing methyl mercury hydroxide and blotted onto DBM paper. When the blots were probed with an MVM c-strand DNA probe no MVM RNA was detected. Hybridization of blots with the v-strand DNA probe demonstrated 4RNAs having sizes of 4.9, 3.3, 3.0, and 1.8 kb. The relative abundance of these four RNAs as a percentage of the total was 10.0, 15, 70, and 1–2% respectively. The 4.9-kb RNA is found predominantly in the nucleus and the other smaller RNAs are predominantly cytoplasmic. The relative abundance of these four RNAs was unchanged between 14 and 26 hr after infection.

The 1.8-kb RNA which is present in very low abundance has not been completely characterized but appears to map to the left of position 20.0 on the genome. The structure of the three larger and more abundant RNAs was analyzed further by the Berk and Sharp procedure (Dadchanji and Ward, unpublished). Analysis of the S1 or mung bean endonuclease digestion products in neutral pH gels revealed three major products of 4.9, 3.3, and 3.0 kb which were analogous to the three species of the same size seen in the Northern blot. The three nuclease digestion products were separately excised from the neutral gels and analyzed in alkaline pH agarose gels. The 4.9-kb band revealed three products in alkali: about 10% of this band was the same size in alkali indicating a small number of transcripts having no splice. Most (about 90%) of the 4.9-kb band yielded in alkali two bands of 2.2 and 2.7 kb, respectively, thus indicating that most of the 4.9-kb RNA contains a very small splice. The 3.0-kb band yielded in alkali a 2.7- and a 0.3-kb band thus also indicating splicing. The 3.3-kb band in alkali yields a 2.6-kb band and two 0.3-kb bands having slightly different sizes.

Further precise mapping of the MVM RNAs was complicated by the inability to use exonuclease VII which is inactive on intact MVM DNA due to the presence of the terminal hairpins. However, the structure of these RNAs has been deduced by additional Northern blotting and Berk and Sharp mapping using specific MVM DNA fragments, by primer extension studies and by analysis of clones of cDNA prepared from copying

TABLE I. Analysis of *in Vivo* RNA Transcripts of AAV and MVM by Northern Blotting and Berk and Sharp Mapping[a]

AAV			MVM		
Berk and Sharp				Berk and Sharp	
Alkaline	Neutral	Northern	Northern	Neutral	Alkaline
4.2	← 4.2	4.2	4.9	4.9 →	[4.9] 10%
1.6 + 2.3	← 3.9	3.9			[2.2 + 2.6] 90%
3.6	← 3.6	3.6	3.3	3.3 →	0.295 + 0.306 + 2.6
1.0 + 2.3	← 3.3	3.3			
2.6	← 2.6	2.6	3.0	3.0 →	0.295 + 2.6
0.05 + 2.25	← 2.3	2.3			

[a] The size in kilobases of RNAs observed in Northern blots or of DNA digestion products observed in Berk and Sharp mapping are indicated. The data are discussed further in the text. The families of RNA of MVM and AAV are arranged to correspond based on the map location of the promoter.

MVM RNA (Dadachanji, Pintel, Tattersall, Merschlinstein, Chow, Astell, and Ward, unpublished). These experiments are summarized in Fig. 8. The fragments obtained in Northern blots and Berk and Sharp analysis are summarized in Table I, which also includes a summary of similar data for AAV RNA.

Comparison of the data in Table I and the maps in Fig. 8 and Fig. 2 illustrates the similarities between AAV and MVM transcription maps. However, MVM does not appear to have an RNA promoted from position 19.0 and there are differences in detail of the splicing. The MVM genome has two introns and AAV only one.

2. Analysis of H-1 RNA

Green *et al.* (1979) reported a preliminary analysis of H-1 RNA transcripts synthesized in infected rat RL5E cells. Liquid hybridization experiments indicated that at both 10 and 24 hr after infection viral RNA transcripts saturated 90% of the virion DNA strand. RNA extracted at 24 hr and analyzed in the Berk and Sharp procedure using ^{32}P vDNA as the probe and analyzed in neutral gels yielded five major digestion products with sizes of 4.7, 3.0, 2.8, 1.45, and 1.3 kb. The three larger species of 4.7, 3.0, and 2.8 kb were also detected in Northern blots of glyoxylated RNA. The three major bands were cut out of neutral Berk and Sharp gels and analyzed in alkaline gels. This showed that all three RNAs were spliced. The 4.7-kb species yielded two products having sizes of 2.2 and 2.6 kb. The 3.0- and 2.8-kb species each yielded a 2.6-kb product and smaller material of only a few hundred nucleotides in size. Analysis in higher-resolution gels indicated that the smaller products appeared to be a doublet of about 170 nucleotides. All of these RNAs contained poly A. Analysis of the cellular location showed that the 4.7-kb species was restricted to the nucleus whereas all the other species were identified in both the nucleus and the cytoplasm. While the total amount of these H-1 RNAs increased from 10 to 24 hr after infection the relative abundance of each species remained unchanged. The relative abundances were 2.8, 3.0, 4.7, 1.45, and 1.3 kb.

Green *et al.* (1979) suggested that the 4.7-kb RNA was the only H-1 transcriptional unit and represented the precursor of both the 3.0- and 2.8-kb RNAs which contained the same size 2.6-kb subunit as the larger RNA. This hypothesis is now known to be incorrect. Further mapping of H-1 *in vivo* RNAs, *in vitro* transcription, and DNA sequence analysis (R. Lebowitz, personal communication) shows that H-1 transcription is very closely analogous to that described for MVM (see Fig. 8) both in the location of promoter sites and RNA splice sites. For both viruses, the RNA transcripts promoted from $p_{4.0}$ have an intron spliced out which spans map position 9.0/38.5. In each case this intron has an acceptor splice site located about 20 bases upstream from the 5′ end of the RNA synthesized from $p_{39.5}$. For H-1, sequence analysis shows that this acceptor splice junction is located midway between the $p_{39.5}$ promoter and the TATA box located 30 bases upstream.

3. KRV RNA

KRV RNA synthesized in rat nephroma cells has also been analyzed. Salzman and Redler (1974) reported a main cytoplasmic species of KRV RNA sedimenting at about 18–20 S in sucrose gradients. Sedimentation analysis of cytoplasmic or polysomal RNA isolated from cells at 12 to 20 hr after infection with KRV revealed a major 21 S species and a minor (less abundant) 12 S species of KRV specific RNA (Mitra, Wali, Valdez, Fabisch, and Salzman; personal communication). Analysis by agarose gel electrophoresis and Northern blotting of the cytoplasmic or polysomal RNA revealed KRV RNAs having size of 4.8, 3.0, 2.5, 1.7, 1.0 kb and a very minor component of about 0.6 kb. The largest species, the 4.8 kb, represents a transcript of about 95% of the virus DNA strand. The most abundant species was the 3.0-kb RNA which appears to comprise the bulk of the polysomal 21 S RNA. This 21 S ^{32}P-RNA was isolated from sucrose gradients of polysomal RNA and mapped both by the electron microscopy of KRV RNA/DNA heteroduplexes and by annealing of the labeled RNA to KRV DNA restriction fragments. These measurements place the 5′ end of the abundant 3.0-kb KRV RNA at about map position 40.0 on the genome map and the 3′ terminus close to the right-hand end of the viral DNA. Again these studies with KRV RNA appear to be quite analogous to the MVM transcription map.

4. BPV RNA

A preliminary analysis of bovine parvovirus RNA has been reported by Bates *et al.* (1981). Poly(A)-containing BPV RNA extracted from infected cells was complementary to the viral strand. Northern blotting experiments detected five RNAs having sizes of 3.8, 3.6, 3.4, 2.6, and 1.8 kb. All of these RNAs appeared simultaneously between 12 and 16 hr after infection. Berk and Sharp mapping indicated that the 3.8-, 3.4-, and 2.6 kb RNAs were not spliced while the 1.8-kb RNA has a single splice and two exons of 1.1 and 0.7 kb. The 2.6- and 1.8-kb RNAs were associated with polysomes. These experiments suggest that the transcription map of BPV may differ, at least in detail, from those of H-1 and MVM.

VI. *IN VITRO* SYNTHESIS OF PARVOVIRUS RNA

A complete description of the mechanism of initiation and control of transcription of eukaryotic genes, including those of parvoviruses, ultimately requires the ability to reconstruct *in vitro* accurate transcription and its appropriate controls. A variety of procedures have been adopted to this end including incubation *in vitro* of isolated cell nuclei or viral transcription complexes extracted from infected cells. More recently two procedures for *in vitro* reconstruction of faithful initiation of transcription have been reported (Weil *et al.*, 1979; Manley *et al.*, 1980). These latter two procedures have resulted in important progress towards defining the structures or recognition sequences required around eukaryotic transcription promoters.

There are two reports of studies of parvovirus transcription in isolated nuclei. Bloom and Rose (1978) analyzed RNA synthesis *in vitro* in cell nuclei prepared 15 hr after infection of KB cells with AAV and adenovirus. About 14% of the total RNA synthesized *in vitro* was AAV-specific. More than 99% of this AAV-specific RNA synthesis was inhibited by concentrations of α-amanitin (1.0 mg/ml) known to inhibit eukaryotic RNA polymerase II. In these isolated nuclei, adenovirus RNA synthesis was preferentially inhibited when compared to synthesis in nuclei prepared from cells infected with adenovirus alone. This probably reflects the inhibition of adenovirus DNA synthesis by AAV coinfection (Carter *et al.*, 1979; Laughlin *et al.*, 1979b), which may result in decreased levels of adenovirus late transcription.

Patton *et al.* (1979) examined RNA synthesis *in vitro* in nuclei prepared from bovine fetal spleen cells infected with bovine parvovirus (BPV). In these nuclei, at 20 hr after infection, about 5% of the RNA synthesized *in vitro* was BPV-specific and all of this RNA appeared to be synthesized by RNA polymerase II as judged by α-amanitin inhibition. The relative levels of the eukaryotic RNA polymerases I, II, and III, as well as the total RNA polymerase level, in the cells was unchanged by BPV infection.

Jay *et al.* (1978) analyzed AAV transcription complexes isolated from infected KB cells. When purified nuclei from infected cells were lysed with sarkosyl in the presence of 0.1 M salt according to the procedure of Gariglio and Mousset (1975) at least 50% of the AAV-specific transcription activity was released from infected cells as a soluble viral transcription complex (VTC) sedimenting in sucrose at 16–26 S. This complex synthesized *in vitro* mostly AAV-specific RNA. Synthesis of this AAV RNA was inhibited with intermediate levels (1–2 μg/ml) of α-amanitin, again indicating that AAV RNA is apparently synthesized by RNA polymerase II. These transcription complexes isolated with sarkosyl elongated, but did not initiate, RNA transcripts as evidenced by the failure of heparin to inhibit RNA synthesis *in vitro*. Furthermore addition of purified RNA polymerase II did not result in stimulation of activity or initiation. These complexes are not useful for studying initiation but may be a useful measure of the level of AAV transcription activity in the cell. Thus inhibition of AAV DNA synthesis with FUdR at 10 hr after infection did not inhibit AAV RNA synthesis and active transcription complexes could still be isolated at 20 hr after infection (C. Laughlin, unpublished). This indicates that AAV RNA synthesis, once begun, is not dependent upon continuing AAV DNA synthesis. However, AAV DNA synthesis is required for AAV RNA synthesis to occur, presumably because of the requirement for production of a double stranded AAV transcription template (Carter *et al.*, 1973; Laughlin *et al.*, 1981).

Since reinitiation did not occur in the AAV transcription complexes isolated with sarkosyl all the observed transcription must represent continuation of RNA synthesis which had been initiated *in vivo*. This was therefore a useful system to determine whether primary transcription was symmetric or asymmetric. Hybridization between separated strands of AAV DNA and AAV-specific RNA synthesized in these complexes showed that RNA was complementary to the AAV DNA minus strand only (C. Laughlin, unpublished). This confirmed the previous *in vivo* studies which indicated that AAV transcription is completely asymmetric (see above).

Isolation of transcribable complexes may be useful as an approach to determining the structure of the viral transcription template. For instance it seems likely that intracellular parvovirus genomes may have a chromatin structure resembling that of cellular chromatin. Indeed MVM nucleoprotein complexes can be leached from infected cell nuclei in the presence of hypotonic buffer (Ben-Asher *et al.*, 1982). Biochemical and electron microscopic analysis of these complexes indicate the presence of cellular histones complexed with MVM RF DNA in an apparent nucleosome structure containing about 26 nucleosomes per genome monomer length. The extract containing these "minichromosomes" supported transcription by endogenous RNA polymerase II.

Initial attempts at reconstructing eukaryotic mRNA transcription *in vitro* by incubating purified RNA polymerase II with purified DNA met with little success. In general, such experiments yielded RNA products which did not exhibit either the strand specificity or the precise genome location of any particular mRNA.

Transcription of purified AAV2 DNA with the RNA polymerase II isolated from uninfected or adenovirus-infected KB cells was studied and the enzyme from either source yielded similar results (Jay *et al.*, 1978). The purified enzyme initiated *in vitro* transcription randomly on either AAV DNA strand. These findings imply that the strand-specific AAV transcription observed *in vivo* requires other factors in addition to RNA polymerase II. The prokaryotic *E. coli.* RNA polymerase also transcribed purified AAV DNA symmetrically (Rose and Koczot, 1971).

The first successful demonstration *in vitro* of faithful and accurate transcription in a eukaryotic system employing RNA polymerase II and which produced RNA having 5′ termini identical to those of the analogous mRNA synthesized *in vivo* was reported by Weil *et al.* (1979). These authors reported that an *in vitro* system consisting of purified RNA polymerase II, a cytoplasmic S100 extract and cloned adenovirus DNA fragments when incubated *in vitro* produced RNA transcripts from the adenovirus major late promoter. These RNA transcripts had exactly the same sequence and structure, including the 5′ methyl cap, as had authentic adenovirus late transcripts synthesized *in vivo*. In the system of Weil *et al.* (1979) adenovirus DNA fragments which contained no known polymerase II promoter sites did not yield any discrete RNA transcripts. Also synthesis was inhibited with α-amanitin at 1.0 μg/ml. This system provided strong evidence that the 5′ terminus (excluding the 5-methyl-cap) of mature eukaryotic mRNA does indeed represent the sequence at which transcription begins.

An alternative *in vitro* system, which also gives accurate transcription initiation, was developed by Manley *et al.* (1980). In this system the purified DNA is incubated with an ammonium sulfate lysate of whole cells from which the chromatin has been removed. This latter procedure has the advantage that it can be rapidly prepared, it does not require addition of purified pol II, and it is much more active (perhaps tenfold) than the system of Weil *et al.* (1979).

Green and Roeder (1980b) showed that the *in vitro* system of Weil *et al.* (1979), when programmed with AAV2 DNA fragments containing the region between 38 and 40 map units, synthesized a discrete AAV specific "runoff" transcript. The size of this runoff transcript was shortened by the expected amount when the AAV DNA fragment was truncated downstream (i.e., to the right) of map position 40. This suggested that the *in vitro* transcript resulted from initiation close to the same position at which the 5′ end of the *in vivo* 2.3- or 2.6-kb AAV RNAs map. This was demonstrated by comparing the 5′ sequence of the *in vitro* transcript with the 5′ sequence of *in vivo* RNA using the variation of the Berk and Sharp procedure described above. A 5′-^{32}P-labeled AAV DNA probe (in which the terminal label was located at the *Hind*III site at nucleotide 1881) was hybridized to the *in vitro* transcript and the hybrids were digested with nuclease S1 and analyzed on a sequencing gel. Both *in vitro* synthesized and *in vivo* AAV RNA resulted in an identical series of ^{32}P-DNA digestion products ranging in length from 33 to 37 nucleotides. This places the 5′ end of both *in vivo* and *in vitro* RNAs at nucleotide 1852 to 1854 in the DNA sequence of Fig. 6. The heterogeneity in the location of the 5′ ends is due to a low degree of nonspecificity of the nuclease S1. When mung bean nuclease was used in similar studies (Green and Roeder, 1980b) both *in vivo* and *in vitro* synthesized AAV transcripts yielded a single predominant ^{32}P-DNA band consistent with transcription beginning at an A opposite the nucleotide (T) at position 1853 (Fig. 6). These experiments show that the *in vitro* system produces RNAs from the AAV2 promoter P_{40} which had identical 5′ termini to *in vivo* RNAs from the same promoter. Thus AAV RNA from this promoter does not appear to exhibit microheterogeneity at the 5′ terminus. Initiation of AAV RNA with an A at nucleotide 1853 would be consistent with the observation that the bulk of *in vivo* synthesized AAV2 mRNA contains m^6A^m as the penultimate nucleotide in the 5′ terminal cap (B. Carter, A. Gershowitz, and B. Moss, unpublished—see below).

If it is assumed that the location of the 5′ terminal cap in a mature enkaryotic mRNA and the point of initiation of *in vitro* RNA synthesis together define the promoter site then the 4.3-, 3.9-, 3.6-, and 3.3-kb AAV RNA must be synthesized from different promoters than that at 40.0 used for the 2.6- and 2.3-kb mRNA. Additional evidence for this has been obtained from experiments using the Manley *in vitro* transcription system. This system, which also recognizes the AAV2 p_{40} promoter (B. Carter, unpublished), produces runoff transcripts from the promoter p_5 and $P_{19.0}$ (K. Berns, personal communication). The 5′ termini of the *in vitro* transcripts from p_5 and $p_{19.0}$ correspond to the termini of the *in vivo* transcripts from these same promoters identified by the primer extension experiments of Lusby and Berns (1981) as described above. At all three AAV promoter sites the *in vivo* or *in vitro* transcripts begin 30 to 32 nucleotides downstream from a TATA box.

MVM transcription has also been analyzed using the Manley *in vitro* system and cloned DNA fragments (D. Pintel and D. Ward, personal communication). There is a strong promoter in MVM yielding *in vitro* runoff products beginning at map position 38.5 which is 30 nucleotides downstream from a TATA box at 38.0. This MVM promoter, p_{38}, which is as strong *in vitro* as the adenovirus major late promoter (one of the strongest known) appears to correspond to the promoter for the major 3.0-kb MVM RNA (Fig. 8). There is a second MVM promoter, which is less active *in vitro*, at map position 4.0 (nucleotide 210–200). Again, this is about 30 nucleotides downstream from the TATA box at nucleotide 185 (Astell *et al.*, 1979). This appears to correspond to the promoter for the less abundant 4.9- and 3.3-kb MVM RNAs (Fig. 8).

Other experiments (R. M. Lebowitz, unpublished) employing *in vitro* transcription systems also indicate, for H-1 DNA, a similar arrangement of RNA polymerase II *in vitro* promoter sites as that observed for MVM.

VII. PROMOTER STRUCTURES

It is now apparent that parvovirus genomes contain more than one RNA promoter site. Furthermore, as the DNA sequences and locations of 5′ mRNA termini are determined it becomes possible to begin identification of the critical elements in these sequences. Before discussing parvovirus promoters it is useful to briefly describe our present knowledge of eukaryotic polymerase II promoter sites. We will give only a general summary because the state of this field is advancing (and consequently changing) rapidly.

Much of the search for eukaryotic RNA polymerase II promoters is based upon prior studies of prokaryotic promoter sites. In general, a great many prokaryotic promoter sites appear to include a DNA sequence (the so-called "Pribnow" box), related to the canonical structure 5′-TATPuATG-3′, which occurs about 10 nucleotides upstream from the short point of RNA synthesis (Pribnow, 1975a,b). In addition, many prokaryotic promoters appear to include a second sequence element about 35 nucleotides upstream from the RNA start point (Rosenberg and Court, 1979). These sequence elements of the prokaryotic promoter are the sites at which RNA polymerase binds (Siebenlist *et al.*, 1980).

The search for the promoter sites of eukaryotic genes transcribed by RNA polymerase II has been advanced enormously in recent years by two major developments. First, the availability of the *in vitro* transcription systems described above (Weil *et al.*, 1979; Manley *et al.*, 1980) has allowed the demonstration that for most genes the 5′ end of a mature *in vivo* synthesized mRNA is identical to the 5′-terminal structure of *in vitro* synthesized transcripts of the same gene. This has provided strong support for the notion that the locations of the 5′ termini of mature mRNAs in eukaryotic DNA sequences do indeed indicate the general location of eukaryotic promoter sites.

The second development has been the generation of a vast and rapidly increasing amount of sequence analysis of regions of eukaryotic cell and viral DNA surrounding potential promoter sites. Inspection of these DNA sequences has identified at least three classes of sequence elements which may be important for defining eukaryotic promoters.

One DNA element, the "capping box" which occurs around the 5′ end of some mRNAs, is similar to the sequence PyCATTCPu or PyPyATTCPu seen in sea urchin histone genes (Sures *et al.*, 1979; Hentschel *et al.*, 1980; Busslinger *et al.*, 1980). Another somewhat related sequence, 5′-GTCAGTTGC-3′, occurs immediately before the mouse β-minor globin gene and certain other cellular and viral class II genes (Konkel *et al.*, 1978; 1979).

A second, more commonly occurring element, is the so-called "Goldberg–Hogness" or "TATA" box which comprises a sequence closely homologous to the canonical structure 5′-TATAAATA-3′ and present about

30 nucleotides (−30) upstream from the 5′ terminus of mRNA (Goldbert, 1978; Gannon *et al.*, 1979). This TATA box may be similar to the prokaryotic Pribnow box.

Other sequences which may occur even further upstream are the CCAAT sequence at about −80 (Efstratiadis *et al.*, 1980) or other upstream "enhancer" regions, such as the SV40 72 base pair repeat, between about −100 and −200 nucleotides (Benoist and Chambon, 1981; Gruss *et al.*, 1981a).

The significance or particular function of any of these sequence elements in eukaryotic class II promoters remains uncertain and in some cases ambiguous. First, not all of these types of sequences occur in every or any particular gene. Secondly, the effect of mutation, deletion, alteration, or normal absence of these sequences upon subsequent transcription varies depending on the assay system. A number of genes which do not have a TATA box are transcribed efficiently *in vivo* (Baker *et al.*, 1979; Haegeman and Fiers, 1978; Grover *et al.*, 1978; Ghosh *et al.*, 1979; Plavell *et al.*, 1980) but only at very low efficiency, in the *in vitro* systems (Lee and Roeder, 1981). Secondly deletion of the TATA box or site-specific mutation to TATA, drastically reduces the *in vitro* transcription efficiency of some genes but shows a much lower reduction on the *in vivo* transcription efficiency (either by injection into *Xenopus* oocytes or transfection of cells with recombinant plasmids). The RNA synthesized from such mutants may have heterogenous 5′ termini (Wasylyk *et al.*, 1980; Gorden *et al.*, 1980; Renoist and Chambon, 1981; Grosschedl and Birnstiel, 1980; Ghosh *et al.*, 1981a; Hu and Manley, 1981; Myers *et al.*, 1981; Grosveld *et al.*, 1981; Grosschedl *et al.*, 1981).

The TATA box may not be essential but may instead control the rate of synthesis and the positioning of the 5′ terminus. Evidence for the latter hypothesis comes from experiments in which cap sites or regions 3′ to the TATA box were deleted. This results in a new 5′ RNA terminus which, however, is still located some 30 nucleotides downstream from the TATA box (Grosschedl and Birnstiel, 1980; Grosveld *et al.*, 1981; Corden *et al.*, 1980; Dierks *et al.*, 1981).

Deletion of upstream enhancer sequences in some cases appears to drastically reduce *in vivo* transcription but appeared to have little effect *in vitro* (Dierks *et al.*, 1981, Benoist and Chambon, 1980; Gruss *et al.*, 1980; Grosveld *et al.*, 1981). When the region between −40 and −100 (including the CCAAT homology at −80) was deleted from the herpes simplex virus thymidine kinase (tk) gene, *in vivo* transcription (as assayed by injection into *Xenopus* oocytes or by transformation of tk^- mouse fibroblasts to tk^+) was reduced 50-fold (McKnight *et al.*, 1981). The remaining 2% level of *in vivo* transcription was abolished by deletion of the TATA box at −30. This residual 2% level *in vivo* is similar to the *in vitro* transcription efficiency of the unaltered tk gene. However, in some cases deletion of sequences around −80 increased *in vivo* transcription (Grosschedl and Birnstiel, 1980). Very recent evidence reported by Grosschedl and Birnstiel (1982) indicates that the enhancer region is required for *in vitro* transcription if the template is circular.

Thus, a gross generalization at present may be that *in vivo*, the upstream sequences are critically important for RNA polymerase entry to the promoter and the TATA box region may be important for defining the precise location of the mRNA 5′ terminus.

In contrast to prokaryotic promoters, eukaryotic promoters may have an additional level of complexity imposed by the chromatin structure. Thus, sequence elements of eukaryotic promoters might be identified by locations about 80 base pairs apart (one turn around a nucleosome), by altered chromatin structure, or by phasing of nucleosomes (Weintraub, 1980).

Some of the class II promoter sequence elements described above can be observed in parvovirus promoters. The sequences around the three AAV promoters are shown in Fig. 6. Both *in vivo* and *in vitro* synthesis of AAV RNAs from all three promoters, p_5, p_{19}, and p_{40} begins about 30 nucleotides downstream from a TATA box (Green and Roeder, 1980b, Lusby and Berns, 1982; and personal communication). Similarly, the two MVM promoters p_4 and p_{38} contain TATA boxes and in each case *in vivo* and *in vitro* transcription begins about 30 nucleotides downstream (D. Ward, personal communication).

Two of the AAV promoters exhibit capping boxes: for p_5, CCATTT occurs at −2 to +4 and for p_{40}, GTCAGTTGC occurs at −3 to +6. The third AAV promoter, p_{19}, does not exhibit either type of capping box. All three AAV promoters have TATA boxes. For both p_5 and p_{19}, the entire sequence GTATTTAAGC is identical and the general region at each site has 16 of 18 bases identical to the adenovirus 5, E1a gene promoter (Lusby and Berns, 1982; Van Ormondt *et al.*, 1978). Also a putative KRV promoter, p_4 at nucleotide 180 (Astell *et al.*, 1979) has the sequence GTATATAAGC which is only one nucleotide different from AAV p_5 and p_{19}. Lusby and Berns (1982) have also noted a remarkable number of short repeats and reverse repeats in the sequence surrounding the p_5 TATA box.

The AAV promoter, p_{19} has the sequence CCAAT ending 88 nucleotides upstream from the RNA 5′ termini. The p_5 promoter has no such sequence but for p_{40} the sequence CAT appears at −89 to −87.

An interesting structural feature of the AAV p_{40} promoter, noted by Green and Roeder (1980b), is a palindromic sequence which could form a stem and loop structure involving the region between the TATA box and the cap box. Similar possible stem and loop structures have been observed for other class II genes (Ziff and Evans, 1978; Nishioka and Leder, 1979; Friedman *et al.*, 1979). Such structures offer attractive possibilities for promoter recognition and control but evidence for such functions awaits experimental verification.

VIII. POST-TRANSCRIPTIONAL MODIFICATION

The metabolism of most eukaryotic mRNAs involves several post-transcriptional modifications including addition of a 5′-terminal methyl cap and a 3′ terminal polyadenylate tail, methylation of internal bases and removal of introns by splicing. These events may alter the stability or coding capacity of individual RNAs and thus contribute to regulation of the level of gene expression.

A. RNA Stability

Early isotopic pulse chase analysis of AAV RNA in KB cells showed that the poly(A)-containing 18–20 S mRNA (i.e., the 2.3-kb mRNA) was stable for at least 6 hr whereas the non-poly(A)-containing heterogenous 4–18 S RNA was much less stable (Carter, 1978). There have been no studies that have measured the half-life of any individual parvovirus mRNA species. For AAV, H-1 and MVM the relative stability of the several overlapping transcripts appears to be similar since the relative amounts of these RNAs does not change between 10 and 24 hr after infection. There does not therefore appear to be any control of parvovirus gene expression based on temporal alterations in mRNA stability. However, there may be some differences in actual stability of different overlapping RNAs (see below).

The significance of the heterogenous unstable 4–18 S AAV RNA transcripts remains unclear. It was suggested (Carter and Rose, 1974; Jay *et al.*, 1979) that these might represent incomplete, nascent transcripts, aberrantly processed transcripts or transcripts of DI genomes. Recent experiments suggested that such RNA does not generally arise from transcription of DI genomes. Isotopic pulse-label experiments show that the relative amount of this unstable RNA is not related directly to the relative proportion of AAV DI genomes in the infected cell (Marcus *et al.*, 1981). Also, this unstable RNA initially accumulates more rapidly than the major 20 S poly(A) containing AAV RNA (Jay *et al.*, 1979). These observations suggest that this heterogenous nuclear RNA may most likely represent nascent AAV transcripts.

These pulse-labeled unstable AAV RNA sequences preferentially accumulate in the supernatant fraction of the nucleus when cells are fractionated according to the Hirt (1967) procedure. This observation is particularly interesting because, as noted above, Marcus *et al.* (1981) observed in the nuclear Hirt pellet fraction two RNAs which appear to be the nonpolyadenylated precursors of the 2.3- and 2.6-kb AAV RNAs. Together with the ready detection of the unspliced AAV RNAs, these observations suggest that AAV may offer a very useful probe to analyse the subcellular (or subnuclear) location of several post-transcriptional processing events for a discrete mRNA.

B. Methylation

No specific studies of parvovirus mRNA methylation have been reported, although preliminary work indicates that AAV mRNA has the general 5′ methylated cap structure expected for eukaryotic mRNAs synthesized by RNA polymerase II (B. Carter, A. Gershowitz, and B. Moss, unpublished; M. Green and R. Roeder, unpublished).

Studies of *in vivo* synthesized mRNAs as well as *in vitro* transcription have led to the general model that for eukaryotic mRNAs synthesized by RNA polymerase II the penultimate nucleotide in the 5′-cap of mature mRNA does indeed represent the point of initiation of RNA synthesis. Specific evidence for this hypothesis was recently reported by Gidoni *et al.* (1981), who analyzed *in vitro* transcription and capping of SV40 mRNAs using β-^{32}P-ATP and β-^{32}P-CTP. It is more common for eukaryotic mRNAs to be initiated with a purine nucleotide, but there are now clear examples of initiation at pyrimidines for cellular (Shatkin, 1976) and viral (Hashimoto *et al.*, 1980; Gidoni *et al.*, 1981) mRNAs.

Carter *et al.* (unpublished) observed that most (50–80%) of the purified cytoplasmic, poly(A)-containing AAV RNA molecules contained, in addition to the added 5′ terminal m^7G, a penultimate nucleotide of m^6A^m. A minority of the molecules contained G^m as the penultimate nucleotide. This is consistent with the expected RNA start points shown in Fig. 6 except that at p_{19}, initiation would occur at the G on either side of the indicated C since there was no evidence for C^m as the penultimate nucleotide of any AAV cap structures. Green and Roeder (unpublished) also observed m^6A^m as the penultimate nucleotide in AAV RNA but in addition observed U^m in this position. This latter observation would be evidence for initiation at a pyrimidine. Complete RNA sequence data will be required before the precise transcription initiation site of AAV RNA can be determined.

Carter *et al.* (unpublished) also observed 4 to 6 residues per AAV mRNA molecule of internally located m^6A. However, neither the precise location nor the function of these internally located methylated residues is known.

C. Polyadenylation

The bulk of the parvovirus specific RNA which accumulates in infected cells contains a 3′-terminal polyadenylate tail. Polyadenylation of AAV RNA has been examined in detail (Carter, 1976). Most of the 20 S AAV RNA contained a polyadenylate tail at the 3′ end, and the lower-molecular-weight, unstable AAV RNA contained little, if any, poly(A). To estimate the size of the poly(A), AAV RNA labeled with [^{3}H]adenosine was purified by hybridizing to AAV DNA and then digested with ribonuclease under conditions in which poly(A) was resistant to cleavage. The size of the poly(A) tract was deduced both from the proportion of [^{3}H]adenosine label which was resistant to ribonuclease digestion and by polyacrylamide gel electrophoresis of the limit digest. The two procedures gave similar results. Cytoplasmic RNA labeled for 4 hr contained a poly(A) sequence about 180–200 nucleotides long, which is about 6% of the total molecular mass. The poly(A) in nuclear AAV RNA was slightly larger than cytoplasmic poly(A) and also decreased in size with the length of the labeling period, as has also been reported for poly(A) in other mRNA molecules (Sheiness and Darnell, 1973).

The location of the 3′ termini, and thus the site of addition of the poly(A) tail, at about map position 96.0 appears to be a common feature of all parvovirus RNAs examined (Laughlin *et al.*, 1979a; Tal *et al.*, 1979; Green *et al.*, 1979; Green and Roeder, 1980a; Ward *et al.*, unpublished). Indeed the poly(A) tail can be observed in parvovirus RNA–DNA heteroduplexes examined by electron microscopy (Tal *et al.*, 1979; Laughlin *et al.*, 1979a) as shown in Fig. 3.

Analysis of the 3′ terminal sequence of a large number of eukaryotic mRNAs has revealed a common unique sequence, 5′-AAUAAA-3′ (Proudfoot and Brownlee, 1974). Analysis of mutants of SV40 by Fitzgerald and Shenk (1981) provided evidence that the AAUAAA sequence is indeed a required part of the recognition signal for addition of the poly(A) sequence. The poly(A) sequence generally appears to be added at a site in which a C, G, or U is followed by an A at a distance between 11 and 30 bases downstream from the hexanucleotide recognition signal. For many viral and eukaryotic mRNAs there is good evidence that transcription proceeds well beyond the poly(A) addition site (Lai *et al.*, 1978; Nevins and Darnell, 1978; Hofer and Darnell, 1981). This then requires an endonucleolytic cleavage at the poly(A) addition site prior to polyadenylation. For some genes transcription may not extend beyond the polyadenylation site (Roop *et al.*, 1980).

AAV2 (K. Berns, personal communication) and H-1 (S. Rhode, personal communication) DNAs each contain polyadenylation recognition signals at the expected map position of 95.0 to 96.0. The H-1 DNA sequence contains several possible polyadenylation sites in the expected region. Furthermore, the 55 base pair sequence which is reiterated in H-1 variant DNA contains a possible polyadenylation site (see Chapter 6). This may provide an interesting model to study poly(A) site selection since not all possible AAUAAA sequences are used for poly(A) addition (Shenk and Fitzgerald, 1981).

There is some suggestive evidence that AAV transcription may proceed beyond the poly(A) addition site at map position 96.0. Jay *et al.* (1979) employed RNA–DNA solution hybridization experiments and observed that more of the AAV DNA *Hinc*II 86/100 fragment was saturated by total RNA present in the Hirt pellet fraction than by the poly(A)-containing AAV RNA. As noted above, Marcus *et al.* (1981) observed, in the Hirt pellet fraction of nuclear RNA, two discrete AAV RNAs having sizes of 2.5 and 2.8 kb that may represent the nonpolyadenylated counterparts of the 2.3- and 2.6-kb RNAs in which transcription has proceeded to the end (map position 100) of the AAV DNA genome.

The biological function of polyadenylation of mRNA is not completely understood. The poly(A) is not required absolutely for translation (Greenberg, 1975) but probably plays a role in mRNA stability in the cytoplasm. Recent evidence supports this latter role (Zeevi *et al.*, 1981). Other possible properties of polyadenylation including a role in RNA splicing or control of the relative abundance of different mRNAs are discussed below.

D. RNA Splicing

The discovery of RNA–RNA splicing in viral and cellular mRNAs (Berget *et al.*, 1977; Chow *et al.*, 1977; Kitchingman *et al.*, 1977; Aloni *et al.*, 1977; Leder *et al.*, 1977; Tonegawa *et al.*, 1978; Breathnach *et al.*, 1977) revealed a much greater potential coding capacity for eukaryotic DNAs as well as possible mechanisms to control the relative level of expression of different proteins. Although RNA splicing is a ubiquitous phenomenon, it is not universal. There are now several examples of mRNAs which are not spliced, including those for histones, interferon, adenovirus protein IX, and herpes virus thymidine kinase (Shaffner *et al.*, 1976; Nagata *et al.*, 1980; Lawn *et al.*, 1981; Alestrom *et al.*, 1980; McKnight, 1980; Wagner *et al.*, 1981; Ghosh *et al.*, 1981b).

Sequence analysis of spliced mRNAs led to the proposal (Breathnach *et al.*, 1978) that all intervening sequences (or introns) have the canonical genome sequences GT at their 5′ end (donor site) and AG at their 3′ end (acceptor site). Seif *et al.* (1979) and Mount (1982) suggested additional canonical features of the sequences bordering introns. As described above, RNA transcripts from both promoters in the autonomous parvoviruses and from all three AAV promoters are spliced. Both introns in H-1 RNAs and MVM and the intron in the AAV 2.3-kb mRNA exhibit canonical splice junction sequences (Green and Roeder, 1980b; R. Lebowitz, personal communication; D. Ward, personal communication). It is important to note that while the AAV 3.9-kb and 3.3-kb RNAs appear to be spliced at the same sites as the AAV 2.3-kb mRNA (Laughlin *et al.*, 1979a; Green and Roeder, 1980a,b; Marcus *et al.*, 1981), this remains to be rigorously proven by direct sequencing of RNA or cDNA.

The mechanism of RNA splicing remains unknown, but two recent possibilities have been suggested. In general poly(A) is added to nuclear RNAs before splicing occurs (Nevins and Darnell, 1978; Loi *et al.*, 1978). In concordance with this, most of the unspliced nuclear 4.2-, 3.6-, and 2.6-kb AAV RNAs contain poly(A) (Laughlin *et al.*, 1979a; Green *et al.*, 1979; Marcus *et al.*, 1981) as do the discrete nuclear transcripts of MVM (Tal *et al.*, 1979) and H-1 (Green *et al.*, 1979). An ingenious model was proposed in which splicing would be facilitated by poly(A) promoted formation of triple-stranded structures within the unspliced RNA precursor (Bina *et al.*, 1980). However, Zeevi *et al.* (1981) recently reported evidence that splicing of adenovirus mRNAs can occur normally when poly(A) addition is prevented with the drug cordycepin (3′ deoxyadenosine). Thus poly(A) addition and RNA splicing appear to be independent events even though poly(A) addition generally precedes splicing.

An alternative model for RNA splicing is based on homology between sequences around splice junctions in mRNA precursors and sequences in small nuclear RNAs (snRNAs) such as U1 (Lerner *et al.*, 1980; Rogers and Wall, 1980). These snRNAs, possibly in the form of ribonucleoprotein (RNP) complexes, may act as templates on which splicing can occur. Sharp (1981) recently suggested that these RNP complexes may form a ribosome-type particle to direct RNA splicing. It has been suggested that the small VA RNAs transcribed from adenovirus may play a similar role in splicing of adenovirus messages (Murray and Holliday, 1979). Indeed Mathews (1980) reported evidence that VA RNA anneals to spliced adenovirus RNAs but not to the same regions in the viral genomic DNA. As noted by R. McPherson (cited by Janik *et al.*, 1981) the adenovirus VA RNA does have some sequence homology with sequences surrounding the splice junctions in AAV RNA. This may suggest a role for adenovirus VA RNA in splicing of AAV RNA. In support of this, Janik *et al.* (1981) reported DNA transfection experiments suggesting that the adenovirus VA genes are essential for the AAV helper function. In contrast, Richardson and Westphal (1981) performed microinjection experiments and concluded that adenovirus VA RNA is not required for the AAV helper function. This point is discussed further by Carter and Laughlin (see Chapter 3). It is also possible that homology between mRNA and small RNAs such as VA RNA might reflect events involved in mRNA stability, transport or translation rather than the mechanism of RNA splicing. Recent evidence (Thimappaya, Weinberger, and Shenk, personal communication) does indeed show that VA RNA is involved in translation of adenovirus late mRNA apparently by promoting ribosome binding.

Other studies on AAV RNA (Laughlin *et al.*, 1979a; Marcus *et al.*, 1981; Myers *et al.*, 1980) provided further suggestive insights into the possible effect of conformation of the primary RNA transcript on the apparent splicing efficiency and to proteins which may control RNA splicing.

In the course of analysing the AAV helper ability of various adenovirus mutants, Myers *et al.* (1980) reported that the mutant adenovirus Ad5*ts*125 failed to help AAV efficiently at the nonpermissive temperature of 40°C. This mutant adenovirus contains a temperature-sensitive lesion in the M_r 72,000 DNA-binding protein (E72K protein) coded by the early region 2A gene (Einsinger and Ginsberg, 1972; Ginsberg *et al.*, 1974; Van der Vliet *et al.*, 1975). The E72K protein has a number of pleiotropic effects on adenovirus gene expression and replication and appears also to have pleiotropic effects on AAV growth. This is discussed more extensively elsewhere in this volume (see Chapter 3).

One of the effects of the Ad E72K protein appears to be on the accumulation of certain AAV spliced RNAs. When AAV was grown in human KB cells at 40°C using the mutant Ad5*ts*125 as the helper, the accumulation in the cytoplasm of the spliced 3.9- and 3.3-kb AAV RNAs was greatly decreased whereas the relative amounts of all the other AAV RNAs were similar to those seen with wild-type adenovirus as the helper (Myers *et al.*, 1980). This is illustrated in Fig. 9, which shows a Berk and Sharp analysis of cytoplasmic AAV RNAs present in KB cells at 40°C using either the mutant or wild-type adenovirus as helper. Remarkably, the AAV 2.3-kb RNA, which contains apparently the same splice as the 3.3- and 3.9-kb RNAs, accumulated in the cytoplasm to a similar level with either mutant or wild-type helper.

Analysis of nuclear AAV RNA revealed normal amounts of the three unspliced AAV RNA transcripts in mutant helped cells (C. Marcus, C. Laughlin, and B. Carter, unpublished). Thus the *ts*125 mutation does not affect the initiation of AAV RNA synthesis. Similarly polyadenylation (Myers *et al.*, 1980) and 5′ methyl capping (B. Carter, A. Geshowitz, and B. Moss, unpublished) of AAV RNA transcripts are not impaired when the mutant helper is used. These observations suggest that the conformation or size of an RNA transcript may have an important effect upon either the efficiency of RNA splicing or the subsequent stability of the spliced transcript.

Marcus *et al.* (1981), who studied nuclear and cytoplasmic AAV RNA, also reported suggestive evidence that the splicing of the AAV 4.2- and 3.6-kb nuclear RNA transcripts may occur much less efficiently than splicing of the 2.6-kb transcript. This decreased efficiency of splicing of the larger AAV RNAs is also implicit in the data reported by Green *et al.* (1980) and indeed may account for the failure of these authors (Green and Roeder, 1980a) to observe the spliced 3.3- and 3.9-kb AAV RNAs. Thus splicing of the two larger AAV RNA transcripts may be very inefficient even when using a wild-type adenovirus as the helper. The lesion in the Ad5*ts*125 mutant may therefore decrease even further, an already inefficient splicing of some AAV RNAs. In this regard, it is interesting that there is evidence that variation in the structure of sequences upstream from the intron may also alter the level of accumulation of the mRNA for SV40 T-antigen. These observations were also interpreted as indicating a role of RNA conformation in RNA splicing (Khoury *et al.*, 1979).

Evidence for an effect of the adenovirus E72K protein on splicing of some adenovirus mRNA per se, rather than on subsequent stability of the spliced RNA, was also independently obtained by Klessig and Chow (1980) as described more fully elsewhere in this volume (see Chapter 3).

In this context it is also interesting that a functional adenovirus 72K DNA-binding protein appears to *decrease* the cytoplasmic stability of

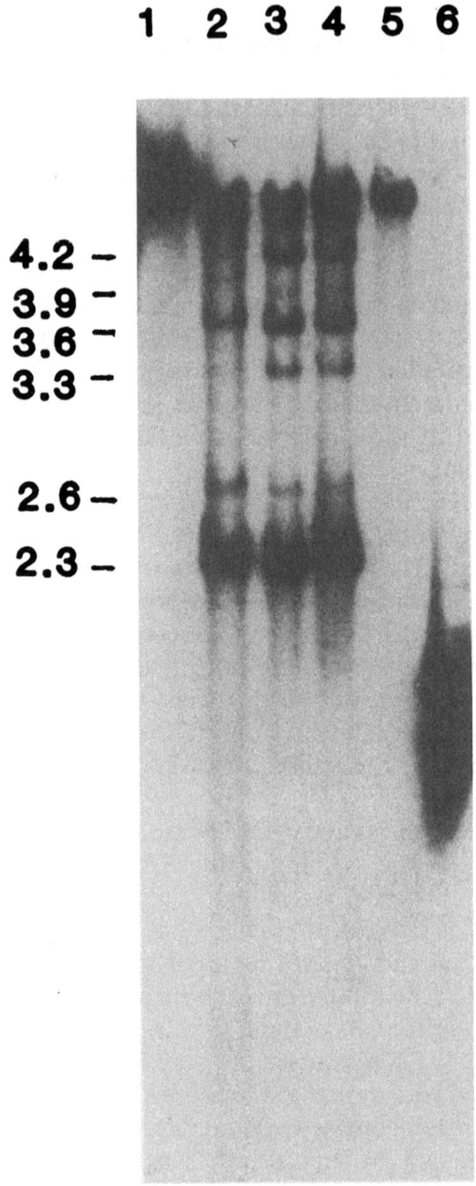

FIGURE 9. Cytoplasmic, poly(A)-containing RNA in cells grown at the nonpermissive temperature (40.5°C) using Ad5*ts*125 or Ad5*wt* as the helper. RNA was annealed with denatured AAV ^{32}P-DNA, digested with endodeoxyribonuclease S1 and electrophoresed at neutral pH in 1.4% agarose gels. Track 2, RNA from *ts*125-helped cells. Tracks 3 and 4, RNA from wt-helped cells. For the reactions in track 4 the amount of nuclease S1 used was one half that in the other reactions. Track 5, control yeast RNA. Molecular weight markers were intact AAV DNA (track 1) and AAV DNA *Hae*II fragments A and B. The individual digestion products are identified at the right by size (in kilobases), which indicates also the AAV RNA species from which they are derived. The molecular weight of the markers are identified at the left. The 4.5-kb digestion product is an artifact from formation of R-loops (Marcus *et al.*, 1981). The 4.7-kb band represents reannealing of the AAV DNA probe. This figure is redrawn from Myers *et al.* (1980).

mRNAs from adenovirus early genes. The stability of these mRNAs is *increased* when the mutant Ad5*ts*125 is grown at the nonpermissive temperature of 40°C (Babich and Nevins, 1981). This observation also suggests that the effect of this mutant on AAV mRNAs may be due to alterations in RNA splicing per se rather than stability. It is of course possible to argue that some product of another adenovirus early gene, which would be produced at an elevated level by Ad5*ts*125 at 40°C might instead decrease the stability of some AAV RNAs.

RNA splicing was suggested to be necessary for mRNA stability or transport of mRNA to the cytoplasm. This suggestion was based on studies with SV40 genes, or globin genes inserted into SV40, in which the intron had been precisely deleted. These deletions resulted in an inability to synthesize stable RNA transcripts (Gruss *et al.*, 1979; Hamer and Leder, 1979). However, the generality of this rule is doubtful since recent analogous work with rat preproinsulin genes inserted into an SV40 vector indicates that, for this mRNA, splicing is not a requirement for transport to the cytoplasm (Gruss *et al.*, 1981b). In addition, as noted above several examples of unspliced cell or viral mRNAs are now known.

Studies with AAV have often revealed unspliced AAV RNAs in the cytoplasm (Laughlin *et al.*, 1979a; Green *et al.*, 1980; Marcus *et al.*, 1981). This may represent evidence that unspliced AAV RNAs can be transported to the cytoplasm. This is even more interesting in view of the AAV2 DNA sequence (Lusby and Berns, 1982) which indicates possible reading frames located well upstream of the intron in the AAV RNAs synthesized from p_5 and p_{19} (see below). However, as noted above (see Section V. 5), it is difficult to exclude leakage of unspliced AAV RNAs from the nucleus during cell fractionation.

IX. CONTROL OF RNA ABUNDANCE

Two important ways to regulate the level of expression of any gene are RNA abundance controls which affect the cytoplasmic level of mRNA and translation controls which affect the efficiency or rate at which the mRNA is translated. For AAV there appear to be translational level controls at least for synthesis of the capsid protein (see Chapter 3). The abundance of a given mRNA in the cell cytoplasm may be regulated at a number of steps including promoter recognition and initiation, chain termination and polyadenylation, splicing, transport from nucleus to cytoplasm, and also mRNA stability. For virus genes in particular (but not exclusively) mRNA abundance might also be controlled by gene dosage resulting from increased copy numbers after replication of the viral genome.

It is not yet clear for any parvovirus gene precisely which mechanisms control the relative abundance of individual mRNAs. Polyadenylation does not appear to be an important abundance control, since for each parvovirus examined thus far, all RNAs appear to have a common polyadenylation site near map position 96.0. This is in contrast to other viruses such as adenovirus where polyadenylation is clearly an important mRNA abundance control (Ziff, 1980; Flint and Booker, 1980; Nevins, 1982).

Splicing of RNA may regulate the level of expression of a particular gene either by the efficiency with which splicing occurs or by altering the reading frame of a mRNA. We have already discussed the suggestive evidence that the efficiency of splicing may differentially regulate the level of AAV spliced RNAs. Based on the available AAV DNA sequence information (Lusby and Berns, 1982) splicing would not change the potential reading frames of RNA transcripts from p_5 and p_{19}, but would change the carboxyl terminus of potential proteins (see Section X below). It is difficult to assess the effect of splicing on the 2.6-kb RNA transcript in the absence of complete AAV DNA sequence data.

Whether there is any specific effect of virus DNA replication upon the abundance of RNA transcripts is difficult to assess since the infecting viral genome is single-stranded and viral RNA synthesis does not begin until viral DNA synthesis occurs. As noted above, for AAV, preliminary evidence suggests that the amount of RNA synthesis is not directly related to the amount of DNA synthesis.

The available evidence, though sparse, suggests that the most important control on parvovirus mRNA abundance may be exerted at the level of promoter recognition or initiation. For instance, as judged by the steady-state amounts of nuclear RNAs, for both AAV and autonomous parvoviruses, the major promoter at map position 38 to 40 appears to be much more active than those in the left hand end (Green *et al.*, 1980; Marcus *et al.*, 1981; Ward *et al.*, unpublished). Whether this truly reflects the relative promoter "strengths" or subsequent RNA stability remains to be determined. The relative half-lives of individual parvovirus mRNAs have not been measured. The use of recombinant DNA technology can now provide a variety of DNA probes to assess, both *in vitro* and *in vivo*, the relative efficiencies of the parvovirus promoters.

X. CODING POTENTIAL OF PARVOVIRUS mRNA

The precise mapping of parvovirus RNA transcripts on the viral genome provides a first step towards determining the codigenic regions of the DNA. Comparison of the RNA maps together with the viral DNA sequence can locate potential open reading frames. Very recent work has now provided precise information on coding regions of both AAV and autonomous parvoviruses.

The DNA sequence of AAV2 between map position 0 and 50 (Lusby and Berns, 1982; Green and Roeder, 1980b) can be divided into three reading frames beginning at nucleotides 1, 2, and 3 respectively. This suggests a number of intriguing possibilities for protein coding regions (see Fig. 2). In reading frame 1 the only significant open coding region begins at the AUG at nucleotide 2203 within the intron, and proceeds at least to nucleotide 2390. The full extent of this reading frame remains to be determined, but it suggests the possibility of a protein coded by the

2.6-kb mRNA beginning 350 nucleotides from the 5′ end. There are, however, two other AUG codons closer to the 5′ end of the 2.6-kb mRNA which might be preferentially recognized.

A second potential open coding region (reading frame 2) for a 9K protein exists in the 2.6-kb mRNA between nucleotides 1919 and 2165. Coding would thus begin 66 nucleotides from the 5′ terminus and would be contained entirely within the intron sequence.

Reading frame 3 is open between nucleotides 321 and 2184. This would allow coding of a 68K protein beginning 34 nucleotides from the 5′ end of the 4.2-kb mRNA. Interestingly, splicing of the 4.2-kb mRNA to yield the 3.9-kb mRNA would remove the two UAA terminators at nucleotides 2184 and 2193 respectively. Since splicing appears to occur in frame this would allow synthesis of a 59K protein exhibiting extensive homology with the 68K protein but having a different carboxyl terminus. Reading frame 3 may also be entered at nucleotide 993 which is 120 bases downstream from the 5′ end of the 3.6-kb and 3.3-kb RNAs. This would allow synthesis of two putative proteins, 44K and 34K, which would have carboxyl region homologies to the putative 68K and 59K proteins, respectively, but would be truncated at the N-terminal region.

The same reading frame 3 can also be entered from an AUG at nucleotide 1937 which is 84 bases from the 5′ end of the 2.6 kb RNA. This would allow synthesis of a 9K protein from within the intron (nucleotides 1937 to 2184) which would be homologous to the carboxyl termini of the putative 68K and 44K proteins. This 9K protein would be coded by almost the same DNA sequence as the putative 9K protein in reading frame 2.

These possibilities indicate that the use of 3 different promoters and one intron might allow production of 5 proteins from 5 individual RNAs using the same reading frame. Together with the additional 9K reading frame 2 protein there are 6 putative AAV proteins possibly synthesized in addition to the three capsid proteins VP1, VP2, and VP3. This would provide a rationale for 6 different AAV transcripts and would imply that nearly all the AAV genome has a coding function.

The only AAV transcript for which a message function appears to be clearly established is the AAV 2.3-kb mRNA. This is the major AAV RNA species in both cytoplasm and polysomes (Laughlin *et al.*, 1979; Green *et al.*, 1980). The coding function of the 2.3-kb mRNA was demonstrated by Jay *et al.* (1981) who isolated cytoplasmic, poly(A) containing RNA from KB cells 20 hr after infection with AAV2 and adenovirus type 5. The RNA was fractionated by velocity sedimentation in a sucrose gradient under denaturing conditions. Aliquots of RNA from individual gradient fractions were analyzed to detect AAV RNA using the Berk and Sharp procedure. Other aliquots were translated in an *in vitro* reticulocyte lysate system. The ^{35}S-labeled proteins synthesized *in vitro* were analyzed by electrophoresis after immunoprecipitation with an antiserum raised against purified AAV2 particles and known to precipitate all three AAV capsid proteins VP1, VP2, and VP3. As shown in Fig. 10, three ^{35}S-labeled

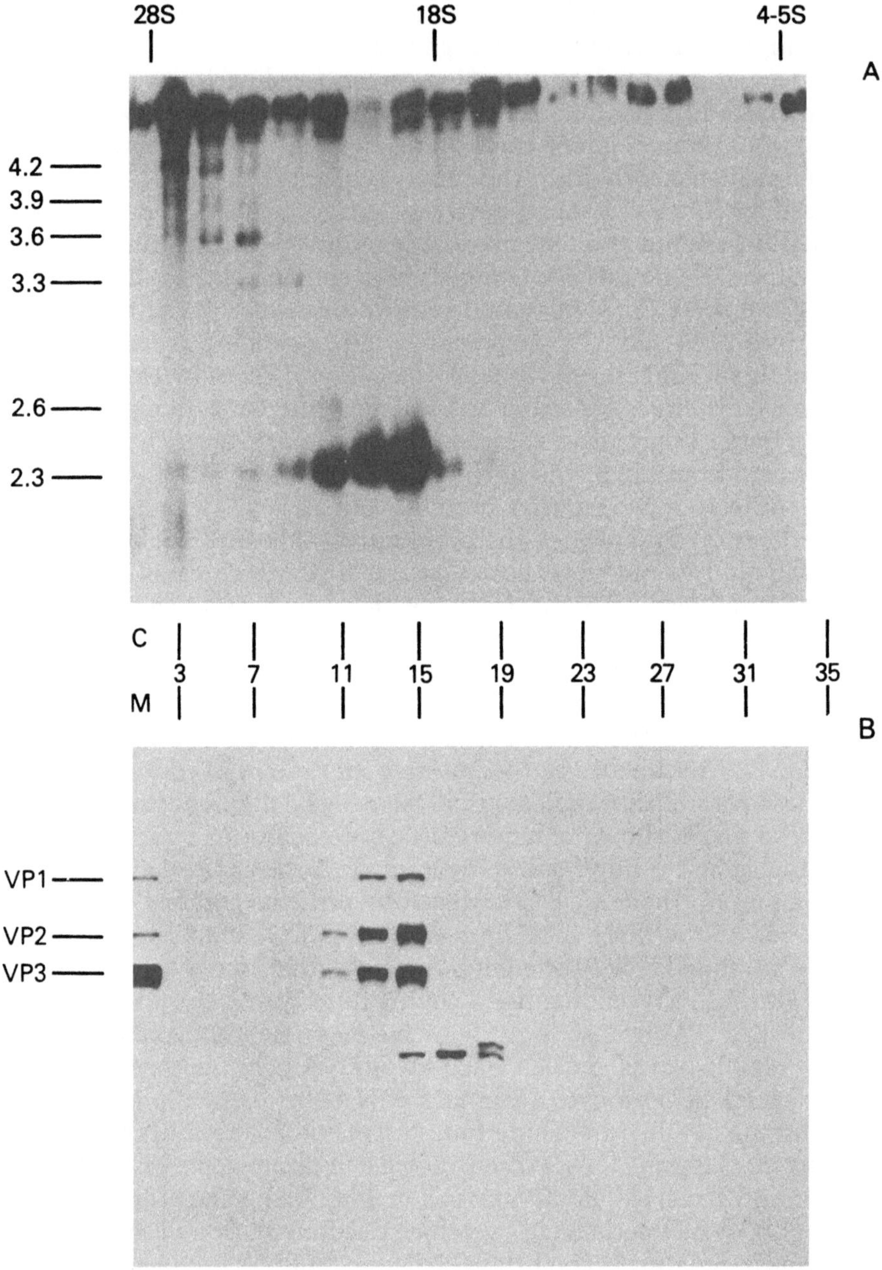

FIGURE 10. Transcription and translation analysis of AAV RNA. Cytoplasmic, poly(A)-containing RNA (100 μg) from KB cells infected with AAV2 (10 i.e./cell) and Ad5wt and grown at 37°C was isolated and sedimented in a neutral sucrose gradient under denaturing conditions. RNA from each fraction was recovered and dissolved in 10 μl of water. Panel A: Aliquots (4 μl) of RNA from odd-numbered fractions were incubated with heat-denatured AAV2 ^{32}P-DNA then digested with endodeoxyribonuclease S1 and electrophoresed at

proteins that coelectrophoresed with the three AAV capsid proteins were specifically immunoprecipitated with the AAV antiserum.

The sedimentation profile distribution of these three *in vitro* synthesized AAV proteins coincided precisely with that of the 2.3-kb AAV mRNA as revealed by the Berk and Sharp mapping (Fig. 10). It was therefore concluded that the AAV capsid proteins VP1, VP2, and VP3, which are known to cross react immunologically (Johnson *et al.*, 1977), are all coded by the spliced 2.3-kb mRNA. This is consistent with, but does not prove, the previous hypothesis that VP2 and VP3 are derived by post-translational cleavage of VP1 (Buller and Rose, 1978). The altered ratios of VP1, VP2, and VP3 seen in the *in vitro* translation (Fig. 10) when compared to the ratios in intact virions or infected cells may indicate a decreased efficiency of post-translational cleavage in the *in vitro* system. It is important to note that the results of Jay *et al.* (1981) indicate only that VP1, VP2, and VP3 are apparently coded by the 2.3-kb RNA species. This does not rigorously prove that each protein is coded by the identical 2.3-kb mRNA. It is probably necessary to sequence a large number of cDNA clones to rigorously eliminate the possibility that VP1, VP2, and VP3 are not derived from 2.3-kb mRNAs which have minor variations in their splicing patterns or 5′ start points.

In earlier studies, *in vitro* translation of AAV mRNA was reported to yield two proteins having sizes of 24,900 and 15,800, respectively, as well as the three capsid proteins VP1, VP2, and VP3 (Buller and Rose, 1978a). These two additional proteins were also observed in AAV infected KB cells and were suggested to represent products from proteolytic cleavages to yield VP2 and VP3 (Buller and Rose, 1978a,b). However, no immunoprecipitations were performed in these studies. These two smaller putative AAV proteins were not observed in *in vitro* translations by Jay *et al.* (1981), who noted that these proteins may be unstable in the *in vitro* translation system, not precipitated by the capsid antiserum used, derived from broken AAV mRNA, or perhaps are not AAV-specific proteins. Whether these two smaller proteins observed by Buller and Rose (1978b) could correspond to any of the other putative reading frames in the left half of AAV DNA is not known.

---

neutral pH in a 1.4% agarose gel according to the Berk and Sharp procedure. The gel was visualized by autoradiography. The sizes of the digestion products (indicated at the left in kilobases) were determined from restriction endonuclease fragments of AAV DNA electrophoresed in the same gel (not shown). Gradient fraction numbers are indicated between the upper and lower panels. Track C contained control yeast RNA. Panel B: Aliquots (2 μl) of RNA from odd-numbered fractions were translated *in vitro* in the reticulocyte lysate. ^{35}S-labeled AAV proteins were detected by immunoprecipitation and electrophoresis in a 9% polyacrylamide gel. Track M contained ^{35}S-labeled capsid proteins from purified AAV particles. The tracks are numbered (3–35) according to the fraction numbers in the sucrose density gradient. The molecular weights shown for the AAV proteins are those determined relative to [^{3}H]methyl proteins obtained from New England Nuclear Corp.

The data of Jay *et al.* (1981) indicate that the AAV VP1, VP2, and VP3 proteins must be coded to the right of the intron in the 2.3-kb mRNA since the 55 nucleotide leader in this mRNA contains no initiator codon (Green and Roeder, 1980b). This places the codigenic region for the AAV capsid proteins between the maximum limits of 48.0/96.0 on the genome map (Laughlin *et al.*, 1979a). This region of AAV contains about 2250 nucleotides equivalent to 750 amino acids which appears only just sufficient to code for a protein (VP1) having a reported molecular weight of about 85,000 (Rose *et al.*, 1971).

An additional consequence of the study reported by Jay *et al.* (1981) is that the 2.3-kb AAV mRNA appears to be under translational control exerted by adenovirus. A similar conclusion can be drawn from studies of monkey cell host-range restrictions for AAV (Buller *et al.*, 1979). This translation control is discussed in more detail elsewhere in this volume (see Chapter 3).

Sequence data and *in vitro* translation studies are now available for several autonomous parvoviruses including MVM and H-1. These studies (D. Ward, personal communication; S. Rhode, personal communication) show that the capsid protein is coded from the right-hand half of the genome; probably from the abundant mRNA transcribed from the promoter at map position 39.0. The left-hand half of the genome has a large open reading frame present in the MVM 4.9-kb RNA. *In vitro* translation of this RNA yields an 84K protein which can be immunoprecipitated with serum from infected animals but not with antivirion antisera. This 84K protein is also found in infected cells and may be the protein found covalently bound at the 5′ end of MVM RF DNA (D. Ward, unpublished). This protein appears to be a *rep* protein. Similar evidence is also available for H-1 (S. Rhode, personal communication). This 84K protein of H-1 or MVM may have functions similar to the putative proteins coded by the left half of the AAV genome.

XI. SUMMARY

From the presently available information, as evidenced by their transcription maps, the defective parvovirus AAV and the autonomous parvoviruses such as MVM and H-1 appear to be more remarkable for the similarities in their genome organization rather than the differences. Both types of virus contain at least two classes of overlapping RNA transcripts which are initiated from either a very active or major promoter near map position 38 to 40 or a less active promoter close to the left-hand end near map position 4.0 to 5.0. Either type of RNA appears to contain an intron located between map position 40 and 48 and to terminate at a polyadenylation site at about map position 96.0. The main difference is that AAV has a third promoter at map position 19.0 but the autonomous viruses do not. In contrast the autonomous parvoviruses apparently produce a second class of RNA from the promoter at the left-hand end by removal of a second intron between map position 9.0 and 38.0. It is possible that this autonomous virus RNA might be functionally analogous to the AAV RNA from p_{19}. Alternatively, these transcriptional differences may reflect the defectiveness of AAV. This latter hypothesis appears perhaps less likely in view of a variety of evidence that the AAV defects requiring helper functions appear to be at the level of DNA replication and mRNA translation or both (see Chapter 3).

The DNA sequences of AAV, H-1 and MVM are complete and all the RNA start points have been accurately mapped. This will now allow further work, particularly employing alterations in the genome sequence by *in vitro* manipulation, to identify functionally important features of sequences surrounding parvovirus promoters. This work will be of significance for several reasons. Firstly, together with many other studies on a large variety of other eukaryotic cell and virus genes, this will contribute to a general definition of the structural and functional principles of eukaryotic gene promoters. More specifically, for individual parvoviruses the several promoters apparently function with significantly different frequency or efficiency. Indeed, for parvoviruses, the initiation or promoter efficiency appears to be a major point at which the relative abundance of individual mRNAs is controlled.

A very important question relevant to obtaining a description of mechanisms of eukaryotic gene promoter functions is the structure of the transcription template particularly in the region of the promoter. It is now apparent that eukaryotic DNA is generally complexed with histones to form a subunit nucleosome structure within the bulk chromatin. The structure of actively transcribed and nontranscribed chromatin is different. Recent studies on AAV (Marcus and Carter, 1982) and MVM (Ben-Asher *et al.*, 1982) indicate that the genomes of both these viruses exist during replication in a chromatin structure containing a nucleosomal subunit configuration analogous to that of cellular chromatin. Because parvovirus genomes are small, they can be readily separated from cellular chromatin. This should facilitate studies of the structure of chromatin on a transcriptional template of known sequence. In addition, the number of gene copies of actively transcribing parvovirus genes in an infected cell is likely to be much higher than for any cellular gene transcribed by RNA polymerase II. These particular properties of parvoviruses should allow bulk amounts of material to be obtained to study specific eukaryotic promoter structures in chromatin. The only other system with analogous properties is that represented by the papovaviruses. Thus, the use of parvoviruses will significantly increase the number of promoters which can be studied in this way. Parvoviruses also differ from papovaviruses in at least two ways. First, parvoviruses have linear rather than circular genomes. Second, there is no evidence that parvovirus promoter functions are controlled by any parvovirus coded protein in the way that papovavirus T-antigens may affect papovavirus promoters. It is likely therefore that parvovirus promoters will be controlled only by cellular factors. Of course for AAV it is possible that an adenovirus coded function could control AAV promoters although so far there is no evidence for this.

The development of transcription maps of parvoviruses together with the determination of the genome sequences has also formed a most important background for prospective studies aimed at developing these viruses as potential eukaryotic cloning vectors. There are a number of properties of parvoviruses which need not be listed here that suggest such vectors would be extremely useful.

Other features of eukaryotic RNA metabolism may also prove to be particularly amenable to study using parvoviruses. We have already noted that processes such as splicing and polyadenylation may occur at different subnuclear locations. For AAV, the relationship between the adenovirus 72K DNA-binding protein and RNA splicing is interesting. Also parvoviruses may offer a useful system in which to obtain biochemically useful amounts of defined, unique unspliced precursor RNAs for studies of RNA splicing *in vitro*.

In summary, molecular biology studies on parvoviruses were begun little more than a decade ago both because of the intrinsic interest of these viruses and because it seemed likely that their small size would offer a level of approachable simplicity for biochemical analysis. To date these viruses remain both intrinsically interesting and approachably simple. In addition these viruses now offer a system for analysis of transcription in chromatin structures which will yield insights of general significance.

ACKNOWLEDGMENT. We thank Joan Mok for preparation of the manuscript. We thank Dave Ward, Ken Berns, Russell Lebowitz, and Lois Salzman for communication of unpublished experiments.

REFERENCES

Alestrom, P., Akusjarvi, G., Perricaudet, M., Mathews, M.B., Klessig, D.C., and Petterson, U., 1980, The gene for polypeptide IX of adenovirus type 2 and its unspliced messenger RNA, *Cell* **19**:671.

Aloni, T., Dhar, R., Lamb, O., Horowitz, M., and Khoury, G., 1977, Novel mechanism for RNA maturation: The leader sequences of simian virus 40 mRNA are not transcribed adjacent to coding sequences, *Proc. Natl. Acad. Sci. USA* **74**:3686.

Alwine, J.C., Kemp, F.J., Parker, B.A., Reiser, J., Renart, J., Stark, G.R., and Wahl, G.M., 1980, Detection of specific RNAs or specific fragments of DNA by fractionation in gels and transfer to diazobenzyloxymethyl paper, *Methods Enzymol.* **68**:220.

Armentrout, R., Bates, R., Berns, K., Carter, B., Chow, M., Dressler, D., Fife, W., Hauswirth, W., Hayward, G., Lavelle, G., Rhode, S., Straus, S., Tattersall, P., and Ward, D., 1978, A standardized nomenclature for restriction endonuclease fragments, in: *Replication of Mammalian Parvoviruses* (D. Ward and P. Tattersall, eds.), pp. 523–526, Cold Spring Harbor Laboratory, New York.

Astell, C.R., Smith, M., Chow, M.B., and Ward, D.C., 1979, Sequence of the 3′ terminus of the genome from Kilham RNA Virus, a non-defective parvovirus, *Virology*, **96**:669.
Babich, A., and Nevins, J.R., 1981, The stability of early adenovirus mRNA is controlled by the viral 72 kd DNA-binding protein, *Cell* **26**:371.
Bailey, J.M., and Davidson, N., 1976, Methylmercury as a reversible denaturing agent for agarose gel electrophoresis, *Anal. Biochem.* **70**:75.
Baker, C.C., Herisse, J., Courtois, G., Galibert, F., and Ziff, E., 1979, Messenger RNA for the Ad2 DNA binding protein: DNA sequences encoding the first leader and heterogeneity of the mRNA 5′ end, *Cell* **18**:569.
Bates, R.C., Patton, J.I., Burd, P.R., and Stout, E.R., 1981, Trancription of the DNA genome of an autonomous parvovirus, *DNA* **1**:80.
Ben-Asher, E., Bratosin, S., and Aloni, Y., 1982, Intracellular DNA of the parvovirus minute virus of mice is organized in a minichromosome structure, *J. Virol.* **41**:1044.
Benoist, C., and Chambon, P., 1980, Deletions covering the putative promoter region of early mRNAs of simian virus 40 do not abolish T-antigen expression, *Proc. Natl. Acad. Sci. USA* **77**:3865.
Berget, S.M., Moore, C., and Sharp, P.A., 1977, Spliced segments at the 5′ terminus of adenovirus 2 late mRNA, *Proc. Natl. Acad. Sci. USA* **74**:3171.
Berk, A.J., and Sharp, P.A., 1977a, Sizing and mapping of early adenovirus mRNAs by gel electrophoresis of S1 endonuclease digested hybrids, *Cell* **12**:721.
Berk, A.J., and Sharp, P.A., 1978, Structure of the adenovirus 2 early mRNAs, *Cell* **14**:695.
Berns, K.I., and Rose, J.A., 1970, Evidence for a single-stranded adenovirus associated virus genome: Isolation and separation of complementary single strands, *J. Virol.* **5**:693.
Berns, K.I., Hauswirth, W.W., Fife, K.H., and Lusby, E., 1978a, Adeno-associated virus DNA replication, *Cold Spring Harbor Symp. Quant. Biol.* **43**:781.
Bina, M., Feldman, R.J., and Deeley, R.G., 1980, Could Poly(A) align the splicing sites of messenger RNA precursors? *Proc. Natl. Acad. Sci. USA* **77**:1278.
Bloom, M.E., and Rose, J.A., 1978, Transcription of adenovirus-associated virus RNA in isolated KB cell nuclei, *Virology* **84**:118.
Breathnach, R., Mandel, J.L., and Chambon, P., 1977, Ovalbumin gene is split in chicken DNA, *Nature (London)* **270**:314.
Breathnach, R., Benoist, C., O'Hare, K., Gannon, F., and Chambon, P., 1978, Ovalbumin gene: Evidence for a leader sequence in mRNA and DNA sequences at the exon-intron boundaries, *Proc. Natl. Acad. Sci. USA* **75**:4853.
Britten, R.J., and Kohne, D.E., 1968, Repeated sequences in DNA, *Science* **161**:529.
Buller, R.M.L., and Rose, J.A., 1978a, Characterization of adenovirus-associated virus-induced polypeptides in vivo and in vitro, in: *Replication of Mammalian Parvoviruses* (D. Ward and P. Tattersall, eds.), pp. 281–288, Cold Spring Harbor Laboratory, New York.
Buller, R.M.L., and Rose, J.A., 1978b, Characterization of adenovirus associated virus-induced polypeptides in KB cells, *J. Virol.* **25**:331.
Buller, R.M.L., Straus, S.E., and Rose, J.A., 1979, Mechanism of host restriction of adenovirus-associated virus replication in African green monkey cells, *J. Gen. Virol.* **43**:663.
Busslinger, M., Portmann, R., Irminger, J.C., and Birnstiel, M.L., 1980, Ubiquitous and gene-specific regulatory 5′ sequences in a sea urchin histone DNA clone coding for histone protein variants, *Nucleic Acids Res.* **8**:957.
Carter, B.J., 1974, Analysis of parvovirus mRNA by sedimentation and electrophoresis in aqueous and nonaqueous solutions, *J. Virol.* **14**:834.
Carter, B.J., 1976, Intracellular distribution and polyadenylate content of adeno-associated virus RNA sequences, *Virology* **73**:273.
Carter, B.J., 1978, Parvovirus Transcription, in *Replication of Mammalian Parvoviruses* (D. Ward and P. Tattersall, eds.), pp. 33–52, Cold Spring Harbor Laboratory, New York.
Carter, B.J., and Khoury, G., 1975, Specific cleavage of adenovirus-associated virus DNA by restriction endonuclease R. *Eco* RI. Characterization of cleavage products, *Virology* **63**:523.

Carter, B.J., and Rose, J.A., 1972, Adenovirus-associated virus multiplication, VIII. Analysis of *in vivo* transcription induced by complete or partial helper viruses, *J. Virol.* **10:**9.

Carter, B.J., and Rose, J.A., 1974, Transcription in vivo of a defective parvovirus: Sedimentation and electrophoretic analysis of RNA synthesized by adenovirus-associated virus and its helper adenovirus, *Virology* **61:**182.

Carter, B.J., Khoury, G., and Denhardt, D.T., 1975, Physical map and strand polarity of specific fragments of adenovirus-associated virus DNA produced by endonuclease, R. *Eco*R1, *J. Virol.* **16:**559.

Carter, B.J., Khoury, G., and Rose, J.A., 1972, Adenovirus-associated virus multiplication IX. Extent of transcription of the viral genome *in vitro, J. Virol.* **10:**1118.

Carter, B.J., Koczot, F.J., Garrison, J., Rose, J.A., and Dolin, R., 1973, Separate helper functions provided by adenovirus for adenovirus associated virus multiplication, *Nature New Biol.* **244:**71.

Carter, B.J., Laughlin, C.A., de la Maza, L.M., and Myers, M., 1979, Adeno-associated virus autointerference, *Virology* **92:**449.

Casey, J., and Davidson, N., 1977, Rates of formation and thermal stabilities of RNA: DNA and DNA:DNA duplexes at high concentrations of formamide, *Nucleic Acids Res.* **4:**1539.

Chow, L.T., Gelinas, R.E., Broker, T.R., and Roberts, R.J., 1977, An amazing sequence arrangement of the 5′ ends of adenovirus 2 messenger RNA, *Cell* **12:**1.

Corden, J., Wasylyk, B., Buchwalder, A., Sassone-Corsi, P., Kedinger, C., and Chambon, P., 1980, Promoter sequences of eukaryotic protein-coding genes, *Science* **209:**1406.

Crotese, R., Harland, R., and Melton, D., 1980, Transcription of tRNA genes *in vivo*: Single-stranded compared to double-stranded templates, *Proc. Natl. Acad. Sci. USA* **77:**4147.

de la Maza, L.M., and Carter, B.J., 1980a, Molecular structure of adeno-associated virus variant DNA, *J. Biol. Chem.* **255:**3194.

Dierks, P., von Oøyen, A., Mantel, N., and Weissman, C., 1981, DNA sequences preceding the rabbit β-globin gene are required for formation in mouse L cells of β-globin RNA with the correct 5′ terminus, *Proc. Natl. Acad. Sci. USA* **78:**1411.

Drake, S., Schaffer, P.A., Esparza, J., and Mayor, M.D., 1974, Complementation of adeno-associated satellite viral antigens and infectious DNA by temperature-sensitive mutants of herpes simplex virus, *Virology* **60:**230.

Efstratiadis, A., Posakony, J.W., Maniatis, T., Lawn, R.M., O'Connell, C., Spritz, R.A., deRiel, J.K., Forgel, B.G., Weissman, S.M., Slightom, J.C., Blechl, A.E., Smithies, O., Baralle, F.E., Shoulders, C.C., and Proudfoot, N.J., 1980, The structure and evaluation of the human β-globin gene family, *Cell* **21:**653.

Ensinger, M.J., and Ginsberg, H.S., 1972, Selection and preliminary characteristics of temperature-sensitive mutants of type 5 adenovirus, *J. Virol.* **10:**328.

Favaloro, J., Treisman, R., and Kamen, R., 1980, Transcription maps of polyoma virus-specific RNA. Analysis by two dimensional nuclease S1-gel mapping, *Methods Enzymol.* **65:**718.

Fitzgerald, M., and Shenk, T., 1981, The sequence 5′-AAUAAA-3′ forms part of the recognition sequence site for polyadenylation of late SV40 mRNAs, *Cell* **24:**251.

Flavell, A.J., Cowie, A., Legon, S., and Kamen, R., 1979, Multiple 5′ terminal cap structures in late polyoma virus RNA, *Cell* **16:**351.

Flavell, A., Cowie, A., Arrand, J., and Kamen, R., 1980, Localization of three major capped 5′ ends of polyoma virus late mRNAs within a single tetranucleotide sequence in the viral genome, *J. Virol.* **33:**902.

Flint, S.J., and Broker, T.R., 1980, Lytic infection by adenoviruses, in: *DNA Tumor Viruses* (J. Tooze, ed.), pp. 443–546, Cold Spring Harbor Laboratory, New York.

Friedmann, T., Esty, A., LaPorte, P., and Deininger, P., 1979, The nucleotide sequence and genome organization of the polyoma early region: Extensive nucleotide and amino acid homology with SV40, *Cell* **17:**715.

Gannon, F., O'Hare, K., Perrin, F., LePennec, J.P., Benoist, C., Cochet, M., Breathnach, R., Royal, A., Garapin, A., Cami, B., and Chambon, P., 1979, Organization and sequences at the 5′ end of a clonal complete ovalbumin gene, *Nature* **278:**428.

Gariglio, P., and Moussett, S., 1975, Isolation and partial characterization of a nuclear RNA polymerase-SV40 DNA complex, *FEBS Lett.* **56**:149.

Gelb, L.D., Kohne, D.E., and Martin, M.A., 1971, Quantitation of SV40 sequences in African green monkey, mouse and virus transformed cell genomes, *J. Mol. Biol.* **57**:129.

Ghosh, P., Reddy, V., Swiascue, J., Lebowitz, P., and Weissman, S., 1979, Heterogeneity and 5′ terminal structures of the late RNAs of simian virus 40, *J. Mol. Biol.* **126**:813.

Ghosh, P.K., Lebowitz, P., Trisque, F.J., and Gluzman, Y., 1981a, Identification of a promoter component involving positioning the 5′ termini of simian virus 40 early mRNAs, *Proc. Natl. Acad. Sci. USA* **78**:100.

Ghosh, P.K., Roy, P., Barkan, A., Mertz, J.E., Weissman, S., and Lebowitz, P., 1981b, Unspliced functional late 19 S mRNAs containing intervening sequences are produced by a late leader mutant of simian virus 40, *Proc. Nat. Acad. Sci. USA* **78**:1386.

Gidoni, D., Kahana, C., Canaani, D., and Groner, Y., 1981, Specific *in vitro* transcription of simian virus 40 early and late genes occurs at various cap nucleotides including cytidine, *Proc. Natl. Acad. Sci. USA* **78**:2174.

Ginsberg, H.S., Ensinger, M.J., Kauffman, R.S., Mayer, A.J., and Undholm, U., 1974, Cell transformation: A study of regulation with types 5 and 12 adenovirus temperature-sensitive mutants, *Cold Spring Harbor Symp. Quant. Biol.* **34**:419.

Goldberg, M., 1978, Ph.D. thesis, Stanford University, Stanford, California.

Green, M.R., and Roeder, R.G., 1980a, Transcripts of the adenovirus-associated virus genome: Mapping of the major RNAs *J. Virol.* **36**:79.

Green, M.R., and Roeder, R.G., 1980b, Definition of a novel promoter for the major adenovirus-associated virus mRNA, *Cell* **22**:231.

Green, M.R., Lebowitz, R.M., and Roeder, R.G., 1979, Expression of the autonomous parvovirus H1 genome: Evidence for a single transcriptional unit and multiple spliced polyadenylated transcripts, *Cell* **17**:967.

Greenberg, J., 1975, Messenger RNA metabolism in animal cells, *J. Cell. Biol.* **64**:269.

Groner, Y., Carmi, P., and Alomi, Y., 1978, Capping structures of simian virus 40 19S and 16S mRNAs, *Nucleic Acids Res.* **4**:3959.

Grosschedl, R., and Birnstiel, M.L., (1980) Identification of regulatory sequences in the prelude sequences of an H2A histone gene by the study of specific deletion mutants in vivo, *Proc. Nat. Acad. Sci. USA* **77**:1432.

Grosschedl, R., and Birnstiel, M.L., 1982, Delimitation of far upstream sequences required for maximal in vitro transcription of an H2A histone gene, *Proc. Natl. Acad. Sci. USA* **79**:297.

Grosschedl, R., Wasylyk, B., Chambon, P., and Birnstiel, M.L., 1981, Point mutation in the TATA box curtails expression of sea urchin H2A histone gene in vivo, *Nature (London)* **294**:178.

Grosveld, G.C., Shewmaker, C.K., Jat, P., and Flavell, R.A., 1981, Localization of DNA sequences necessary for transcription of the rabbit β-globin gene *in vitro*, *Cell* **25**:215.

Gruss, P., Lai, C-J., Dhar, R., and Khoury, G., 1979, Splicing as a requirement for biogenesis of functional 16S mRNA of simian virus 40, *Proc. Natl. Acad. Sci. USA* **76**:4317.

Gruss, P., Dhar, R., and Khoury, G., 1981a, Simian virus 40 tandem repeated sequences as an element of the early promoter, *Proc. Natl. Acad. Sci. USA* **78**:943.

Gruss, P., Efstratiadis, A., Karathanasis, S., Konig, M., and Khoury, G., 1981b, Synthesis of stable unspliced mRNA from an intronless simian virus 40-rat preproinsulin gene recombinant, *Proc. Natl. Acad. Sci. USA* **78**:6091.

Haegeman, G., and Fiers, W., 1978, Localization of the 5′ terminus of late SV40 mRNA, *Nucleic Acids Res.* **5**:2359.

Hamer, D.H., and Leder, P., 1979, Splicing and the formation of stable RNA, *Cell* **18**:1299.

Handa, H., Shiroki, K., and Shimojo, H., 1975, Complementation of adeno-associated virus growth with temperature-sensitive mutants of human adenovirus types 12 and 5, *J. Gen. Virol.* **29**:239.

Hashimoto, S., Pursley, M.H., Wold, W.S.M., and Green, J., 1980, Characterization of distinct 5′-terminal cap structures of adenovirus type 2 early messenger ribonucleic acid and KB cell messenger ribonucleic acid, *Biochemistry* **19**:294.

Hentschel, C., Irminger, J-C., Bucher, P., and Birnsteil, M.L., 1980, Sea urchin histone mRNA termini are located in gene regions downstream from putative regulatory sequences, *Nature* **285:**147.

Hirt, B., 1967, Selective extraction of polyoma DNA from infected mouse cell cultures, *J. Mol. Biol.* **26:**365.

Hofer, E., and Darnell, J.E. Jr., 1981, The primary transcription unit of the mouse β-major globin gene, *Cell* **23:**585.

Hu, S.L., and Manley, S.L., 1981, DNA sequence required for initiation of transcription in vitro from the major late promoter of adenovirus 2, *Proc. Natl. Acad. Sci. USA* **78:**820.

Ishibashi, M., and Ito, M., 1971, The potentiation of type 1 adeno-associated virus by temperature-sensitive conditional lethal mutants of CELO virus at the restrictive temperature, *Virology* **45:**317.

Ito, M., and Susuki, E., 1970, Adeno-associated satellite virus growth supported by a temperature-sensitive mutant of human adenovirus, *J. Gen. Virol.* **9:**243.

Janik, J.E., Huston, M.M., and Rose, J.A., 1981, Locations of adenovirus genes required for the replication of adenovirus-associated virus, *Proc. Natl. Acad. Sci. USA* **78:**1925.

Jay, F.T., Laughlin, C.A., de la Maza, L.M., Carter, B.J., and Cook, W.J., 1978, Adeno-associated-virus RNA synthesis in vivo and in vitro, in: *Replication of Mammalian Parvoviruses* (D. Ward and P. Tattersall, eds.), pp. 385–397, Cold Spring Harbor Laboratory, New York.

Jay, F.T., de la Maza, M., and Carter, B.J., 1979, Parvovirus RNA transcripts containing sequences not present in mature mRNA: A method for isolation of putative mRNA precursor sequences, *Proc. Natl. Acad. Sci. USA* **76:**625.

Jay, F.T., Laughlin,C.A., and Carter, B.J., 1981, Eukaryotic translational control: Adeno-associated cirus protein synthesis is affected by a mutation in the adenovirus DNA binding protein, *Proc. Natl. Acad. Sci. USA* **78:**2927.

Johnson, F.B., Thomson, T.A., Taylor, P.A., and Vlazny, D.A., 1977, Molecular similarities among the adenovirus-associated virus polypeptides and evidence for a precursor protein, *Virology* **82:**1.

Khoury, G., Gruss, P., Dhar, R., and Lai, C-J., 1979, Processing and expression of early SV40 mRNA: A role for RNA conformation in splicing, *Cell* **18:**85.

Khoury, G., and Martin, M., 1977, A comparison of SV40 transcription *in vivo* and *in vitro*, *Nature New Biol.* **238:**4.

Kitchingman, G.R., Lai, S.P., and Westphal, H., 1977, Loop structures in hybrids of early RNA and the separated strands of adenovirus DNA, *Proc. Natl. Acad. Sci. USA* **74:**4392.

Klessig, D.F., and Chow, L.T., 1980, Incomplete splicing of several late viral mRNAs in monkey cells infected by human adenovirus type 2, *J. Mol. Biol.* **139:**221.

Konkel, D.A., Tilghman, S.M., and Leder, P., 1978, The sequence of the chromosomal mouse β-globin major gene: Homologies in capping, splicing and poly(a) sites, *Cell* **15:**1125.

Konkel, D.A., Maizel, J.V., Jr., and Leder, P., 1979, The evaluation and sequence comparison of two recently diverged mouse chromosomal β-globin genes, *Cell* **18:**865.

Lai, C-J., Dhar, R., and Khoury, G., 1978, Mapping the spliced and unspliced late lytic SV40 RNAs, *Cell* **14:**971.

Laughlin, C.A., Westphal, H., and Carter, B.J., 1979a, Spliced adenovirus-associated virus RNA, *Proc. Natl. Acad. Sci. USA* **76:**5566.

Laughlin, C.A., Myers, M.W., Risin, D.L., and Carter, B.J., 1979b, Defective-interfering particles of the human parvovirus adeno-associated virus, *Virology* **94:**162.

Lawn, R.M., Adelman, J., Franke, A.E., Houck, C.M., Gross, M., Nagarian, R., and Goeddel, D.V., 1981, Human fibroblast interferon gene lacks introns, *Nucleic Acids Res.* **9:**1045.

Leder, P., Tilghman, S.M., Tiemeier, D.C., Polsky, F.I., Seidman, J.G., Edgell, M.H., Enquist, L.W., Leder, A., and Norman, B., 1977, The cloning of mouse globin and surrounding gene sequences in bacteriophage λ, *Cold Spring Harbor Symp. Quant. Biol.* **42:**915.

Lee, D.C., and Roeder, R.G., 1981, Transcription of adenovirus type 2 genes in a cell-free system: Apparent homogeneity of initiation at some promoters, *Molecular and Cellular Biology* **1:**635.

Legon, S., Flavell, A.J., Lowie, A., and Kamen, R., 1979, Amplification in the leader sequence of late polyoma virus mRNAs, *Cell* **16**:20.

Lenk, R., Storch, I., and Maizel, J.V., Jr., 1980, Cell architecture during adenovirus infection, *Virology* **105**:19.

Lerner, M.R., Boyle, J.A., Mount, S., Wolin, S., and Steitz, J.A., 1980, Are snRNPs involved in splicing? *Nature* **283**:220.

Lusby, E.W., and Berns, K.I., 1982, Mapping the 5′-termini of two adeno-associated virus RNAs in the left half of the genome, *J. Virol.* **41**:518.

Manley, J.L., Fire, A., Cano, A., Sharp, P.A., and Gefter, M.A., 1980, DNA-dependent transcription of adenovirus genes in a soluble whole cell extract, *Proc. Natl. Acad. Sci. USA* **77**:3855.

Marcus, C.J., and Carter, B.J., 1982, Nucleosome formation by adeno-associated virus DNA, *Fed. Proc. Abs.* **515**.

Marcus, C.J., Laughlin, C.A., and Carter, B.J., 1981, Adeno-associated virus RNA transcription in vivo, *Europ. J. Biochem.* **121**:147.

Mathews, M.B., 1980, Binding of adenovirus VA RNA to mRNAs: A possible role in splicing? *Nature* **285**:575.

Mayor, H.D., and Ratner, J., 1972, Conditionally defective helper adenovirus and satellite virus replication, *Nature New Biol.* **239**:20.

McDonnell, M.W., Simon, M.N., and Studier, F.W., 1977, Analysis of restriction fragments of T7 DNA and determination of molecular weights by electrophoresis in neutral and alkaline gels, *J. Mol. Biol.* **146**:119.

McKnight, S.L., 1980, The nucleotide sequence and transcript map of the herpes simplex virus thymidine kinase gene, *Nucleic Acids Res.* **8**:5949.

McKnight, S.L., Gavis, E.R., Kingsburg, R., and Axel, R., 1981, Analysis of transcriptional regulating signals of the HSV thymidine kinase gene: Identification of an upstream control region, *Cell* **25**:385.

McMaster, G.K., and Carmichael, G.G., 1977, Analysis of single- and double-stranded nucleic acids on polyacrylamide and agarose gels by using glyoxal and acridine orange, *Proc. Natl. Acad. Sci. USA* **74**:4835.

Mount, S.M., 1982, A catalogue of splice junctions, *Nucleic Acids Res.* **10**:459.

Murray, V., and Holliday, R., 1979, Mechanism for splicing of gene transcripts, *FEBS Lett.* **106**:5.

Muster, C.J., Lee, Y.S., Newbold, J., and Leis, J., 1980, Physical mapping of adenoassociated virus serotype 4 DNA, *J. Virol.* **35**:653.

Myers, M.W., Laughlin, C.A., Jay, F.T., and Carter, B.J., 1980, Adenovirus helper function for growth of adeno-associated virus: Effect of temperature-sensitive mutations in adenovirus early gene region 2, *J. Virol.* **35**:65.

Myers, R.M., Rio, D.C., Robbins, A.K., and Tjian, R., 1981, SV40 gene expression is modulated by the cooperative binding of T-antigen to DNA, *Cell* **25**:373.

Nagata, S., Mantei, N., and Weissman, 1980, The structure of one of the eight or more distinct chromosomal genes for human interferon-α, *Nature (London)* **287**:401.

Nevins, J.R., 1982, Adenovirus gene expression: Control at multiple steps of mRNA biogenesis, *Cell* **28**:1.

Nevins, J.R., and Darnell, J.E., Jr., 1978, Steps in the processing of Ad2 mRNA: Poly(A)+ nuclear sequences are conserved and poly(A) addition precedes splicing, *Cell* **15**:1477.

Nishioka, Y., and Leder, P., 1979, The complete sequence of a chromosomal mouse α-globin gene reveals elements conserved throughout vertebrate evolution, *Cell* **18**:875.

Ostrove, J., Duckworth, D.H., and Berns, K.I., 1981, Inhibition of adenovirus transformed cell oncogenicity by adeno-associated virus, *Virology* **113**:521.

Patton, J.T., Stout, E.R., and Bates, R.C., 1979, Transcription of the bovine parvovirus genome in isolated nuclei, *J. Virol* **30**:917.

Pribnow, D., 1975a, Nucleotide sequence of an RNA polymerase binding site at an early T7 promoter, *Proc. Natl. Acad. Sci. USA* **72**:784.

Pribnow, D., 1975b, Bacteriophage T7 early promoters: Nucleotide sequences of two RNA polymerase binding sites, *J. Mol. Biol.* **99**:419.

Proudfoot, N., and Brownlee, G.G., 1974, Sequence at the 3′ end of globin mRNA shows homology with immunoglobin light chain mRNA, *Nature* **252**:359.

Rhode, S.L., 1973, Replication process of the parvovirus H-1.I. Kinetics in a parasynchronous cell system, *J. Virol.* **11**:856.

Richardson, W.R., and Westphal, H., 1981, A cascade of adenovirus early functions is required for expression of adeno-associated virus, *Cell* **27**:133.

Rigby, P.W.J., Dieckman, M., Rhodes, C., and Berg, P., 1977, Labeling of deoxyribonucleic acid to high specific activity *in vitro* by nick translation with DNA polymerase I., *J. Mol. Biol.* **113**:237.

Roeder, R.C., 1976, Eukaryotic nuclear RNA polymerases, in: *RNA Polymerase* (R. Losick and M. Chamberlain, eds.), p. 285, Cold Spring Harbor Laboratory, Cold Spring Harbor, New York.

Rogers, J., and Wall, R., 1980, A mechanism for RNA splicing, *Proc. Natl. Acad. Sci. USA* **77**:1877.

Roop, D.R., Tsai, M-J., and O'Malley, B.W., 1980, Definition of the 5′ and 3′ ends of transcripts of the ovalbumin gene, *Cell* **19**:63.

Rose, J.A., and Koczot, F.J., 1971, Adenovirus-associated virus multiplication. Base composition of the deoxyribonucleic acid strand species and strand-specific *in vivo* transcription, *J. Virol.* **8**:771.

Rose, J.A., Maizel, J.V., Jr., Inman, J.K., and Shatkin, A.J., 1971, Structural proteins of adenovirus-associated viruses, *J. Virol.* **8**:766.

Rosenberg, M., and Court, D., 1979, Regulatory sequences involved in the promotion and termination of RNA transcription, *Ann. Rev. Genetics* **13**:319.

Salzman, L.A., and Redler, B., 1974, Synthesis of virus-specific RNA in cells infected with parvovirus Kilham rat virus, *J. Virol.* **14**:434.

Salzman, L.A., White, W.L., and L. McKerlie, 1972, Growth characteristics of Kilham rat virus and its effect on cellular macromolecular synthesis, *J. Virol.* **10**:573.

Sanger, F.G., Air, G.M., Barrell, B.G., Brown, N.L., Coulson, A.R., Fiddes, T.C., Hutchison, C.A., III, Slocombe, P.M., and Smith, M., 1977, Nucleotide sequence of bacteriophage φX174, *Nature* **265**:687.

Schaffner, W., Groos, K., Telford, J., and Birnstiel, M., 1976, Molecular analysis of the histone gene cluster of *Psammechinns miliaris*: II. The arrangement of the five histone-coding and spacer sequences, *Cell* **8**:471.

Seif, I., Khoury, G., and Dhar, R., 1979, BKV splice sequences based on analysis of preferred donor and acceptor sites, *Nucleic Acids Res.* **6**:3387.

Sharp, P.A., 1981, Speculations on RNA splicing, *Cell* **23**:643.

Shatkin, A.J., 1976, Capping of eukaryotic mRNAs, *Cell* **9**:645.

Sheiness, D., and Darnell, J.E., 1973, Polyadenylic acid segment in mRNA becomes shorter with age, *Nature New Biol.* **241**:265.

Siebenlist, U., Simpson, R.B., and Gilbert, W., 1980, *E. coli* RNA polymerase interacts homologously with two different promoters, *Cell* **20**:269.

Siegl, G., and Gautschi, M., 1973, The multiplication of parvovirus LuIII in a synchronized culture system II. Biochemical characteristics of virus replication, *Arch. Gesamte Virusforsch.* **40**:119.

Southern, E.M., 1975, Detection of specific sequences among DNA fragments separated by gel electrophoresis, *J. Mol. Biol* **98**:503.

Straus, S.E., Ginsberg, H.S., and Rose, J.A., 1976a, DNA-minus temperature-sensitive mutants of adenovirus type 5 help adenovirus-associated virus replication, *J. Virol.* **17**:140.

Sures, I., Levy, S., and Kedes, L.H., 1980, Leader sequences of *Strongylocentrotus purpuratus* histone mRNAs start at a unique heptanucleotide common to all five histone genes, *Proc. Natl. Acad. Sci. USA* **77**:1265.

Sutton, W.D., 1971, A crude nuclease preparation suitable for use in DNA reassociation experiments, *Biochem. Biophys. Acta.* **240**:522.

Tal, J., Ron, D., Tattersall, P., Bratosin, S., and Aloni, Y., 1979, About 30% of minute virus of mice RNA is spliced out following polyadenylation, *Nature* **279**:649.
Tattersall, P., 1972, Replication of parvovirus MVM. I. Dependence of virus multiplication and plaque formation on cell growth, *J. Virol.* **10**:586.
Tennant, R.W., and Hand, R.E., Jr., 1970, Requirement of cellular DNA synthesis for Kilham rat virus replication, *Virology* **42**:1054.
Tonegawa, S., Maxam, A.N., Tizard, R., Bernhard, O., and Gilbert, W., 1978, Sequences of a mouse germ-line gene for a variable region of an immunoglobulin light chain, *Proc. Natl. Acad. Sci. USA* **75**:1485.
VanderVliet, P.C., Levine, A.J., Ensinger, M.J., and Ginsberg, H.S., 1975, Thermolabile DNA binding proteins from cells infected with a temperature-sensitive mutant of adenovirus defective in viral DNA synthesis, *J. Virol.* **15**:348.
Van Ormondt, H., Moat, J., deWaard, A., and Van der Eb, A.J., 1978, The nucleotide sequence of the transforming *HpaI*-E fragment of adenovirus type 5 DNA, *Gene* **4**:309.
Vogelstein, B., and Gillespie, D., 1977, RNA-DNA hybridization in solution without DNA reannealing, *Biochem. Biophys. Res. Commun.* **75**:1127.
Wagner, M.J., Sharp, J.A., and Summer, W.C., 1981, Nucleotide sequence of the thymidine kinase gene of herpes simplex type 1, *Proc. Natl. Acad. Sci. USA* **78**:1441.
Wasylyk, B., Oerbyshire, R., Guy, A., Molko, D., Roget, A., Teoule, R., and Chambon, P., 1980, Specific in vitro transcription of conalbumin gene is drastically decreased by a single point mutation in the TATA box homology sequence, *Proc. Natl. Acad. Sci. USA* **77**:7024.
Weaver, R.F., and Weissman, C., 1979, Mapping of RNA by a modification of the Berk and Sharp procedure: The 5′ termini of 15S β-globin MRNA precursor and mature 10S β-globin mRNA have identical map coordinates, *Nucleic Acids Res.* **7**:1175.
Weil, P.A., Luse, D.S., Segall, J., and Roeder, R.G., 1979, Selective and accurate initiation of transcription at the Ad2 major late promoter in a soluble system dependent on purified RNA polymerase II and DNA, *Cell* **18**:469.
Weintraub, H., 1980, Recognition of specific DNA sequences in eukaryotic chromosomes, *Nucleic Acids Res.* **8**:4745.
Weinmann, R., Jaehning, J.A., Raskas, H.J., and Roeder, R.G., 1976, Viral RNA synthesis and levels of DNA-dependent RNA polymerases during replication of adenovirus 2, *J. Virol.* **17**:114.
Zeevi, M., Nevins, J.R., and Darnell, J.E., Jr., 1981, Nuclear RNA is spliced in the absence of poly(A) addition, *Cell* **26**:39.
Ziff, E.B., 1980, Transcription and RNA processing by the DNA tumour viruses, *Nature (London)* **287**:491.
Ziff, E.B., and Evans, R.M., 1978, Coincidence of the promoter and capped 5′ terminus of RNA from the adenovirus 2 major late transcription unit, *Cell* **15**:1463.

CHAPTER 6

Variant and Defective Interfering Parvoviruses

Barrie J. Carter

I. INTRODUCTION

Growth of viruses in eukaryotic cells frequently results in production of variant viruses having altered or deleted genome sequences. Detailed study of these variants provides important information about virus growth and virus–host interactions. Variant genomes may arise through aberrant replication or recombination. In many cases, these variant genomes are encapsidated into virus particles. The physical structures and biological properties of such variants have important implications for the involvement of *cis-* or *trans-*acting functions of regions of the virus genome sequence in replication and particle assembly. Understanding the mechanisms by which variant genomes are generated may be a productive approach to analysis of some aspects of eukaryotic recombination.

Encapsidated variant genomes often are able to interfere with virus replication in subsequent growth cycles. Such variants were named defective interfering (DI) particles by Huang and Baltimore (1970). These same authors also pointed out that such DI particles may play an important role in the evolution of viral infections.

Variant virus particles have been observed for most parvoviruses. When lysates from parvovirus infected cells are banded to equilibrium in CsCl bouyant density gradients the infectious parvovirus particles band at a density of approximately 1.41 g/cm^3. A minor, less infectious pop-

BARRIE J. CARTER • Laboratory of Cell Biology and Genetics, National Institute of Arthritis, Diabetes, and Digestive and Kidney Diseases, National Institutes of Health, Bethesda, Maryland 20205.

ulation of particles bands at about 1.45 g/cm^3. When these infectious particles are viewed by electron microscopy using negative staining they are not penetrated by the stain and thus appear "full." Parvovirus "empty" particles, which are penetrated by a negative stain, band at a buoyant density of about 1.31–1.32 g/cm^3. In addition, there are often present particles which band in the density region between "empty" and "full" particles and show partial penetration of an electron-dense, negative stain. These intermediate density particles are observed for both the helper-dependent, adeno-associated virus, AAV (Torakai *et al.*, 1970), and the autonomous (helper-independent) parvoviruses (reviewed by Siegl, 1976).

These intermediate-density particles, which are now known to contain genomes of varying molecular weight or sequence, are the subject of this chapter. It was quickly realized that these variant particles contained smaller DNA molecules (Torakai *et al.*, 1970; Crawford, 1966; Johnson *et al.*, 1974; Siegl, 1972, 1973) but the first systematic studies on the physical structure and biological properties of parvovirus variants were not reported until the parvovirus meeting held at Cold Spring Harbor in 1977 (de la Maza and Carter, 1978; Siegl *et al.*, 1978; Rhode, 1978b). Since that time a great deal of detailed information has been obtained and this constitutes most of the work which I shall review here. I will first review these particles collectively as variant particles and then discuss their defective interfering properties. Prior to discussion of either of these aspects I will clearly define certain aspects of nomenclature and terminology as I will use it here.

I will not review here those variations which appear to be of the deletion-substition type occurring in localized regions of some autonomous parvovirus genomes and which may alter the tissue and cell tropism of these viruses. This subject is discussed elsewhere in this volume (see Chapter 8).

II. NOMENCLATURE

A variety of different, inconsistent, and sometimes confusing terms have been used in the literature in reference to parvovirus particles, proteins, and DNA. It is important, therefore, to define clearly at the beginning of this review certain terms and the sense in which they will be used.

The terminology used in reference to parvovirus particles has been particularly confusing. The intermediate-density particles (or the genomes which they contain) variously have been referred to as "incomplete", "D," "defective," "light," or "variant" particles. Many of these terms are unsatisfactory for general use. The term "incomplete" has been used to refer to assembly intermediates. Also, broken genomes are incomplete. "Defective" and "D" were used to denote noninfectivity of variant particles but are unwieldy for use with AAV which is a helper-dependent (defective) virus. The term "viable defective" to denote variants of H-1 virus which can replicate in the absence of standard virus is also confusing (Rhode, 1978b). The term "light" also has been used to refer to the major type of infectious particles banding in CsCl at 1.41 g/cm^3 in contrast to the "heavy" infectious particles that band at 1.45 g/cm^3. I will use the following four terms as a general nomenclature.

1. *Standard* virus particles, as defined by Huang and Baltimore (1970), have a normal complement of structural proteins and contain a homologous infectious genome capable of independent replication. This definition is extended to include standard AAV particles. Although AAV is a helper-dependent virus the helper is a heterologous virus. Indeed, AAV can be viewed semantically as a nondefective virus for which the host range is an adenovirus-infected or herpes virus-infected cell as discussed elsewhere in this volume (see Chapter 3). Standard particles variously have been referred to as wild-type, mature, or infectious virus or full particles.

2. *Variant* particles contain a normal complement of structural proteins and contain a homologous genome that is lacking part of, or has a different conformation than, the standard genome. These particles include most of the intermediate-density particles mentioned above but do not include assembly intermediates such as those described for AAV (Myers and Carter, 1980; de la Maza and Carter, 1980). These assembly intermediates can be identified because the genome associated with them is available for digestion by deoxyribonucleases. The DNA in mature standard or DI particles is completely resistant to nuclease digestion.

3. *Defective interfering* (*DI*) particles were defined by Huang and Baltimore (1970) as variant particles which interfere specifically with intracellular replication of nondefective, homologous (standard) virus (see Section VIII.B).

4. *Empty* particles contain normal structural proteins but have no DNA. It should be noted that for the autonomous parvoviruses the ratio of three polypeptides differs between empty particles and standard virus as a consequence of a proteolytic cleavage during assembly (Tattersall, 1978).

The above nomenclature scheme relies only on the structure of the encapsidated genome and can be used to describe the purified DNA as well as whole particles. The use of "defective" will be restricted to the description of DI particles. Thus, AAV will be referred to as a helper-dependent virus. The autonomous parvoviruses can then logically be described as helper independent.

It is frequently useful to classify virus particles in terms of their bouyant density at equilibrium in CsCl gradients. Operationally, this is usually the first step in purification and fractionation of the various species of parvovirus particles. In this classification system (Carter *et al.*,

1979) the virus name is followed by the bouyant density of the particle in parentheses. Thus, AAV particles banding at 1.41 g/cm^3 are called AAV (1.41) particles. When it is clear which parvovirus is being referred to, the virus name is omitted, e.g., 1.41 particles, 1.41 DNA.

The nomenclature of parvovirus structural protein unfortunately is inconsistent and no general consensus has been reached. Most parvoviruses contain three size classes of structural polypeptides when analyzed by gel electrophoresis in denaturing conditions (Tattersall, 1978). Most groups have designated these proteins as A, B, C or VP1, VP2, VP3 where A or VP1 is the highest species. However, one group has labeled the AAV proteins VP3, VP2, and VP1, where VP3 is the highest and VP1 is the lowest molecular weight. I will use the single nomenclature system, VP1, VP2, and VP3 where VP1 is the highest-molecular-weight species (Bachmann *et al.*,1979).

Some variant genomes have a conformation similar to that of certain intracellular replicating forms of parvovirus DNA. These molecules consist of complementary strands which are covalently joined at one terminus by a hairpin structure formed by virtue of the palindromic terminal sequence 126 nucleotides in length. These molecules at neutral pH spontaneously renature with zero-order kinetics to form duplex structures. I will refer to such molecules as hairpin molecules. They have been variously referred to as spontaneously renaturing, terminally cross-linked, or snap-back molecules.

For description of parvovirus genome and map locations the nomenclature system derived by Armentrout *et al.* (1978) will be used. Briefly, the genome is divided into 100 map units (1 map unit is approximately 45 to 50 nucleotides). Restriction fragments (or RNA transcripts) are designated by their map coordinates, e.g., *Eco*R1 38.1/42.6.

III. DETECTION OF VARIANT PARVOVIRUSES

Parvovirus variants can be detected in a number of ways dependent upon either physical or biological properties. As already mentioned, these variants were first detected because they banded to equilibrium in CsC1 gradients at intermediate bouyant densities. It was then demonstrated that in general these variant particles were noninfectious and contained DNA molecules which were smaller than standard genomes (Torakai *et al.*, 1970; Crawford, 1966; Johnson *et al.*, 1974; Siegl, 1972, 1973; de la Maza and Carter, 1978; Muller *et al.*, 1978; Faust and Ward, 1979; Laughlin *et al.*, 1979).

Parvovirus variants have also been detected by analysis of intracellular DNA from infected cells. Thus, analysis of isotopically labeled intracellular AAV DNA molecules extracted from infected cells using the method of Hirt (1967) also revealed variant genomes (Hauswirth and Berns, 1978, 1979; Carter *et al.*, 1979). In addition to the standard and replicating form (RF) AAV genomes there was a population of smaller variant AAV DNA molecules. Similar intracellular AAV variant genomes, which were described as "dead-end" molecules, were observed earlier by Straus *et al.* (1976). These intracellular variant molecules appear to be the counterparts of encapsidated variant genomes (see below). Variant genomes of H-1 obtained from intracellular DNA have also been analyzed by Rhode (1978a,b,c).

In some cases parvovirus variants have been recognized because of their biological properties of interference. Early studies of AAV growth indicated that some AAV stocks exhibited a multiplicity-dependent autointerference (Torakai *et al.*, 1970; Rose and Koczot, 1972), and this phenomenon was later shown to be associated with the presence of variant genomes (Carter *et al.*, 1979; Laughlin *et al.*, 1979a). Rhode (1978a,b,c) obtained H-1 variants by successive high multiplicity passage and assay of the resulting virus for loss of infectious yield and for interference. More recently, it was suggested that Aleutian disease virus of mink (ADV) may generate variant or interfering particles since repeated undiluted passage resulted in loss of virus titer (Bloom *et al.*, 1980).

IV. PHYSICAL PROPERTIES OF VARIANT PARTICLES

The physical properties of variant particles from several parvoviruses are summarized in Table I. Two features are immediately apparent from the collected data:

i. There is a remarkable similarity in the properties of variant particles from the several different viruses.

ii. There is a direct relationship between the molecular weight of variant DNA and the bouyant density in CsCl or the sedimentation value of the particle. This relationship is also seen for AAV when virus is labeled under steady state conditions with both [^{3}H]thymidine and [^{35}S]methionine. The purified virus particles exhibit a ratio of $^{35}S:^{3}H$ which decreases linearly with CsCl bouyant density (M. Myers and B. Carter, unpublished). These observations imply that each variant particle contains only one molecule of DNA of the size indicated in Table I. This conclusion is further supported for AAV variant particles by several analyses which show that these particles have exactly the same proportions of the three capside proteins as do empty (1.32) particles and standard (1.41) or (1.45) particles (Johnson *et al.*, 1975; de la Maza and Carter 1980b; F. Jay, L. de la Maza, and B. J. Carter, unpublished). Furthermore, all the AAV particles exhibit the same degree of charge microheterogeneity in each of the three capsid proteins when analyzed in the O'Farrell (1975) two-dimensional gel electrophoresis system (F. Jay *et al.*, unpublished).

The above relationships do not account for the difference in properties of the standard 1.45 and 1.41 particles. For AAV this appears to be due to some difference in the protein conformation rather than any dif-

TABLE I. Physical Properties of Parvovirus Particles[a]

Particle[b]	Sedimentation value[c] Particle	Sedimentation value[c] DNA	DNA mol. wt. ($\times 10^{-6}$)	Percent genome	Infectivity ratio particles/infectious unit
AAV(1.45)	109	15.5	1.4	100	1600
AAV(1.41)	111	15.5	1.4	100	40–100
AAV(1.37)	82	11.0	0.63	45	$>10^7$
AAV(1.35)	78	8.5	0.35	20	$>10^7$
AAV(1.32)	66	5.0	0.10	7	$>10^8$
LuIII(1.41)	110	16.0	1.6	100	Infectious
LuIII(1.37)	94	11.4	0.69	42	Noninfectious
LuIII(1.36)	90	9.6	0.45	28	Noninfectious
LuIII(1.34)	78	7.7	0.26	16	Noninfectious
LuIII(1.33)	69	6.1	0.14	9	Noninfectious
MVM(1.41)	110	16.0	1.4	100	Infectious
MVM(1.38)	n.d.[d]	11.5	0.84	60	n.d.
MVM(1.36)	n.d.	10.0	0.63	45	n.d.
MVM(1.35)	n.d.	9.0	0.43	31	n.d.
MVM(1.33)	n.d.	8.0	0.35	25	n.d.
MVM(1.32)	n.d.	6.0	0.17	12	n.d.
FPV(1.45)	n.d.	16.6	1.7	100	Infectious
FPV(1.41)	n.d.	16.6	1.7	100	Infectious
FPV(1.36)	n.d.	12.0	0.78	46	Noninfectious

[a] In general, these data were obtained by pooling virus particles from contiguous regions of CsCl buoyant density gradients. In the data shown here, the ranges for the CsCl density, sedimentation values, and DNA molecular weight were obtained by taking averages of the ranges reported in the literature.

[b] The data were compiled from the following references: AAV2 (de la Maza and Carter, 1978, 1980a, 1980b; Laughlin *et al.*, 1979a; Myers and Carter, 1980), LuIII (Muller *et al.*, 1978), MVM (Faust and Ward, 1979), FPV (Johnson *et al.*, 1974).

[c] The DNA sedimentation values are those measured at pH 13.0. Particle sedimentation values were measured at neutral pH.

[d] n.d.: not done.

ference in protein type, amount, or charge or in the packaged DNA (de la Maza and Carter, 1980b). For helper-independent viruses, the 1.45 particle contains capsid proteins A and B as do empty particles whereas in 1.41 particles the B protein is cleaved to yield protein C (Clinton and Hayashi, 1975, 1976; Tattersall *et al.*, 1976, 1977; Kongsvik *et al.*, 1978). This proteolytic cleavage reflects the precursor product relationship between MVM (1.45) and (1.41) particles (Tattersall *et al.*, 1976; Richards *et al.*, 1977). For AAV such a precursor–product relationship could not be demonstrated (Myers and Carter, 1980; de la Maza and Carter, 1980b). The proteins of helper-independent variant particles have not been studied in detail but would be of particular interest. Richards *et al.* (1978) noted that encapsidation of the MVM genome into the protein shell has profound effects on surface properties of the particle and results in significant changes in the particle hemagglutination activity and the accessibility of the VP2 protein to proteolytic processing both *in vivo* and *in vitro*. It may be useful to determine the effect on these properties of encapsidation of a shorter variant genome.

Standard (1.41) parvovirus particles have a UV absorbance spectrum with a maximum at 260 nm and a $A_{260}:A_{280}$ ratio of approximately 1.40 whereas empty (1.32) particles have a maximum at 280 nm and an $A_{260}:A_{280}$ ratio of approximately 0.67 (Tattersall *et al.*, 1976; Laughlin *et al.*, 1979a). AAV variant particles exhibit an intermediate UV spectrum with a broad maximum between 260 and 280 nm and an $A_{260}:A_{280}$ ratio of approximately 1.0.

As noted in Table I, variant particles are essentially noninfectious and any residual infectivity associated with them is probably accounted for by contamination with standard particles.

For LuIII and FPV, it has not yet been shown that the DNA in the variant particles is in fact virus specific. For AAV, MVM, and H-1 this has been documented in the course of extensive studies on the physical structure and sequence arrangement of the variant DNA as described in the next section.

V. PHYSICAL STRUCTURE OF VARIANT DNA

A. Encapsidated AAV DNA

1. Fractionation of Variant Genomes

The physical structure and sequence arrangement of variant DNA from AAV2 particles has been extensively analyzed by de la Maza and Carter (1978, 1980a,b).

DNA was extracted from purified virus particles by sedimentation in 5–20% alkaline sucrose gradients. Most of the DNA from 1.41 particles sedimented as a discrete 15.5 S species designated DNA 1.41. Approximately 5%–10% of the total DNA from the 1.41 particles which sedimented at 11–12 S was pooled separately and designated DNA 1.41b. The 1.35 particles yielded adenovirus DNA which pelleted to the bottom of the tube and a discrete 10 S AAV species which was pooled and designated DNA 1.35. The 1.32 particles yielded very small DNA molecules designated DNA 1.32.

The four classes of DNA obtained from alkaline gradients were further purified by reassociation and sedimentation in neutral sucrose gradients. The 1.41 DNA yielded 14.5 S standard duplexes and a small amount of slower sedimenting 12–13 S DNA, designated DNA 1.41a. The reassociated 1.35 DNA molecules exhibited a heterogenous sedimentation profile in neutral sucrose: about one-half sedimented at around 8–10 S while the remainder sedimented faster and appeared to be oligomeric structures. These oligomeric structures reflect intermolecular reannealing due to the "cohesive-end" properties of AAV DNA termini (Gerry *et al.*, 1973; Carter and Khoury, 1975).

The DNA 1.41a, 1.41b, 1.35, and 1.32 molecules were previously referred to as DNA I, II, III, and IV respectively, by de la Maza and Carter

TABLE II. Size and Kinetic Complexity of AAV Variant DNA[a]

	Molecular weight determined in:					
	Alkaline sucrose		Neutral sucrose			
DNA	$M_r \times 10^{-6}$	Percent standard DNA	$M_r \times 10^{-6}$	Percent standard DNA	M_r (neutral) / M_r (alkaline)	Relative reassociation rate
1.41	1.55	100	3.1	100	2.0	1.0
1.41a	1.55	100	1.47	47	0.95	1.7
1.41b	0.90	58	0.94	30	1.04	3.3
1.35	0.31	20	0.56	18	1.80	3.9
1.32	0.05	3.2	0.05	1.6	1.0	n.d.[b]

[a] The data were taken from de la Maza and Carter (1980a).
[b] n.d.: not done.

(1978, 1980a). This previous nomenclature, which is less descriptive, will not be used here to avoid confusion with that used by Faust and Ward (1979) in a different sense (see Section V.C below).

2. Physical Mapping

a. Molecular Weight

Sedimentation analysis of the variant DNAs suggested the important feature that many of these molecules form hairpin structures at neutral pH. For a normal linear DNA duplex the ratio of the molecular weight of the duplex DNA in neutral sucrose to that of the single-strand in alkaline sucrose is 2.0. For a palindromic molecule which could form a hairpin structure by spontaneous self-renaturation the ratio would approach 1.0. As shown in Table II, DNA 1.41a, 1.41b, and 1.32 appeared to be mainly hairpin molecules. DNA 1.35 had a ratio of 1.8 and was a mixture of linear duplexes and hairpin molecules.

b. Genetic Complexity

Because they are smaller than standard genomes, individual variant molecules must have a lower genetic complexity. Two lines of evidence showed that the entire population of AAV variants generally had a lower genetic complexity. The rate of reassociation of variant DNA was inversely proportional to the molecular weight (Table II). Secondly, restriction endonuclease digestion of variant DNAs revealed that only certain regions of the AAV genome sequence were present in variant genomes. One example of a restriction endonuclease digest, of AAV variant DNA using *Hha*I is shown in Fig. 1.

These two analytical techniques also showed that variant DNA is composed predominantly of AAV genome sequences and no evidence for

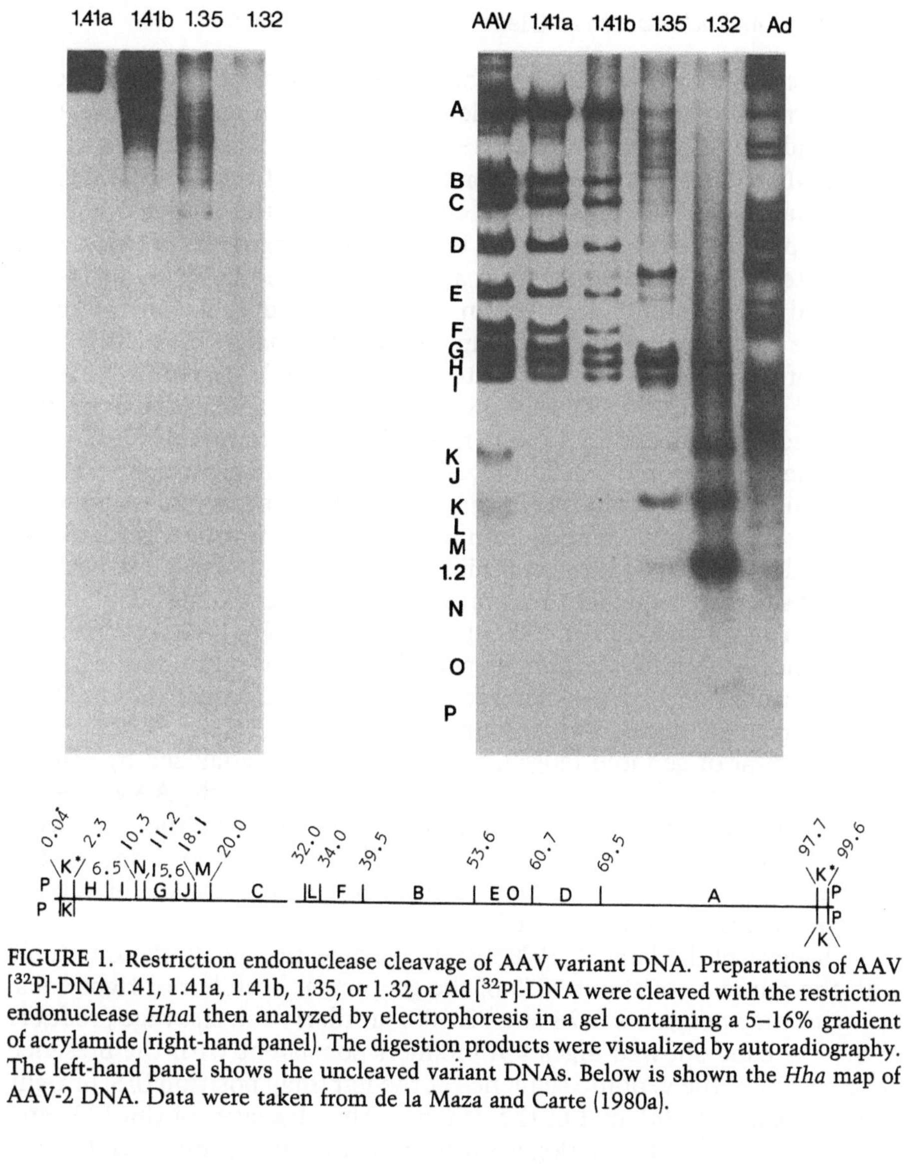

FIGURE 1. Restriction endonuclease cleavage of AAV variant DNA. Preparations of AAV [^{32}P]-DNA 1.41, 1.41a, 1.41b, 1.35, or 1.32 or Ad [^{32}P]-DNA were cleaved with the restriction endonuclease *Hha*I then analyzed by electrophoresis in a gel containing a 5–16% gradient of acrylamide (right-hand panel). The digestion products were visualized by autoradiography. The left-hand panel shows the uncleaved variant DNAs. Below is shown the *Hha* map of AAV-2 DNA. Data were taken from de la Maza and Carte (1980a).

the presence of adenovirus or host-cell sequences was obtained. Thus, reassociation of ^{32}P-labeled variant DNA was accelerated to the expected degree by addition of precisely measured concentrations of unlabeled AAV DNA but not by adenovirus or cell DNA (de la Maza and Carter, 1980a). Extensive restriction endonuclease digestion with many different enzymes also revealed no evidence for adenovirus or host-cell sequences in AAV variants. Most of the cleavage products correspond to known AAV DNA fragments. It remains possible that a very small fraction of the AAV variant DNA might include some host-cell sequence as has been suggested for MVM variants (Faust and Ward, 1979).

c. Restriction Endonuclease Cleavage of Variant DNA

An extensive analysis of restriction endonuclease cleavage patterns showed that the variant DNAs are deleted for internal regions of the AAV genome (de la Maza and Carter, 1980a). In general, as the size of the variant DNA decreases, there is a progressive loss of the internal regions of the genome with a concomitant enrichment for the terminal regions. For example, as shown in Fig. 1, the internal *Hha* fragments between map coordinates 20 to 70 are present in a low molar ratio in DNA 1.41a and 1.41b and are completely absent in DNA 1.35 and 1.32. On the other hand, the *Hha* fragments that map in the terminal regions (0/15 and 97/100) of the genome are represented in the DNA 1.41a and 1.41b at an increased molar ratio. In DNA 1.35, *Hha* 11.2/15.6 and 5.6/10.3 are also mostly deleted and even 2.3/6.5 is mostly deleted from DNA 1.32.

In the population of variant DNAs both left- and right-hand termini were retained. This could imply that individual variant molecules were internally deleted but retained both left and right termini. Alternatively, individual molecules might contain only one terminus and a variable length of adjacent sequence. This problem was addressed in the following way (de la Maza and Carter, 1980a). ^{32}P-labeled DNA 1.35 was fractionated by electrophoresis in agarose gels. This yielded prominent discrete species rather than a general heterogenous range. Individual variant DNA bands corresponding to DNA of four size classes equivalent to 28%, 12%, 9.5%, and 8% of genome length, respectively, were analyzed by restriction endonuclease cleavage. Most internal fragments of the AAV genome were not present in the digests and fragments closer to the ends were progressively lost as the size of the band decreased. More importantly, both left and right-hand terminal fragments were present in the digests of each specific band of variant DNA.

Cleavage of the largest (28%) band gave digestion products consistent with the presence of about 28% of the genome at both left or right ends, i.e., with map positions 0/28 and 72/100. Similarly the cleavage products of the 8% band showed digestion products consistent with the presence of about 8% of the genome sequence at the left (map position 0/8) or right (map position 92/100) end of the genome. The cleavages of the 12% and 9.5% DNA bands gave analogous results. Furthermore, for each discrete size class of DNA there was a gradient of relative molar amounts of the left- or right-hand termini. Fragments nearer to the terminus were present in greater molar amounts than fragments mapping further inboard.

The gradients of fragment representation in discrete size classes of DNA are not readily accounted for by size heterogeneity within individual DNA 1.35 bands. This observation is only explained if most of the variant DNAs contain left and right terminal regions joined in molecules having internal deletions. If the precise location of the internal deletion is variable within the population of molecules, then even for discrete size classes of variant DNA, a gradient of fragment molarities is generated at both left- and right-hand ends (see below).

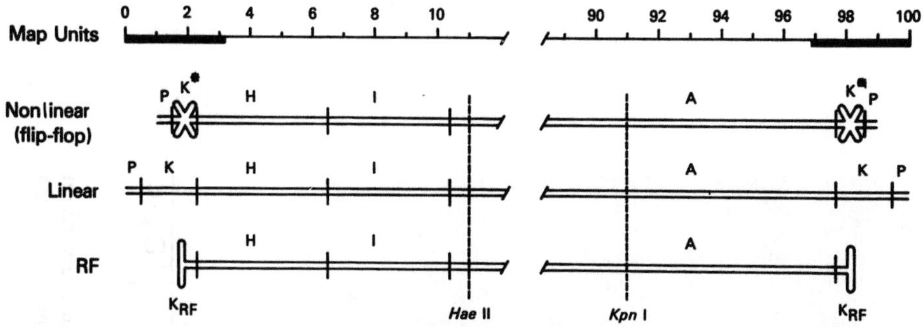

FIGURE 2. Schematic representation of the structural configuration of AAV2 DNA duplex termini. The fragments derived by *Hha* cleavage are designated P, K, K*, K_{RF}, H, I, and A. The *Hae*II and *Kpn*I cleavage sites are indicated by broken lines. The sites have been mapped (see Fig. 3). The regions with map coordinates 0/3.1 and 96.9/100, indicated by the solid bar (upper line), contain the terminal repeated, palindromic sequences which can fold into a T-shaped, base-paired hairpin. Nucleotides 44–83 (map positions 1/2 and 98/99) occur in two orientations (i.e., flip-flop). Two forms of terminal duplexes occur: linear termini and flip-flop termini. The flip-flop configuration results in one of several nonlinear, base-paired cruciform structures, shown here schematically. From de la Maza and Carter (1980a).

3. Terminal Fragments of Variant DNA

The two possible orientations (i.e., flip-flop) of the terminal repetitious sequence result in two types of termini in AAV DNA duplex as illustrated in Fig. 2 (Berns *et al.*, 1975, 1978; Denhardt *et al.*, 1976; Carter *et al.*, 1976; de la Maza and Carter, 1977; Spear *et al.*, 1977). Restriction endonucleases that do not cleave AAV DNA within the terminal repetition yield two types of terminal structure; the α component which corresponds to the cruciform duplex and the γ component corresponding to the linear duplex structure (Fig. 4). For *Hha*, which cleaves within the terminal repetition, the terminal fragment P is unique since this region of the terminal repetition is identical in either orientation. However the *Hha*-K and K* fragments contain the flip-flop region; K corresponds to the linear duplex, and K* the cruciform structure (Figs. 1, 3).

A third type of AAV DNA terminal structure, designated RF, can be formed by self-renaturation of the terminal sequence within a single DNA strand. Also, as shown in Fig. 2, the RF structure is the configuration that would be expected during replication of AAV DNA if this occurs by a Cavalier-Smith, hairpin transfer mechanism (Cavalier-Smith, 1974; Berns *et al.*, 1978; Straus *et al.*, 1976; Tattersall and Ward, 1976).

All three types of terminal structures shown in Fig. 2 were observed in the population of variant DNA molecules (de la Maza and Carter, 1980a). The presence of α, γ, and RF termini is indicated in Figs. 1 and 3. Comparison with cleavages of DNA 1.41 showed that in the variant DNAs there is a decreased amount of the α components relative to the γ components. Concomitant with the decrease in α components, an equivalent amount of the RF component appeared (Fig. 3). The RF con-

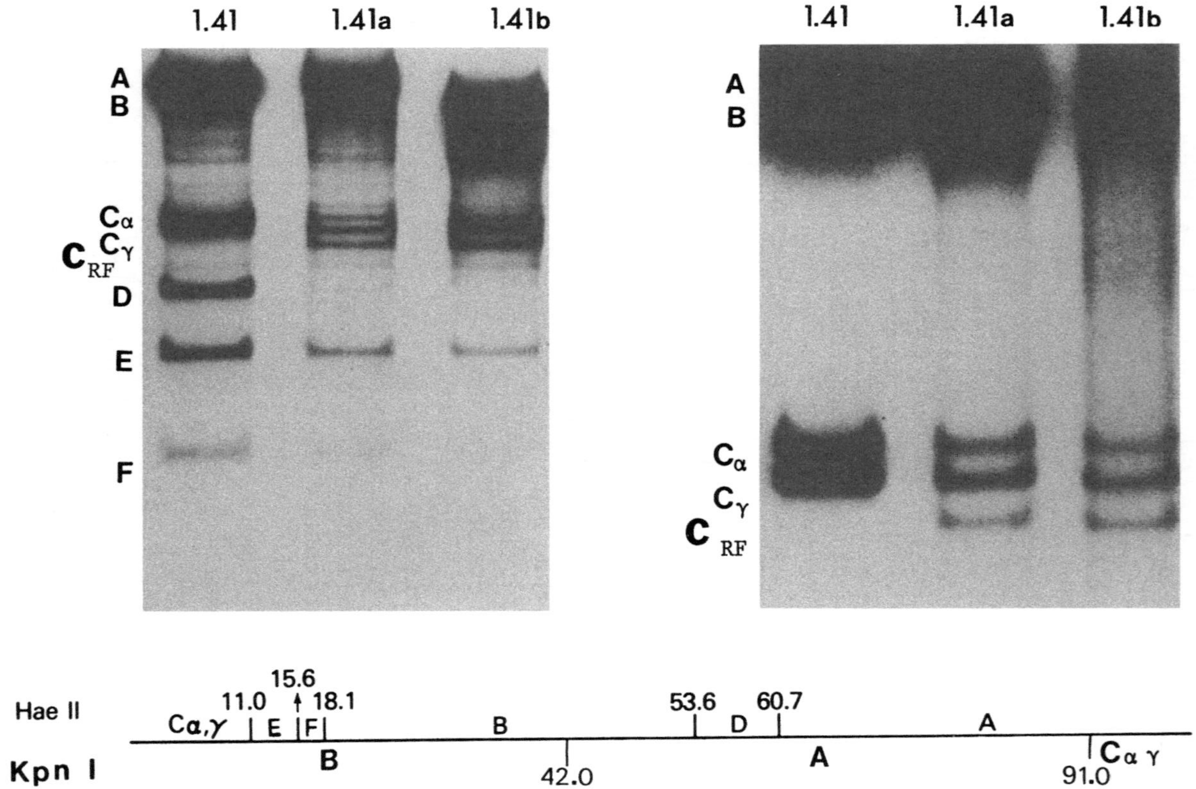

FIGURE 3. ^{32}P-DNA 1.41, 1.41a, or 1.41b cleaved with *Hae*II (left panel) or *Kpn*I (right panel) and analyzed on 3–10% acrylamide gels. The AAV2 cleavage maps for *Hae*II and *Kpn*I are shown. Redrawn from de la Maza and Carter (1980).

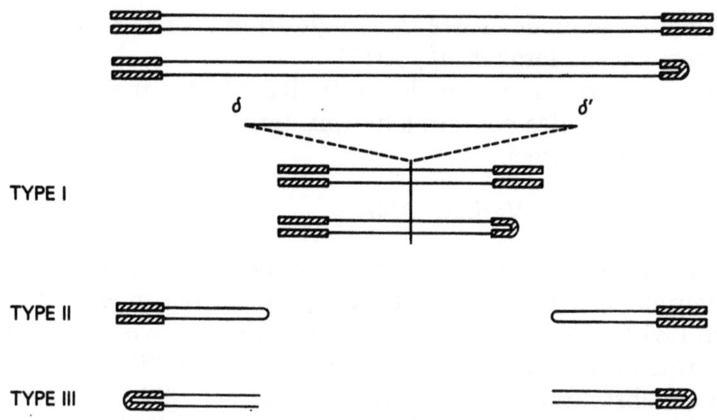

FIGURE 4. General models for AAV variant DNA. The AAV genome is represented in its duplex form. Boxed regions represent the terminal palindrome in its linear or RF form. δ and δ' represent the boundaries of an internally deleted region. The models are discussed further in the text. Redrawn from de la Maza and Carter (1980a).

figuration present in the variant DNA is also reflected in the *Hha*-K_{RF} fragment (Fig. 1) which has an apparent size of about 50 base pairs. This *Hha*-K_{RF} fragment, previously designated *Hha*-1.2 (de la Maza and Carter, 1978) is the most predominant fragment present in DNA 1.32.

Several lines of evidence supported the interpretation of the structure of the RF fragments (de la Maza and Carter, 1980a):

a. The RF component is about 60 nucleotides shorter than the γ component as expected.

b. None of the RF components could be labeled using polynucleotide kinase and ^{32}P-γ-ATP. This suggests that the RF fragment contained no single-stranded gaps although it is difficult to eliminate the possibility that there were gaps containing blocked 5′-termini.

c. Digestion of duplex DNA 1.41 with *E. coli* exonuclease III, to the extent that about 5% of the DNA was rendered acid-soluble, resulted in increased amounts of *Hha*-K_{RF} and a concomitant decrease in K* and K. Similar results were obtained when preparations of DNA 1.41b or 1.35 were digested with exonuclease III. This experiment confirmed that *Hha*-K_{RF} located within 5% of the genome termini and could be formed when the 5′ end of the DNA strand spontaneously renatured after exonuclease III digestion of the 3′ ended strand.

d. When a *Hha* digest of AAV DNA 1.32 was treated with alkali immediately prior to electrophoresis in a nondenaturing gel at room temperature, the K and K* fragments were abolished and the K_{RF} fragment remained. Alkali treatment of undigested DNA 1.32 did not alter its migration rate. This was interpreted as indicating denaturation of K and K* followed by spontaneous renaturation to yield K_{RF} in the nondenaturing conditions of the gel. This is exactly the behavior predicted from the AAV

DNA nucleotide sequence since the single strands of K, K* and K_{RF} have identical sequences (Berns *et al.*, 1978).

In summary, the above evidence indicated that many of the AAV variant DNA molecules contain a hairpin terminus analogous to that in AAV RF DNA molecules.

4. Structure of Hairpin Variant DNA

As noted above (Section V.A.2.a) the velocity sedimentation analysis of AAV variant DNA suggested that at neutral pH most of the 1.41a, 1.41b, and 1.32 variant DNAs were hairpin molecules while the 1.35 DNA was a mixture of normal linear and hairpin duplexes (Table II). This conclusion was substantiated (de la Maza and Carter, 1980a) by fractionating these DNAs on hydroxylapatite then digesting them, both before and after heat denaturation, with the single-strand specific endonuclease S1. This confirmed that most of DNA 1.41a, 1.41b, and 1.32 hairpin molecules contained no appreciable proportion of single-strand region.

That DNA 1.41a was mostly hairpin molecules was not surprising because it was precisely this property which allowed its separation from DNA 1.41 (Section V.A.1). Similarly, DNA 1.32 was a hairpin structure because the mapping experiments showed that it contained primarily the terminal palindromic sequence. DNA 1.35 is a mixture of both normal duplexes (69% of the total) and hairpin duplexes (15% of the total). The remainder of DNA 1.35 (16% of the total) appeared to represent heteroduplex molecules which consisted of plus and minus strands containing internal deletions at different sites in each strand and thus behaved as partly single-strand, nonhairpin molecules.

The location of the crosslink in the variant hairpin molecules was also determined (de la Maza and Carter, 1980a). A preparation of DNA 1.41b that had a mean molecular size estimated from gel electrophoresis of about 50% of genome length was cleaved jointly with the restriction endonucleases *Kpn* and *Bam*H1. This yielded the left-terminal *Bam*-$B_{\alpha,\gamma}$, RF fragments (0/22) and the right-terminal fragments *Kpn*-$C_{\alpha,\gamma}$, RF (91/100) and one new fragment, approximately 19%–20% of genome length. Therefore, most of this DNA 1.41b contained internal deletions of about 50% of the genome. To enrich for fragments containing the hairpin region, the cleaved DNA was heat-denatured and quick-cooled, then fractionated on hydroxylapatite. The DNA which still eluted at 0.4 M phosphate was highly enriched for the *Kpn*-C_{RF} and *Bam*-B_{RF} fragments. This observation showed that the RF fragments accounted for the hairpin properties of the variant DNA molecules. Thus, the cross-linked region in the variant DNA appears to be at the molecular termini with respect to the complete genome.

5. The Overall Structure of Encapsidated AAV Variant DNA

The simplest general structure which accounts for all the physical properties of encapsidated variant DNA is the type I structure shown in

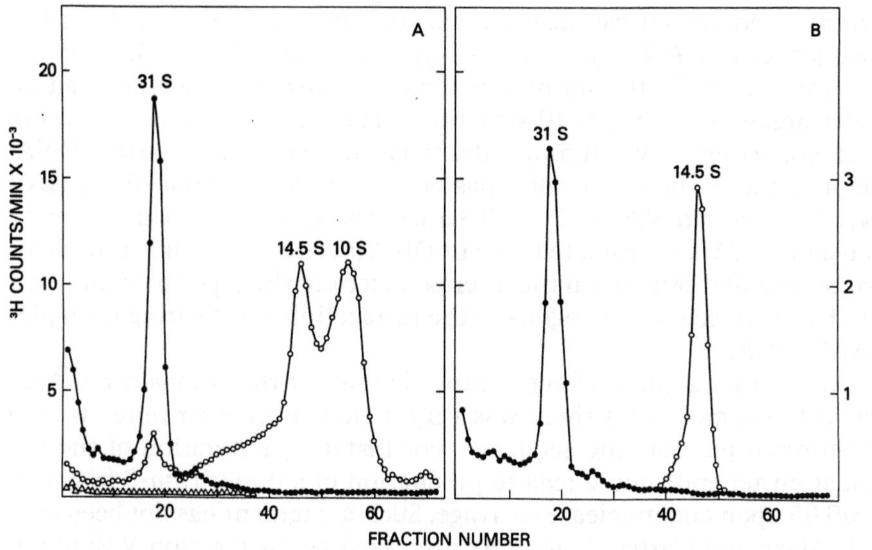

FIGURE 5. Sedimentation in neutral sucrose gradients of intracellular DNA from AAV-infected human KB cells. Panel A: ^{3}H-labeled Hirt supernatant DNA from cells infected with adenovirus 2 at 5 p.f.u. per cell (●) or adenovirus 2 plus AAV(1.41) at 50 infectious units (i.e., 2000 particles) per cell (○), or mock infected cells (△). Panel B: ^{3}H-labeled Hirt supernatant DNA from cells infected with adenovirus 2 and AAV (1.35) at 2000 particles per cell (●). Duplex AAV ^{32}P-DNA was added as a molecular weight marker (○). Reprinted from Laughlin *et al.*(1979a).

Fig. 4 as proposed by de la Maza and Carter (1980a). The variant DNAs are shown as molecules containing left and right molecular termini and having an internal deletion. The size of the deletion is variable. Also, the location of the deletion, even for a given size class of variant DNA molecules, is variable, i.e., in the model, for a given size of deletion, the distance δ–δ′ is constant but the locations of positions δ and δ′ are variable. The hairpin molecules consist of those in which one terminus (left or right) is in the covalent hairpin RF conformation. For the linear duplex variants, the proportion of linear (γ) or flip-flop (α) terminal fragments would be approximately equal as for the normal AAV genome. For the hairpin variant molecules the γ and RF conformations would be equivalent. Thus, for variant DNA populations there would be an inverse relationship between the relative amounts of RF and α terminal fragments as is observed.

Alternative possible conformations for the hairpin molecules are those shown in type II or type III (Fig. 5) in which individual molecules contain sequences from left or right-hand termini only. This model is less likely because for a discrete size class of variant DNA there would not be a gradient of molar amounts of fragments for left- or right-hand fragments. Also, this model would not readily explain the inverse relationship between the α and RF terminal fragments. In addition, double restriction endonuclease cleavage experiments such as that described above (Section V.A.4) argue against type II or type III models. The observation that only the normal genome terminal or fragments can be labeled argues against type III structures. There may, however, be a very small proportion of type II molecules in encapsidated AAV variant DNA. The presence of such molecules has been claimed for intracellular AAV DNA (see Section V.B.2). Type II structures have been clearly demonstrated for MVM encapsidated variant DNA (Section V.C.). It is important to note here that other than the internal deletion, the type I hairpin structure (Fig. 5) is exactly analogous to the intracellular replicating form (RF) of AAV DNA.

An additional point of significance is that for the encapsidated AAV variant DNA molecules there was very little evidence for reiteration of the terminal palindromic sequence. For instance, reiteration of the termini at either end should lead to production of a dimer *Hha*-P fragment 99.5/0.05 upon endonuclease cleavage. Such a fragment has not been seen (de la Maza and Carter, 1980a). As discussed below (Section V.E) reiteration of the terminal sequence of H-1 variant DNA has been observed (Rhode, 1978a).

B. Intracellular AAV DNA

Variant AAV DNA was also be detected in the population of intracellular DNA. When viral DNA was isotopically labeled for one to four hours late (i.e., 16 to 20 hr) after infection then selectively extracted using the detergent-lysis and high-salt precipitation procedure of Hirt (1967) variant DNA molecules were identified by velocity sedimentation in neutral sucrose (Hauswirth and Berns, 1979; Carter *et al.*, 1979). These variant molecules sedimented more slowly than standard 14.5 S AAV DNA in a broad zone having a peak at about 10S equivalent to a size about 20% that of the standard genome (Fig. 5). Although there was a clear majority of 10 S DNA the population contained molecules ranging from less than 10% of genome length to almost full-length strands.

The intracellular variant DNA was heterogenous with respect to topology as well as size. As judged by sedimentation in neutral and alkaline sucrose gradients, by resistance to the single-strand-specific endonuclease S1, and by chromatography on hydroxylapatite, the variant DNA was a mixture of single-stranded and duplex molecules (Carter *et al.*, 1979; Hauswirth and Berns, 1979). The amount of apparently single-stranded variant DNA varied in different experiments and with different virus stocks, ranging from a high of about 50% (Hauswirth and Berns, 1979) to a low of about 10% (L.M. de la Maza and B.J. Carter, unpublished). The structure of this single-stranded DNA has not been further investigated but may represent annealing of strands containing only the most terminal sequences in common and having deletions of different extents or different regions as mentioned above (Section V.A.4) for similar molecules obtained from encapsidated AAV DNA.

The intracellular duplex variant DNA contained a large proportion of hairpin molecules. Several features of the physical structure of the duplex variants were revealed by cleavage of the molecules with restriction endonucleases. First, the variant DNA was predominently AAV-specific. Second, only AAV sequences from map-position 0/20 or 80/100 were prominently represented. Third, in addition to the enrichment for terminal genome sequences there was a gradient of fragment molarity. Fragments nearest the termini were present in higher molar amounts than those at more internal locations. These properties suggest that these variant molecules are the intracellular counterparts to the encapsidated variant molecules having a general type I structure (Fig. 5) as described above. However, the intracellular AAV variants have been characterized in much less detail and therefore their assignment as type I structures is less certain and indeed more controversial.

Hauswirth and Berns (1979) reached the conclusion that these molecules were predominantly of the type II structure (Fig. 4) having one normal left- or right-hand genome terminus and one "internal" hairpin terminus. These workers digested AAV variant DNA with restriction endonuclease *Taq1* and showed enrichment for terminal sequences. At the right-hand end, their digestions yielded *Taq*-D the normal terminal duplex fragment (90.4/100) and a similar amount of a fragment, *Taq*-E1, that was not present in normal standard duplexes. *Taq*-E1 was shown to be exactly analogous to the *Kpn*-C_{RF} fragment described by de la Maza and Carter (1980a) and shown in Figs. 2 and 3. In a second experiment the 10 S intracellular DNA molecules were cut with *Taq*1 then heat-denatured, quick-cooled, and chromatographed on hydroxylapaptite. Most of the digestion products still did not elute from hydroxylapaptite at 0.15 *M* phosphate suggesting that most fragments still had a region capable of forming a duplex structure. Since *Taq*1 cuts AAV2 DNA at map positions 15.5, 30.7, 76, 89, and 90.4 (as well as seven additional sites between 30.7 and 76.0), each intracellular 10 S molecule (20% of genome length) should have received on average only one cut. One half of the fragments should contain a normal genome duplex which after heat-denaturation and quick cooling yields a hairpin terminal palindrome structure that is retained on hydroxylapatite at 0.15 M phosphate. Because on average the other half of the fragments were also retained on hydroxylapatite, Hauswirth and Berns (1979) argued that these contained "internal" hairpins. However, it seems more likely that these represented the *Taq*-E1 fragment from the right end or the analogous fragment from the left end. For this reason, the weight of available evidence suggests that the bulk of these intracellular variant molecules may more likely be of the type I structure than the type II structure. A critical resolution of this discrepancy requires analysis of precise size classes of this DNA to determine if a gradient of fragment molarities is observed. Alternatively electron microscopy of heteroduplexes or even molecular cloning of individual molecules may resolve this issue. A resolution of the structure of these variant DNAs is critical to determining the probable mechanism by which these molecule are generated.

C. Encapsidated MVM DNA

The DNA obtained from variant MVM particles which banded in CsCl at buoyant densities between 1.33 and 1.38 g/cm^3 (see Table I) was extensively characterized by Faust and Ward (1979). In these studies variant particles were obtained by propogating MVM(T) virus in Ehrlich ascites cells either in a single-cycle, high-multiplicity infection or after several rounds of undiluted passage. The encapsidated variant DNA, isolated by sedimentation of particles in alkaline sucrose gradients, was very heterogenous but had an average size of about 12S which corresponds to a single strand length of about 2000 nucleotides or 40% of the MVM genome length (Table I). Analysis of this variant DNA with the single-strand-specific endonuclease S1 or by hydroxylapatite chromatography revealed two types of molecule. About 50% of the variant DNA was resistant to nuclease S1 digestion and therefore contained extensive duplex regions. Moreover, a large proportion of this DNA remained resistant to nuclease S1 even after heat denaturation and quick cooling. Chromatography of [^{32}P]-DNA on hydroxylapatite showed that the variant DNA was comprised of approximately equal amounts of single-stranded (type I) or double-stranded (type II) molecules. The double-stranded molecules appeared to be hairpin structures.

1. Structure of MVM Type I Variant DNA

The type I variant MVM DNA obtained by hydroxylapatite chromatography was characterized by Faust and Ward (1979). Standard MVM DNA contains hairpin duplexes at both termini which resist digestion with single-strand-specific nucleases (Chow and Ward, 1978; Bourguignon *et al.*, 1976; Astell *et al.*, 1979). After nuclease digestion the remaining 3′ terminal hairpin is 115 nucleotides, and the 5′ terminal hairpin is 130 nucleotides, in length. Digestion of type I variant DNA with the single-strand-specific mung bean nuclease yielded a 115-base-pair 3′ hairpin but the 5′hairpin was about 10 to 15 base pairs longer than that from standard MVM DNA.

The sequence organization of the single-stranded type I MVM DNA was examined by Faust and Ward (1979) by conversion of the V strand to a duplex RF structure by *in vitro* synthesis of the C strand using *E. coli* DNA polymerase and the self-priming properties of the 3′ terminal hairpin (Bourgingnon *et al.*, 1976). The duplex RF [^{32}P]-DNA was cleaved with restriction endonucleases. The results obtained were analogous to

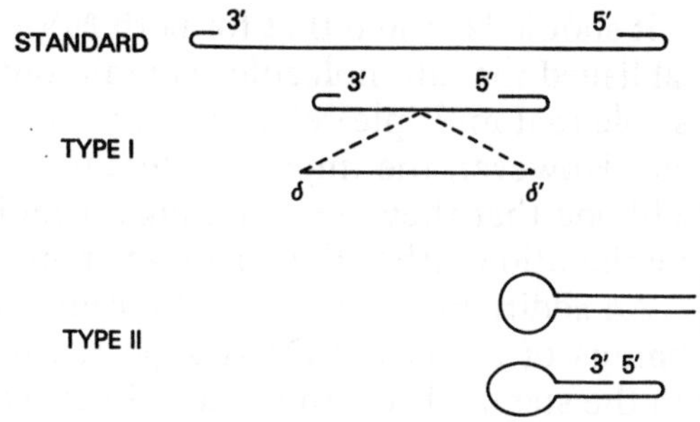

FIGURE 6. Schematic diagrams of structures of type I and type II MVM variant DNA. The internal deletion in type I molecules is represented by the line δ, δ′. Type II structures are shown with the 5′ viral strand (right-hand) palindrome in either the extended or RF form. The models are based on those of Faust and Ward (1979).

those observed with AAV variant genomes. Although all the genome fragments were observed, fragments near the genome termini were present in much higher abundance than internally located fragments. The same result was also obtained when the variant DNA was first annealed to ^{32}P-labeled C-strand DNA then cleaved with restriction enzymes. In both cases the contiguous region, 17.8/86.5, on the physical map was clearly underrepresented in the digest.

The sequence organization of individual type I variant DNA molecules was examined in two ways. First, type I DNA was fractionated by size in alkaline sucrose gradients prior to *in vitro* conversion to RF DNA. The individual size classes were then digested with restriction enzymes. This showed that (1) the sequence complexity decreased with DNA size and that (2) the decrease in complexity was due to progressive loss of internal genome regions. The smallest variant DNA contained only the 3′ terminal genome region 0/9.8 and the 5′ terminal region 91.5/100. These results again are analogous to those described above for AAV.

To prove that the type I DNA contained both 3′ and 5′ genome termini in the same molecule Faust and War (1979) performed the following analysis. Restriction fragments representing the 3′ half or the 5′ half of the MVM genome were separately annealed with preparations of type I variant DNA and the resulting hybrids were purified first by hydroxylapatite chromatography and then by BD-cellulose chromatography. The type I DNA strands from the hybrids were then converted *in vitro* to RF duplexes and cleaved with restriction enzymes. The same result was observed whether the type I DNA had been selected by annealing to 3′ or 5′ terminal genome fragments. In each case both 5′ and 3′ termini were present in the type I DNA and the extent of the deletion was similar.

The observations of Faust and Ward (1979) suggested that MVM type I DNA has the structure shown in Fig. 6. This structure is exactly analogous to the proposed for the bulk of encapsidated AAV variant DNA by de la Maza and Carter (1980a) as described in Fig. 4. Apparently, MVM type I molecules are recombinant genomes retaining both 3′ and 5′ genome termini but having an internal deletion which has a variable position and extent in individual molecules.

It should be noted that for both MVM and AAV type I DNA it is not established that all molecules contain only a single simple deletion. It is possible that multiple deletions or rearrangements of the sequence might occur. However, the important feature of type I molecules is the strong likelihood that they are recombinant molecules which arise initially by recombination rather than aberrant replication.

An additional point noted by Faust and Ward (1979) was that the 5′ terminus of the type I DNA appeared to be 10 to 15 base pairs longer than the standard genome 5′ terminus. This variation is not resolved but might indicate that there is reiteration of the genome terminal sequence as in some H-1 variants (see Section V.E below). Alternatively it might represent the presence of a protein known to be covalently bound to the 5′ termini of RF DNA of several autonomous parvoviruses including H-1 (Revie *et al.*,1978) and MVM (M. Chow and D. Ward, personal communication; Chow, 1981) and perhaps AAV (B. Stillman, personal communication).

2. Structure of MVM Type II Variant DNA

The type II MVM DNA appeared to consist of hairpin molecules and the structure of these was also studied further by Faust and Ward (1979). Type II DNA was purified as double-stranded DNA on hydroxylapatite. These molecules were analyzed by single-strand-specific nuclease digestion or hydroxylapatite chromatography after a variety of denaturation and renaturation protocols. These studies showed that 85% of the DNA spontaneously renatured. This spontaneous renaturation was prevented by digestion with single-stranded specific nuclease prior to denaturation. At least 75% of the type II DNA was virus specific. This DNA appeared to contain complementary strands covalently associated by relatively large single-stranded loops as illustrated in Fig. 6. Further analyses by electron microscopy and restriction enzyme digestion substantiated this conclusion.

When total MVM variant DNA was viewed by electron microscopy, about 30% of the molecules consisted of a duplex "panhandle" and a single-strand loop. The length of both the panhandle (up to 1000 base pairs) and the loop (300 to 1000 nucleotides) was variable. Analysis of the type II DNA with restriction enzymes showed that the duplex region was derived only from the 5′ (85/100) terminal right-hand region of the V strand. No evidence for 3′ terminal left-hand type II MVM DNA was obtained. Some of the type II DNA contained the 5′ genome terminus present in the hairpin, RF configuration.

The structure of type II MVM DNA is summarized in Fig. 6. These molecules appear to be analogous to the type II AAV molecules proposed by Berns and Hauswirth (1979) and described in Fig. 4. One difference, however, is that any type II AAV molecules that exist do not appear to have significantly sized single-stranded loops as judged by single-strand-specific nuclease digestion studies (de la Maza and Carter, 1980a; Hauswirth and Berns, 1979). This point is significant in discussing the mechanism by which variant molecules are generated (Section VII.A).

D. Intracellular MVM DNA

There are no detailed studies on intracellular MVM variant DNA but in this context the report of Bratosin *et al.*(1979) deserves mention. These workers lysed infected cell nuclei with the detergents triton X-100 and sarkosyl and removed the cellular chromatin by centrifugation. When the supernatant fraction was viewed by electronmicroscopy about 5% of the total MVM DNA molecules had a "noose" and "stem" or "lasso" structure. The sum of the lengths of the stem and the noose was equivalent to one genome size. Both the stem and noose appeared to be double stranded and the structure did not depend upon either protein or RNA linkers for its conformational integrity. Furthermore, the noose appeared to be located at the right-hand genome end (i.e., the 5′ terminus of the viral DNA strand).

Bratosin *et al.* (1979) suggested that in these "lasso" structures, the noose slides along the stem (i.e., towards the 3′ end of the viral strand) and this might be a mechanism for circularization of the DNA. However, this model remains highly speculative. For instance, whether this is a static structure or a kinetically sliding molecule is completely unknown. Also, one major prediction implied by the proposed structure is that the entire lasso should have no free ends and upon denaturation should yield a covalently closed single-stranded circle of two genome equivalents in length. This point was not addressed.

Rhode (1978) also saw similar lasso or "lollipop" structures among the intracellular H-1 DNA isolated from semipermissive NB cells but not from permissive hamster embryo cells. Rhode suggested that these lollipops were formed by initiation of DNA replication on a covalently closed hairpin loop of the 5′ end of viral DNA possibly by RNA priming.

The relevance of these "lasso" or "lollipop" structures to variant DNA is not established but might provide a basis for production of type II molecules that would originate only from the right-hand or 5′ end of viral DNA as observed by Faust and Ward (1979). This is discussed further below (Section VI).

E. Intracellular H-1 DNA

Variant genomes of H-1 and H-3 have been obtained and studied extensively by Rhode (1978a,b,c). Variants of H-1 were obtained by growth of H-1 standard virus (H-1 wt) in secondary hamster embryo cells. This virus was then passaged 20 times at high multiplicity in NB cells (simian virus 40 transformed human newborn kidney cells). At each passage after the first, additional H-1 wt was added at a multiplicity of 10 to act as a helper. The virus preparation obtained after 20 passages was called H-1-DI-wt. From this stock a viable variant virus, DI-1, was isolated. Two other virus stocks were then made by taking the H-1-DI-wt stock from passage level 14 and passaging an additional six times in NB cells using an added helper in each passage either DI-1 or the H-1 mutant, *ts*14. (The H-1 *ts*14 mutant is temperature sensitive for plaque production at 39.5°C but also has a *cis*-dominant RF rep$^-$ phenotype at both permissive and nonpermissive temperatures such that *ts*14 RF replication is only 3–7% that of H-1 wt.) These variant stocks are called H-1-DI-wt-*ts*14 and H-1-DI-wt-DI-1. Since all of these stocks were DI (defective interfering) particles (see Section VIII.E below), the nomenclature used for each stock was to record after the DI prefix the names of the helper viruses in the order used. A stock of the closely related parvovirus H-3 was also prepared by 10 passages at high multiplicity. The interfering properties of these virus stocks are described below. It should be noted that the H-1 DI-1 virus is not strictly speaking a defective or DI particle since it is viable and grows, albeit poorly, to produce very small plaques. For this reason Rhode (1978a) called DI-1 a viable defective. Because this term is confusing in a general context (Section II, above) I have avoided it but have retained the mutant name DI-1.

The structures of the variant genomes in the several H-1 and H-3 stocks were examined by Rhode (1978a) by isolating intracellular RF duplex molecules after labeling *in vivo* with [^{32}P]orthophosphate. This provided convenient substrates for mapping with restriction enzymes. The general structure of the H-1 variants showed several characteristics which differed from those of AAV and MVM variants characterized thus far:

1. Most H-1 variants had a size of 90% to 100% of that of a standard genome.
2. Specific regions of the H-1 genome were often reiterated to high frequency.
3. There were corresponding deletions elsewhere in the genome.

However all the H-1 variants apparently retained both the 5′ and 3′ palindrome termini. Also, there were some exceptions to the above generalizations. For instance reiteration occurred without any corresponding deletion and similarly deletions occurred without any corresponding insertion. However, in each of these cases, the deletions and insertions were of rather small portions (1–5%) of the genome length.

The H-1-DI-wt virus analyzed at passage 20 (Rhode 1978a) contained a preponderance of molecules which were enriched for regions in the left-hand molecular end (0/20) and at the extreme right end (95/100) whereas some sequences between map positions 36 and 91 were clearly deficient. The deficiency in this region of the genome appears to be due to two sets of deletions. One set of deletions occurs in the left-hand side in the localized region between map position 32 to 44 and apparently nearly always deleting a *Hha* site at position 38.0. The deletions in the right-hand side of the genome were less well characterized but apparently are localized just to the left of the *Hpa*I cleavage site at 91.0. The DI-wt-*ts*14 and the DI-wt-DI-1 stocks showed variant molecules with similar general characteristics for the DI-wt variants.

One particular deletion which occurs in H-1 variants is that occurring in the DI-1 mutant (Rhode, 1978a). This virus has a 58 base pair insertion in the *Hind*II 91.3/95.7 fragment. This insertion is just in board from the right-hand terminal palindrome and results in an increase in the size of the *Hind*II 91.3/95.7 fragment. This region of H-1 DI-1 has now been sequenced (S. Rhode, personal communication) as has the corresponding region of the parent H-1 wt virus and the closely related virus H-3. The results are summarized in Fig. 7. In this region H-1 contains two copies of a 55 nucleotide pair repeated sequence. Each of the repeats is extremely AT rich (45 of 55 bases are either A or T) and the two repeats are separated by the trinucleotide 5′-TTA-3′. H-3 wt virus contains only one copy of the 55 nucleotide sequence and does not contain the TTA trinucleotide. Interestingly, the DI-1 mutant contains a third copy of the 55 base sequence and this is also separated by the trinucleotide TTA. The DI-1 mutant has no other known alterations in the genome sequence.

The H-1 variant preparations described above all showed evidence of insertions of approximately 60 nucleotides in the *Hind*II 91.3/95.7 fragment. Interestingly all three preparations, DI-wt, DI-wt-DI-1, and DI-wt-ts14, respectively, showed evidence for multiple insertions of 60 base pairs in this region (Rhode 1978a). This is illustrated in Fig. 8. Cleavage of H-1 with *Hpa*I results in a single cut at 91.3 to yield the fragment 91/100. The variant preparations all show a decrease or loss of this band and the appearance of multiple bands differing from each other by about 60 bases. This appears to represent reiteration of the same region as that sequenced in the mutant DI-1 (S. Rhode, personal communication). Extensive passage of these virus stocks showed an accumulation of variants with increased reiteration of this region. Furthermore, the use of H-1 *ts*14 as a helper tends to yield variants with a higher average reiteration number than did a H-1 wt helper.

When the H-3-DI-wt stocks were analyzed after 10 passages by cleavage with *Hpa*I an analogous phenomenon was observed (Rhode, 1978a). In this case the *Hpa*I cleavage site is at 92.4 and in the variant stock there is apparently increasing reiteration of a 55- to 60-nucleotide sequence (Fig. 8).

The extensive reiteration of the 55-nucleotide sequence is of intense interest because it is close to, and may even constitute part of the origin of replication for viral strand synthesis. It is thus of further interest that a revertant of *ts*14, that is RF rep$^-$ and much less temperature sensitive for plaque formation, also has reiterations of the same genome region (Rhode, 1978a).

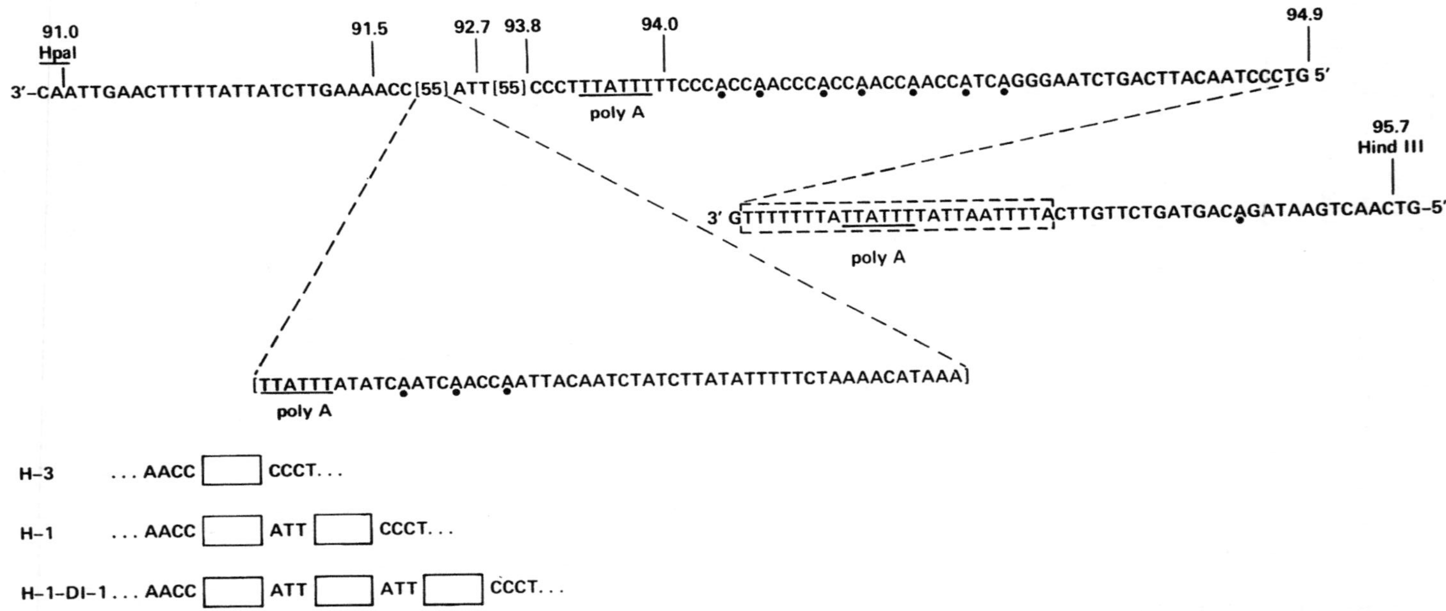

FIGURE 7. Sequence of the H-1 DNA viral (minus) strand in the region 91.0/95.7. The square brackets at map position 92.7 indicate an exact repeat of 55 nucleotides. The structures of the H-3, H-1 and H-1-DI-1 genomes in this region are shown schematically in the lower portion of the figure where the dot represents the 55-base pair sequence. Also indicated (by underlining and solid spot) are possible polyadenylation recognition sites (Fitzgerald and Shenk, 1981) and a 25-base-pair AT sequence (dashed box). The data were obtained by S. Rhode (unpublished experiments).

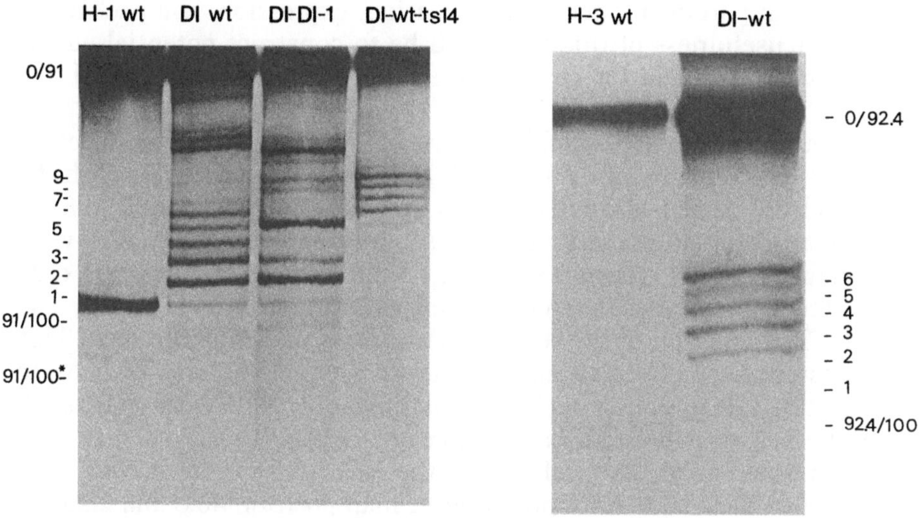

FIGURE 8. The structure of H-1 and H-3 variant DNA grown in human NB cells. ^{32}P-labeled intracellular RF DNAs of H-1 wt, H-1-DI-wt (passage 30), H-1-DI-DI-1 (passage 10), H-1-DI-wt-*ts*14 (passage 30), H-3-wt, or H-3-DI-wt (passage 10) were digested with endonuclease *Hpa*I, electrophoresed in an acrylamide agarose gel, and autoradiographed. The left-hand panel shows H-1 DNA and the right-hand panel shows H-3 DNA. The *Hpa*I 91/100 and 92.4/100 fragments are the right-hand end of H-1 and H-3 DNA, respectively. The 91/100* fragment is the same fragment as 91/100 but in the RF conformation. The bands indicated by numbers 1–9 represent DNA with successive insertions of a 55 base pair fragment. Redrawn and reprinted with permission from Rhode (1978a).

As noted above it is possible to obtain variants of H-1 containing deletions but having no reiteration. One such variant is H-1 *dl*-1 which has deleted 350 nucleotides at map position 38.5 (S. Rhode, personal communication). This virus is a DI particle and produces no capsid proteins. If transcription and translation of H-1 is similar to that of AAV as seems likely (see Chapter 5 of this volume) it is possible that the deletion in *dl*-1 interferes with a major H-1 promoter or with splicing of H-1 RNA.

In summary, the H-1 genome appears to have two regions highly susceptible to rearrangement. One region is close to the origin of viral strand DNA synthesis and is frequently reiterated. The other region is near the probable location of an H-1 promoter and is often deleted.

VI. *IN VITRO* CONSTRUCTION OF VARIANT GENOMES

With the advent of *in vitro* molecular recombination technology it has become possible to construct specific variants of genomes. Application of this approach to parvovirus genomes has only very recently begun but can be expected to yield important results with several implications. Firstly, this is a useful approach to define *cis*-acting sequences which are important for genome replication, expression, and packaging. A broader usefulness of this work will be to construct potential eukaryotic vectors suitable for cloning, amplification, or integration of any gene.

Cheung and Berns (unpublished) investigated the effect of deletion of the 38.1/42.6 region of AAV 2 DNA by cleaving at these sites with the endonuclease *Eco*R1 and religating to yield A (42.6/100) and B (0/38.1) fragments in the combinations AB, AA, and BB. These molecules were then transfected into human KB cells. All three types of molecules only replicated when standard AAV, as well as adenovirus helper, was also added. Thus deletion of the region 38.1/42.6 prevented replication. This is consistent with observations that the AAV type I variants (de la Maza and Carter, 1980a) do not replicate unless standard AAV (as well as adenovirus helper) is present (Laughlin *et al.*, 1979a).

It is now known that deletion of the 38.1/42.6 region of AAV 2 removes both the major AAV promoter at map position 40.0 and also the probable donor splice junction at 41.0 which is required for all the spliced AAV RNA species (Laughlin *et al.*, 1979b; Green and Roeder, 1980; see also Chapter 5 of this volume). This deletion probably blocks production of the major AAV 2.3-kb spliced RNA promoted from map position 40.0 which specifies the AAV capsid proteins (Jay *et al.*,1981) and may prevent splicing of the other AAV RNAs. At least some AAV-coded proteins are required for AAV replication (Laughlin *et al.*, 1979a; Myers and Carter, 1981; Jay *et al.*, 1981).

Deletion mutants of H-1 wt and H-1-DI-1 have also been constructed *in vitro* (S. Rhode, personal communication). These mutants were deleted in map regions 42/67, 34/51, or 20/79. When transfected into permissive cells together with helper H-1-wt virus all these variants were unstable and undergo rearrangements to yield a size which is 0.9 to 0.96 that of the standard wt virus. These rearrangements do not involve recombination with the helper virus. This behavior is analogous to that of the *in vivo* generated H-1 variants described above. However, the *in vivo* rearrangements of the *in vitro* generated variants appear to promote their replication because they rapidly predominate over the wt virus within two or three passages.

VII. *IN VIVO* GENERATION AND REPLICATION OF VARIANT DNA

A. Generation of Variant DNA

It is not possible to observe the original molecular event which generated a particular variant genome. The mechanism of such events can only be inferred from the structure of variant molecules as they are observed after several rounds of replication or several cell passages. Obviously such observations may be biased by selective pressures occurring after the original event. Selection may be influenced by many factors including passage level, the host cell, the helper virus, and relative efficiency of replication or packaging of variant DNA. However, based on the observed structures of variant DNA and the current models for parvovirus DNA replication (see Chapter 1) several possible mechanisms for the generation of variant genomes can be proposed. These models fall into two classes: recombinational mechanisms which would give rise to type I variants, and aberrant replication mechanisms which would yield type II models.

It seems likely that the type I AAV or MVM variants (de la Maza and Carter, 1980a; Faust and Ward, 1979) and the variants of H-1 (Rhode, 1978a) arise by a recombination mechanism resulting in internal deletions. Whether such recombination is an intermolecular or intramolecular event is not clear. However, such recombination events may be facilitated by a strand displacement replication mechanism and the very large number of parvovirus genomes accumulated in the cell nucleus. For AAV about 10^6 genomes copies are accumulated per infected cell (Rose and Koczot, 1972; Laughlin *et al.*, 1982) and only about one third of them are packaged (Myers and Carter, 1981, and unpublished results).

It is of some interest that there may be certain preferred sites for the occurrence of deletions. For instance, it was noted above that H-1 may have a high frequency of deletion near a probable promoter site at map position 38.0 (Rhode, 1978a, and unpublished data). It is possible that recombination might be enhanced at such sites due to pausing of replication caused by the presence of RNA polymerase, to preferential melting of promoter regions or differential chromatin structure. In this respect it is important to note that at least for adenovirus DNA the processes of replication, transcription, and recombination apparently are not compartmentalized in the cell nucleus (Wolgemuth and Hsu, 1980; Matsuguchi *et al.*, 1979). Indeed, Wolgemuth and Hsu (1980) isolated intracellular adenovirus DNA and showed by electron microscopy that all three processes could occur simultaneously on a single molecule.

The type II variants probably are the result of aberrant replication and several models can be proposed as detailed in Fig. 9. The simplest schemes are those in Figs. 9A, 9B, and 9C which were suggested by de la Maza and Carter (1978) and Faust and Ward (1979). In the model of Figs. 9A and 9B, breaks occur either in the parental input strand or in displaced progeny strands. The newly created 3′ end might then act as a primer by a random looping over or illegitimate base pairing to produce type II molecules with the loop structure shown in Fig. 9B. For MVM these molecules would contain the right end of the genome only. This type of model has also been proposed by Daniell (1976) to explain the generation of variant adenovirus genomes. Breakage of the genome would also yield the left-hand portion of the strand containing the 3′ genome terminus which could yield the structure shown in Fig. 9C. This type of molecule

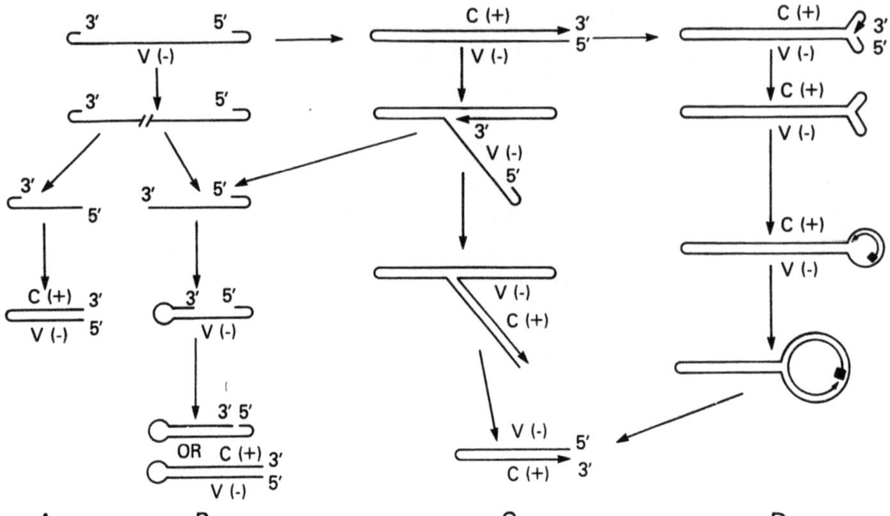

FIGURE 9. Possible replication models for the derivation of certain types of variant DNA. The overall scheme is based on various models suggested by de la Maza and Carter (1978); Rhode (1978b); Faust and Wart (1979); and Bratosin *et al.* (1979). The model is discussed further in the text.

has not been observed in virion DNA of AAV or MVM nor clearly distinguished in intracellular DNA. However, if breakage occurred during displacement replication of the strand then the left-hand portion may not be generated at all. An alternative model to produce type II variants shown in Fig. 9C suggests a template strand switch during displacement replication. It is difficult to choose between models B and C. Faust and Ward (1979) favored model B for production of MVM type II variants since, as they pointed out, pathway C would appear to be less likely to produce molecules with large loops.

Other more complicated schemes (Fig. 9D) can be proposed which are based on the observation of "lollipop" or "lasso" replicating molecules of H-1 (Rhode, 1978b) or MVM (Bratosin *et al.*, 1980). The model shown in Fig. 9D is based on the suggestion of Rhode (1978b) that covalent cross linking at the 5′ terminus of the v strand and melting out of the cross-linked DNA could allow replication around the loop perhaps by RNA priming.

All of the models shown in Fig. 9 could provide type II molecules containing the 5′ viral strand terminus (right-hand molecular end of the genome) for autonomous parvovirus as such as MVM or either the left or right terminus for AAV. For MVM this is consistent with the observations of Faust and Ward (1979).

B. Effect of Passage Level, Host Cell, or Helper Virus

A number of factors, including the multiplicity of infection, passage level of virus stocks, and the helper virus or host cell used may influence production of variant genomes and particles. In most cases it is not clear whether these factors influence the generation or subsequent amplification of variants. However, a number of these effects have been documented.

In general, the effect of increased multiplicity of infection with virus stocks is to increase the number of variants produced. This is readily understandable in view of the *rep*$^-$ properties of most variant genomes (see Section VII.C below). Thus, increased multiplicity may provide increased helper functions from standard genomes. Of course, increasing the multiplicity sufficiently may result in interference with multiplication of standard virus and a consequent overall decrease in production of both standard and variant virus (see Section VIII).

Continued passage of virus stocks, particularly at high multiplicity, may result in appearance of new variant genomes, amplification of existing variants, or even loss of some types of variants. All these situations have been observed for parvovirus variants.

The effect of multiple passage on variants of H-1 or H-3 has been discussed already (Section V.E). In general these variants accumulated successive reiterations of a region near the 5′ viral strand terminus and progressively deleted regions elsewhere in the genome (Rhode, 1978a).

Faust and Ward (1979) described both type I and type II variant genomes for MVM. Serial, undiluted passage of these virus stocks resulted in a gradual loss of the type II variants. In a single passage yield approximately equal numbers of type I and type II genomes were encapsidated. After ten successive passages the variant population contained almost entirely type I molecules. Two other changes also occurred in the type I variants with passage. There was a decrease in average size from about 2 to 1.2 kb due to progressive deletion of internal genome regions and selective enrichment of the left- and right-hand molecular termini. In addition, restriction enzyme mapping of the type I genomes indicated that they increasingly accumulated rearrangements in the genome sequence or incorporated cellular DNA sequences. These results may be interpreted in several ways. The type II molecules may arise by aberrant replication but may not subsequently be replicated. Alternatively these molecules may be replicated more slowly or packaged less efficiently.

Serial passage of AAV at high multiplicity also results in accumulation of variant molecules and a progressive decrease in size (M. Myers, C. Laughlin and B. Carter, unpublished). These AAV variants from different passage levels have not yet been analyzed in detail but AAV may be similar to MVM in accumulating preferentially type I molecules. Most of the encapsidated AAV variant genomes examined by de la Maza and Carter (1980) were type I and were obtained from virus stocks that had been serially passaged at high multiplicity. If AAV generates type II variants, as proposed by Hauswirth and Berns (1979) in their analysis of AAV intracellular DNA, these molecules also must be replicated or packaged inefficiently.

It appears to be a general phenomenon that most of the variants of H-1, MVM, and AAV which preferentially accumulate with passage appear to be internally deleted type I recombinant molecules with progressively larger deletions. However, for H-1 there appears to be an additional selection for molecules which have generated multiple reiterations near the 5′ viral strand replication origin. Whether this is due to intrinsic differences in the H-1, MVM, and AAV replication mechanism, to differences in the various host cells used, or to the experimental procedure is not clear. It may simply reflect the fact that the H-1 stocks were passaged more extensively (10–30 passages) than the AAV or MVM stocks. One important point may be that, in the H-1 or H-3 passages, standard helper virus was added at each passage (Rhode, 1978a), whereas for AAV (de la Maza and Carter, 1980a; M. Myers, C. Laughlin, and B. Carter, unpublished) or MVM (Faust and Ward, 1979) this was not done.

In studies by Muller *et al.* (1978), LuIII was serially passaged undiluted in HeLa cells 30 times beginning from a purified stock of standard 1.41, 110 S particles. During the subsequent passages, incorporation of [^{3}H]thymidine into standard 1.41, 110 S particles or variant 70–90 S particles was analyzed. This first passage yielded an appreciable proportion of variants. In subsequent passages there was a cyclic variation in the production of "high" or "low" amounts of variants but a progressive and constant decline in infectious titer from 10^7 p.f.u./ml at passage 1 to 10^3 p.f.u./ml at passage 25. This cycling of variant particle production has been reported for many viruses and probably is related to interference by the variants (Huang and Baltimore, 1970, 1977; see also Section VIII).

The type of standard helper virus used may have an effect on production of variants. The effect of *wt* or mutant H-1 helpers on the passage of H-1 variants (Rhode, 1978a) has already been described above (Section V.E). The main difference appears to be on the rate at which the H-1 variants appeared and accumulated reiterations.

Differences in variant multiplication have been noted when different host cells are used. Growth of an unfractionated stock of AAV at a multiplicity of 5.0 was compared in primary human embryonic kidney (HEK) cells, the human KB cell line, and monkey Vero cells (M. Myers and B. Carter, unpublished). Viral DNA or particles were analyzed by labeling the cultures jointly with [^{3}H]thymidine and [^{35}S]methionine. In all three cell lines the ratio of standard to variant genomes, determined by sucrose gradient sedimentation of intracellular viral DNA, was approximately 2. When purified particles were examined this ratio was increased to 5, 14, or 25 in KB, HEK, and Vero cells, respectively. This may suggest some preferential packaging of standard genomes depending on the cell line. For LuIII (Muller *et al.*, 1978) the ratio of standard particles to variant particles was 4.0 for HeLa cells and 0.5 for NB cells. Furthermore, production of LuIII variants was multiplicity dependent in Hela cells but not in NB cells.

C. Replication of Variant DNA

Most of the variant particles of AAV, MVM, H-1, and LuIII which have been characterized are noninfectious. An exception to this are the

H-1 viable variants such as DI-1 discussed above. The lack of infectivity reflects the DNA *rep*⁻ properties of the variant genomes.

Infection of cells with variant AAV, H-1, or LuIII particles alone did not lead to detectable duplex RF or progeny single-strand DNA synthesis (Laughlin *et al.*, 1979a; Rhode, 1978a; Muller *et al.*, 1978). This is shown for AAV in Fig. 5. It must be noted that these experiments were performed using *in vivo* isotopic labeling. This should be reinvestigated using blotting experiments and high specific activity *in vitro* labeled probes to rigorously eliminate even low levels of replication. AAV variants do apparently replicate when both adenovirus and standard AAV particles are added as helpers. Under these conditions there was a large increase in the synthesis of AAV variant DNA over that observed in infections using only standard AAV and adenovirus (Laughlin *et al.*, 1979a). An important point was that the helper function for AAV variant particles required both adenovirus and standard AAV and could not be provided by either virus alone. Furthermore UV-inactivated, standard AAV also was not a helper for variant replication.

One initially puzzling feature of AAV (Laughlin *et al.*, 1979a) was that infection with purified preparations of standard 1.41 particles yielded in a one-step growth experiment large amounts of variants with an average size of 10S when sedimented in neutral sucrose (see Fig. 5). Also the production of these variants was multiplicity-dependent. It was not clear if this was due to preferential accumulation of infecting variants genomes or to *de novo* generation of new variants. Based on the structural studies it now seems likely that this is accounted for by the presence in the 1.41 particles of variant genomes with the hairpin RF structure of type I DNA (de la Maza and Carter, 1980; see also Fig. 5). These molecules upon replication can readily yield single strands which are not hairpinned. Furthermore, it is possible that some of the RF type I variant genomes encapsidated in 1.41 particles may be oligomers of smaller variant molecules.

Support for this idea also comes from pulse-labeling experiments (Hauswirth and Berns, 1978), in which infected KB cells were pulse-labeled with [^{3}H]thymidine for 5 min at 18 hr after infection. In addition to standard genomes and faster sedimenting RF molecules, the pulse-labeled AAV DNA contained two variant species equivalent to about 55% and 30% of genome length. After a 1 hr chase most of the large molecules had apparently been converted to the smaller species.

Experiments with LuIII variant particles in Hela cells produced a different result than with AAV. Infection with LuIII variant particles alone yielded no detectable DNA synthesis or particle production whereas addition of standard LuIII to the variant infection did yield production of LuIII variants (Muller *et al.*, 1978). However, in contrast to the results with AAV, there was no net increase in the amount of variant DNA over that produced by infection with standard virus alone. It is therefore not clear if the added light density LuIII variant particle genomes actually replicated, and the structure of these genomes has not been studied (Muller *et al.*, 1978). It is possible that the LuIII variant genomes produced in HeLa cells replicate inefficiently as suggested for MVM type II molecules by Faust and Ward (1979).

D. Growth Cycle Parameters

Growth kinetics of variant particles have been examined under the conditions of single cycle growth both for LuIII in human Hela cells and AAV in human KB cells.

When Hela cells were synchronized for DNA synthesis then infected with LuIII at the time of release from the synchronization block, accumulation of virus particles began about 10 hr after infection (Siegl and Gautschi, 1973 a,b). Pulse-labeling with [^{3}H]thymidine revealed significant differences in the rate of accumulation of standard and variant virus (Muller *et al.*, 1978). At 11 to 13 hr the amount of label in variant particles represented only 10% of that in standard particles. By 17 to 19 hr the amount of label in variant particles was more than 50% of that in standard virus. This suggests that the rate of accumulation of LuIII variants increased during the later stages of the growth cycle.

Different results were seen when one-step growth of AAV variants was examined. The maximum rate of accumulation of AAV particles in human KB cell monolayers using adenovirus as the helper occurs between 12 and 24 hr after infection (Myers and Carter, 1980) whereas DNA replication begins at about 6 hrs (Carter *et al.*, 1973). In one set of studies (Myers and Carter, 1980) cultures were pulse-labeled with [^{3}H]thymidine for 30 min beginning at 18 hr after infection then chased with unlabeled thymidine. Under these conditions (Fig. 10) DNA chases into standard 1.41 particles at the same rate as into 1.37 and 1.35 DI particles. The relative amounts of ^{3}H-labeled 1.41, 1.37, or 1.35 particles were the same over the 4-hr chase period. Similar results were obtained when the pulse was begun at 14 hr after infection. Further experiments (M. Myers and B. Carter, unpublished) were performed in which viral DNA in the infected cultures was labeled for 1 hr periods then chased for 1 hr. At the end of the chase period the virus DNA was extracted and sedimented in neutral sucrose gradients to determine the amount of ^{3}H-labeled standard or variant genomes. These experiments showed that the relative rates of accumulation of standard and variant genomes were unchanged between 10 and 20 hr. Thus AAV variant genomes and particles do not occur preferentially later during the period of progeny strand synthesis, which is in contrast to the results obtained with LuIII (Muller *et al.*, 1980).

For AAV, these observations suggest that interference with standard genome replication by variants may occur early during RF replication (see Section IX).

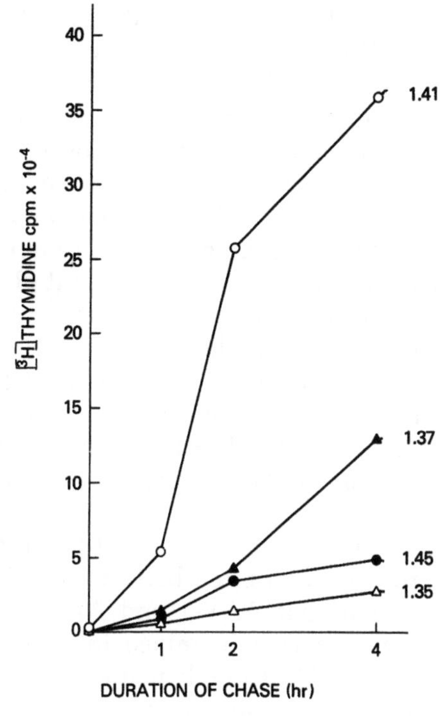

FIGURE 10. Pulse-chase kinetics of AAV particle assembly. KB monolayer cells were infected with AAV 2 (3 infectious units/cell) and adenovirus helper (5 p.f.u/ml). At 17 hr after infection, the cultures were labeled with [^{3}H]thymidine for 30 min then chased with unlabeled thymidine. During the chase the amount of labeled DNA incorporated into standard particles (1.41) or (1.45) or variant particles banding in CsCl at 1.37 or 1.35 g/cm^3 was determined as described by Myers and Carter (1980). The data were redrawn from Myers and Carter (1980).

VIII. INTERFERING PROPERTIES OF VARIANTS

A. Viral Interference

Some parvovirus variants have been shown to have the properties of defective interfering (DI) particles. Before discussing the DI phenomenon which is concerned with a specific type of viral interference it is useful to summarize the general phenomenon of viral interference. Many viruses interfere with each other when replicating in the same host cell and several types of interference can be classified.

Heterologous interference may occur between viruses of completely different groups. Examples of this type of interference are the interactions between AAV and adenovirus or SV40. AAV interferes with replication of its helper adenovirus in a multiplicity-dependent fashion (Hoggan *et al.*, 1966; Smith *et al.*, 1966; Casto *et al.*, 1967a,b; Parks *et al.*, 1967, 1968). This does not appear to be related to interferon since AAV does not produce an interferonlike substance (Casto *et al.*, 1967; Parks *et al.*, 1968) and adenovirus and AAV multiplication are very resistant to interferon action (H. Handa and B. Carter, unpublished). The interference of AAV with adenovirus apparently occurs early in the adenovirus replication cycle at a step after absorption and uncoating and is greatly decreased if cells are preinfected with adenovirus for 8 to 10 hr prior to

AAV (Parks *et al.*, 1968; Myers and Carter 1981), In support of this Carter *et al.* (1979) showed that AAV inhibits adenovirus DNA replication in a way which is dependent on the multiplicity of both adenovirus and AAV. AAV also interferes with growth of SV40 virus as shown by inhibition of SV40 plaque formation (Costa *et al.*, 1967b).

Heterotypic interference may occur between viruses of the same family or genus which may have varying degrees of genome sequence homology. One example is the interference between the helper-independent parvoviruses H-1, H-3, and LuIII (Rhode, 1978c). H-1 and H-3 are closely related based on restriction enzyme mapping whereas LuIII shows rather less homology. The mutant of H-1, *ts*14, is apparently a temperature-sensitive rep^- mutant which is deficient in RF DNA replication. In mixed infections, *ts*14, unlike H-1, caused a multiplicity-dependent inhibition of H-3 but not LuIII RF replication. In contrast, H-1 inhibits capsid protein synthesis of both the closely related H-3 virus and the less closely related LuIII.

A more extreme case of heterotypic interference is the *serotypic* interference exhibited between AAV serotypes. Torikai and Mayor (1969) reported that AAV-1 and AAV-4 exhibited mutual interference when grown together in mixed infections in primary African green monkey kidney (AGMK) cells using simian adenovirus SV15 as the helper. The degree of interference was directly proportional to the multiplicity of either serotype and also appeared to occur relatively early in the growth cycle, i.e., within the first 12 hr after infection. In view of the subsequent demonstration of AAV DI particles (Carter *et al.*, 1979; Laughlin *et al.*, 1979a) it remains to be demonstrated that this serotypic interference was due to interference by standard virus genomes rather than DI genomes.

Homologous interference has also been observed within a single serotype of AAV. This type of interference is manifested as a multiplicity-dependent autointerference (Rose and Koczot, 1972; Carter *et al.*, 1979) and is now known to be due to AAV DI particles (Laughlin *et al.*, 1979).

B. Defective Interfering Particles

As animal virus research progressed during the last 30 years it became increasingly apparent that production of biologically active defective particles was a frequent occurrence. The original observation was made by von Magnus (1954), who described incomplete or defective particles of influenza virus which arose during high multiplicity passage and which interfered with growth of standard influenza virus. Since then this phenomenon has been recorded for many virus families. This phenomenon was reviewed by Huang and Baltimore (1970), who designated the particles which cause it as defective interfering (DI) particles. These same authors suggested that DI particles may play a role in the evolution of viral diseases.

The general properties of DI particles were codified by Huang and Baltimore (1970, 1977) and Huang (1973): DI particles (1) contain normal structural proteins, (2) contain a part of the virus genome, (3) reproduce only in the presence of helper virus, (4) interfere specifically with the intracellular growth of homologous, standard virus, and (5) are able to increase their proportion in the yield from coinfections with standard virus in successive passages.

Although DI particles are deletion mutants, not all deletion mutants are DI particles. For instance the H-1 DI-1 virus is not a DI particle because it is viable. Similarly, Daniell (1976) reported that adenovirus synthesizes and encapsidates aberrant genomes with structures analogous to those of the MVM type II DNA analyzed by Faust and Ward (1979). However, these particles do not interfere with replication of standard adenovirus.

The encapsidated variant particles of H-1, AAV, and LuIII described in preceding sections are DI particles. That their properties satisfy criteria 1, 2, 3, and 5 of Huang and Baltimore has already been noted. Their interfering properties, which satisfy criterion 4, are described in the succeeding sections. The interfering properties of MVM variants have not been studied. The type I and type II variant of MVM each satisfy criteria 1 and 2, whereas type I but not type II satisfies criterion 5 (Faust and Ward, 1979). Thus, MVM type I variants may be DI particles whereas type II variants probably are not.

As for variant particles in general, the production and accumulation of DI particles may vary depending on the host cell used. The level of interference may also be dependent upon some host-cell factors. This has already been discussed for AAV and for LuIII (Section VII.B).

C. Interference by DI Particles

As well as the general criteria for DI particles proposed by Huang and Baltimore (1970), an additional set of criteria for interference with growth of standard virus by DI particles were defined by Huang (1973). Thus inhibition of standard virus growth caused by DI particles (1) is specific for homologous virus, (2) is prevented by UV irradiation of DI particles, (3) is an intracellular phenomenon, and (4) results in alteration of intracellular viral nucleic acid synthesis. Implicit in these general criteria is the idea that interference results from competition for some rate-limiting factor which is required for viral genome replication or encapsidation and which is, or is controlled by, a virus-coded function. This virus-coded function is deleted in the DI genome (see Section IX).

According to these criteria, DI particles do not interfere with adsorption, penetration, or uncoating of standard virus. The homologous interference may be extremely stringent and restricted to a homotypic serotype within a single genus. The effects of DI particles should also be distinguished from interference due to interferon production which is usually host-cell specific rather than virus species specific.

D. Interfering Properties of AAV Variants

AAV variant particles have been shown to exhibit the interfering properties of DI particles but also exhibit certain other (non-DI) properties of interference with adenovirus.

1. DI Properties of AAV Variants

AAV interference was analyzed by Carter *et al.* (1979). When human KB cells were infected with adenovirus and increasing amounts of AAV (1.41) or AAV (1.45) the maximum yield of AAV (about 10^4 infectious units/cell) was obtained at a multiplicity of 10–15 infectious units/cell. Above this multiplicity there was a linear, inverse relationship between the multiplicity of infection and yield of infectious AAV. In similar experiments, analysis of intracellular AAV DNA synthesis showed a corresponding increase in the amount of variant DNA and a concomitant decrease in synthesis of standard AAV DNA. This inhibition was ascribed (Carter *et al.*, 1979) to the variant molecules present in AAV (1.41) or AAV (1.45) preparations (de la Maza and Carter, 1980, 1981). These variant genomes appear therefore to have DI properties, and it can be shown that AAV (1.41) preparations obtained from high multiplicity passage stocks contain a higher proportion of such interfering particles than similar preparations from stocks grown at low dilution (M. Myers and B. Carter, unpublished).

This interference by AAV preparations is not due to inhibition of adenovirus (see below) and consequent depression of some helper function because it is still observed even when cells are infected with adenovirus for 10 hr prior to AAV (D. Risin and B. Carter, unpublished).

DI properties could be more directly demonstrated for lighter AAV particles such as AAV (1.35) or AAV (1.32) since these can be purified from AAV (1.41) particles (Laughlin *et al.*, 1979a). Addition of AAV (1.35) particles to an infection with AAV (1.41) particles inhibited production of infectious progeny AAV at low particle ratios (Table III).When intracellular DNA was analyzed in similar experiments (Fig. 11) there was progressive inhibition of standard AAV DNA (and increased synthesis of 10S variant DNA) as the amount of added 1.35 particles was increased. Again the inhibition occurred at low ratios of 1.35 : 1.41 particles. At a ratio of 10 : 1 there was a 50% inhibition of standard DNA. Similar inhibition was obtained wit AAV (1.32) particles, but the efficiency of inhibition was less (i.e., a relatively higher proportion of 1.32 particles was required) due to the presence of empty particles in AAV (1.32) preparations. For the experiment shown in Fig. 11 it was shown that the amount

TABLE III. Effect of Added AAV(1.35) Particles on the Yield of Infectious AAV[a]

Added AAV(1.35) particles per cell	Ratio of added AAV(1.35) particles to AAV(1.41) particles	Yield of infectious AAV: Infectivity/ml	Yield of infectious AAV: Percentage control
None	—	4.0×10^8	100.0
200	1	4.0×10^8	100.0
2,000	10	0.45×10^8	11.2
6,000	30	0.15×10^8	3.8
20,000	100	0.19×10^8	4.9

[a] KB monolayer cells (5×10^5 cells) were infected with Ad2 (5 pfu/cell) and AAV(1.41) particles at 5 infectious units/cell (i.e., 200 particles per cell). AAV(1.35) particles in the amount shown were also added at the time of infection. The data are taken from Myers and Carter (1980).

of 1.35 or 1.32 particles added did not inhibit the adsorption of ^{32}P-AAV (1.41) particles (Laughlin *et al.*, 1979a). Ultraviolet irradiation of variant particles prior to infection abolished their interfering properties (Laughlin *et al.*, 1979a) and the irradiation dose required was qualitatively propor-

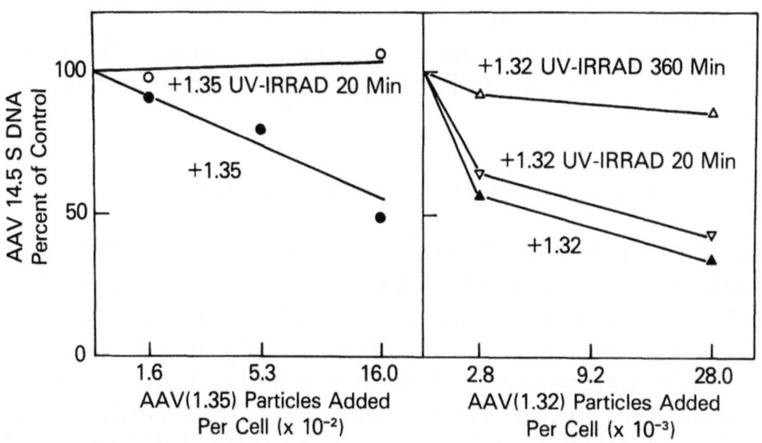

FIGURE 11. Effect of UV irradiation on AAV DI inhibition. KB cells in monolayers were infected with Ad2 at a multiplicity of 5 p.f.u./cell together with AAV (1.41) particles at a multiplicity of 3 infectious units (i.e., 150 particles) per cell and varying amounts of AAV(1.35) or (1.32) particles. DNA was labeled at 16 to 20 hr after infection with ^{3}H-thymidine and extracted and analyzed on neutral sucrose gradients. The total amount of 14.5 S (standard) AAV DNA in gradients was calculated and plotted as a percentage of the amount (2.9×10^4 counts/min) synthesized in the absence of added DI particles. Left-hand panel: Inhibition by AAV(1.35) particles not irradiated (●) or AAV(1.35) particles UV-irradiated for 20 min (○). Right-hand panel: An analogous experiment except that AAV(1.32) particles were used in place of AAV(1.35) particles. (▲) AAV(1.32) particles not irradiated; (▽) AAV(1.32) irradiated for 20 min; (△) AAV(1.32) irradiated for 360 min. Redrawn from Laughlin *et al.* (1979a).

tional to the variant particle genome target size (see Fig. 11). These experiments showed that inhibition was not mediated at the level of cellular adsorption but required replication of the variant genome.

Thus AAV variant particles appear to satisfy all the criteria of DI particles. Since individual variants have not been cloned, it is not know if all the variant particles in the population have DI properties.

2. AAV Interference with Adenovirus Replication

It was noted above (Section VIII.A) that AAV inhibits adenovirus DNA replication and viral assembly. When KB cells were coinfected with purified AAV DI particles and adenovirus there was no measurable inhibition of adenovirus DNA replication or particle assembly (Laughlin *et al.*, 1979a). In the same cells there was no detectable replication of the AAV DI genomes even when 2×10^5 AAV particles were added per cell. Thus, interference by AAV DI particles is quite specific for AAV replication and has no apprent effect upon adenovirus replication. Furthermore, the interference upon production of adenovirus type 2 DNA and particles by coinfection of KB cells with standard AAV (1.41) was progressively relieved if the cells were also coinfected with increasing amounts of AAV (1.35) or (1.32) DI particles (Laughlin *et al.*, 1979a). This abrogation of the interference with adenovirus by AAV (1.41) is presumably due to inhibition of the AAV (1.41) replication by the AAV DI particles. These results imply that DI particles compete with AAV (1.41) particles for an AAV gene product required for AAV replication but not for Ad replication.

3. AAV Interference with Adenovirus Tumorigenesis

Tumor production in newborn hamsters by oncogenic strains of human adenovirus such as adenovirus type 12, can be inhibited by concomitant injection of AAV (Kirchstein *et al.*, 1968; Mayor *et al.*, 1973) or H-1 (Toolan and Ledinko, 1968). More recently de la Maza and Carter (1981) have demonstrated that this inhibition of adenovirus oncogenesis can be mediated by injection of purified AAV-2 DNA. Furthermore, AAV (1.35) or AAV (1.32) DI particles, or even the purified DNA extracted from AAV DI particles can inhibit the Ad12 oncogenesis as efficiently as does standard AAV. These observations are in contrast to the inability of AAV DI genomes to inhibit lytic replication of adenovirus in permissive cells (Carter *et al.*, 1979, Laughlin *et al.*, 1979a). This suggests that inhibition of adenovirus oncogenesis does not require replication of AAV DNA. It is also consistent with the evidence that malignant transformation (or oncogenesis) does not require replication of the adenovirus genome (Doerfler and Lundholm, 1979; Doerfler, 1977; Graham *et al.*, 1974; Ginsberg *et al.*, 1974). Thus AAV DI genomes may allow a functional distinction between inhibition of adenovirus DNA replication and adenovirus oncogenesis. AAV may be a useful probe for analysis of certain adenovirus early gene functions (see Chapter 3).

E. Interfering Properties of H-1 Variants

Preparation of stocks of variant H-1 particles designated as DI-wt, DI-wt-*ts*14, or DI-wt-DI-1 and physical analysis of their genomes was described above (Section V.E). The defective-interfering properties of these viruses were demonstrated by Rhode (1978a) and shown to be those of DI particles.

Several experiments (Rhode, 1978a) indicated that H-1 DI virus interfered with replication of H-1 wt in a manner that was intracellular, was multiplicity dependent, and did not apparently depend on interferon production.

A preparation of H-1 DI particles added to NB cells infected with H-1 wt at a multiplicity of 3.0 p.f.u./cell resulted in a 90%–95% inhibition in the yield of infectious H-1 particles or H-1 hemagglutination activity. The effect of interference was the same when the DI particles were absorbed to the cells either 4 hr before or 4 hr after the H-1 wt virus. The inhibition was reduced to 50% when the multiplicity of H-1 wt was increased to 50 p.f.u./cell.

Interference by DI particles was also very sensitive to inactivation by UV irradiation. A dose that reduced the infectivity of H-1 by 10^3-fold eliminated all interfering activity from H-1 DI particles.

The interference by H-1 DI particles did not appear to be due to induction of interferon. Although helper-independent parvoviruses are sensitive to interferon they are very poor interferon inducers (Kilham *et al.*, 1968). The stocks of H-1 DI particles apparently did not contain interferon and there was no detectable interferon produced when NB cells infected with H-1 DI particles were assayed 6–12 hr after infection.

H-1 DI preparations also inhibited production of hemagglutinin (viral capsid protein) synthesis of the heterotypic viruses H-3 and LuIII (Rhode 1978a). This is analogous to the multiplicity-dependent, heterotypic cross-interference among wt H-1, H-3 and LuIII for viral protein synthesis (Rhode, 1978c).

Preparations of H-1 DI particles were deficient in production of the viral capsid protein VP1 and VP2 as determined by immunofluorescent staining or by labeling of infected cultures with ^{35}S-methionine followed by gel electrophoresis (Rhode, 1978a). This is in accord with deletions in the DI genomes which may remove a major transcription promoter (see Section V.E). In this respect it is of interest that the H-1 mutant *dl*-1, having only a 350 nucleotide deletion at map position 38.0 and no other insertions, is also an interfering genome (S. Rhode, personal communication).

Cells jointly infected by H-1 wt and H-1 DI particles were examined by immunofluorescent staining. Both the number of cells producing viral antigen and the amount of antigen per cell were decreased (Rhode 1978a). Interestingly the DI interference delayed or prevented development of viral cytopathology. This suggests that H-1 DI particles might be able to aid in establishment of persistent infections (see Section VII.G below).

F. Interfering Properties of LuIII

The variant particles of LuIII analyzed by Muller *et al.* (1978) appeared to contain predominantly single strand DNA which may be analogous to the type I MVM genomes analyzed by Faust and Ward (1979). The LuIII variants showed no ability to replicate in the absence of standard LuIII but interfered with replication of LuIII DNA in single cycle growth experiments. Also, production of LuIII variants showed cyclic variation and interference during continued passage. Therefore, although competition at the level of cellular absorption has not been rigorously eliminated it seems likely that the LuIII variants can also be classified as DI particles.

G. DI Particles and Persistent Infection

There are many examples of viruses which can establish persistent infections in some cell lines (Youngner and Preble 1980). Many explanations have been offered for this phenomenon including the involvement of interferon or the evolution of temperature-sensitive virus variants. Huang and Baltimore (1970) suggested that DI particles may play a role in establishing or maintaining persistent infections, and there is now good evidence for this in some cases (Holland *et al.*, 1980).

It was noted that H-1 DI particles might be capable of establishing persistent infections (Rhode 1978a). The helper-independent parvovirus Aleutian disease virus (ADV) causes a persistent infection in mink (Porter and Cho 1980). Replication of naturally isolated ADV is temperature sensitive in cell culture. Also, ADV produces large amounts of virus, containing DNA, with a CsCl buoyant density of 1.33–1.35 g/cm^3. The virus titer is attenuated by successive passage in Crandal feline kidney cells as shown by Bloom *et al.* (1980), who suggested that the variant ADV may be DI particles. It has been suggested also (Bloom *et al.*, 1980; Porter and Cho, 1980) that such DI particles may play a role in establishment of persistent infections by ADV although there is, as yet, no direct evidence for this.

IX. MECHANISM OF INTERFERENCE

The mechanism of interference by parvovirus DI particles is not clearly understood but must be considered in relation to the general parvovirus replication cycle. A variety of experiments from a number of laboratories has led to the formulation of a general scheme of parvovirus DNA replication. Their work is reviewed in detail elsewhere in this volume.

Briefly, parvovirus replication can be envisaged to occur in the following steps. After absorption and uncoating the parental single-strand genome is converted to a double-stranded parental RF DNA. This process, called *parental strand RF replication,* may occur by self-priming from the base-paired 3′ terminal hairpin and for helper-independent viruses may require a cell product present in late S-phase. In the next stage, the *parental RF* is replicated through a number of rounds to amplify the pool of *progeny RF* molecules which may be oligomeric. In the third stage, progeny RF molecules are the templates for displacement synthesis of *progeny single-strand* (SS) DNA which is packaged into a preformed capsid.

Two lines of evidence indicate that accumulation of progeny SS DNA apparently requires concomitant synthesis of capsids or capsid protein. Temperature-sensitive mutants of H-1, such as *ts*-1, fail to produce H-1 capsid proteins and also fail to accumulate progeny SS DNA (Rhode, 1976). For AAV, capsid formation can be inhibited either by the arginine analog, L-canavanine, which inhibits production of the major VP1 capsid protein (Buller and Rose, 1978) or by use of the Ad5 mutant, Ad5*ts*125, which prevents translation of the 2.3-kb RNA coding for the AAV capsid proteins (Myers *et al.*, 1980; Jay *et al.*, 1981). In either case inhibition of AAV capsids is accompanied by inhibition of AAV progeny SS DNA accumulation (Myers and Carter, 1981). Evidence that AAV DNA is packaged into preformed capsids comes from isotopic pulse-chase studies of AAV assembly (Myers and Carter, 1980) and from a consideration of the physical structure of parvovirus DI particles (see Section X).

For replication of progeny RF molecules of either H-1 or AAV, current evidence suggests that the parvovirus genome may supply a necessary *rep* product (see below, Section X). The first stage of parental strand RF replication may require a cell product (or for AAV, perhaps a helper virus coded product) other than or in addition to, the mammalian cell DNA polymerases. In spite of its self-priming 3′ terminal hairpin structure, parvovirus DNA is not utilized as a substrate *in vitro* by purified mammalian DNA polymerases (D. Ward, personal communication; Jagadeeswaran and Lavelle, 1979) but is replicated by nuclear lysates (Kollek *et al.*, 1980; Pritchard *et al.*, 1981).

Interference by DI particles could occur at any of the several stages in the parvovirus replication cycle. There is very little direct evidence, but several lines of indirect evidence argue that the interference may occur primarily at a relatively early stage of RF replication.

Since DI genomes in general probably do not make capsid proteins they must compete for capsids produced from standard genomes. However, this is probably not the primary level of interference for several reasons. First, in most parvovirus infections, there appears to be a significant excess of empty capsids produced (Myers and Carter 1980; Tattersall 1978). This excess of capsid protein is apparently related to the abundance of the mRNA coding for it (see Chapter 5) and the probability that the pool of templates utilized for transcription may be limited (see below). Of course, in the presence of a very large number of DI genomes it is possible that there would be sufficient interference with standard RF replication to reduce the number of transcription templates, and consequently that number of capsids to a rate-limiting level. At this stage additional interference might occur during encapsidation. It also must be recognized that there is no evidence that all of the empty capsids in the apparently excess pool are in fact available for use in packaging DNA. The pool of utilizable capsids might be rate limiting.

Additional evidence that most of the interference does not occur at the level of progeny strand synthesis or encapsidation comes from studies using the adenovirus mutant, Ad5*ts*125, as a helper for AAV at the nonpermissive temperature of 40°C in human KB cells. Under these conditions AAV RF replication is normal but synthesis of AAV capsid proteins and accumulation of AAV progeny SS DNA does not occur (Myers *et al.*, 1980; Myers and Carter, 1981; Jay *et al.*, 1981). Using this mutant helper, the same amount of DI genomes are replicated (Myers and Carter, 1981) and the same degree of interference with replication of standard RF DNA is observed (B. Carter, unpublished) as when adenovirus wild-type helper is used.

AAV DI particles do not seem to interfere simply by additionally inhibiting adenovirus and decreasing the level of adenovirus helper functions. As noted, above, AAV DI particles alone apparently do not interfere with adenovirus functions expressed during a lytic infection. Furthermore, infection of cells with adenovirus for 10 hr, to allow expression of adenovirus helper functions prior to infection with AAV, still resulted in extensive interference (D. Risin and B. Carter, unpublished).

Interference by AAV DI particles does not appear to occur primarily at the level of transcription and DI genomes do not give rise to a significant number of cytoplasmic transcripts. First, most of the AAV DI genomes are deleted for the major promoter region at map position 40.0 and also for the next strongest promoter at 18.0 but do retain the promoter at 5.0 (Laughlin *et al.*, 1979b; de la Maza and Carter, 1980a; Green and Roeder, 1980). Analysis of cytoplasmic RNA by electron microscopy of DNA–RNA hybrids revealed no significant amount of AAV RNA molecules that appeared to be transcripts of DI molecules although some rare AAV RNAs were seen with additional deletions located at varying map positions (Laughlin *et al.*, 1979b).

Marcus *et al.* (1981) analyzed quantitatively the steady state AAV RNA species present in the nucleus and cytoplasm of infected KB cells under conditions in which the proportion of DI genomes ranged from less than 5% to more than 90% of the total intracellular AAV DNA. In these experiments, the amount of standard DNA synthesis varied by more than 10-fold, but there was no significant variation in the amount of type of the standard AAV RNAs which accumulated and there was no significant accumulation of DI transcripts.

These experiments, as well as indicating that interference at transcriptional levels is not significant, also imply that the pool of active transcription templates may be only a small function of the total RF DNA. This would be consistent with the proposal that a parvovirus *rep* product is required for RF replication. Thus transcription may begin from the earliest RF molecules produced. Rhode (1976) has also interpreted experiments with BUdR inhibition as indicating that the pool of H-1 transcription templates is small.

The main point at which parvovirus DI interference occurs appears to be at an early stage of infection at the level of parental or progeny RF DNA replication. Some evidence in support of this has been obtained (B. Carter, unpublished) by isolating intracellular DNA, labeled with [^{3}H]thymidine between 10 and 20 hr after infection, under conditions in which SS DNA does not reanneal (Straus *et al.*, 1976b; Myers and Carter, 1981). In these experiments, as the amount of DI genomes was increased in the infecting virus innoculum there was a concomitant decrease in the amount of duplex RF DNA replication. The decreased accumulation of SS DNA was proportional to that of RF DNA. Also, as noted above (Section VII.D), the rates of accumulation of AAV standard or DI genomes and particles were equivalent between 12 and 20 hr after infection (M. Myers and B. Carter, unpublished). Finally, the interference between AAV serotypes observed by Torikai and Mayor (1969) occurred within the first 12 hr after infection.

The precise mechanism of interference with RF DNA replication is not clear, but several possibilities can be identified. There may be competition for a rate-limiting produce involved in RF replication, such as the proposed *rep* product. Amplification of the region near the 5′ replication origin in the H-1 DI molecules may offer a selective advantage for initiating replication (Rhode, 1978a, and unpublished experiments). For AAV or MVM, once oligomeric RFs are produced, those from DI genomes may replicate faster because of the selective enrichment for replication origins. Whether AAV or MV variant genomes can accumulate more readily by replicating faster simply because they are shorter molecules is not clear. It seems, however, that the rate-limiting steps in replication are more likely to involve initiation or termination rather than elongation.

Retention of a replication origin at either end of a variant genome appears to be important for amplification. Type I variant genomes of MVM containing both genome termini are amplified selectively whereas type II genomes containing only the 5′ terminus are not (Faust and Ward, 1979). For AAV this same situation might not necessarily apply since the replication origins at either end of the genome are identical (Berns *et al.*, 1978). However, AAV type II variant genomes are relatively rare in encapsidated DNA (de la Maza and Carter, 1980). If indeed these type II molecules are present in intracellular AAV DNA as suggested by Hauswirth and Berns (1979) they apparently do not interfere or compete well.

One interesting possibility is that the parvovirus DI genomes which are apparently deleted for the major promoter at map position 38.0 to 40.0 might have an advantage in the early rounds of replication because they are less likely to sequestered as transcription templates (B. Carter, unpublished; S. Rhode, unpublished).

Rhode (personal communication) has suggested that, at least for H-1, interference may occur at the very early stage of conversion of input genomes to parental RF. His study (unpublished) of the kinetics of initiation of infection as a function of input multiplicity suggests that parental RF synthesis is a very rate-limiting step and highly dependent upon input DNA concentration. If it is also hypothesized that appearance of capsid proteins in the cell nucleus inhibits further parental RF production (by encapsidating any free input genomes), the virus genome which replicated earliest would have a selective advantage.

In summary, although the precise mechanism is not known, the basis for the interference by parvovirus DI genomes and their selective amplification appears to lie in early events in the intracellular replication cycle.

X. BIOLOGICAL INFERENCES

The structure and properties of the parvovirus variants have a number of important biological implications. Analysis of the sequence retained or amplified in variant DNA molecules can help to delineate *cis*-acting elements. A number of studies (summarized elsewhere in this volume) have identified the parvovirus genome terminal palindrome regions as having important functions in the replication origin and in associated events such as site-specific nicking and perhaps "hairpin" transfer. The retention of these palindromes as the minimal conserved unit in AAV or MVM variant DNA confirms their importance as replication origins. Similarly, the 55-nucleotide sequence that is just inboard of the H-1 viral strand 5′ palindrome and which is reiterated in the variant genomes may have an important *cis*-acting function for H-1 replication.

As already discussed above, the requirement of a helper virus for multiplication of the deleted variants implied the existence of a *trans*-acting parvovirus product which may be necessary for RF replication. Furthermore, this *rep* product may not by synonymous with the capsid proteins which are required for SS DNA accumulation but not for RF replication. Perhaps one *rep* product is the protein that may be covalently bound to the 5′ terminus of intracellular parvovirus DNA (Revie *et al.*, 1979; Chow, 1981; B. Stillman, personal communication). For AAV the capsid proteins are encoded by the right-hand half of the genome (Jay *et al.*, 1981). Thus a potential *rep* function may be coded by the left-hand half of the genome which has two open reading frames (Lusby and Berns, 1981).

The structure of variant DNA also has a number of implications for the mechanism of encapsidation. Since generally only one AAV or MVM variant DNA molecule is packaged per particle, a full-length strand is not required for formation or stability of variant particles. This is consistent with evidence that AAV is assembled into preformed capsids rather than by condensation of capsid subunits around the DNA (Myers and Carter, 1980). Also, production of progeny SS DNA from concatemeric intermediates does not appear to involve a length-measuring mechanism such as the head-full mechanism for phage T4 DNA (Streisinger *et al.*, 1964, 1967).

Parvovirus SS DNA replication is believed to occur by strand displacement from concatemeric RF molecules which requires a sequence-specific cleavage at both the 3′ and 5′ ends of the strand at sites defined by the terminal palindromic sequence. If the AAV DI type I RF structures shown in Fig. 5 are packaged according to the general model, then as the DNA strand is displaced from its 5′ end cleavage does not always occur at the first terminal palindromic sequence reached. Thus, AAV appears to be unique among viruses in its ability to encapsidate an RF structure. If these structures are viewed as head-to-head (or tail-to-tail) oligomers of single stranded variants then they are analogous to the reiterated variants of SV40 DNA described by Davoli and Fareed (1975).

Although RF molecules much larger than a normal genome are found in the intracellular DNA there is no evidence that any parvovirus DNA molecules longer than genome length are encapsidated in particles. This presumably indicates a restriction on the size of the DNA that can be packaged.

Current evidence suggests that the structures of H-1 particles are very similar to those of MVM. However, H-1 variants tend to accumulate molecules which approach standard genome length whereas MVM does not. This probably is not a reflection of selective packaging of large H-1 variants and is more likely related to intracellular replication and the tendency for certain region of H-1 to reiterate at high frequency.

It was suggested on the basis of the structure of AAV and MVM variant genomes (de la Maza and Carter, 1980) that if there are any *cis*-acting DNA sequences required for DNA encapsidation, these sequences must exist in or near the terminal palindromes. There is evidence for such *cis*-acting regions in adenovirus DNA (Daniell and Mullenbach, 1978). Furthermore, the existence of such sequences in parvovirus DNA together with the fact that AAV has identical 3′ and 5′ termini while autonomous parvoviruses do not, might explain why the autonomous viruses package only minus (viral) strands. However, this idea requires reinterpretation in the light of recent evidence that autonomous parvoviruses such as LuIII or KRV can encapsidate efficiently plus, as well as minus, DNA strands depending on the host cell used (G. Siegl, personal communication; R. Bates, C. Snyder, S. Mitra, unpublished experiments cited in Mitra, 1980).

The analysis of variant molecules has yielded considerable insight into the replication of parvovirus genomes. Further analysis of these molecules, and particularly the deleted, recombinant genomes, may yield insight into cellular recombination mechanisms. One particularly important aspect of parvoviruses which is highlighted by the study of the variant genomes, as noted by Faust and Ward (1979), is their use as potential cloning vectors for propagation of genetic elements in mammalian cells.

ACKNOWLEDGMENTS. Much of the experimental and conceptual framework of studies on AAV variants from my own laboratory is due to the contributions of Luis de la Maza, Catherine Laughlin, Maureen Myers, Carol Marcus, Francis Jay, and Deborah Risin. I have also derived much stimulation from rewarding discussions with Solon Rhode, Emmanuel Faust, David Ward, Ken Berns, William Hauswirth, Peter Tattersall, and Günter Siegl. I also thank Joan Mok and Mary Demory for typing the manuscript.

REFERENCES

Armentraut, R., Bates, R., Berns, K., Carter, B., Chow, M., Dressler, D., Fife, W., Hauswirth, W., Hayward, G., Lavelle, G., Rhode, S., Straus, S., Tattersall, P., and Ward, D., 1978, A standardized nomenclature for restriction endonuclease fragments, in: *Replication of Mammalian Parvoviruses* (D. Ward and P. Tattersall, eds.) pp. 523–526, Cold Spring Harbor Laboratory, New York.

Astell, C.R., Smith, M., Chow, M.B., and Ward, D.C., 1979, Structure of the 3′ hairpin termini of four rodent parvovirus genomes: Nucleotide sequence homology at the origins of DNA replication, *Cell* **17**:691.

Bachmann, P.A., Hoggan, M.D., Kurstak, E., Melnick, J.L., Pereira, H.G., Tattersall, P., and Vago, C., 1979, Parvoviridae: Second report, *Intervirology*, **11**:248.

Berns, K.I., Kort, J., Fife, K.H., Grogan, E.W., and Spear I., 1975, Study of the fine structure of adeno-associated virus DNA with bacterial restriction endonucleases, *J. Virol.* **16**:712.

Berns, K.I., Hauswirth, W.W., Fife, K.H., and Lusby, E., 1978, Adeno-associated virus DNA replication, *Cold Spring Harbor Symp. Quant. Biol.* **43**:781.

Bloom, M.E., Race, R.E., and Wolfinbarger, J.B., 1980, Characterization of Aleutian disease virus as a parvovirus, *J. Virol.* **35**:836.

Bourguignon, G.J., Tattersall, P., and Ward, D.C., 1976, DNA of minute virus of mice: Self-priming, non-permuted, single-stranded genome with a 5′ terminal hairpin duplex, *J. Virol.* **20**:290.

Bratosin, S., Laub, O., Tal, J., and Aloni, Y., 1979, Mechanism for circularization of linear DNAs: Circular parvoviral MVM DNA is formed by a "noose" sliding in a "lasso"-like DNA structure, *Proc. Natl. Acad. Sci. USA* **76**:4289.

Buller, R.M.L., and Rose, J.A., 1978, Characterization of adenovirus associated virus-induced polypeptides in KB cells, *J. Virol.* **25**:331.

Carter, B.J., and Khoury, G., 1975. Specific cleavage of adenovirus-associated virus DNA by restriction endonuclease R. *Eco* RI. Characterization of cleavage products, *Virology* **63**:523.

Carter, B.J., Koczot, F.J., Garrison, J., Rose, J.A., and Dolin, R., 1973. Separate helper functions provided by adenovirus for adenovirus-associated virus multiplication, *Nature New Biol.* **244:**71.

Carter, B.J., Fife, K.H., de la Maza, L.M., and Berns, K.I., 1976, Genome localization of adeno-associated virus RNA, *J. Virol.* **19:**1044.

Carter, B.J., Laughlin, C.A., de la Maza, L.M., and Myers, M., 1979, Adeno-associated virus autointerference, *Virology* **92:**449.

Casto, B.C., Atchison, R.W., and Hammon, W.McD., 1967a, Studies on the relationship between adeno-associated virus type I (AAV-1) and adenoviruses. I. Replication of AAV-1 in certain cell cultures and its effect on helper adenovirus, *Virology* **32:**52.

Casto, B.C., Armstrong, J.A., Atchison, R.W., and Hammon, W.McD., 1967b, Studies on the relationship between adeno-associated virus type I (AAV-1) and adenoviruses. II. Inhibition of adenovirus plaques by AAV; its nature and specificity, *Virology* **33:**452.

Cavalier-Smith, T., 1974, Palindromic base sequences and replication of eukaryotic chromosome ends, *Nature* **250:**467.

Chow, M.B., 1981, DNA Replication of Minute virus of mice, a parvovirus, Ph.D. thesis, Yale University, Connecticut.

Chow, M.B., and Ward, D.C., 1978, Comparison of the terminal nucleotide structures in the DNA of non-defective parvoviruses, in: *Replication of Mammalian Parvoviruses* (D. Ward and P. Tattersall, eds.), pp. 205–217, Cold Spring Harbor Laboratory, New York.

Clinton, G.M., and Hayashi, M., 1975, The parvovirus MVM: Particles with altered structural proteins, *Virology* **66:**261.

Clinton, G.M., and Hayashi, M., 1976, The parvovirus MVM: A comparison of heavy and light particle infectivity and their density conversion in vitro, *Virology* **74:**57.

Crawford, L.V., 1966, A minute virus of mice, *Virology* **29:**605.

Daniell, E., 1976, Genome structure of incomplete particles of adenovirus, *J. Virol.* **19:**685.

Daniell, E., and Mullenbach, T., 1978, Synthesis of defective viral DNA in Hela cells infected with adenovirus type 3, *J. Virol.* **26:**61.

Davoli, D., and Fareed, G.C., 1975, Formation of reiterated simian virus 40 DNA, *Cold Spring Harbor Symp. Quant. Biol.* **39:**137.

de la Maza, L.M., and Carter, B.J., 1977, Adeno-associated virus DNA structure. Restriction endonuclease maps and arrangement of terminal sequences, *Virology* **82:**409.

de la Maza, L.M., and Carter, B.J., 1978, DNA structure of incomplete adeno-associated virus particles, in: *Replication of Mammalian Parvoviruses* (D. Ward and P. Tattersall, eds.), pp. 193–204, Cold Spring Harbor Laboratory, New York.

de la Maza, L.M., and Carter, B.J., 1980a, Molecular structure of adeno-associated virus variant DNA, *J. Biol. Chem.* **255:**3194.

de la Maza, L.M., and Carter, B.J., 1980b, Heavy and light particles of adeno-associated virus, *J. Virol.* **33:**1129.

de la Maza, L.M., and Carter, B.J., 1981, Inhibition of adenovirus tumorigenicity by adeno-associated virus, *J. Nat. Cancer Inst.* **67:**1323.

Denhardt, D.T., Eisenberg, S., Bartok, K., and Carter, B.J., 1976, Multiple structures of adeno-associated virus DNA. Analysis of terminally labeled molecules with endonuclease R. *Hae* III, *J. Virol.* **18:**672.

Doerfler, W., 1977, Animal virus-host genome interactions, in: *Comprehensive Virology* (H. Fraenkal-Conrat and R. Wagner, eds.), Vol. 10, pp. 279–399, Plenum Press, New York.

Doerfler, W., and Lundholm, U., 1970, Absence of replication of the DNA of adenovirus type 12 in BHK21 cells, *Virology* **40:**754.

Faust, E.A., and Ward, D.C., 1979, Incomplete genomes of the parvovirus minute virus of mice: Selective conservation of genome termini, including the origin for DNA replication, *J. Virol.* **32:**276.

Fitzgerald, M., and Shenk, T., 1981, The sequence 5′-AAUAAA-3′ forms part of the recognition site for polyadenylation of late SV40 mRNAs. *Cell* **24:**251.

Gerry, H.W., Kelly, T.J., and Berns, K.I., 1973, Arrangement of nucleotide sequences in adeno-associated virus DNA, *J. Mol. Biol.* **79:**207.

Ginsberg, H.S., Ensinger, M.J., Kauffman, R.S., Mayer, A.J., and Lundholm, U., 1974, Cell transformation: A study of regulation with types 5 and 12 adenovirus temperature-sensitive mutants, *Cold Spring Harbor Symp. Quant. Biol.* **39:**419.

Graham, F.L., van der Eb, A.J., and Heijneker, H.L., 1974, Size and location of the transforming region in human adenovirus type 5 DNA, *Nature* **251:**687.

Green, M.R., and Roeder, R.G., 1980, Definition of a novel promoter for the major adenovirus-associated virus mRNA, *Cell* **22:**231.

Hauswirth, W.W., and Berns, K.I., 1978, Initiation and termination of adeno-associated-virus DNA replication, in: *Replication of Mammalian Parvoviruses* (D. Ward and P. Tattersall, eds.), pp. 257–267, Cold Spring Harbor Laboratory, New York.

Hauswirth, W.W., and Berns, K.I., 1979, Adeno-associated virus DNA replication: Non-unit length molecules, *Virology* **93:**57.

Hirt, B., 1967, Selective extraction of polyoma DNA from infected mouse cell cultures, *J. Mol. Biol.* **26:**365.

Hoggan, M.D., Blacklow, N.R., and Rowe, W.P., 1966, Studies of small DNA viruses found in various adenovirus preparations: Physical, biological, and immunological characteristics, *Proc. Natl. Acad. Sci. USA* **55:**1467.

Holland, J.J., Kennedy, S.I.T., Semler, B., Jones, C.L., Raux, L., and Grabou, E.A., 1980, Defective interfering RNA viruses and the host-cell response, in: *Comprehensive Virology* (H. Fraenkal-Conrat and R. Wagner, eds.), Vol. 10, pp. 137–192, Plenum Press, New York.

Huang, A.S., 1973, Defective interfering viruses, *Ann. Rev. Microbiol.* **27:**101.

Huang, A.S., and Baltimore, D., 1970, Defective Viral Particles and Viral Disease processes, *Nature* **226:**325.

Huang, A.S., and Baltimore, D., 1977, Defective interfering animal viruses, in: *Comprehensive Virology* (H. Fraenkal-Conrat and R.W. Wagner, eds.), Vol. 10, pp. 73–116, Plenum Press, New York.

Jagadeeswaran, P., and Lavelle, G., 1979, In vitro replication of single stranded DNA of Kilham rat virus, *Fed. Proc. Abs.* **38:**689.

Jay, F.T., Laughlin,C.A., de la Maza, L.M., Carter, B.J., and Cook, W.J., 1978, Adeno-associated virus RNA synthesis in vivo and in vitro, in: *Replication of Mammalian Parvoviruses* (D. Ward and P. Tattersall, eds.), pp. 385–397, Cold Spring Harbor Laboratory, New York.

Jay, F.T., Laughlin, C.A., and Carter, B.J., 1981, Eukaryotic translational control: Adeno-associated virus protein synthesis is affected by a mutation in the adenovirus DNA binding protein, *Proc. Natl. Acad. Sci. USA* **78:**2927.

Johnson, F.B., Whitaker, C.W., and Hoggan, M.D., 1975, Structural components of adenovirus-associated top component, *Virology* **65:**196.

Johnson, R.H., Siegl, G., and Gautschi, M., 1974, Characteristics of feline panleucopaenia virus strains enabling definitive classification as parvoviruses, *Arch. Gesamte Virusforsch.* **46:**315.

Kilham, L., Buckler, C.E., Ferm, V.H., and Baron, S., 1968, Production of interferon during rat virus infection, *Proc. Soc. Exp. Biol. Med.* **129:**274.

Kirchstein, R.L., Smith, K.O., and Peters, E.A., 1968, Inhibition of adenovirus 12 oncogenicity by adeno-associated virus, *Proc. Soc. Exp. Biol. Med.* **128:**670.

Kollek, R., Tseng, B.Y., and Goulian, M., 1980, Studies on the effect of inhibitors of DNA polymerases on synthesis of H-1 DNA, *J. Supramolec. Struct.* **Suppl 4:**322.

Kongsvik, J.R., Hopkins, M.S., and Ellem, K.A.O., 1978, Two populations of infectious virus produced during H-1 infection of synchronized transformed cells, in: *Replication of Mammalian Parvoviruses* (D. Ward and P. Tattersall, eds.), pp. 505–520, Cold Spring Harbor Laboratory, New York.

Laughlin, C.A., Myers, M.W., Risin, D.L., and Carter, B.J., 1979a, Defective-interfering particles of the human parvovirus adeno-associated virus, *Virology* **94:**162.

Laughlin, C.A., Westphal, H., and Carter, B.J., 1979b, Spliced adenovirus-associated virus RNA, *Proc. Natl. Acad. Sci. USA* **76:**5566.

Laughlin, C.A., Jones, N.C., and Carter, B.J., 1982, The effect of deletions in adenovirus early region 1 genes upon replication of adeno-associated virus, *J. Virol.* **41:**868.

Lusby, E., and Berns, K.I., 1981, Mapping the 5′ termini of two adeno-associated virus RNAs in the left half of the genome, *J. Virol* **41:**518.

Marcus, C.J., Laughlin, C.A., and Carter, B.J., 1981, Adeno-associated RNA transcription in vivo, *European J. Biochem.* **121:**147.

Matsuguchi, M., Puvion-Dutilleul, F., and Moyne, G., 1979, Late transcription and simultaneous replication of simian adenovirus 7 DNA as revealed by spreading lytically infected cell cultures, *J. Gen. Virol.* **42:**443.

Mayor, H.D., Houlditch, G.S., and Mumford, D.M., 1973, Influence of adeno-associated satellite virus on adenovirus-induced tumors in hamsters, *Nature New Biol.* **241:**44.

Mitra, S., 1980, DNA replication in viruses, *Ann. Rev. Genet.* **14:**347.

Muller, H.P., Gautschi, M., and Siegl, G., 1978, Defective particles of parvovirus LuIII, in: *Replication of Mammalian Parvoviruses* (D. Ward and P. Tattersall, eds.), pp. 231–240, Cold Spring Harbor Laboratory, New York.

Myers, M.W., and Carter, B.J., 1980, Assembly of adeno-associated virus, *Virology* **102:**71.

Myers, M.W., and Carter, B.J., 1981, Adeno-associated virus replication. The effect of L-canavanine or a helper virus mutation on accumulation of viral capsids and progeny single-stranded DNA, *J. Biol. Chem.* **256:**567.

Myers, M.W., Laughlin, C.A., Jay, F.T., and Carter, B.J., 1980, Adenovirus helper function for growth of adeno-associated virus: Effect of temperature-sensitive mutations in adenovirus early gene region 2, *J. Virol.* **35:**65.

O'Farrell, P.H., 1975, High resolution two dimensional gel electrophoresis, *J. Biol. Chem.* **246:**6159.

Parks, W.P., Melnick, J.L., Rongay, R., and Mayor, H.D., 1967, Physical assay and growth cycle studies of a defective adeno-satellite virus, *J. Virol.* **1:**171.

Parks, W.P., Casazza, A.M., Alcott, J., and Melnick, J.L., 1968, Adeno-associated satellite virus interference with the replication of its helper adenovirus, *J. Exp. Med.* **127:**91.

Porter, D.D., and Cho, H.J., 1980, Aleutian disease of mink: A model for persistent infection, in: *Comprehensive Virology* (H. Fraenkel-Conrat and R. Wagner, eds.), Vol. 10, pp. 233–256, Plenum Press, New York.

Pritchard, C.R., Stout, E.R., and Bates, R.C., 1981, Replication of parvoviral DNA. I. Characterization of a nuclear lysate system, *J. Virol.* **37:**352.

Revie, D., Tseng, B.Y., Grafstrom, B.H., and Goulian, M., 1979, Covalent association of protein with replicative form DNA of parvovirus H-1, *Proc. Natl. Acad. Sci. USA* **76:**5539.

Rhode, S.L.,III, 1976, Replication process of the parvovirus H-1. V. Isolation of and characterization of temperature-sensitive H-1 mutants defective in progency DNA synthesis, *J. Virol.* **17:**659.

Rhode, S.L., III, 1978a, Defective interfering particles of parvovirus H-1, *J. Virol.* **27:**347.

Rhode, S.L., III, 1978b, H-1 DNA synthesis, in: *Replication of Mammalian Parvoviruses* (D. Ward and P. Tattersall, eds.), pp. 279–296, Cold Spring Harbor Laboratory, New York.

Rhode, S.L., III, 1978c, Replication process of the parvovirus, H-1. X. Isolation of a mutant defective in replicative-form DNA replication, *J. Virol.* **25:**215.

Richards, R., Linser, P., and Armentraut, R.W., 1977, Kinetics of assembly of a parvovirus, minute virus of mice, in synchronized brain cells, *J. Virol.* **22:**778.

Richards, K., Linser, P., and Armentraut, R.W., 1978, Maturation of minute-virus-of-mice particles in synchronized rat-brain cells, in: *Replication of Mammalian Parvoviruses* (D. Ward and P. Tattersall, eds.), pp. 447–458, Cold Spring Harbor Laboratory, New York.

Rose, J.A., and Koczot, F.J., 1972, Adenovirus-associated virus multiplication. VII. Helper requirement for viral deoxyribonucleic acid and ribonucleic acid synthesis, *J. Virol.* **10:**1.

Seigl, G., 1972, Parvoviruses as contaminants of permanent human cell lines. V. The nuclei acid of KBSH virus, *Arch. Gesante. Virusforsch.* **37**:267.

Siegl, G., 1973, Physiochemical characteristics of the DNA of parvovirus LuIII, *Arch. Gesamte. Virusforsch.* **43**:334.

Siegl, G., 1976, *The Parvoviruses*, Springer-Verlag, Wien.

Siegl, G., and Gautschi, M., 1973a, The multiplication of parvovirus LuIII in a synchronized culture system. I. Optimum conditions for virus replication, *Arch. Gesamte Virusforsch.* **40**:105.

Siegl, G., and Gautschi, M., 1973b, The multiplication of parvovirus LuIII in a synchronized culture system. II. Biochemical characteristics of virus replication, *Arch. Gesamte Virusforsch.* **40**:119.

Smith, K.O., Gehle, W.D., and Thiel, J.F., 1966, Properties of a small virus associated with adenovirus type 4, *J. Immunol.* **97**:754.

Spear, I., Fife, K.H., Hauswirth, W.W., Jones, C.J., and Berns, K.I., 1977, Evidence for two nucleotide sequence orientations within the terminal repetition of adeno-associated virus DNA, *J. Virol.* **24**:627.

Straus, S.E., Sebring, E.D., and Rose, J.A., 1976, Concatemers of alternating plus and minus strands are intermediates in adenovirus-associated virus DNA synthesis, *Proc. Natl. Acad. Sci. USA* **73**:742.

Streisinger, G., Edgar, R.S., and Denhardt, G.H., 1964, Chromosome structure in phage T4, I. Circularity of the linkage map, *Proc. Natl. Acad. Sci. USA* **51**:775.

Streisinger, G., Emrich, J., and Stahl, M.M., 1967, Chromosome structure in phage T4. III. Terminal redundancy and length determination, *Proc. Natl. Acad. Sci. USA* **57**:292.

Tattersall, P., 1978, Parvovirus protein structure and virion maturation, in: *Replication of Mammalian Parvoviruses* (D. Ward and P. Tattersall, eds.), pp. 53–72, Cold Spring Harbor Laboratory, New York.

Tattersall, P., and Ward, D.C., 1976, The rolling hairpin: A model for replication of parvovirus and linear chromosomal DNA, *Nature* **263**:156.

Tattersall, P., Cawte, P.J., Shatkin, A.J., and Ward, D.C., 1976, Three structural polypeptides coded for by minute virus of mice, *J. Virol.* **20**:273.

Tattersall, P., Shatkin, A.J., and Ward, D.C., 1977, Sequence homology between the structural polypeptides of minute virus of mice, *J. Mol. Biol.* **111**:375.

Toolan, H.W., and Ledinko, N., 1968, Inhibition by H-1 virus of the incidence of tumores produced by endovirus type 12 in hamsters, *Virology* **35**:475.

Torikai, K., and Mayor, H.D., 1969, Interference between two adeno-associated satellite viruses: A three component system, *J. Virol.* **3**:484.

Torikai, K., Ito, M., Jordan, L.E., and Mayor, H.D., 1970, Properties of light particles produced during growth of type 4 adeno-associated satellite virus, *J. Virol.* **6**:363.

von Magnus, P., 1954, Incomplete forms of influenza virus, *Adv. Virus Res.* **2**:59.

Wolgemuth, D.J., and Hsu, M-T., 1980, Visualization of genetic recombination intermediates of human adenovirus type 2 DNA from infected Hela cells, *Nature* **287**:168.

Youngner, J.S., and Preble, O.T., 1980, Viral persistance: Evolution of viral populations, in: *Comprehensive Virology* (H. Fraenkal-Conrat and R. Wagner, eds.), Vol. 10, pp. 73–135, Plenum Press, New York.

CHAPTER 7

Parvovirus Proteins

F. BRENT JOHNSON

I. INTRODUCTION

Division of the parvovirus group classically has been based upon the requirement for coinfection with a helper virus. Thus, some parvoviruses are regarded as defective and require the assistance of another virus for complete replication. These defective viruses will be referred to as helper-dependent viruses. The adenovirus-associated viruses (AAV) are the prototype viruses of the helper-dependent group. The other major group of parvoviruses is composed of the autonomously replicating viruses or helper-independent viruses.

When parvovirus particles from infected cells are purified on isopycnic cesium chloride gradients, typically four main density classes are noted. A minor proportion of the infectious particles band at 1.45–1.47 g/cm^3. The major proportion of the infectious particles band at 1.39–1.42 g/cm^3. The particles which band at 1.35–1.37 g/cm^3 are heterogeneous. They represent a mixed population of variant particles containing DNA molecules shorter than genome length. The particles banding at 1.30–1.32 g/cm^3 are empty capsids. They not only appear to be empty in negatively stained preparations in the electron microscope, but do not label with radioactive thymidine. For the sake of simplicity and consistency these different types of particles will be referred to by the virus designation followed by the relative specific density in cesium chloride enclosed in parentheses, e.g., AAV(1.40).

Despite the recognition of most parvoviruses as members of this group at the time of the analysis of their structural proteins, nonuniform systems of nomenclature have been used for designation of the proteins.

F. BRENT JOHNSON • Department of Microbiology, Brigham Young University, Provo, Utah 84601.

At the parvovirus conference at Cold Spring Harbor in 1977 it was suggested that a uniform nomenclature system be adopted. The system simply was that the viral proteins be designated by the abbreviation "VP" and that the proteins be numbered (1,2,3,etc.) beginning with the protein of highest molecular weight. Although this system is somewhat flawed, it is gaining wider acceptance and will be used in this chapter.

Several review articles have appeared on parvovirus structure, replication, and biology. The serious reader is referred to those reviews (Berns, 1974; Berns and Hauswirth, 1979; Henry, 1973; Hoggan, 1970; Hoggan, 1971; Kilham and Margolis, 1975; Margolis and Kilham, 1975; Porter and Larsen, 1974; Porter and Cho, 1980; Porter *et al.*, 1980; Rose, 1974; Siegl, 1976; Tattersall, 1978; Toolan, 1972; Young and Mayor, 1979a) in addition to these in the current collection.

II. ADENOVIRUS-ASSOCIATED VIRUS

A. Composition of the Capsid

1. Capsid Morphology

The parvovirus virion appears to be icosahedral, nonenveloped, with a diameter which has been estimated to range from 15 to 28 nm (Hoggan, 1971; Atchison *et al.*, 1965; Hoggan *et al.*, 1966; Mayor *et al.*, 1965; Archetti and Bocciarelli, 1964, 1965; Crawford *et al.*, 1969). Several parvoviruses observed in the same laboratory in a comparative study had diameters ranging from 19 to 24 nm (Hoggan, 1971). The number of capsomers and their arrangement in the virion is uncertain. Estimates of the number of capsomers have ranged from 12 for AAV (Mayor *et al.*, 1965), 20 or 32 for AAV (Archetti and Bocciarelli, 1964, 1965), to 32 for Rat Virus and several other parvoviruses (Karasaki, 1966) to 42 for DNV (Kurstak and Cote, 1969). One group of workers (Smith *et al.*, 1966) reported that the individual capsomers of AAV were not readily distinguishable and that the surface of the particle might best be represented by a reticulum or netlike arrangement of protein fibers.

Chemical studies have shown the AAV virion consists of protein (80%) and DNA (20%) (Parks *et al.*, 1967). The relatively high DNA-to-protein ratio gives the virion a density in cesium chloride of about 1.4 g/cm^3 which allows ready separation from the helper adenovirus (density = 1.35–1.36 g/cm^3) on cesium chloride gradients.

2. Number of Structural Proteins

The capsid proteins of three AAV serotypes (AAV-1, AAV-2, and AAV-3) were compared by SDS gel electrophoresis by Rose *et al.* (1971).

TABLE I. AAV Structural Polypeptides

Virus	Particle type[a]	Virus protein	Estimated mol. wt. ($\times 10^{-3}$)	Fraction of virion protein (%)	Reference
AAV-1	AAV(1.40)	VP1	83.5–87	8–10	Rose *et al.* (1971)
		VP2	70.3–73	5–12	Salo and Mayor (1977)
		VP3	56.3–62	78–86	Salo and Mayor (1977)
AAV-2	AAV(1.39)	VP1	87	8	Rose *et al.* (1971)
		VP2	73	5	
		VP3	62[b]	86	
AAV-3	AAV(1.34)	VP1	91.6	11	Johnson *et al.* (1975)
		VP2	79.3	10	
		VP3	65.9	79	
AAV-3	AAV(1.40)	VP1	87–91.6	8–11	Rose *et al.* (1971)
		VP2	73–79.3	5–10	Johnson *et al.* (1971)
		VP3	62–65.9	79–86	Johnson *et al.* (1971)
AAV-3	AAV(1.467)	VP1	92.5	7	Johnson *et al.* (1971)
		VP2	80.3	7	
		VP3	68.6	85	
AAV-4	AAV(1.43); AAV(1.30)	VP1	83.4	7	Salo and Mayor (1977)
		VP2	71.1	10	
		VP3	58.3	83	

[a] Designated by the relative specific density of the particle in cesium chloride.
[b] AAV-2 VP3 molecular weight reported to be slightly less than AAV-1 VP3 in comparative gels (Rose *et al.*, 1971).

Three structural proteins were found in each virus (refer to Table I). The smallest of the three proteins was found in the greatest abundance in the virion. Among the three serotypes homologous components were electrophoretically similar except AAV-2 VP3 was slightly smaller than the VP3 molecule of AAV-1 or AAV-3. Similar results were reported by Johnson *et al.* (1971), who found three proteins in the capsid of AAV-3. They also compared the protein of AAV(1.40) with AAV(1.46) particles. These two types of particles contained similar proteins; however, the amount of VP1 and VP2 seemed to be reproducibly less in the AAV(1.46) particles when scanning the Coomassie Blue stained bands under conditions of densitometric linearity. These protein differences seemed to account for the greater density of AAV(1.46) particles in cesium chloride density gradients; however, incorporation of radiolabeled amino acids into the proteins does not consistently confirm the findings observed by staining.

Salo and Mayor (1977) reported studies on AAV type 4. Two minor proteins and one major protein were found. The protein composition of AAV-4 empty particles and AAV-4(1.43) particles was essentially identical (Table I). In other studies on empty capsids, AAV-3(1.34) particles were shown to contain the same proteins in similar distribution to the infectious AAV(1.40) particles (Johnson *et al.*, 1975). These particles completely lacked the radioactive thymidine which was included in the infected cells. The findings on the composition of the empty capsids suggested that AAV does not contain a core protein which is lacking when the DNA is absent.

3. Hemagglutinin

One property of the parvoviruses which has been used to characterize them is their varied ability to hemagglutinate different erythrocytes. Of the four major AAV serotypes, hemagglutinating activity is associated with only AAV type 4 (Ito and Mayor, 1968). AAV types 1, 2, and 3 do not hemagglutinate either at 4 or 37°C (Hoggan, 1971) but type 4 agglutinates human, guinea pig, and sheep erythrocytes at 4 but not at 37°C (Ito and Mayor, 1968). Salo and Mayor (1978) noted that infectious virions of AAV type 4 were more dependent than empty capsids upon temperature for hemagglutination. When the virus was tested for hemagglutination with either guinea pig or human erythrocytes at 4 and 22°C, the HA titers were nearly maintained in the case of empty capsids but were almost totally lost with complete particles at 22°C.

Attempts to determine which AAV protein possesses hemaglutinating activity were unsuccessful as purification and isolation of the proteins resulted in the loss of all hemagglutinating activity (Salo and Mayor, 1977). In addition, type 4 AAV grown in arginine deficient medium lacked the hemagglutinin normally associated with the type 4 virion (Mayor and Gorman, 1975).

4. External Orientation of the Polypeptides in the Capsid

Earlier studies suggested the absence of a basic AAV core protein (Rose *et al.*, 1971; Johnson *et al.*, 1975), but it still was unknown whether any of the three AAV proteins was exclusively internally oriented. Lubeck and Johnson (1977) reported studies wherein sepharose-bound (solid-state) lactoperoxidase was used to radioiodinate the externally oriented polypeptides. This technique has found wide usage to identify external virion proteins as it employs low levels of relatively weak oxidizing agents and labels only those residues which are exposed to and have contact with the enzyme. It was determined that both AAV(1.40) and AAV(1.46) particles had externally oriented VP1, VP2, and VP3. An offshoot of the study was that when AAV(1.40) particles or solubilized proteins were iodinated the labeling approximated the stoichiometric ratios of the three polypeptides as found by other methods in the intact virion, indicating a similar susceptibility of the three polypeptides to iodination.

Salo and Mayor (1978) reported that AAV-4, as well as X14 particles, migrated to a pH of 2.6 in sucrose-stabilized isoelectric focusing gradients.

Empty capsids and complete capsids had the same isoelectric point. The authors suggested this extremely acidic isoelectric point indicates that acidic amino acids are on the surface of the capsid and the basic amino acids are aligned internally. The isoelectric points of the individual purified capsid proteins were reported by Johnson *et al.* (1978) to fall near to one another at pH 5.0–5.3. These results correlate with the conclusion of Salo and Mayor that the acidic portions of the molecule are oriented outward, while the basic residues orient inward because the overall isoelectric point is much less acidic than when the proteins are assembled into the capsid structure.

B. Relationships among the Structural Proteins

1. Amino Acid Labeling

AAV was grown in the presence of individual radiolabeled amino acids, purified and the structural proteins separated on polyacrylamide gels (Johnson *et al.*, 1977). None of the proteins was specifically enriched with leucine, lysine, or aspartic acid, suggesting that no one polypeptide is significantly more basic or acidic than any other and that, with respect to these amino acids, the amino acid content is similar between VP1, VP2, and VP3.

2. Immunochemical Relationships

Antisera produced in guinea pigs against purified SDS treated VP1, VP2, and VP3 reacted in immunofluorescence tests with AAV antigens that appeared in infected cells (Johnson *et al.*, 1972). These infected cell antigens were viral specific protein(s) that were reactive with the antisera prior to the acquisition by the protein of its final tertiary conformation. Likewise, the capsid proteins could be denatured by SDS treatment which restored their immunoreactivity to these antisera. The combination of SDS and heat dissociates the protein and unfolds each polypeptide chain to form an SDS–polypeptide complex. In this complex the polypeptide chain interacts with SDS molecules in such a way that renaturation of the protein into its native conformation is virtually impossible. When this polypeptide chain serves as an immunogen, antibody molecules are made to the numerous immunologic determinants exposed along the molecule. The SDS molecule was not immunogenic and the antibodies were directed against the polypeptides and not against the bound detergent (Johnson *et al.*, 1977). Each antiserum was composed of a heterogeneous population of antibodies that was directed against many immunologic determinants along the polypeptide molecules.

Direct radioimmunoprecipitation assays using fragmented polypeptides revealed that VP1, VP2, and VP3 share a large number of immunogenic determinants (Johnson *et al.*, 1977). These same authors showed in immunocompetition assays that VP1, VP2, and VP3 share antigens that compete for the antibodies in the reagent antisera.

Not only do VP1, VP2, and VP3 share antigens within a given serotype, but some of these polypeptide immunologic determinants are shared with AAV of other serotypes (Johnson *et al.*, 1972) even though the capsid antigens are serotypically unique or nearly unique.

3. Isoelectric Points

AAV capsid proteins dissociated in urea and focused on polyacrylamide isoelectric focusing gels appeared in the pH gradient at pH 5.0 to 5.3 (Johnson *et al.*, 1978). All three polypeptides had similar acidic isoelectric points indicating their similarity in amino acid content and an enrichment with acidic amino acids (first noted by Rose *et al.*, 1971).

4. Peptide Maps

Peptide maps of the individual AAV capsid proteins were produced by Lubeck *et al.* (1979). The proteins were radiolabeled with ^{125}I, individually purified and digested with trypsin or chymotrypsin. The digests were separated on cellulose thin layer plates by electrophoresis and chromatography. A large number of common peptides were observed which were shared by all three capsid proteins, suggesting that all three polypeptides are encoded in the same sequence of virus DNA. The tryptic and chymotryptic maps clearly demonstrated three to four peptides that were unique to VP1 (the largest polypeptide). The data suggested that the VP3 amino acid sequences were largely contained within VP2 and the VP2 sequences were contained within VP1.

5. Amino Acid Analysis

Amino acid analyses of the purified capsid polypeptides of AAV-3, AAV-2 (1.40), and AAV-2(1.46) were reported by Lubeck *et al.* (1979). The amino acid composition of these proteins was similar, implying a common origin. The amino acid composition of VP3 was contained within VP2 which was contained within VP1. The most common amino acids in these proteins are aspartic acid, glycine, then glutamic acid which accounts for the acidic isoelectric point of the proteins.

C. Protein Biosynthesis

Normally a cell must be coinfected with AAV and its adenovirus helper for complete AAV replication to occur. Partial replication occurs in the presence of herpes viruses but assembly is incomplete. In the absence of helper viruses AAV penetrates to the nucleus and is uncoated there (Berns *et al.*, 1975) but there is no transcription (Rose and Koczot, 1972). Even in the adenovirus coinfected cell, AAV DNA synthesis precedes RNA synthesis, suggesting the absence of any AAV "early" function (Handa *et al.*, 1976). That AAV has the ability to establish a latent infection in cells in the absence of helper (Hoggan *et al.*, 1972; Berns *et al.*, 1975) implies that the cell provides all necessary functions for the establishment of latency and the maintenance of the persistent infection without AAV protein synthesis.

1. Herpes Viruses as Incomplete Helpers

A variety of herpes viruses have been shown to provide incomplete helper functions for AAV replication (Atchison, 1970; Blacklow *et al.*, 1970; Blacklow *et al.*, 1971; Dolin and Rabson, 1973). These partial helpers permitted AAV antigen synthesis (detected as immunofluorescent stainable antigen) but infectious AAV or AAV particles were not produced. Cells coinfected with herpes simplex virus produced infectious AAV DNA which remained unencapsidated (Boucher *et al.*, 1971) suggesting the inability of herpes viruses to help the AAV assembly process. Cells transformed by adenoviruses (Hoggan, 1970) or herpes virus (reported by Mayor and Drake, 1974) were not capable of supporting AAV replication without superinfection by either an adenovirus or herpes virus. However, herpes virus transformed cells coinfected with a helper virus demonstrated abundant cytoplasmic AAV antigen while the bulk of AAV antigen is nuclear in the normal cell counterpart (Mayor and Drake, 1974). In contrast to the above results, Blacklow (1975) reported that immunofluorescent stainable antigen appeared in 333-8-9 cells infected with AAV-3 alone. These infected cells were stained with sera to VP1, VP3 and whole virion antisera. The results suggested that 333-8-9 cells synthesized at least two of the three AAV polypeptides without exogenous helper; that this incomplete potentiation might be related to HSV-2 genetic material possessed by the cells. However, in view of the strong cross-reactions between the VP1, VP2, and VP3 antisera and the extensive sharing of immunologic determinants on the three polypeptides, it is not clear why the VP2 antiserum failed to stain these cells.

2. Intracellular Location of AAV Protein and Transport

Following penetration of the AAV particles to the nucleus of the infected cell and uncoating, DNA replication and transcription occurs. The messenger RNA(s) are transported to the cytoplasm where AAV proteins first appear (Johnson *et al.*, 1972; Salo and Mayor, 1979). These cytoplasmic proteins are largely immature as the final conformation and antigenic structure of the proteins is first found in the nucleus after transport from the cytoplasm (Mayor *et al.*, 1967; Johnson *et al.*, 1972). The mechanism of transport or the form of the protein during transport remains unclear. Assembled complete particles as well as empty particles are first found in the nucleus (Atchison *et al.*, 1966; Mayor *et al.*, 1967; Torpier *et al.*, 1971). Cells that have a high concentration of intranuclear AAV usually have scarce adenovirus particles and usually form separate inclusions (Mayor *et al.*, 1967). Under conditions of arginine deprivation potentially infectious AAV particles were synthesized, although fewer particles were produced and the particles appeared to lack structural stability (Mayor and Gorman, 1975). The observation that AAV particles were found in both the nucleus and the cytoplasm of cells infected under conditions of arginine starvation indicates that in AAV systems, in contrast to herpes simplex, migration of viral constituents from cytoplasm to nucleus and vice versa was not inhibited.

3. Translation of AAV Proteins

Although the subject of transcription is treated elsewhere in this volume it seems appropriate briefly to review some of the information on RNA forms and synthesis to obtain a clearer view of AAV translation. The thymidine-rich or minus strand of AAV DNA is transcribed *in vivo* (Carter *et al.*, 1972). Approximately 50% of the particles in a normal virus population contain minus strand DNA and the other 50% of particles contain positive strand DNA. No reports convincingly show that the positive-stranded particles are noninfectious. The observation that transcription occurs only from the minus strand implies that if the positive strand is infectious a minus copy must be formed in order to allow transcription to proceed. Early on, Carter (1974) reported finding a single RNA transcript having a molecular weight of about 0.9×10^6 to 1.0×10^6. Later refined observations demonstrated several AAV RNA forms in infected cells.

Green *et al.* (1980) reported finding four prominent viral transcripts containing 4.3, 3.6, 2.6, and 2.3 kilobases (kb). All species were polyadenylated and found in nuclear as well as whole cell preparations. The 2.3-kb RNA was the predominant form, appeared to be spliced, and was the only form found on polysomes. The 4.3-kb RNA (90% of the genome length) and the 3.6-kb RNA were confined to the nucleus. The 5′ termini of the three larger transcripts map to distinct positions, while the 3′ termini of all the RNAs map to a common location at position 96 on the genome. The abundant 2.3-kb RNA is spliced and is composed of a short leader sequence of about 50 nucleotides ligated to a body of about 2250 nucleotides. The 2.3-kb RNA appeared to be derived from the 2.6-kb transcript (Green and Roeder, 1980a). Further, Green and Roeder (1980b) proposed the following model for the synthesis of the 2.3-kb mRNA: transcription is initiated near position 39, the RNA chain is elongated to position 96, then polyadenylated producing the unspliced 2.6 kb RNA. A single splice then generates the mature 2.3-kb RNA which is then transported to the cell cytoplasm. The 4.3 and 3.6-kb RNAs are not found on polysomes and appear not to be obligatory precursors to the functional 2.3-kb RNA.

In other reports cytoplasmic AAV RNA exists as a family of overlapping spliced RNAs of 2.3, 3.3, 3.6, and 3.9 kb (Laughlin *et al.*, 1979b). The 2.3-kb RNA was the most abundant, accounting for at least 50% of all the cytoplasmic AAV transcripts. The RNA's were coterminal at the 3′ end. Jay *et al.* (1981) reported the 2.3-kb RNA codes for all three proteins VP1, VP2, and VP3 when translated in an *in vitro* reticulocyte lysate showing that the capsid proteins are coded by the genome sequence between map positions 48.0 and 96.0. When a temperature-sensitive helper (Ad5*ts*125) was used at the nonpermissive temperature the accumulation of the AAV capsid proteins decreased by 50-fold. Normal amounts of 2.3-kb mRNA were made, however, and this RNA could be translated *in vitro* to VP1, VP2, and VP3 suggesting that control is exerted upon the 2.3-kb mRNA at the translational level and that this control can be influenced by mutations in adenovirus. Post-translational processing of the AAV protein may occur since a single mRNA encodes all three capsid proteins. Whether the processing is catalyzed by a cellular enzyme, or by a helper virus product, or is autocatalytic is not known. It also remains unknown whether processing occurs at the amino terminal, the carboxyterminal, or both termini. It is clear that the capsid proteins possess blocked amino termini that preclude dansylation under normal conditions (Johnson *et al.*, unpublished results), but the amino acid sequences near the amino termini remain unknown.

Further evidence for post-translational processing of AAV protein was provided by Buller and Rose (1978a), and Myers and Carter (1981). The arginine analog, L-canavanine, inhibited the accumulation of the major protein (VP3), empty or full AAV capsids or a nonstructural 24,000 polypeptide. In contrast, accumulation of VP1 and VP2 was not affected, therefore L-canavanine apparently inhibits a post-translational cleavage required to produce VP3. Also, the appearance of the three capsid proteins in infected cells is in the same molar proportions as found in the virion (Johnson *et al.*, 1977; Salo and Mayor, 1979; Buller and Rose, 1978).

D. Nonstructural Proteins

AAV-induced noncapsid proteins appear in infected cells and in *in vitro* translation systems. Buller and Rose (1978a) demonstrated the presence of small AAV-specific peptides of molecular weight 24,900 and 15,800. Other AAV peptides of molecular weights 75,000, 61,400, 44,500, 30,200, 22,400, and 13,200, in addition to VP1, VP2, and VP3, appeared in *in vitro* syntheses directed by AAV mRNA (Buller and Rose, 1978b). In work reported by Johnson *et al.* (1977) immune sera precipitated AAV polypeptides of molecular weights 56,000, 100,000, 110,000, and 120,000.

The origin and nature of all of these protein forms is not yet known; however, it is presumed that many of them represent cleavage products from the posttranslational processing of the primary gene product and perhaps catabolic intermediates in the degradation of unused protein molecules. Some of these forms probably represent multimeric forms of the structural or nonstructural molecules.

E. Assembly and Protein Modification

1. Assembly

Results described by Myers and Carter (1980) suggested a general model for AAV assembly. In this model DNA progeny strands are supplied by strand displacement replication from duplex replicating intermediates. Newly synthesized AAV protein is rapidly assembled (by an as yet unidentified mechanism) into empty 1.32 capsids which then associate with progeny DNA strands concomitant with, or soon after, displacement. These intermediate structures are converted (in the absence of DNA synthesis) over a period of several hours to mature particles containing either a standard or a DI genome. DNA replication can occur in the absence of capsid assembly and the virus "maturation" process occurs in the absence of DNA replication. Assembly of the proteins into the 1.32 particle requires modification of the proteins, the appearance of new antigens, and requires a function in HSV coinfections that is sensitive to 2-deoxy-D-glucose, an inhibitor of protein glycosylation (Young and Mayor, 1979b).

A question relating to assembly of the AAV virion is whether AAV(1.45) particles have a precursor relationship with AAV(1.41) particles or if these two particle types are assembled in separate pathways. Early after infection the two particle types are assembled in separate pathways. Early after infection the two particle types appear in about equal amounts, but later on the 1.41 particles become the predominate type (De La Maza and Carter, 1980). In pulse-chase experiments a precursor–product relationship could not be demonstrated, indicating independent assembly of the two particle types.

2. Biological Function of the AAV Proteins

Aside from the structural function of the AAV proteins and the ability of the AAV type 4 capsid to hemagglutinate erythrocytes little is known of other functions. No unequivocal enzymatic activities have yet been identified. However, Myers and Carter (1981) showed that inhibition of AAV capsid accumulation was accompanied by inhibition of accumulation of AAV single-stranded DNA but not accumulation of RF DNA. This observation suggested that appearance of the single-stranded progeny DNA is linked to the accumulation of the AAV capsid proteins (see also Laughlin *et al.*, 1979a).

3. Modification

AAV type 1 proteins produced in cells labeled with [^{32}P]orthophosphate or [^{14}C]glucosamine were examined by gel electrophoresis (Salo and Mayor, 1979). Even though control adenoviral proteins were phosphorylated the AAV proteins were not phosphorylated. In cells coinfected with either adenovirus or herpes simplex virus the AAV polypeptides were not glycosylated while control proteins were glycosylated. The AAV proteins are, however, highly labeled with [^{14}C]acetate suggesting acetylation of the protein (Johnson and Hoggan, 1979) and possibly accounting for the *N*-terminal block.

III. ALEUTIAN DISEASE VIRUS

A. Composition and Properties of the Capsid

Aleutian disease, a virus-induced immune complex disease of mink, is characterized by glomerulonephritis, arteritis, generalized plasmacytosis, and hypergammaglobulinemia with high titers of antiviral antibody. Negatively stained aleutian disease virus (ADV) appears icosahedral in morphology by electron microscopy and has a diameter of 23–25 nm (Cho and Ingram, 1973; Chesebro *et al.*, 1975; Porter *et al.*, 1977). The virion appears to contain 32 capsomers which are hollow tubular-shaped structures 4.5 nm in diameter with a 1–2-nm central hole. The virus contains single-stranded DNA (mol. wt. = 1.2×10^6) and is stable to ether, chloroform, fluorocarbon, pH 3, proteases, nucleases, and heating at 56°C for 30 min (Porter and Cho, 1980). On CsC1 gradients the virus bands at buoyant densities of 1.295, 1.332, and 1.405–1.416 g/cm^3 (Cho, 1977). Most of the infectivity bands at 1.405–1.416 g/cm^3.

Two reports have appeared describing the ADV protein. Shahrabadi *et al.* (1977) isolated ADV particles from infected mink by stepwise fluorocarbon extraction, CsCl density gradient centrifugation and velocity sedimentation through sucrose. Four polypeptides were found in these particles with molecular weights of 30,000, 27,000, 20,500, and 14,000 (See Table II). In contrast, Bloom *et al.* (1980) reported a strain of ADV adapted to grow in cell culture contained two structural polypeptides. The virus was purified on CsCl gradients without fluorocarbon extraction. The two polypeptides had molecular weights of 89,100 and 77,600.

B. Synthesis of Viral Antigen

A strain of ADV (the Utah-1 strain) has been adapted to grow in cell culture producing viral antigen and infectious virus in a continuous line of feline kidney cells (CRFK cells) or in primary feline kidney cells (Porter

TABLE II. Structural Polypeptides of Helper-Independent Parvoviruses

Virus	Particle type[a]	Virus protein	Estimated mol. wt. ($\times 10^{-3}$)	Fraction of virion protein (%)	Reference
Aleutian disease virus	ADV(1.41)	VP1	30	42	Shahrabadi *et al.* (1977)
		VP2	27	12	
		VP3	20.5	42	
		VP4	14	4	
	ADV(?)	VP1	89.1	—	Bloom *et al.* (1980)
		VP2	77.6	—	
Bovine parvovirus	BPV(1.40); BPV(1.30)	VP1	85.5	10	Johnson and Hoggan (1973)
		VP2	77	8	
		VP3	67	82	
Densonucleosis virus	DNV(1.40); DNV(1.30); DNV(1.44)	VP1	98–109	1–15[b]	Tijssen *et al.* (1976)
		VP2	69–71	15–28	Kelly *et al.* (1980)
		VP3	58–61	15–24	Kelly *et al.* (1980)
		VP4	42–49	27–69	Kelly *et al.* (1980)
Feline panleukopenia virus	FPV(1.41)	VP1	73.1	10	Johnson *et al.* (1974)
		VP2	60.3	86	
		VP3?	39.6	3–6	
H-1	H-1(1.43)	VP1	88–92	25	Kongsvik *et al.* (1974)
		VP2′	68–72	75	Paradiso (1981)
		VP2	65–69	(75)[c]	Paradiso (1981)
Lu III	Lu III(1.41)	VP1	75	15.5	Gautschi and Siegl (1973)
		VP3	62	84.5	Gautschi and Siegl (1973)
	Lu III(1.31) Lu III(1.35)	VP1	75	—	Gautschi and Siegl (1973)
		VP2	69	—	
		VP3	62	—	
	Lu III(1.30–1.32)	VP1	82	V[d]	
		VP2	67	V	
	Lu III(1.33–1.38)	VP1	82	V	

		VP2	67	V	Majaniemi, Tratschin, and Siegl (personal communication)
		VP3	64	V	
	Lu III(1.39–1.42)	VP2	67	V	
		VP3	64	V	
	Lu III(1.43–1.46)	VP3	64	V	
MVM	MVM(1.30)	VP1	83.3–92	14–18	Clinton and Hayashi (1975)
		VP2	64.3–72	79–85	Tattersall *et al.* (1976)
		(VP3)	62–69	7	Tattersall *et al.* (1976)
	MVM(1.42)	VP1	83.3–92	12–18	
		VP2	64.3–72	Variable (e.g., 27)	
		VP3	61.4–69	Variable (e.g., 61)	Peterson *et al.* (1978)
	MVM(1.47)	VP1	92	12–18	
		VP2	72	86	
		VP3	69	2	
RV	RV(1.41)	VP1	72	13.1	Salzman and White (1970)
		VP2	62	75.5	
		VP3	55	11.4	
	RV(1.30)	VP1	81	9.7	Peterson *et al.* (1978)
		VP2	63.5	87.1	
		VP3	59.5	3.2	
	RV(1.42)	VP1	81	10.2	
		VP2	63.5	30.3	
		VP3	59.5	59.5	
X-14	X-14(?) (Complete virions and empty capsids)	VP1	82.4	13.7	Salo and Mayor (1977)
		VP2	67.5	86.3	

[a] Particle type referred to by its relative specific density in CsCl.
[b] The variance in the relative amounts of the proteins present in DNV is apparently due to the different serotypes tested and the age of the larvae in which the virus is grown.
[c] The concentration of VP2 varies inversely with VP2′; it arises by proteolytic cleavage of VP2′ during virion maturation. VP2 is not present in empty capsids.
[d] V = variable. The concentrations of the polypeptides shift with the density of the virion. As the density increases, VP3 appears then becomes the predominate component in the densest particles.

et al., 1977). The virus exhibits a degree of temperature sensitivity at usual incubation temperatures. The temperature optimum is about 31.8°C (Porter *et al.*, 1980). Because the virus does not produce plaques in cell culture, it is usually quantitated by an immunofluorescence focus assay. ADV requires rapidly dividing cells for replication, a property shared with other parvoviruses. Infected mink cells in culture produce viral antigen but little infectious virus. Virus antigen has been found in macrophages (perhaps phagocytized immune complexes) and in an intranuclear location in infected mink cells two days after infection with small amounts of antigen appearing in the cytoplasm by 3 or 4 days postinfection (Porter and Larsen, 1974). Intranuclear ADV antigen appears transiently in infected CRFK cells (Hahn *et al.*, 1977). The antigen fades with 2–4 days after reaching maximum fluorescence. The number of antigen positive cells varies depending on the amount of inoculum but rarely exceeded 20% of the cells.

IV. BOVINE PARVOVIRUS

A. Properties and Composition of the Virus

Abinanti and Warfield (1961) isolated a virus from the gastrointestinal tract of calves that hemagglutinated both human and guinea pig erythrocytes. Hemadsorption of guinea pig erythrocytes onto infected bovine embryonic kidney cells appeared in a characteristic manner. The virus was referred to as the hemadsorbing enteric virus (HADEN). The virus was ether resistant, stable at 56°C for 8 hr and at −20°C for more than 6 months. The diameter of the particle was less than 30 nm as determined by Gradocol filtration. Approximately 86% of the adult cattle in tested herds had hemagglutination inhibition antibody titers. Spahn *et al.* (1966b) observed that calves inoculated with the virus developed a mild respiratory illness and diarrhea. In other studies the frequency of infection varied between 14% and 100% of the animals in infected herds (Storz *et al.*, 1972). These authors also reported bovine parvovirus antibody in cynomolgus monkeys, guinea pigs, goats, dogs, and horses. No antibody was detected in sheep. Experimental infection in calves led to a cytocidal intestinal tract infection with a transient viremia and infection in the lymphatic tissues, thymus, adrenal glands, and brain (Storz and Bates, 1973; Storz and Leary, 1979). The virus has been isolated from cattle in various regions of the United States (Abinanti and Warfield, 1961; Bates *et al.*, 1972), Algeria (Vincent, 1971), and Japa (Inaba *et al.*, 1973). Spahn *et al.* (1966a) reported the virus to be heat stable and resistant to ether, sodium desoxycholate, and pH 2 for 30 min. The viral genome is DNA (Storz and Warren, 1970; Bachmann, 1971).

Electron microscopic examination of BPV reveals the particle to have an average diameter of 22 nm (Hoggan, 1971). Virus particles are found to band at various densities in cesium chloride gradients. Bands are found at 1.30–1.31 g/cm^3 (consists mainly of empty particles as observed by negative staining in the electron microscope), 1.40–1.42 g/cm^3 (contains infection particles), and a dense band at a density greater than 1.45 g/cm^3. BPV(1.40) and BPV(1.30) particles were purified and their structural proteins separated on polyacrylamide gels (Johnson and Hoggan, 1973). Three polypeptides were identified in both types of particles (See Table II). The major polypeptide (VP3), accounting for 75%–83% of the virion protein, has a molecular weight of about 67,000. VP1 has a molecular weight of 85,500 and VP2, 77,000. When the BPV polypeptides were compared with the AAV structural polypeptides they were found to have almost coelectrophoresing VP3s but the molecular weights of VP1 and VP2 of BPV were lower than the VP1 and VP2 of AAV.

B. Replication of Bovine Parvovirus

The host range of the virus is relatively narrow with optimal virus replication occurring in bovine embryonic kidney cells (Abinanti and Warfield, 1961(or in bovine fetal lung or bovine fetal spleen cells (Bates and Storz, 1973). Actively dividing cells are required. Some viral replication occurs in bovine cells derived from various tissues such as testicle, intestine, and adrenal (Bates and Storz, 1973) and in buffalo lung fibroblasts and embryonic bovine tracheal cells (Lubeck and Johnson, 1976). Limited replication occurs in fetal lamb kidney cells (Jacob and Johnson, unpublished results), but nonbovine mammalian cells and chicken cells are nearly uniformly refractory (Bates and Storz, 1973).

Studies on the replication kinetics of the virus showed an eclipse period of 16 hr (Bates and Storz, 1973). Most of the virus remained cell-associated. Release of some virus was detected from 30 to 36 hr after inoculation, depending on the host cell type. Leary and Storz (1980) described a number of nucleopathic changes associated with BPV replication. Nucleolar stranding occurred; increased basophilia and degranulation of the nucleoplasm were detected 12 hr postinfection. By 18 hr there was a replacement of the nucleoplasm with a fine-grained homogeneous material. Cowdry type A inclusions and other novel multifocal inclusions (empty virus capsids and virions) developed beginning by 18–24 hr. Patton *et al.* (1979) demonstrated that BPV transcription beginning 8–12 hr postinfection proceeds to 20 hr. By 20 hr postinfection 5% of the total RNA synthesized in infected nuclei was viral specific. In synchronized cells titers of infectious virus increased at 8 hr postinfection and the maximum titer was reached by 20 hr. Viral DNA synthesis was observed 2 hr before progeny virus appeared. Cellular RNA and protein synthesis was reduced but did not affect the rate of progression of the cells through S phase (Parris and Bates, 1976).

It was shown by Pritchard *et al.* (1978b) that BPV (as well as H-1 virus, LuIII virus, and Kilham rat virus) does not contain a virion-associated DNA polymerase. Levels of cellular DNA polymerases vary during infection, with levels of DNA polymerase alpha activity closely paralleling viral DNA synthesis and the production of progeny virus (Pritchard *et al.*, 1978a). Other work confirmed the requirement of DNA polymerase alpha for synthesis of BPV DNA (Pritchard *et al.*, 1981).

Utilizing immune sera prepared against the SDS-treated polypeptides of the virus Croft *et al.* (1974) determined the presence of BPV proteins in the cytoplasm of infected cells 2 hr before the appearance of completed virus antigen in the nucleus. Formation of the tertiary protein structure appeared to begin in the paranuclear area of the infected cell. The protein was transported to the nucleus where final protein maturation and virus assembly takes place. Further, Bates *et al.* (1974) found evidence of progeny particles by electron microscopy first in the nucleus and in association with smooth endoplasmic reticulum. Some particles were found on the cytoplasmic side of the nuclear membrane. Egress of the virus was visualized as occurring primarily by degeneration of the cell and rupture of the nuclear membrane late in infection.

V. CANINE PARVOVIRUS

Beginning in 1978 a pandemic of fatal myocarditis and enteritis in puppies occurred. Epidemics of this disease occurred virtually simultaneously in the United States, Canada, England, France, Scotland, and Australia (Appel *et al.*, 1979; Burtonboy *et al.*, 1979; Johnson and Spradbrow, 1979; Kelly, 1978; Moraillon, 1980; Thomson and Gagnon, 1978). In studies reported by Levin *et al.* (1979) and Levin (1980) most affected puppies are between 3 weeks and 3 months of age. There was no evidence of hereditary, breed, or sex predisposition. The animals appeared robust and healthy until a few minutes or hours before they died. Signs included hyperpnea, dyspnea, crying, and cyanosis. Gross lesions often included pale streaking of the myocardium, dilation of the left ventricle, and sometimes marked thickening of the endocardium of the left ventricle and atrium. The lungs were usually edematous. In many cases basophilic intranuclear inclusion bodies were found in cardiac myofibers and parvovirus particles identified in the inclusion bodies by electron microscopy.

The clinical and pathological picture of the disease closely resembles feline panleukopenia and mink enteritis. The causative agent of this disease, in agreement with the clinical findings, was found to be a parvovirus which proved to be antigenically closely related to feline panleukopenia virus (FPV) (Carmichael *et al.*, 1980; Johnson and Spradbrow, 1979; Lenghaus and Studdert, 1980). Mink enteritis virus (MEV) is closely related or even identical to feline panleukopenia virus and is frequently referred to as a strain of FPV (Burger *et al.*, 1963; Johnson, 1967; Johnson *et al.*, 1974). In a comparative study, the canine parvoviral genome was compared to the MEV genome by restriction enzyme mapping (McMaster *et al.*, 1981b). The two viruses were found to share 86% of the mapped sites. Thus all three viruses, CPV, FPV, and MEV are closely related viruses; in fact, immunization with FPV vaccines appears to protect dogs, cats, and mink against infection by these viruses.

The origin of CPV is unclear. Dog sera collected before 1978 proved to be free of antibodies to the virus (Carmichael *et al.*, 1980; Johnson and Spradbrow, 1979). But antibodies to another parvovirus of dogs, the minute virus of canines, were present in dogs before 1978 (see Siegl, 1976). This virus has no serologic relationship with FPV but has a density of 1.39 g/cm^3 (Williams, 1980). Because no antibodies appeared in dogs before 1978, the disease was not endemic in dogs before this time. It has been suggested that wild-type FPV was adapted to canine cells either deliberately during the production of attenuated live FPV vaccine or accidentally by contamination of canine cell cultures in laboratories (Johnson and Spradbrow, 1979; Lenghaus and Studdert, 1980). After mutating to dog virulence, the virus then could have been spread worldwide with any vaccine produced in canine cells. Alternatively, Moraillon *et al.* (1980) suggested that CPV developed from mink enteritis virus.

VI. DENSONUCLEOSIS VIRUS

A. Biophysical Properties

The densonucleosis viruses (DNV) have been isolated from insects. DNV I is the causative agent of the densonucleosis disease of *Galleria mellonella* (the greater wax moth) larvae. DNV 2 has been isolated from the Trinidadian buckeye caterpillar (*Junonia coenia*). The two viruses are very closely related as determined by DNA hybridization studies (Kelly *et al.*, 1977). The virus contains a single-stranded DNA genome having a molecular weight of about 1.6×10^6 to 1.9×10^6 and, as pointed out elsewhere, like AAV contains separately encapsidated complementary DNA strands (Kelley *et al.*, 1977; Kurstak *et al.*, 1971; Kurstak *et al.*, 1973; Mayor and Kurstak, 1974). The double-stranded DNA created upon extraction of the DNA from viral particles exists predominately as linear monomers, although circular monomers and concatamers and other less well defined structures were observed by electron microscopy and agarose gel electrophoresis (Kelly and Bud, 1978). There appeared to be an inverted terminal repetition in the DNA molecule. The size of the DNV particles has been reported by different authors to be in the range of 19 to 24 nm (Kurstak, 1972). The capsid appears to have 42 capsomers as observed in the electron microscope (Kurstak and Côté, 1969; Lowff and Tournier, 1971). Molecular studies on the protein do not necessarily confirm this as Tijssen *et al.* (1976) and Tijssen and Kurstak (1979b) suggested the capsid may have 12 capsomers. Studies on the buoyant density of infectious DNV particles revealed infectious particles of two densities, 1.40 g/cm^3 and 1.44 g/cm^3 (Hoggan, 1971; Truffant *et al.*, 1967). The 1.44 particles appear to have less protein (Tijssen *et al.*, 1977). The infectious particles contain the polyamines putrescine, spermidine, and spermine which are absent in the empty particles (Kelly and Elliott, 1977). The virus is pH stable (Kelly *et al.*, 1980). Morphogenesis of the virus occurs in intranuclear locations in the infected cell (Garzon and Kurstak, 1976).

B. Structural Proteins

The polypeptide composition of purified DNV particles has been analyzed by a variety of methods in polyacrylamide gels by several authors. Tijssen *et al.* (1976) reported four structural proteins in the DNV capsid. These four proteins had molecular weights of 49,000, 58,000, 69,000, and 98,000. The 49,000 protein was the major protein accounting for 69% of the protein mass. [This may vary to as low as 40% or lower depending on the stage or instar of the larvae in which the virus is grown (Tijssen and Kurstak, 1981).] The 58,500 and 69,000 protein accounted for 15% each of the virion protein and the 98,000 protein accounted for only 1% of the virion protein. This report was in contrast to an earlier report (Tinsley and Longworth, 1973) in which it was noted that three proteins were found in 5% polyacrylamide gels but four were found in 10% gels. Further, the molecular weights (72,000, 57,000, 53,000, 46,000) were considerably lower than the molecular weights reported by Tijssen *et al.* (1976), who could not confirm the latter results. Kelley *et al.* (1980) reported a greater spread in molecular weight values for the two serotypes of DNV with the molecular weights ranging from 109,600 (DNV-2 VP1) to 41,900 (DNV-2 VP4). These authors further reported that the top component (empty particles) appeared to have a protein composition identical to that of DNV(1.40) particles and was a more stable structure.

Early on, Tijssen *et al.* (1976) suggested VP1 (p98) was a dimer of VP4 (p49) because the molecular weight of VP1 is double that of VP4 and because in different experiments the percentage of VP1 fluctuated, depending on salt conditions, but the sum of VP1 plus VP4 was always constant. In later reports (Tijssen and Kurstak, 1979a; Tijssen and Kurstak, 1981) these authors give peptide mapping evidence showing VP1 not to be a dimer of VP4 as VP1 and VP2 share peptides not found in VP4. Furthermore, peptide mapping results show that the DNV proteins share extensive amino acid sequences in addition to antigenic homologies (Tijssen and Kurstak, 1981). Regarding location of the proteins in the capsid, labeling with periodate-oxidized glycoprotein supported the hypothesis that 60 molecules of VP4 aggregate into a dodecahedron with 12 pentamers and VP2 and VP3 may have a stabilizing function and are located on the outer surface of the particle (Tijssen and Kurstak, 1979b). The acidic DNA charges are believed to be neutralized by cations other than basic amino acid side chains of the capsid protein (Kelley *et al.*, 1980).

VII. FELINE PANLEUKOPENIA VIRUS

As noted above in Section V, feline panleukopenia virus (FPV) is closely related to mink enteritis virus and canine parvovirus. These viruses may represent simply strain variations or host range mutants of the original parent virus. For an excellent review of FPV see Siegl (1976).

The virus has a linear single-stranded DNA genome having a molecular weight of 1.7×10^6 daltons. The particle is 20–24 nm in diameter. Infectious particles band at buoyant densities of 1.44 and 1.41 g/cm^3. The virus is heat resistant, stable under storage conditions at 4°C for many months, and resistant to chloroform, trypsin treatment, and pH 3.

The virion protein was separated into two major polypeptides with molecular weights of 73,100 (VP1, comprising 10% of the virion protein) and 60,300 (VP2, comprising 86% of the virion protein) (R.H. Johnson *et al.*, 1974). A third minor protein was detected. It had a molecular weight of 39,600 and comprised only 3–6% of the protein. It was not clear if this protein was an FPV VP3 or merely a protein contaminant of the purified virus.

VIII. H-1 PARVOVIRUS

A. Composition of the Virion

H-1 virus was originally isolated from the transplantable human tumor HEp-1 after serial transplantation in rats. The virus was also recovered from various tissues of cancer patients (Toolan, 1961a,b). This virus along with several other parvoviruses have been termed the "hamster osteolytic virus" (see Siegl, 1976) because of the characteristic pathogenesis in hamsters. The viruses affect the replicating cells of developing bone, causing in the hamsters, among other things, dwarfism and malformation of teeth and bones (Toolan, 1972). The virions are about 22 nm in diameter, contain single-stranded DNA, and are stable to heating within a broad pH range, stable to ether, chloroform, and butanol, and survive storage conditions for very long periods of time.

The structural proteins of H-1 virus were originally reported to be comprised of three proteins with molecular weights of 92,000, 72,000, and 56,000 daltons accounting for 15%, 75%, and 10% of the virion protein, respectively (Kongsvik and Toolan, 1972). Kongsvik *et al.* (1974) later reported the synthesis of two viral proteins in infected cells and in purified virions that were propagated in synchronous cell cultures (Kongsvik *et al.*, 1974). The two proteins were VP1 (molecular weight 92,000) and VP2′ (molecular weight 72,000). The 56,000-dalton protein originally found associated with H-1 virions apparently is not a virus-coded structural protein. The most likely structure of H-1 virus is, then, an icosahedron composed of 60 VP2 and 12 VP1 molecules arranged in 12 pentamers of VP2 (VP2′) with a molecule of VP1 situated at the vertex of each pentamer (Kongsvik *et al.*, 1974). The full, DNA-containing virus particle appears to contain the three polypeptides VP1, VP2′, and VP2. The empty capsids contain only VP1 and VP2′ (Tattersall, 1978). VP1 and VP2′ are viral products possessing overlapping amino acid sequences such that all of the sequences of VP2′ are contained within VP1 (Tattersall *et al.*, 1977). VP2 is a proteolytic cleavage product of VP2′ with cleavage occurring only in the full virus after the particle is assembled in a manner similar to MVM (Clinton and Hayashi, 1975; Tattersall *et al.*, 1976; Paradiso, 1981). The conversion of VP2′ to VP2 can be achieved *in vitro* by treating full capsids with various proteases. Empty capsids, however, are resistant to proteolytic cleavage indicating structural differences between empty and full capsids (Clinton and Hayashi, 1976; Kongsvik *et al.*, 1978).

Infectious H-1 particles band in two density classes in CsCl: 1.47 g/cm^3 and 1.43 g/cm^3. The relative amounts of these two viral forms appear to be partially dependent on the time post-infection at which the virus is harvested (Kongsvik *et al.*, 1978). At early times H-1(1.47) particles predominate and at later times H-1(1.43) particles predominate. Similarly, at least for MVM, the protein content of the virus purified at various times postinfection appears to be time-related in that VP2′ is predominate at early times and VP2 predominates at later times (Tattersall *et al.*, 1976). Thus, several authors have suggested that the major viral protein is different in H-1(1.47) particles from that in H-1(1.43) particles, that is, VP2′ would be the major protein in H-1(1.47) and VP2 the major protein in H-1(1.43) particles (Clinton and Hayashi, 1975; Kongsvik *et al.*, 1978; Kongsvik *et al.*, 1979; Richards *et al.*, 1978). The suggestion was made, then, that the conversion of 1.47 particles to 1.43 particles is related to the proteolytic conversion of VP2′ to VP2. However, *in vitro* conversion of 1.47 to 1.43 particles by proteolytic cleavage has not been accomplished.

H-1 virus grown in two different cell lines (NB and THK) were analyzed for their protein content (Paradiso, 1981). In contrast to the previously reported results, it was found the ratio of VP2′ to VP2 was the same in 1.47 and 1.43 particles derived from the same cell line but differed significantly between the two hosts. In studies on the fate of infectious virus using radioiodinated H-1, the conversion of VP2′ to VP2 occurred in a time-dependent manner up to 24 hr postinfection. There was a proteolytic cleavage with a simultaneous shift in density (1.47 particles to 1.43 particles); however, there was no enrichment of VP2 in the 1.43

particles. It was thus interpreted that the cleavage of VP2′ to VP2 was not responsible for the shift in density.

B. Viral Replication

Ultrastructural studies on H-1 replication have revealed many interesting features of the replication cycle (Singer and Toolan, 1975; Singer, 1975; Singer, 1976; Singer and Rhode, 1977a; Singer and Rhode, 1977b; Singer and Rhode, 1978). In the early stages of viral infection (12 hr postinfection) there was fragmentation and loss of nucleolar fibrous components. High concentrations of empty H-1 particles formed compact linear arrays on the chromatinlike fibers which remained in the rarefied areas. By 18–36 hr empty particles were found in association with nucleolar-derived spherules. Complete virions were found in extranucleolar chromatin fibers. After 36 hr the chromatin fibers appeared to condense and release the previously attached virus. Intranuclear crystalline arrays of virus particles are seen, and also changes in the cellular deoxyribonucleoprotein. In a study of H-1 temperature-sensitive mutants immunospecific staining patterns indicated that the H-1 capsid proteins formed two separate intranuclear antigens. One of these was a thermostable chromatin-associated antigen present in proteins that have not formed capsids and are concentrated on heterochromatin and nucleolar-associated chromatin. The second was a thermolabile inclusion-associated antigen found in the proteins of assembled empty capsids (composing H-1 inclusions). In a study on DNA replication (Singer and Rhode, 1978) DNA synthesis was found to begin at localized euchromatic or nucleolar sites and then spread outward. The foci became larger with increasing cellular damage, forming a limited number of H-1 DNA synthetic centers in the euchromatin. Each islandlike focus was surrounded by tufts of heterochromatin containing high concentrations of unassembled H-1 capsid proteins. It was suggested that H-1 proteins along with cellular cofactors associated with the fibrillar component of the nucleolus and the euchromatin may play a role in the regulation of H-1 DNA synthesis.

Studies on the transcriptional events in H-1 infection have revealed the most abundant viral transcript, present in both nucleus and cytoplasm, to be a 2.8-kb transcript representing about 56% of the viral genome. Less abundant transcripts of 3.0, 1.45, and 1.30 kb were also found. In contrast, a prominant 4.7-kb transcript which corresponds to 95% of the viral DNA was found only in the nucleus, suggesting that the H-1 genome may function as a single transcription unit. Further, evidence indicated the 4.7-, 3.0-, and 2.8-kb RNA's were spliced (Green *et al.*, 1979).

Replication of H-1 involves the synthesis and replication of a doublestranded replicative form DNA (Rhode, 1974a). Progeny viral strand DNA is then synthesized with the complementary strand DNA in the RF as template. The viral hemagglutinin, viral antigen, and the proteins

VP1 and VP2 have been shown to depend on prior DNA synthesis (i.e., formation of no "early" proteins). Viral DNA synthesis takes place near the end of S phase of the infected cell (Kongsvik *et al.*, 1974; Rhode, 1973; Rhode, 1974b). Viral protein synthesis and the initiation of RF DNA replication occur concomitantly shortly after hemagglutinating-DNA synthesis. Initiation of RF DNA synthesis was shown to require protein synthesis; however, studies on *ts* mutants of H-1 reveal that a viral capsid protein is required for synthesis of single-stranded progeny DNA but not for RF DNA replication (Rhode, 1976). In fact, viral protein may be inhibitory to RF DNA replication (Rhode, 1974).

Defective interfering particles of H-1 virus have been described (Rhode, 1978). The particles were produced by serial high-multiplicity passage and were shown to interfere with the synthesis of capsid proteins and standard infectious virus.

IX. LuIII VIRUS

In a long-term study (Hallauer *et al.*, 1971; Siegl, 1976) 43 permanent cell strains were subjected to gentle extraction of viral hemagglutinins by means of an alkaline buffer. Of the 43 "noninfected" cell strains 36 were shown to be contaminated with parvoviruses belonging to four serotypes designated I to IV. Three reference isolates were designated KBSH, TVX, and LuIII. The KBSH virus proved to be similar, if not identical, to porcine parvovirus. This virus could have been inadvertently introduced into the cell cultures by using contaminated trypsin derived from the pancreas of porcine parvovirus infected pigs (Siegl, 1976). The biophysical properties, structure, and replication of LuIII virus have been extensively studied, although its natural animal host, if any, remains obscure. The LuII virus has, however, been adapted to grow in newborn and fetal hamsters, causing a systemic infection with massive intestinal hemorrhage, transplacental infection, abortion, and lesions in the heart, liver, kidney, and CNS (Soike *et al.*, 1976).

The virus possesses cubic morphology with a diameter of 19–21 nm (Siegl *et al.*, 1971). The capsid probably contains 32 capsomers. They buoyant denisty in CsCl of the infectious virus particle is 1.395 g/cm^3. Empty particles band at 1.31 g/cm^3 while an intermediate band at 1.35 g/cm^3 contained hemagglutinin and some infectivity. A dense band is found at 1.44 g/cm^3 (Siegl, 1973). The virus was also resistant to ether, chloroform, freon, sodium deoxycholate, trypsin, pepsin, papain, RNase, DNase, and heating at 75°C for 1 hr reduced infectivity and hemagglutination titers only one hundred-fold. The virus contains single-stranded DNA (Siegl *et al.*, 1971; Siegl, 1973; Siegl and Gautschi, 1976; Gautschi and Reinhard, 1978). The DNA is linear and has a molecular weight of about 1.59 × 10^6 daltons. As noted above for H-1 virus, synthesis of detectable amounts of ds DNA coincides with the display of the cellular helper function late in S phase. Data reported by Siegl and Gautschi (1976) suggested the same is true for the synthesis of the RF form of LuIII ds DNA in synchronized HeLa cells. Further work revealed the presence of at least one additional nucleic acid species which is a linear double-stranded molecule bearing single-stranded branches of various length and presumed to be a replicative intermediate form in the synthesis of progeny viral DNA.

In studies on the replication of LuIII virus (Siegl and Gautschi, 1973a,b) detection of immunofluorescent stainable cytoplasmic antigen was possible by 3 to 4 hr postinfection. The appearance of this antigen was independent of the physiologic state of the infected cell. Intranuclear antigen appeared at 8 to 10 hr postinfection. Appearance of this antigen was strictly linked to cellular events. Viral replication was complete by 16–18 hr postinfection. LuIII virus, then, is absorbed, penetrates, is uncoated, and cytoplasmic antigen develops without obvious control by cellular events. Nuclear antigen, on the other hand, appears only after the infected cell arrives at a status of "physiologic competence" probably identical with late S phase. It was not entirely clear if the so-called early cytoplasmic antigen was distinctly identified as a newly synthesized protein or simply represented accumulation of input viral capsid antigen modified with the loss of hemagglutinating activity; but unpublished investigations of Siegl, Gautschi, and Trachsel led to the conclusion that development of the cytoplasmic antigen was nothing more than the uptake of excess viral antigen by the infected cell (Siegl, 1976).

Regarding the protein composition of the LuIII virion, Gautschi and Siegl (1973) reported early on that the protein content was somewhat related to the relative specific density of the particle. Infectious LuIII virions (1.41 g/cm^3) contained two proteins: a major protein with a molecular weight of 62,000 daltons amounting to 84.5% of the virion protein, and a minor protein of 75,000 daltons accounting for 15.5% of the virion protein. Particles banding at 1.35 or 1.31 g/cm^3 contained an additional third protein with a molecular weight of 69,000 daltons, but it was not clear whether this protein was of viral or cellular origin. In a follow-up study Gautschi *et al.* (1976) reported that the 62- and 74-kilodalton proteins were detected in the nuclei of infected cells at 11–12 hr postinfection. Most of the intranuclear viral polypeptides, at this time point, were associated with the chromatin acidic proteins. By 13–14 hr postinfection about one third of the virus protein could be extracted with the nuclear sap proteins. About 90% of the chromatin-associated viral polypeptides were in the form of empty viral capsids.

As of this writing a somewhat different picture of the protein content of LuIII virus is emerging. I. Majaniemi, J. D. Tratschin and G. Siegl (personal communication; Abstract, Fifth International Congress of Virology) have found different amounts of viral protein in particles of the various densities (see Table II). The proteins are of different molecular weights than originally reported, but this probably can be accounted for by the different techniques employed. The three proteins have molecular weights of 82,000, 67,000, and 64,000 daltons. For the sake of consistency, in this review, these will be referred to as VP1 (82 kilodalton), VP2 (67 kilodalton), and VP3 (64 kilodalton). Empty capsids (banding at 1.30–1.32 g/cm^3) contain predominately VP2 with a trace of VP1. Particles banding at 1.33–1.38 g/cm^3, containing less than genome length DNA, contain VP1, VP2, and VP3. VP1 is present at trace levels with VP2 and VP3 accounting for most of the protein. As the density of the particles increases from 1.33 to 1.37 g/cm^3 the distribution of VP2 and VP3 shifts from nearly all VP2 to about equal amounts of VP2 and VP3. As further density shifts occur from 1.40 to 1.42 a further shift in VP2 and VP3 is noted such that in 1.42 particles the predominate protein is VP3 with virtually no VP2 or VP1. The 1.43 to 1.46 particles contain only VP3.

X. MINUTE VIRUS OF MICE

During a study by Crawford (1966) of mouse adenovirus, a contaminant virus was detected which became known as minute virus of mice (MVM). The virus contains single-stranded DNA and is about 20–28 nm in diameter. A feature that this virus has in common with other parvoviruses is the appearance of particles of four density classes: empty particles (1.30–1.32 g/cm^3), particles containing less than genome length DNAs (1.35–1.37 g/cm^3), full infectious particles (1.42 g/cm^3) and dense infectious particles (1.47 g/cm^3). Furthermore, the virus requires host-cell functions that are not present in resting cells (Tattersall, 1972).

Regarding the protein structure of the virus, Tattersall *et al.* (1976) reported that the empty capsid contains two polypeptides while infectious particles contain three polypeptides. The empty particle contains a protein of molecular weight 83,300 (VP1) representing 15–18% of the protein mass and a protein of molecular weight 64,300 (VP2) comprising 82–85% of the protein mass. The full particle contains three polypeptides: VP1 (again at 15–18% of the protein mass), VP2, and a third protein, VP3, with a molecular weight of 61,400. The concentrations of VP2 and VP3 vary inversely in different preparations of full particles. These were found to be viral-specific polypeptides, not found in uninfected cells, and their molecular weights and molar ratios were independent of the species of the host cell in which the virus was grown. Moreover, experiments involving sequential harvesting of nuclei of cells under one cycle growth conditions revealed an increase in the proportion of VP3 in full particles as the infection progressed, suggesting that VP3 was derived from VP2. Clinton and Hayashi (1975) also reported on the protein structure of MVM. In contrast to Tattersall *et al.* (1976), these authors reported the presence of VP3 in empty capsids albeit at small and highly variable amounts. VP2 was the major protein component of the empty capsid as well as the 1.47 particle. VP3 was the major component of the 1.42 particle (see Table II). Further, these authors reported that 1.47 particles were converted to 1.42 particles after one round of infection when left in the infected culture. The conversion in density appeared to be due to conversion of VP2 to VP3. An MVM variant has been isolated (McMaster *et al.*, 1981a) which is immunosuppressive. Even though its genome is slightly different from prototype MVM, the structural proteins appeared to be identical.

In a comparative study, Peterson *et al.* (1978) reported on the structural proteins of MVM, H-1, H-3, and Kilham rat virus (KRV). The empty capsids of all four viruses contained only two types of polypeptides, VP1 and VP2. All full-virus preparations contained all three structural polypeptides, VP1, VP2, and VP3. VP1 represented about 15% of the virion protein mass in all of the viruses but VP2 and VP3 varied significantly in the different viruses. VP2 of all four viruses was selectively susceptible to proteolytic cleavage, suggesting that these viral capsids are similar in their conformational properties and may undergo similar maturation events. Other results presented by Peterson *et al.* (1978) demonstrated that all three species of viral proteins can exist as phosphoproteins with microheterogeneity reflecting subpopulation of proteins phosphorylated to different levels. The proteins can also be labeled with [^{3}H]adenosine.

That the MVM proteins are structurally related and are presumably encoded within the same nucleotide sequences was shown by Tattersall *et al.* (1977). Tryptic and chymotryptic fingerprinting was done after *in vitro* radioiodination of the proteins. VP2 and VP3 digests were almost identical. Both digests showed a unique polypeptide in VP3. In addition to the close amino acid sequence relationship between VP2 and VP3, all of the sequence of VP2 was present within VP1, the largest polypeptide (about 16% of the virion protein). The VP1 polypeptide also contained additional peptides, comprising about 20% of the total, which were not found in either VP2 or VP3. The fingerprints of VP2 and VP3 supported the precursor–product relationship between them suggested by the previous studies of Clinton and Hayashi (1975) and Tattersall *et al.* (1976). The occurrence of VP2 and VP3 within VP1 also suggests either their derivation from VP1 by post-translational processing or by production of RNA transcripts with overlapping sequences.

Studies on the proteolytic digestion of MVM particles *in vitro* (Tattersall *et al.*, 1977; Clinton and Hayashi, 1976) revealed, in the case of intact full particles, that the cleavage of VP2 *in vivo* (or infected-cell-conditioned medium) could be closely mimicked by trypsin and to a lesser extent by chymotrypsin, but this did not change their density in CsCl. In contrast, the VP2 polypeptide in the empty virion was resistant to cleavage by either enzyme, a finding suggesting that this polypeptide adopts a different conformation in each particle type. The VP1 molecule was resistant to cleavage by either enzyme, in either particle type. A maturation scheme was thus devised by these authors in which VP1 and VP2, independently synthesized, condensed to form the VP1, VP2 empty particle. This particle was then the functional procapsid for packaging of progeny viral DNA strands forming the VP1, VP2, DNA "heavy full" (1.46) particle. This particle in turn was then converted to the VP1, VP3, DNA "light full" (1.42) particles, a finding confirmed by Richards *et al.* (1977, 1978).

XI. PORCINE PARVOVIRUS

Porcine parvovirus (PPV) isolates have been identified in various geographical regions of the world (Bachmann, 1970; Cartwright and Huck, 1967; Cartwright *et al.*, 1969; Johnson, 1969; Mayr *et al.*, 1968; Mangeling, 1972; Bachmann *et al.*, 1975). A high incidence of infection of swine, more than 40%, is noted in some serological surveys (Mengeling, 1972). The virus is heat stable, resistant to lipid solvents, and contains DNA. In addition to isolations from swine, the virus has been isolated from preparations of commercial 1:250 trypsin (Croghan *et al.*, 1973), and Hallauer *et al.* (1971) reported that trypsin, commercially prepared from hog pancreas, might be involved in the widespread contamination of stable cell lines of nonswine origin with parvoviruses.

PPV infections in swine may result in reproductive failure. Sixty percent of litters of infected animals were affected in one study (Mengeling *et al.*, 1980). Viral antigen could be identified in the affected embryos. Some of the embryos were dead and in various stages of decomposition and resorption. Antigen was found in some live embryos but these tended to be next to the infected dead littermates. Indicating intrauterine spread of the infection. Other reports have shown prenatal infection after maternal exposure to the virus (Mengeling, 1978; Joo *et al.*, 1976; Bachmann *et al.*, 1975). Several inactivated PPV vaccines have been found to induce an antibody response following administration to susceptible pigs (Joo and Johnson, 1977; Suzuki and Fujisake, 1976; Mengeling, 1977) and effective prevention of transplacental infection and fetal death has been shown (Mengeling *et al.*, 1979).

PPV has been isolated from buffy-coat leukocytes (Mengeling, 1972) and PPV antigens have been detected in lymphoid tissues (Cutlip and Mengeling, 1975; Mengeling and Cutlip, 1976). Further, it has been shown that mitogen-stimulated T and B lymphocytes were able to support PPV replication, while peripheral blood monocytes and peritoneal macrophages phagocytized the virus but did not support virus replication (Paul *et al.*, 1979).

Little information appears in the published literature on PPV structure or replication. In a personal communication, S. Belak, D. E. Gutekunst, and W. L. Mengeling point out that three particle bands appear on CsCl gradients, a top band with a density of 1.30–1.31 g/cm^3, and intermediate band at a density of 1.39–1.41 g/cm^3 and a lower band at 1.46 g/cm^3. Particles at 1.31–1.35 were also found. The PPV (1.39–1.41) particles were found to contain three structural proteins (VP1, VP2, VP3) having molecular weights of 79,000, 62,000, and 55,000. PPV (1.31–1.35) particles contained VP1 and VP2, but no VP3. Also these particles contained two additional proteins, believed to be precursor proteins, with molecular weights of 68,000 and 60,500.

XII. RABBIT PARVOVIRUS

The first report characterizing a virus isolate from rabbits as a parvovirus appeared in 1977 (Matsunaga *et al.*, 1977). The virus was isolated from rabbit feces. The virus replicated in rabbit kidney cell cultures but the susceptibility of primary or secondary cell cultures was very low. One cell strain was highly susceptible between passages 8 and 30. The virus agglutinated human group O erythrocytes at 4°C, it was stable at acid pH, and was resistant to chloroform and heat treatment. Virus replication was inhibited by 5-iodo-2-deoxyuridine and results of acridine orange staining suggested the virions contain single-stranded DNA. The relative specific density of the virus was 1.41 to 1.44 g/cm^3 in CsCl. The infectious particles had cubic symmetry and were 27 to 28 nm in diameter. In a serological survey a number of rabbits (42 of 90 sera) from a commercial source were found to contain hemagglutination-inhibiting antibody to this virus.

Preliminary work has been done on the protein structure of this rabbit parvovirus (Y. Matsunaga, personal communication). The following polypeptides were found in the infectious particles: VP1, a minor polypeptide, molecular weight 96,000; VP2, a minor polypeptide, molecular weight 85,000; VP3, the major structural polypeptide, molecular weight 75,000 accounting for about 73% of the virion protein. Another minor protein (molecular weight 70,000) was found in infectious particles but it was thought that this protein could be a cleavage product of VP3. The polypeptide content of empty particles was the same as the RPV virion except the minor 70,000 protein was not present; that is, the empty particle contains VP1, VP2, and VP3. In experiments on the kinetics of infection, cell-associated virus was detected at 15 hr postinfection, reaching a plateau by 36 hr. The released virus lagged 1 to 1.5 days behind the cell-associated virus. VP3 could be found in both the cytoplasm and nuclear fractions at 15 hr, while VP1 and VP2 were detected at 18 hr. The minor, 70,000 molecular weight peptide could not be found in infected cells through the observation period of 44 hr.

XIII. RAT VIRUS

The Kilham rat virus (KRV) or simply rat virus (RV) was discovered during investigations of Kilham and Olivier (1959) wherein a search was underway to identify oncogenic rat papovaviruses. This isolate multiplied in embryonic rat cells, agglutinated both guinea pig and rat erythrocytes, was resistant to ether treatment as well as heating to 80°C for 2 hr. Screening for specific antibodies in sera collected from both wild and laboratory rats yielded positive results in about 80% of the tested samples, reflecting an extremely high prevalence of RV infections in rat colonies (Moore and Nicastri, 1965; Kilham, 1966). Apparently RV can establish a persistent or latent infection (Robey *et al.*, 1968), and when an animal develops a tumor or is immunosuppressed (El Dadah *et al.*, 1967) the virus replicates to high titer due to the presence of replicating S-phase cells (Tennant and Hand, 1970) or lack of immunomodulation.

Rat virus contains a well-characterized linear single-stranded DNA genome. The capsid is proposed to have an icosahedral morphology, 20 to 29 nm in diameter, containing 32 capsomers (see Siegl, 1976). The rat virus capsid was shown to be composed of three structural polypeptides (Salzman and White, 1970). The molecular weights of the three proteins were 72,000 (VP1), 62,000 (VP2), and 55,000 (VP3). VP2 comprised about 75.5% of the total protein. A slightly different picture of the structural proteins of RV was seen by Peterson *et al.* (1978). The empty capsid had little or no VP3 while the full particle had all three proteins, but VP3 was the most abundant (see Table II).

A virus that is serologically similar, if not identical, to rat virus is the X-14 virus. Salo and Mayor (1977) described the composition of X-14 virus. The capsid of the virus consisted of only two polypeptides. VP1 (molecular weight 82,400) and VP2 (67,500) represented 13.7 and 86.3 percent of the total protein mass, respectively. The naturally occurring empty capsids had proteins indistinguishable in size and concentration from those of complete virions.

XIV. DISCUSSION

It appears, at this stage of investigation of parvovirus protein, that its major function is that of capsid structural material. There appear to be no nonstructural proteins synthesized in parvovirus infected cells that could normally be called "early" proteins with enzymatic or regulatory function. At the same time, it is not yet entirely clear whether the structural proteins possess any enzymatic or regulatory activity. There are some viral non-structural polypeptides that appear in the infected cell after DNA synthesis. Some of these polypeptides could be immature protein forms and cleavage fragments, but none are known to have biological functions.

As far as the capsid proteins are concerned, it seems clear that the AAV virion contains three structural polypeptides. The major one of these accounts for 80–85% of the virion protein. The molecular weight of the major component (VP3) is 56,000 to 68,000 daltons, depending partially on AAV serotype and virion density. The molecular weight of the major protein in the helper-independent parvoviruses varies from 58,000 to 77,000 daltons with a wide diversity of concentrations of the protein found in the particle. Most helper-independent parvoviruses have been shown to contain three structural polypeptides, but some have two proteins and some have four. In some cases the number of structural proteins found in the capsid is a function of the maturation state of the particle, with, in general, a trend toward smaller proteins and released cleavage products as the particle matures to its final inert state. As maturation of the virion occurs shifts in relative specific density occur, in that density is a function of protein and DNA content. Conformational shifts in the capsid protein due to posttranslational (even postassembly) modifications such as cleavages, acetylation, and phosphorylation result in density changes, but not necessarily changes in infectivity. Some particles, banding at lower densities, contain subgenomic lengths of DNA. These apparent DI particles contain the normal structural polypeptides as far as has been determined.

The mechanism of biosynthesis of parvoviral proteins remains somewhat obscure. Insofar as has been studied the capsid proteins within a given virus share extensive areas of amino acid sequence, suggesting a common genetic origin. How these common nucleotide sequences are expressed and where on the genetic map the proteins begin and end are not yet clear. Moreover, further information should be acquired on what role post-translational modifications play in virion assembly, stabilization, and maturation.

REFERENCES

Abinanti, F.R., and Warfield, M.S., 1961, Recovery of a hemadsorbing virus (HADEN) from the gastrointestinal tract of calves, *Virology* **14**:288–289.

Appel, M.J.G., Scott, F.W., and Carmichael, L.E., 1979, Isolation and immunization studies of a canine parvo-like virus from dogs with haemorrhagic enteritis, *Vet. Rec.* **105**:156–159.

Archetti, I., and Bocciarelli, D.S., 1964, On the structure of a small simian virus, Proceedings Third European Regional Conference Electron Microscopy, Prague, p. 343.

Archetti, I., and Bocciarelli, D.S., 1965, Biological structural characteristics of a still untyped small virus, *Ann. Ist. Super. Sanita* **1**:103–106.

Atchison, R.W., 1970, The role of herpes viruses in adenovirus-associated virus replication *in vitro, Virology* **42**:155–162.

Atchison, R.W., Casto, B.C., and Hammon, W. McD., 1965, Adenovirus-associated defective virus particles, *Science* **149**:754–756.

Atchison, R.W., Casto, B.C., and Hammon, W. McD., 1966, Electron microscopy of adenovirus-associated virus (AAV) in cell cultures, *Virology* **29**:353–357.

Bachmann, P.A., 1970, Parovoviren bein Schwein, *Zentralbl. Veterinaermed.* **17**:192–194.

Bachmann, P.A., 1971, Properties of a bovine parvovirus, *Zentralbl. Veterinaermed.* **18**:80–85.

Bachmann, P.A., Sheffy, B.E., and Vaughan, J.T., 1975, Experimental in utero infection of fetal pigs with a porcine parvovirus, *Infect. Immun.* **12**:455–460.

Bates, R.C., and Storz, J., 1973, Host cell range and growth characteristics of bovine parvoviruses, *Infect. Immun.* **7**:398–402.

Bates, R.C., Storz, J., and Reed, D.E., 1972, Isolation and comparison of bovine paroviruses, *J. Inf. Dis.* **126**:531–536.

Bates, R.C., Storz, J., and Doughri, A.M., 1974, Morphogenesis of bovine parvoviruses and associated cellular changes, *Exp. Molec. Pathol.* **20**:208–215.

Berns, K.I., 1974, Molecular biology of the adeno-associated viruses, *Curr. Top. Microbiol. Immunol.* **65**:1–20.

Berns, K.I., and Hauswirth, W.W., 1979, Adeno-associated viruses, *Adv. Virus Res.* **25**:407–449.

Berns, K.I., Pinkerton, T.C., Thomas, G.F., and Hoggan, M.D., 1975, Detection of adeno-associated virus (AAV)—Specific nucleotide sequences in DNA isolated from latently infected Detroit 6 cells, *Virology* **68**:556–560.

Blacklow, N.R., 1975, Potentiation of an adenovirus-associated virus by herpes simplex virus type-2-transformed cells, *J. Natl. Cancer Inst.* **54**:241–244.

Blacklow, N.R., Hoggan, M.D., and McClanahan, M.S., 1970, Adenovirus-associated viruses: Enhancement by human herpes viruses, *Proc. Soc. Exp. Biol. Med.* **134**:952–954.

Blacklow, N.R., Dolin, R., and Hoggan, M.D., 1971, Studies of the enhancement of an adenovirus-associated virus by herpes simplex virus, *J. Gen. Virol.* **10**:29–36.

Bloom, M.E., Race, R.E., and Wolfinbarger, J.B., 1980, Characterization of aleutian disease virus as a parvovirus, *J. Virol.* **35**:836–843.

Boucher, D.W., Melnick, J.L., and Mayor, H.D., 1971, Nonencapsidated infectious DNA of adeno-satellite virus in cells coinfected with herpesvirus, *Science* **173**:1243–1245.

Buller, R.M.L., and Rose, J.A., 1978a, Characterization of adenovirus-associated virus-induced polypeptides in KB cells, *J. Virol.* **25**:331–338.

Buller, R.M.L., and Rose, J.A., 1978b, Characterization of adeno-associated-virus polypeptides synthesized in vivo and in vitro, in: *Replication of Mammalian Parvoviruses* (D.C. Ward and P. Tattersall, eds.), pp. 399–410, Cold Spring Harbor Laboratory, Cold Spring Harbor, New York.

Burger, D.J., Gorham, J.R., and Ott, R.L., 1963, Protection of cats against feline panleucopaenia following mink enteritis vaccination, *Small Anim. Clinician* **3**:611–617.

Burtonboy, G., Coignoul, F., Pastoret, P.P., and Delferrière, N., 1979, Canine hemorrhagic enteritis: Detection of viral particles by electron microscopy, *Arch. Virol.* **61**:1–11.

Carmichael, L.E., Joubert, J.C., and Pollock, R.V.H., 1980, Hemagglutination by canine parvovirus: Serologic studies and diagnostic application, *Am. J. Vet. Res.* **41**:784–791.

Carter, B.J., 1974, Analysis of parvovirus mRNA by sedimentation and electrophoresis in aqueous and nonaqueous solution, *J. Virol.* **14**:834–839.

Carter, B.J., Khoury, G., and Rose, J.A., 1972, Adenovirus-associated virus multiplication IX. Extent of transcription of the viral genome *in vivo*, *J. Virol.* **10**:1118–1125.

Cartwright, S.F., and Huck, R.A., 1967, Viruses isolated in association with herd infertility, abortions and stillbirths in pigs, *Vet. Rec.* **81**:196–197.

Cartwright, S.F., Lucas, M., and Huck, R.A., 1969, A small haemagglutinating porcine DNA virus. I. Isolation and properties, *J. Comp. Path.* **79**:371–377.

Chesebro, B., Bloom, M., Hadlow, W., and Race, R., 1975, Purification and ultrastructure of aleutian disease virus of mink, *Nature (London)* **254**:456.

Cho, H.J., 1977, Demonstration of heavy and light density populations of aleutian disease virus, *Can. J. Comp. Med.* **41**:215.

Cho, J.H., and Ingram, D.G., 1973, Isolation, purification and structure of aleutian disease virus by immunological techniques, *Nature (London) New Biol.* **243**:174.

Clinton, G.M., and Hayashi, M., 1975, The parvovirus MVM: Particles with altered structural proteins, *Virology* **66**:261–267.

Clinton, G.M., and Hayashi, M., 1976, The parvovirus MVM: A comparison of heavy and light particle infectivity and their density conversion *in vitro*, *Virology* **74**:57–63.

Crawford, L.V., 1966, A minute virus of mice, *Virology* **29**:605–612.

Crawford, L.V., Follett, E.A.C., Burdon, M.G., and McGeoch, D.J., 1969, The DNA of a minute virus of mice, *J. Gen. Virol.* **4:**37–46.

Croft, G.F., Hoggan, M.D., and Johnson, F.B., 1974, Production and reactivity of immune sera specific for HADEN virus polypeptide antigens, *J. Virol.* **13:**608–613.

Croghan, D.L., Matchett, A., and Koski, T.A., 1973, Isolation of porcine parovirus from commercial trypsin, *Appl. Microbiol.* **26:**431–433.

Cutlip, R.C., and Mengeling, W.L., 1975, Experimentally induced infection of neonatal swine with porcine parvovirus, *Am. J. Vet. Res.* **36:**1179–1182.

De La Maza, L.M., and Carter, B.J., 1980, Heavy and light particles of adeno-associated virus, *J. Virol.* **33:**1129–1137.

Dolin, R., and Rabson, A.S., 1973, Herpesvirus saimiri: Enhancement of adenovirus-associated virus, *J. Natl. Cancer Inst.* **50:**205–208.

ElDadah, A.N., Nathanson, N., Smith, K.O., Squire, R.A., Santos, G.W., and Melby, E.C., 1967, Viral hemorrhagic encephalopathy of rats, *Science* **156:**392–394.

Garzon, S., and Kurstak, E., 1976, Ultrastructural studies on the morphogenesis of the densonucleosis virus (parvovirus), *Virology* **70:**517–531.

Gautschi, M., and Reinhard, P., 1978, Synthesis of a mammalian parvovirus in Brij-58-lysed cells, *J. Virol.* **27:**453–456.

Gautschi, M., and Siegl, G., 1973, Structural proteins of parvovirus Lu III., *Arch. Gesamte Virusforsch.* **43:**326–333.

Gautschi, M., Siegl, G., and Kronauer, G., 1976, Multiplication of parvovirus Lu III in a synchronized culture system IV. Association of viral structural polypeptides with the host cell chromatin, *J. Virol.* **20:**29–38.

Green, M.R., and Roeder, R.G., 1980a, Transcripts of the adeno-associated virus genome: Mapping of the major RNAs, *J. Virol.* **36:**79–92.

Green, M.R., and Roeder, R.G., 1980b, Definition of a novel promoter for the major adenovirus-associated virus mRNA, *Cell* **22:**231–242.

Green, M.R., Lebovitz, R.M., and Roeder, R.G., 1979, Expression of the autonomous parvovirus H1 genome: Evidence for a single transcriptional unit and multiple spliced polyadenylated transcripts, *Cell* **17:**967–977.

Green, M.R., Straus, S.E., and Roeder, R.G., 1980, Transcripts of the adenovirus-associated virus genome: Multiple polyadenylated RNAs including a potential primary transcript, *J. Virol.* **35:**560–565.

Hahn, E.C., Ramos, L., and Kenyon, A.J., 1977, Expression of aleutian mink disease antigen in cell culture, *Infect. Immun.* **15:**204–211.

Hallauer, C., Kronauer, G., and Siegl, G., 1971, Parvoviruses as contaminants of permanent human cell lines I. Virus isolations from 1960–1970, *Arch. Gesamte Virusforsch.* **35:**80–90.

Handa, H., Shimojo H., and Yamaguchi, K., 1976, Multiplication of adenoassociated virus type 1 in cells coinfected with a temperature-sensitive mutant of human adenovirus type 31, *Virology* **74:**1–15.

Henry, C.J., 1973, Adenovirus-associated (satellite) viruses, *Progr. Exp. Tumor Res.* **18:**273–293.

Hoggan, M.D., 1970, Adenovirus associated viruses, in: *Progress in Medical Virology* (J.L. Melnick, ed.), pp. 211–239, S. Karger, New York.

Hoggan, M.D., 1971, Small DNA viruses, in: *Comparative Virology* (K. Maramorosch and E. Kurstak, eds), pp. 43–79, Academic Press, Inc., New York.

Hoggan, M.D., Blacklow, N.R., and Rowe, W.P., 1966, Studies of small DNA viruses found in various adenovirus preparations: physical, biological, and immunological characteristics, *Proc. Natl. Acad. Sci. USA* **55:**1467–1474.

Inaba, Y., Kurogi, H., Takahashi, E., Sato, K., Tanaka, Y., Goto, Y., Omori, T., and Matumoto, M., 1973, Isolation and properties of bovine parvovirus type 1 from Japanese calves, *Arch. Gesamte Virusforsch.* **42:**54–66.

Ito, M., and Mayor, H.D., 1968, Hemagglutination of type 4 adeno-associated satellite virus, *J. Immunol.* **100:**61–68.

Jay, F.T., Laughlin, C.A., and Carter, B.J., 1981, Eukaryotic translational control: Adeno-associated virus protein synthesis is affected by a mutation in the adenovirus DNA binding protein, *Proc. Natl. Acad. Sci. USA* **78:**2927.

Johnson, F.B., and Hoggan, M.D., 1973, Structural proteins of HADEN virus, *Virology* **51:**129–137.

Johnson, F.B., and Hoggan, M.D., 1979, Labeling of adenovirus-associated virus proteins with radioactive acetate, Abstract of the Annual Meeting of the American Society for Microbiology, p. 243.

Johnson, F.B., Ozer, H.L., and Hoggan, M.D., 1971, Structural proteins of adenovirus-associated virus type 3, *J. Virol.* **8:**860–863.

Johnson, F.B., Blacklow, N.R., and Hoggan, M.D., 1972, Immunological reactivity of antisera prepared against the sodium dodecyl sulfate treated structural polypeptides of adenovirus-associated virus, *J. Virol.* **9:**1017–1026.

Johnson, F.B., Whitaker, C.W., and Hoggan, M.D., 1975, Structural polypeptides of adenovirus-associated virus top component, *Virology* **65:**196–203.

Johnson, F.B., Thomson, T.A., Taylor, P.A., and Vlazny, D.A., 1977, Molecular similarities among the adenovirus-associated virus polypeptides and evidence for a precursor protein, *Virology* **82:**1–13.

Johnson, F.B., Vlazny, D.A., Thomson, T.A., Taylor, P.A., and Lubeck, M.D., 1978, Adenovirus-associated virus polypeptides: Molecular similarities, in: *Replication of Mammalian Parvoviruses* (D.C. Ward and P. Tattersall, eds.), pp. 411–421, Cold Spring Harbor Laboratory, Cold Spring Harbor, New York.

Johnson, R.H., 1967, Feline panleucopaenia virus—in vitro comparison of strains with a mink enteritis virus, *J. Small Anim. Pract.* **8:**319–322.

Johnson, R.H., and Spradbrow, P.B., 1979, Isolation from dogs with severe enteritis of a parvovirus related to feline panleucopaenia virus, *Aust. Vet. J.* **55:**151.

Johnson, R.H., Siegl, G., and Gautschi, M., 1974, Characteristics of feline panleucopaenia virus strains enabling definitive classification as parvoviruses, *Arch. Gesamte Virusforsch.* **46:**315–324.

Joo, H.S., and Johnson, R.H., 1977, Serological responses in pigs vaccinated with inactivated porcine parvovirus, *Aust. Vet. J.* **53:**550–552.

Joo, H.S., Donaldson-Wood, C.R., and Johnson, R.H., 1976, Observations on the pathogenesis of porcine parvovirus infection, *Arch. Virol.* **51:**123–129.

Karasaki, S., 1966, Size and ultrastructure of the H-viruses as determined with the use of specific antibodies, *J. Ultrastr. Res.* **16:**109–122.

Kelly, D.C., and Bud, H.M., 1978, Densonucleosis virus DNA: Analysis of fine structure by electron microscopy and agarose gel electrophoresis, *J. Gen Virol.* **40:**33–43.

Kelly, D.C., and Elliott, R.M., 1977, Polyamines contained by two densonucleosis viruses, *J. Virol.* **21:**408–410.

Kelly, D.C., Barwise, A.H., and Walker, I.O., 1977, DNA contained by two densonucleosis viruses, *J. Virol.* **21:**396–407.

Kelly, D.C., Moore, N.F., Spilling, C.R., Barwise, A.H., and Walker, I.O., 1980, Densonucleosis virus structural proteins, *J. Virol.* **36:**224–235.

Kelly, W.R., 1978, An enteric disease of dogs resembling feline panleucoaenia, *Aust. Vet. J.* **54:**593.

Kilham, L., 1966, Viruses of laboratory and wild rats, *Natl. Cancer Inst. Monogr. No.* **20:**117–135.

Kilham, L., and Margolis, G., 1975, Problems of human concern arising from animal models of intrauterine and neonatal infections due to viruses: A review I. Introduction and virologic studies, *Progr. Med. Virol.* **20:**113–143.

Kilham, L., and Olivier, L.J., 1959, A latent virus of rats isolated in tissue culture, *Virology* **7:**428–437.

Kongsvik, J.R., Gierthy, J.R., and Rhode, S.L., III, 1974, Replication process of the parvovirus H-1 IV. H-1-specific proteins synthesized in synchronized human NB kidney cells, *J. Virol.* **14:**1600–1603.

Kongsvik, J.R., and Toolan, H.W., 1972, Capsid Components of the parvovirus H-1, *Proc. Soc. Exp. Biol. Med.* **139:**1202–1205.

Kongsvik, J.R., Hopkins, M.S., and Ellem, K.A.O., 1978, Two populations of infectious virus produced during H-1 infection of synchronized transformed cells, in: *Replication of Mammalian Parvoviruses* (D.C. Ward and P. Tattersall, eds.), pp. 505–520, Cold Spring Harbor Laboratory, Cold Spring Harbor, New York.

Kongsvik, J.R., Hopkins, M.S., and Ellem, K.A.O., 1979, Subfractionation of CsCl-purified H-1 parvovirus on metrizamide gradients, *Virology* **96:**646–651.

Kurstak, E., 1972, Small DNA densonucleosis virus (DNV), *Adv. Virus Res.* **17:**207–241.

Kurstak, E., and Côté, J.R., 1969, Proposition de classification du virus de la densonucléose (VDN) basée sur l'etude de la structure moleculaire et des propriétés physicochimiques, *C. R. Acad. Sci. Paris* **268:**616–619.

Kurstak, E., Vernoux, J.P., Niveleau, A., and Onji, P.A., 1971, Visualisation du DNA du virus de la desonucléose (VDN) a chaînes monocaténaires complémentaires de polarités inverses plus et moins, *C. R. Acad. Sci. Paris* **272:**762–765.

Kurstak, E., Vernoux, J.P., and Brakier-Gingras, L., 1973, Etude biophysique de l'acide désoxyribonucléique du virus de la desonucléose (VDN). II. Extraction du DNA viral et mise en évidence de la présence de chaînes polynucléotidiques complémentaires, encapsidés séparément dans les virions VDN, *Arch Gesamte Virusforsch.* **40:**274–284.

Laughlin, C.A., Myers, M.W., Risin, D.L., and Carter, B.J., 1979a, Defective-interfering particles of the human parvovirus adeno-associated virus, *Virology* **94:**162–174.

Laughlin, C.A., Westpahl, H., and Carter, B.J., 1979b, Spliced adenovirus-associated virus RNA, *Proc. Natl. Acad. Sci. USA* **76:**5567–5571.

Leary, J.J., and Storz, J., 1980, Nucleopathic changes in parvovirus-infected, cultured cells, *Exp. Molec. Pathol.* **32:**188–200.

Lenghaus, C., and Studdert, M.J., 1980, Relationship of canine panleucopaenia (enteritis) and myocarditis parvoviruses to feline panleucopaenia virus, *Aust. Vet. J.* **56:**152–153.

Levin, S., 1980, Parvovirus: A possible etiologic agent in cardiomyopathy and endocardial fibroelastosis, *Human Pathol.* **11:**404–405.

Levin, S., Taylor, R.F., Lowrie, P.M., and Roberts, A.W., 1979, Viral myocarditis in puppies, American Veterinary Laboratory Diagnostics 22nd Annual Proceedings, pp. 395–404.

Lubeck, M.D., and Johnson, F.B., 1976, Multiplication of bovine parvovirus in two cell strains, *Infect. Immun.* **13:**1289–1292.

Lubeck, M.D., and Johnson, F.B., 1977, Radioiodination of adenovirus-associated virus external structural proteins, *Virology* **83:**453–457.

Lubeck, M.D., Lee, H.M., Hoggan, M.D., and Johnson, F.B., 1979, Adenovirus-associated virus structural protein sequence homology, *J. Gen Virol.* **45:**209–216.

Lwoff, A., and Tournier, P., 1971, Remarks on the classification of viruses, in: *Comparative Virology* (K. Maramorosch and E. Kurstak, eds.), pp. 1–42, Academic Press, Inc., New York.

Margolis, G., and Kilham, L., 1975, Problems of human concern arising from animal models of intrauterine and neonatal infections due to viruses: A review II. Pathologic studies. *Progr. Med. Virol.* **20:**144–179.

Matsunaga, Y., Matsuno, S., and Mukoyama, J., 1977, Isolation and characterization of a parvovirus of rabbits, *Infect. Immun.* **18:**495–500.

Mayor, H.D., and Drake, S., 1974, Complementation of adeno-associated satellite virus antigens in cells transformed by human herpes virus, *Microbios* **11A:**37–46.

Mayor, H.D., and Gorman, C., 1975, Growth of adeno-associated satellite virus under conditions of arginine deprivation, *Virology* **66:**25–35.

Mayor, H.D., and Kurstak, E., 1974, Viruses with separately encapsidated complementary DNA strands, in: *Viruses, Evolution and Cancer* (E. Kurstak, and K. Maramerosch, eds.), pp. 55–78, Academic Press, Inc., New York.

Mayor, H.D., Jamison, R.M., Jordan, L.E., and Melnick, J.L., 1965, Structure and composition of a small particle prepared from a simian adenovirus, *J. Bacteriol.* **90:**235–242.

Mayor, H.D., Ito, M., Jordan, L.E., and Melnick, J.L., 1967, Morphological studies on the replication of a defective satellite virus and its helper adenovirus, *J. Natl. Cancer Inst.* **38**:805–820.

McMaster, G.K., Beard, P., Engers, H.D., and Hirt, B., 1981a, Characterization of an immunosuppressive parvovirus related to the minute virus of mice, *J. Virol.* **38**:317–326.

McMaster, G.K., Tratschin, J.-D., and Siegl, G., 1981b, Comparison of canine parvovirus with mink enteritis virus by restriction site mapping, *J. Virol.* **38**:368–371.

Mengeling, W.L., 1972, Porcine parvovirus: Properties and prevalence of a strain isolated in the United States, *Am. J. Vet. Res.* **33**:2239–2248.

Mengeling, W.L., 1977, Diagnosing porcine parvovirus-induced reproductive failure, *Proc. Am. Assoc. Vet. Lab. Diag.* **20**:237–244.

Mengeling, W.L., 1979, Prenatal infection following maternal exposure to porcine parvovirus on either the seventh or fourteenth day of gestation, *Can. J. Comp. Med.* **43**:106–109.

Mengeling, W.L., Brown, T.T., Paul, P.S., and Gutekunst, D.E., 1979, Efficacy of an inactivated virus vaccine for prevention of porcine parvovirus-induced reproductive failure, *Am. J. Vet. Res.* **40**:204–207.

Mengeling, W.L., and Cutlip, R.C., 1976, Reproductive disease experimentally induced by exposing pregnant gilts to porcine parvovirus, *Am. J. Vet. Res.* **37**:1393–1400.

Mengeling, W.L., Paul, P.S., and Brown, T.T., 1980, Transplacental infection and embryonic death following maternal exposure to porcine parvovirus near the time of conception. *Arch. Virol.* **65**:55–62.

Moraillon, A., 1980, Canine parovirus: Safety and efficacy of attenuated feline panleucopenia vaccine, *Vet. Rec.* (June 14, 1980) (letter).

Moraillon, A., Moraillon, R., Person, J.M., and Parodi, A.L., 1980, Parvovirose canine: l'ingestion d'organes de vison atteint d'entérite à virus déclenche clex le chien une maladie identique à la maladie spontanée, *Rec. Med. Vet.* **156**:539–548.

Moore, A.E., and Nicastri, A.D., 1965, Letnal infection and pathological findings in A X C rats inoculated with H-virus and RV., *J. Natl. Cancer Inst.* **35**:937–947.

Myers, M.W., and Carter, B.J., 1980, Assembly of adeno-associated virus, *Virology* **102**:71–82.

Myers, M.W., and Carter, B.J., 1981, Adeno-associated replication, the effect of L-canavanine or a helper virus mutation on accumulation of viral capsids and progeny single-stranded DNA, *J. Biol. Chem.* **56**:567–570.

Paradiso, P.R., 1981, The infectious process of the parvovirus H-1: Correlation of the protein content, particle density and viral infectivity, *J. Virol.* **39**:800–807.

Parks, W.P., Green, M., Pina, M., and Melnick, J.L., 1967, Physicochemical characterization of adeno-associated satellite virus type 4 and its nucleic acid, *J. Virol.* **1**:980–987.

Parris, D.S., and Bates, R.C., 1976, Effect of bovine parvovirus replication on DNA, RNA, and protein synthesis in S phase cells, *Virology* **73**:72–78.

Patton, J.T., Stout, E.R., and Bates, R.C., 1979, Transcription of the bovine parvovirus genome in isolated nuclei, *J. Virol.* **30**:917–922.

Paul, P.S., Mengeling, W.L., and Brown, T.T., Jr., 1979, Replication of porcine parvovirus in peripheral blood lymphocytes, monocytes, and peritoneal macrophages, *Infect. Immun.* **25**:1003–1007.

Peterson, J.L., Dale, R.M.K., Karess, R., Leonard, D., and Ward, D.C., 1978, Comparison of parvovirus structural proteins: Evidence for post-translational modification, in: *Replication of Mammalian Parvoviruses*, (D.C. Ward and P. Tattersall, eds.), pp. 431–445, Cold Spring Harbor Laboratory, Cold Spring Harbor, New York.

Porter, D.D., and Cho, H.J., 1980, Aleutian disease of mink: A model for persistent infection, in: *Comprehensive Virology*, (H. Fraenkel-Conrat and R.R. Wagner, eds.), Vol. 16, pp. 233–255, Plenum Press, New York.

Porter, D.D., and Larsen, A.E., 1974, Aleutian disease of mink, *Progr. Med. Virol.* **18**:32–47.

Porter, D.D., Larsen, A.E., Cox, N.A., Porter, H.G., and Suffin, S.C., 1977, Isolation of aleutian disease virus of mink in cell culture, *Intervirology* **8**:129–144.

Porter, D.D., Larsen, A.E., and Porter, H.G., 1980, Aleutian disease of mink, *Adv. Immunol.* **29**:261–286.

Pritchard, C., Bates, R.C., and Stout, E.R., 1978a, Levels of cellular DNA polymerases in synchronized bovine parvovirus-infected cells, *J. Virol.* **27**:258–261.

Pritchard, C., Patton, J.T., Bates, R.C., and Stout, E.R., 1978b, Replication of nondefective parvoviruses: Lack of a virion-associated DNA polymerase, *J. Virol.* **28**:20–27.

Pritchard, C., Stout, E.R., and Bates, R.C., 1981, Replication of parvoviral DNA I. Characterization of a nuclear lysate system, *J. Virol.* **37**:352–362.

Rhode, S.L., 1973, Replication process of the parvovirus H-1. I. Kinetics in a parasynchronous cell system, *J. Virol.* **11**:856–861.

Rhode, S.L., 1974a, Replication process of the parvovirus H-1, II. Isolation and characterization of H-1 replicative form DNA, *J. Virol.* **13**:400–410.

Rhode, S.L., 1974b, Replication process of the parvovirus H-1. III. Factors affecting H-1 RF DNA synthesis, *J. Virol.* **14**:791–801.

Rhode, S.L., 1976, Replication process of the parvovirus H-1 V. Isolation and characterization of temperature-sensitive H-1 mutants defective in progeny DNA synthesis, *J. Virol.* **17**:659–667.

Rhode, S.L., 1978, Defective interfering particles of parvovirus H-1, *J. Virol.* **27**:347–356.

Richards, R., Linser, P., and Armentrout, R.W., 1977, Kinetics of assembly of a parvovirus, minute virus of mice, in synchronized rat brain cells, *J. Virol.* **22**:778–793.

Richards, R., Linser, P., and Armentrout, R.W., 1978, Maturation of minute-virus-of-mice particles in synchronized rat brain cells, in: *Replication of Mammalian Parvoviruses* (D.C. Ward and P. Tattersall, eds.), pp. 447–458, Cold Spring Harbor Laboratory, Cold Spring Harbor, New York.

Robey, R.E., Woodman, D.R., and Hetrick, F.M., 1968, Studies on the natural infection of rats with the Kilham rat virus, *Am. J. Epidemiol.* **88**:139–143.

Rose, J.A., 1974, Parvovirus reproduction, in: *Comprehensive Virology* (H. Fraenkel-Conrat and R.R. Wagner, eds.), Vol. 3, pp. 1–61, Plenum Press, New York.

Rose, J.A., Maizel, J.V., Jr., Inman, J.K., and Shatkin, A.J., 1971, Structural proteins of adenovirus-associated viruses, *J. Virol.* **8**:766–770.

Salo, R.J., and Mayor, H.D., 1977, Structural polypeptides of parvoviruses, *Virology* **78**:340–345.

Salo, R.J., and Mayor, H.D., 1978, Isoelectric focusing of parvoviruses, *Intervirology* **10**:87–93.

Salo, R.J., and Mayor, H.D., 1979, Adenovirus-associated virus polypeptides synthesized in cells coinfected with either adenovirus or herpesvirus, *Virology* **93**:237–245.

Salzman, L.A., and White, W.L., 1970, Structural proteins of Kilham rat virus, *Biochem. Biophys. Res. Commun.* **41**:1551–1556.

Shahrabadi, M.S., Cho, H.J., and Marusyk, R.G., 1977, Characterization of the protein and nucleic acid of aleutian disease virus, *J. Virol.* **23**:353–362.

Siegl, G., 1973, Physicochemical Characteristics of the DNA of parvovirus Lu III, *Arch. Gesamte Virusforsch.* **43**:334–344.

Siegl, G., 1976, The parvoviruses, in: *Virology Monographs* (S. Gard and C. Hallauer, eds.), Vol. 15, pp. 1–109, Springer-Verlag, New York.

Siegl, G., and Gautschi, M., 1973a, The multiplication of parvovirus Lu III in a synchronized culture system I. Optimum conditions for virus replication, *Arch. Gesamte Virusforsch.* **40**:105–118.

Siegl, G., and Gautschi, M., 1973b, The multiplication of parvovirus Lu III in a synchronized culture system II. Biochemical characteristics of virus replication, *Arch. Gesamte Virusforsch.* **40**:119–127.

Siegl, G., and Gautschi, M., 1976, Multiplication of parvovirus Lu III in a synchronized culture system III. Replication of viral DNA, *J. Virol.* **17**:841–853.

Siegl, G., Hallauer, C., Novak, A., and Kronauer, G., 1971, Parvoviruses as contaminants of permanent human cell lines II. Physicochemical properties of the isolated viruses, *Arch. Gesamte Virusforsch.* **35**:91–103.

Singer, I.I., 1975, Ultrastructural studies of H-1 parvovirus replication. II. Induced changes in the deoxyribonucleoprotein and ribonucleoprotein components of human NB cell nuclei, *Exp. Cell Res.* **95:**205–217.

Singer, I.I., 1976, Ultrastructural studies of H-1 parvovirus replication III. Intracellular location of viral antigens with immunocytochrome c. *Exp. cell Res.* **99:**346–356.

Singer, I.I., and Rhode, S.L., 1977a, Ultrastructural studies of H-1 parvovirus replication IV. Crystal development and structure with the temperature-sensitive mutant ts1, *J. Virol.* **24:**343–352.

Singer, I.I., and Rhode, S.L., III, 1977b, Ultrastructural studies of H-1 parvovirus replication V. Immunocytochemical demonstration of separate chromatin-associated and inclusion-associated antigens, *J. Virol.* **24:**353–362.

Singer, I.I., and Rhode, S.L., III, 1978, Ultrastructural studies of H-1 parvovirus replication VI. Simultaneous autoradiographic and immunochemical intranuclear localization of viral DNA synthesis and protein accumulation, *J. Virol.* **25:**349–360.

Singer, I.I., and Toolan, H.W., 1975, Ultrastructural studies of H-1 parvovirus replication I. Cytopathology produced in human N. B. epithelial cells and hamster embryo fibroblasts, *Virology* **65:**40–54.

Smith, K.O., Gehle, W.D., and Thiel, J.F., 1966, Properties of a small virus associated with adenovirus type 4, *J. Immunol.* **97:**754–766.

Soike, K.F., Iatropoulis, M., and Siegl, G., 1976, Infection of newborn and fetal hamsters induced by inoculation of Lu III parvovirus, *Arch. Virol.* **51:**235–241.

Spahn, G.J. Mohanty, S.B., and Hetrick, F.M., 1966a, Characteristics of hemadsorbing enteric (HADEN) virus, *Can. J. Microbiol.* **12:**653–661.

Spahn, G.J., Mohanty, S.B., and Hetrick, F.M., 1966b, Experimental infection of calves with hemadsorbing enteric (HADEN) virus, *Cornell Vet.* **56:**377–386.

Storz, J., and Bates, R.C., 1973, Parvovirus infections in calves, *J. Am. Vet. Med. Assoc.* **163:**884–886.

Storz, J., and Leary, J.J., 1979, Bovine parvoviruses: Epidemiology and host–virus relationships, Les Colloques de l'INSERM Vol. 90, pp. 63–80.

Storz, J., and Warren, G.S., 1970, Effect of antimetabolites and actinomycin D on the replication of HADEN, a bovine parvovirus, *Arch. Gesamte Virusforsch.* **30:**271–274.

Storz, J., Bates, R.C., Warren, G.S., and Howard, T.H., 1972, Distribution of antibodies against bovine parvovirus 1 in cattle and other animal species, *Am. J. Vet. Res.* **33:**269–272.

Suzuki, H., and Fujisaki, Y., 1976, Immunizing effects of inactivated porcine parvovirus vaccine on piglets, *Natl. Inst. Anim. Health Q.* (*Tokyo*) **16:**81.

Tattersall, P., 1972, Replication of the parvovirus MVM I. Dependence of virus multiplication and plaque formation on cell growth, *J. Virol.* **10:**586–590.

Tattersall, P., 1978, Parvovirus protein structure and virion maturation, in: *Replication of Mammalian Parvoviruses* (D.C. Ward and P. Tattersall eds.), pp. 53–72, Cold Spring Harbor Laboratory, Cold Spring Harbor, New York.

Tattersall, P., Cawte, P.J., Shatkin, A.J., and Ward, D.C., 1976, Three structural polypeptides coded for by minute virus of mice, a parvovirus, *J. Virol.* **20:**273–289.

Tattersall, P., Shatkin, A.J., and, Ward, D.C., 1977, Sequence homology between the structural polypeptides of minute virus of mice, *J. Mol. Biol.* **111:**375–394.

Tennant, R.W., and Hand, R.E., 1970, Requirement of cellular synthesis for kilham rat virus replication, *Virology* **42:**1054–1063.

Thompson, G.W., and Gagnon, A.N., 1978, Canine gastroenteritis associated with a parvovirus-like agent, *Can. Vet. J.* **19:**346.

Tijssen, P., and Kurstak, E., 1979a, A simple and sensitive method for the purification and peptide mapping of proteins solubilized from densonucleosis virus with sodium dodecyl sulfate, *Anal. Biochem.* **99:**97–104.

Tijssen, P., and Kurstak, E., 1979b, Studies on the structure of the two infectious types of densonucleosis virus, *Intervirology* **11:**261–267.

Tijssen, P., and Kurstak, E., 1981, Biochemical, biophysical, and biological properties of densonucleosis virus (parvovirus) III. Common sequences of structural proteins, *J. Virol.* **37**:17–23.

Tijssen, P., van den Hurk, J., and Kurstak, E., 1976, Biochemical biophysical, and biological properties of densonucleosis virus I. Structural proteins, *J. Virol.* **17**:686–691.

Tijssen, P., Tijssen-van der Slikke, T., and Kurstak, E., 1977, Biochemical, biophysical, and biological properties of densonucleosis virus (parvovirus) II. Two types of infectious virions, *J. Virol.* **21**:225–231.

Tinsley, T.W., and Longworth, J.F., 1973, Parvoviruses, *J. Gen. Virol.* **20**(Suppl):7–15.

Toolan, H.W., 1961a, A virus associated with transplantable human tumors, *Bull. N.Y. Acad. Med.* **37**:305–310.

Toolan, H.W., 1961b, Studies on a viral agent associated with human tissues, *Proc. Amer. Assoc. Cancer Res.* **3**:273–274.

Toolan, H.W., 1972, The parvoviruses, *Progr. Exp. Tumor Res.* **16**:410–425.

Torpier, G., D'Halluin, J.C., and Boulanger, P., 1971, Electron microscopic observations on KB cells infected with adeno-associated satellite virus, *J. Microscopie* **11**:259–264.

Truffant, N., Berger, G., Niveleau, A., May, P., Bergoin, M., and Vago, C., 1967, Recherches sur l'acide nucléique du virus de la densonucléose du lépidoptére *G. mellonella, Arch. Gesamte Virusforsch.* **21**:469–474.

Vincent, J., 1971, Isolement en Algérie de quatre souches de parvovirus bovis, *Ann. Inst. Pasteur* **121**:811–814.

Williams, F.P., Jr., 1980, Astrovirus-like, coronavirus-like, and parvovirus-like particles detected in the diarrheal stools of beagle pups, *Arch. Virol.* **66**:215–226.

Young, J.F., and Mayor, H.D., 1979a, Adeno-associated virus—An extreme state of viral defectiveness, *Prog. Med. Virol.* **25**:113–132.

Young, J.F., and Mayor, H.D., 1979b, Studies on the defectiveness of adeno-associated virus (AAV) I. Effects of phosphonoacetic acid and 2-deoxy-D-glucose on the replication of AAV, *Virology* **94**:323–341.

CHAPTER 8

Biology and Pathogenicity of Autonomous Parvoviruses

GÜNTER SIEGL

I. INTRODUCTION

During the past ten years parvoviruses have attracted considerable interest. This is in part due to their unique molecular organization as well as to the fact that they provide an excellent experimental tool to study the replication of a small, single-stranded viral DNA genome and to probe into the synthesis of eucaryotic cell DNA. On the other hand, parvoviruses were shown to be associated with various economically important diseases of animals. At lease since the very recent world-wide epidemic of canine parvovirus enteritis they are no longer regarded as constituting a mere laboratory problem.

In parallel to this increasing importance of parvoviruses as causative agents of specific diseases, several excellent articles on individual parvoviruses have been published (Toolan, 1972; Siegl, 1976; Porter and Cho, 1980; Porter *et al.*, 1980; Mengeling, 1981). These reviews list the available information on distinct characteristics of the viruses in great detail. The present article, on the contrary, traces the properties in common to the various autonomous parvoviruses. It places emphasis on those characteristics which allow rapid and unambiguous identification of individual members of this genus and attempts to provide a basis for the general understanding of this group of viruses.

GÜNTER SIEGL • Institute for Hygiene and Medical Microbiology, University of Bern, CH 3010 Bern, Switzerland.

A. History of Isolation

Isolation of autonomous parvoviruses can be dealt with under two distinct headings, i.e., the isolation by chance and isolation as the result of an intensive search for the causative agent of a specific infectious disease. By far the most parvoviruses have been isolated accidentally. Rat virus (RV), the type species of the genus *Parvovirus*, owes its discovery to the search for an anticipated papovavirus of rats. In the course of these investigation, Kilham and Olivier (1959) inoculated tumor tissue homogenates from three rats into rat embryonic cell cultures and observed the replication of a small, hemagglutinating virus of outstanding stability. In subsequent years similar viruses were frequently recovered from tissues of tumor-bearing rats as well as from tissue of rats treated with carcinogenic or immunosuppressive chemicals (Zhdanov and Merekalova, 1962; Kilham and Molony, 1964; Spencer, 1967). Thus, X-14 virus was isolated in rat embryonic cell cultures inoculated with mammary tumor homogenates of rats X-irradiated and treated with methyl cholanthrene (Payne *et al.*, 1963). The LS agent was recovered by similar techniques from a rat bearing a chloroleukemic tumor (Lum and Schreiner, 1963), and the "hemorrhagic encephalopathy of rats" (HER) agent was found to be responsible for an acute paralysis of the hind limbs of rats which developed after the animals were treated with high concentrations of cyclophosphamide (El Dadah *et al.*, 1967). All these viruses as well as those recovered by Kilham and Molony (1964), Spencer (1967), and Zhdanov and Merekalova (1962) are serologically closely related.

At about the same time as Kilham and Olivier (1959) isolated RV, Toolan (1960) and Toolan and co-workers (1960) observed the development of a rather complex syndrome after inoculation of fractions or cell-free filtrates of transplantable human tumors into newborn hamsters. These animals developed dwarfism, microcephalic domed head, fragility of bones, and malformation of teeth. The causative virus was identified in samples originating from the human tumor HEp-1. It shared many characteristics with RV, yet proved to be serologically distinct from this virus. Because of its source it was named H-1 virus. Under almost identical experimental conditions H-3 virus was recovered from the human tumor HEp-3 (Dalldorf, 1960) and injection of various tissues of cancer patients (Toolan, 1961a,b) as well as of human embryos and placentas (Toolan *et al.*, 1962) into hamster neonates led to the repeated isolation of H-1 virus and the detection of the serologically related HT virus. The still scarcely characterized HB virus also originated from such studies (Toolan, 1964).

A constant and apparently inexhaustable source for the isolation of viruses with the physicochemical characteristics of parvoviruses are primary and permanent cultures of cells from various animals and man. Hallauer and Kronauer (1962) became aware of latent infections of permanent human cell lines with parvoviruses when uninfected cultures subjected to treatment with the alkaline glycine buffer (Hallauer and Kronauer, 1960) yielded a then unidentified hemagglutinating agent. Between 1960 and 1970 a total of 43 strains of permanent human cell lines such as HeLa, KB, HEp-2, FL-Amnion, Lu106 have yielded a total of 38 isolates (Hallauer *et al.*, 1971). The great majority (29) of these isolates, represented by the prototype KBSH virus, proved to be serologically related to porcine parvovirus (PPV). Eight isolates represented by the prototype TVX (= tumor virus X) and LuIII virus (isolated only once from Lu106 cells) were antigenically unrelated to any other known parvovirus. Circumstantial evidence (Hallauer *et al.*, 1971) and direct isolation (Croghan *et al.*, 1973) support the view that the viruses serologically related to PPV had been transmitted to the cultures by the use of trypsin derived from the pancreas of PPV-infected pigs. The origin of TVX and LuIII-type viruses, on the other hand, is still unknown.

Hallauer *et al.* (1971) also reported on the isolation of a serologically unique parvovirus (RTV) from a permanent line of rat cells (AT) originally established by P. Tournier. In contrast, the agents isolated by Melnick *et al.*(1971) and Berquist *et al.* (1972) from Detroit 6 cells (neoplastic human bone marrow) and HEL (human embryonic lung) cells, respectively, had the antigenic characteristics of H-3 virus.

At the present time, it is well known that parvoviruses may persist in their natural hosts in spite of the presence of an impressive humoral immunity. Cell cultures prepared from tissues of infected animals therefore are frequently found to yield these viruses (Mayr *et al.*, 1968; O'Reilly and Whitaker, 1969; Wozniak and Hetrick, 1969). Consequently, it is impossible to decide whether parvoviruses such as the minute virus of mice (MVM) or porcine parvovirus (PPV) which have come to our notice as contaminants in mouse adenovirus stocks (Crawford, 1966) or hog cholera virus (Horzinek *et al.*, 1967; Mayr *et al.*, 1968), respectively, had been already present in the original isolates or were picked up during passage of the latter viruses in latently infected cultures of mouse and pig cells. In any case, cell cultures prepared from tissues of conventionally raised, apparently healthy animals are always prone to yield parvoviruses. The latter fact is emphasized by the observation that apparently healthy animals may shed parvoviruses in their feces (Bouillant and Hanson, 1965; Csiza *et al.*, 1971c; Storz and Leary, 1979). Three distinct parvoviruses—the bovine parvovirus (BPV) (Abinanti and Warfield, 1961), the minute virus of canines (MVC) (Binn *et al.*, 1970), and the lapine parvovirus (LPV) (Matsunaga *et al.*, 1977)—were isolated under these conditions.

In striking contrast to the frequency and readiness with which parvoviruses were recovered as contaminants from various biological materials, their identification generally posed a considerable problem whenever they were involved in clinically overt disease. Thus, both the syndromes of feline ataxia and feline enteritis had been known for almost a hundred years (Herringham and Andrewes, 1888; Zschokke, 1900) and had been studied in much detail during the first half of this century. Nevertheless, the feline parvovirus (FPV) responsible for both syndromes was isolated and characterized in the mid-1960s only (Johnson, 1964, 1965; Kilham and Margolis, 1966a; Johnson and Cruickshank, 1966; Johnson *et al.*, 1967). Similarly, the syndrome of Aleutian disease of mink has been described by Hartsough and Gorham in 1956 and then was thought to be of genetic origin. In 1962, however, its transmissible nature was recognized almost simultaneously by two groups (Karstad and Pridham, 1962; Henson *et al.*, 1962). In spite of the growing knowledge on the properties of parvoviruses, the causative virus was isolated in cell cultures in recent years only (Porter *et al.*, 1977; Hahn *et al.*, 1977a,b) and the definite physicochemical characteristics of the virion enabeling unambiguous classification of ADV with the parvoviruses were reported by Bloom *et al.* in 1980!

Comparable difficulties were encountered with goose parvovirus (GPV). The fatal disease resulting from infection with this virus is confined almost exclusively to goslings aged between 8 and 30 days. It was first described in Poland in 1962 by Wachnik and Nowaki. When the disease caused by the virus swept through Europe between 1962 and 1972 the related syndrome became known as "goose influenza" in Hungary, "virus hepatitis of geese" in Germany, "goose plague" in the Netherlands and "ascitic-hepato-nephritis" in France due to varying clinical symptoms and postmortem findings. Losses of up to 95% in affected geese flocks were reported (Kraus, 1965; van Cleef and Miltenberg, 1966; Derzy, 1967; Schettler, 1971a; Dannacher *et al.*, 1974). Isolation *in vitro* and subsequent detailed characterization of the agent, however, indicated that all the different syndromes were caused by the same parvovirus (Schettler, 1973; Dannacher *et al.*, 1974; Kisary, 1974; Kisary and Derzsy, 1974).

Finally, the very recent isolation and characterization of a parvovirus from dogs either suffering from severe enteritis or dying from acute heart failure (myocarditis) has to be mentioned. This virus—generally referred to as canine parvovirus (CPV)—is unrelated to the minute virus of canines (MVC) isolated previously by Binn *et al.* (1970). As described elsewhere in more detail (see Chapter 9), it represents a new, second, host-range variant of feline parvovirus. The first host-range variant of FPV—mink enteritis virus (MEV)—appeared in Canada more than 30 years ago (Schofield, 1949). Since then it has spread throughout the world and is now considered a major threat to mink-breeding colonies.

B. Host Range of Parvoviruses

In spite of the observations with the host-range variants of FPV the history of isolations of the various parvoviruses is suggestive for their host range and, hence, for their grouping within the genus within the

TABLE I. The Origin of Autonomous Parvoviruses

Virus	Vernacular name	Source of first isolation	Natural host
RV	Rat virus	Rat neoplastic tissues	Rats
RTV	—	Rat cell line	Rats (?)
H-1	—	Human tumors[a]	Rats, man (?)
HB	—	Human fetal tissues[a]	Man (?)
MVM	Minute virus of mice	Mouse adenovirus stock	Mice
FPV:			
FPLV	Feline parvovirus: feline panleukopenia virus	Feline tissues	Cats
MEV	Mink enteritis virus	Mink tissues	Mink
CPV	Canine parvovirus	Canine feces	Dogs
MVC	Minute virus of canines	Canine feces	Dogs
BPV	Bovine parvovirus	Bovine feces	Cattle
PPV	Porcine parvovirus	Primary porcine cell cultures	Swine
GPV	Goose parvovirus	Goose tissues	Geese
ADV	Aleutian disease virus	Mink tissues	Mink
LPV	Lapine parvovirus	Rabbit feces	Rabbits
TVX	—	Human permanent cell line	Rats (?)
LuIII	—	Human permanent cell line	Unknown

[a] Isolated via inoculation of tissue homogenates into newborn hamsters.

helper-independent, autonomous parvoviruses. However, additional factors such as, for example, the incidence of antibodies in populations of the various species and a thorough comparison of viral antigens are necessary to obtain a reliable picture. The synopsis of such an analysis is given in Table I. It clearly indicates that, with very few exceptions which have to be discussed later, individual antigenically distinct parvoviruses apparently are limited to one natural host.

II. ANTIGENIC STRUCTURE AND SEROLOGIC RELATIONSHIP

Parvoviruses are rather simple particles consisting of not more than three structural proteins (Johnson, Chapter 7 of this volume). These proteins also share overlapping peptide sequences (Tattersall *et al.*, 1977). Some of the polypeptides probably are present in precursors of the mature virion only and there is preliminary evidence (Majaniemi *et al.*, 1981) that the capsid of the infectious particle of parvovirus LuIII might consist of multiple copies of one single polypeptide. It may be expected, therefore, that the mature parvovirus particle is of rather homogeneous antigenicity.

The antigen of the mature virion (110S, $\rho = 1.41$ g/ml in CsCl) can be detected and identified by neutralization tests, by hemagglutination and hemagglutination inhibition, as well as by complement fixation. The extent to which the mature virion is antigenically related to the "dense" or "heavy" infectious virion (110S; ρ = 1.44–1.47 g/ml), to the defective or incomplete particles (70–90S; ρ = 1.34–1.39 g/ml), or to empty capsid structures (70S; ρ = 1.32 g/ml), is still unknown. All these types of particles react also in hemagglutination inhibition and complement fixation tests. Whether they can be distinguished serologically like D and L antigens of the infectious picornavirus particle and its corresponding empty capsid has not been investigated.

A. Hemagglutination and Hemadsorption

With exceptions of Aleutian disease virus (ADV) (Porter and Cho, 1980) and goose parvovirus (GPV) (Schettler, 1973; Kisary and Derzsy, 1974) the ability to agglutinate red blood cells of at least one animal species is a common characteristic of all agents listed in Table I. Due to the simple structure of the virus particle the hemagglutinating entity is identical with the virus capsid structure. Empty capsids, incomplete particles, as well as the two density variants of the infectious virion therefore are able to agglutinate suitable erythrocytes. Because the polypeptide composition of these particles differs (cf. Johnson, Chapter 7 of this volume), it could be assumed that the ability to hemagglutinate might also vary. In fact, it has been reported that the "heavy" (1.44–1.46 g/ml) variant of the infectious virion of several parvoviruses has a lower specific hemagglutinating activity than the empty capsid or the "light" mature virion (Clinton and Hayashi, 1976; Tattersall, 1978a; Kongsvik *et al.*, 1978). Interpretation of such results, however, depends on parameters like the purity of virus suspensions, the absence of virus aggregates, and the concentration of particles, none of which are easily controlled. The available results therefore are still far from being conclusive.

The parvovirus hemagglutinin is as resistant as the infectious virion. For various parvoviruses it proved to be stable from pH 2 through pH 11 when the particles were incubated at 4, 25, and 37°C (Green, 1965; Siegl *et al.*, 1971). With increasing temperature, however, a gradual fall of HA titers of the treated virus suspensions could be observed. This became particularly evident at extreme pH values. At pH 5 to 7, on the other hand, the active hemagglutinin could still be demonstrated after incubation at 80°C for 30 min to 2 hr. Treatment with organic solvents such as ether and chloroform, or with various alcohols also does not affect the hemagglutinin (for reference see Siegl, 1976). Similarly, proteolytic enzymes (e.g., trypsin, chymotrypsin, and papain) proved without destructive effect (Green, 1964; Kongsvik and Toolan, 1972b; Siegl *et al.*, 1971; Siegl, 1976). On the contrary, a brief treatment of H-1 virus with papain (Kongsvik and Toolan, 1972) or with ficin (Kongsvik *et al.*, 1974) even enhanced hemagglutination. These results were obtained with unpurified virus suspension of undefined particle composition. Therefore, they cannot be related to the results of the more recent biochemical studies which show that part of the polypeptides contained in the two buoyant density variants of the infectious parvovirus particle are in fact cleaved by both trypsin and chymotrypsin *in vitro* (Kongsvik, 1972b; Clinton and Hayashi, 1976; Tattersall 1978a; Kongsvik *et al.*, 1978).

The effect of formaldehyde, glutaraldehyde, and β-propiolactone on the hemagglutinins of parvoviruses (Kongsvik *et al.*, 1974; Scott, 1980; Siegl, unpublished) mostly depend on the temperature of reaction. Under carefully controlled conditions none of these agents demages the hemagglutination or the antigenicity of the particles even though infectivity is readily abolished.

The counterpart of the viral hemagglutinin—the receptor at the surface of the agglutinated red blood cell—has been assumed to be a glycoplipid structure containing *N*-acetylneuraminic acid linked with carbohydrates and fatty acids (Cocuzza and Russo, 1969). This view is supported by the findings that pretreatment of erythrocytes with influenza virus or Newcastle disease virus, with receptor destroying enzyme (RDE) of bacterial origin, with lipid solvents, or with periodate rendered them nonagglutinable for parvoviruses (Moore, 1962a; Payne *et al.*, 1964; Cocuzza and Russo, 1969; Hallauer *et al.*, 1972). Formaldehyde, glutaraldehyde, and trypsin, on the other hand, proved to be without effect in these studies. Somewhat contradictory results were reported more recently by Mochizuki *et al.* (1978). These authors observed that the receptor for FPV on porcine erythrocytes was almost completely inactivated by trypsin, KIO_4, α-amylase, and, to some extent, also by formaldehyde. In consequence, the receptor structure was supposed by these authors to have the characteristics of a glycoprotein.

As far as the reaction between hemagglutinin and receptors is concerned, the reversible sensitivity of H-1 hemagglutination to *p*-chloromercurybenzoate suggests that virus-bound sulfydryl groups could play an important role (Castro *et al.*, 1971a,b; Kongsvik and Toolan, 1972b). The most critical parameters in the reaction of parvoviruses with suitable red blood cells, however, are temperature and pH. In some instances skillful control of these parameters even can be used to distinguish between antigenically closely related viruses. For example, feline panleukopenia virus (FPLV) preferentially agglutinates pig erythrocyes at pH 6.5, whereas CPV, the canine variant of FPV, shows considerable hemagglutination activity at pH 7.2 (Carmichael *et al.*, 1980; Tratschin *et al.*, 1982). Likewise, FPLV and the mink variant MEV agglutinate pig erythrocytes almost exclusively at 4°C. CPV again reacts effectively with the same type of cells also at room temperature (Tratschin *et al.*, 1982). A reaction temperature around +4°C was also found indispensable for hemagglutination by the minute virus of canines (MVC) (Binn *et al.*, 1970) as well as for the lapine parvovirus (Matsunaga *et al.*, 1977). For all other hemagglutinating parvoviruses the reaction apparently is completed within 2–4 hr

TABLE II. The Most Appropriate Conditions for the Detection and Quantitation of Parvoviruses by Hemagglutination

Virus	Erythrocytes: species of origin	pH	Temperature (°C)
RV	Guinea pig and human O	7.2[a]	20[b]
RTV	Guinea pig and human O	7.2	20
H-1	Guinea pig and human O	7.2	20
HB	Guinea pig	7.2	20
MVM	Guinea pig	7.2	20
FPV:			
FPLV	Pig	6.5	4
MEV	Pig	6.5–7.0	4
CPV	Pig and rhesus monkey	7.2	4–20
MVC	Rhesus monkey	7.2	4
BPV	Guinea pig and human O	7.2	20
PPV	Guinea pig and human O	7.2	20
LPV	Guinea pig	7.2	4
TVX	Guinea pig and human O	7.2	20
LuIII	Guinea pig and human O	7.2	20
GPV	No hemagglutination demonstrable		
ADV	No hemagglutination demonstrable		

[a] Buffers in the range of between pH 6.8–7.4 were used by various authors. According to personal experiences pH 7.2 is a good compromise.
[b] In general, reaction was allowed to proceed at "room temperature." In some instances higher titers may be obtained during incubation at 4°C.

at room temperature. Nevertheless, slightly higher titers are frequently recorded after incubation in the refrigerator overnight (Mayr *et al.*, 1968; Bachmann, 1971; Bates *et al.*, 1972; Mengeling, 1972; Inaba *et al.*, 1973a,b).

Table II summarizes the most convenient hemagglutination conditions for the various parvoviruses. It is evident that the great majority of these viruses agglutinates guinea pig erythrocytes at room temperature. Rhesus monkey or African green monkey red blood cells, on the other hand, are suitable for MVC and CPV, whereas FPLV and MEV are easily detected with pig erythrocytes. All the latter viruses hemagglutinate preferentially at 4°C. They elute from the red blood cells as soon as the temperature is raised to 20 or 37°C; yet, the erythrocytes remain agglutinable and the reaction is reversible upon reincubation at refrigerator temperature.

For many of the viruses listed in Table II erythrocytes of as many as 19 different animal species have been tested for suitability in the heagglutination reaction (Toolan, 1976; Siegl, 1976; Porter and Cho, 1980; Mengeling, 1981; Siegl, Chapter 9 of this volume). In some of these investigations it was also proposed that particular isolates of antigenically closely related viruses can be distinguished on the basis of their hemagglutination pattern. It was realized, however, that reliable results can only be obtained under strictly controlled test conditions. In this respect, minimal prerequisites are standardization of virus suspensions and the use of a single batch of erythrocytes in all comparative experiments (Toolan, 1967b; Hallauer *et al.*, 1972). The importance of the latter parameter is sufficiently illustrated by the frequent observations in routine diagnostic laboratories that hemagglutination titers of one and the same virus suspension may vary with the donor animal of the erythrocytes (e.g., Carmichael *et al.*, 1980). In the case of PPV it was found that the extent to which this virus reacts with chicken erythrocytes depends on both the genetic composition (Cartwright *et al.*, 1969; Morimoto *et al.*, 1972; Pini, 1975; Ruckerbauer *et al.*, 1978) and the age of the donor animal.

Closely related to the ability of parvoviruses to agglutinate red blood cells of various species is the phenomenon of hemadsorption. The respective results, however, are rather inconsistent (Kilham and Olivier, 1959; Portella, 1964; Cartwright *et al.*, 1969; Hallauer *et al.*, 1972; Hinaidy *et al.*, 1979). Portella (1964) reported that in tissue cultures infected with RV, H-1, or H-3 virus only those cells proved to be infected to which guinea pig erythrocytes attached. According to electron micrographs the attachment of the erythrocytes to the cellular membranes was directly mediated by virus particles. Portella therefore recommended hemadsorption as a suitable and rapid means to detect virus infected cells. In contrast, rather slight hemadsorption was observed with BPV in cell cultures exhibiting low degree CPE (Hinaidy *et al.*, 1979) and binding of guinea pig red blood cells to cell cultures infected with KBSH virus took place only after the cell associated virus had reached rather high HA titers (Hallauer *et al.*, 1972). These discrepancies can be explained on the basis of the replication behavior of parvoviruses. Whereas enveloped hemagglutinating viruses mature by budding from the cell membrane, parvoviruses are replicated within the nucleus of an infected cell. The great majority of progeny virions are released upon the final disintegration of the host cell only. They then associate with unoccupied receptor structures both at the surface of intact cells and of cell debris and thus provide the substrate for the adsorption of erythrocytes.

The binding of progeny parvovirus particles to the cell detritus is a well-known phenomenon. During purification of the viruses this fact is allowed for by the use of receptor destroying enzyme (Salzman and Jori, 1970; Siegl *et al.*, 1972; Tattersall *et al.*, 1976) or treatment with buffer solutions of pH 8.0 to 9.0 (Hallauer and Kronauer, 1965; Gierthy *et al.*, 1974) to dissociate the virus–receptor complex.

B. Hemagglutination Inhibition

Hemagglutination inhibition (HI) is the most convenient means of identifying the great majority of autonomous parvoviruses. This serological technique has been extensively used to investigate the antigenic relationship between individual viruses and to screen sera for the presence of antiviral antibodies as diagnostic indicator of a past infection. In both cases, however, the respective results can only be relied upon if hemagglutination conditions are sufficiently standardized (Hallauer *et al.*, 1972; Joo *et al.*, 1976b; Carmichael *et al.*, 1980). Moreover, HI-tests are extremely sensitive to the presence of nonspecific inhibitors. These substances have been demonstrated in sera of humans and of almost all animal species so far tested (Moore, 1962b; Lum, 1970a,c; Hallauer *et al.*, 1972; Joo *et al.*, 1976b; Siegl and Kronauer, 1980). They may be present in extraordinarily high concentration and sometimes were still found in serum dilutions of 1:640! The nature of these inhibitors is largely unresolved. They resist heating to 56°C for 30 min. Pretreatment of sera with RDE, trypsin, sodium periodate, ether, or $MnCl_2$ was found to yield inconsistent results. The most reliable technique of removing the substances from sera consists in extraction with Kaolin (Clarke and Casals, 1958). Even under these conditions, however, HI-titers of up to 1:80 can still be due to the presence of "inhibitory substances" rather than reflecting a low antibody titer. A second parallel extraction of the serum sample under investigation or more subtle techniques such as column chromatography and velocity sedimentation then can be used to reach an unambiguous decision.

To the best of our knowledge, the serum inhibitors dealt with so far are nonspecific, i.e., they indiscriminately inhibit hemagglutination by various parvoviruses. This is apparently not true for that inhibitor of H-1 hemagglutination which has been isolated from human placentas (Toolan 1964, 1968). The substance was purified and characterized by Usatequi-Gomez (1965), Karasaki *et al.* (1965), and Usatequi-Gomez and Morgan (1968) as a glycoprotein with many properties of a macroglobulin. It interfered specifically with hemagglutination by H-1 virus. Hemagglutination by the antigenically closely related HT virus was only occasionally inhibited and hemagglutination by the serologically unrelated HB, RV, and H-3 viruses was not affected at all.

On the basis of HI-tests, the hemagglutinating autonomous parvoviruses can be separated into 12 distinct groups (Table III). These groups apparently share no HI-antigen in common. A one-sided serologic relationship between individual groups (e.g., between TVX and H-1) has been occasionally reported (cf. Lum and Schreiner, 1963; Hallauer *et al.*, 1971). It cannot be excluded, however that such observations reflect a contamination of the antigens used to raise the antisera rather than a true antigenic relationship between the viruses concerned. Contamination of antigens thereby may have resulted from the use of primary cell cultures prepared from the tissues of latently infected animals (Mayr *et al.*, 1968; O'Reilly and Whitaker, 1969) or from cell cultures contaminated in the laboratory (Wozniak and Hetrick, 1969; Hallauer *et al.*, 1971). It is therefore also conceivable that the "new" serotype of BPV described so far only by Inaba *et al.* (1973a) and compared exclusively to BPV reference strains represents an already well-known parvovirus (PPV?).

TABLE III. Serologic Relationship between Parvoviruses as Revealed by Hemagglutination Inhibition[a]

Antisera to		Parvoviruses																	
		RV	H-3	X-14	RTV	H-1	HT	HB	MVM	FPLV	MEV	CPV	MVC	BPV	PPV	KBSH	LuIII	TVX	LPV
RV		160	80	80	–[b]	–	–	–	–	–	–	–	–	–	–	–	–	–	–
	H-3	640	1280	1280	–	–	–	–	–	–	–	–		–	–	–	–	80	–
	X-14	80	160	160	–	–	–	–	–	–	–		–	–	–	–	–	–	–
RTV		–	–	–	2560	–			–	–	–		–	–	–	–	80	–	–
H-1		–	–	–	–	20480	+	–	–	–	–	–	–	–	–	–	–	320	–
	HT	–	–	–	–	+	+	–											
HB		–	–	–	–	–	–	+	–	–	–				–	–	–	–	
MVM		–	–	–	–	–			10240	–	–	–	–		–	–	–	–	–
FPV:	FPLV	–	–	–	–	–			–	640	640	640	–		–	–	–	–	
	MEV	–	–	–	–	–			–	320	160	160	–		–	–	–	–	–
	CPV	–	–	–	–	–			–	1280	640	2560	–		–	–	–	–	–
MVC		–	–	–	–	–			–	–	–	–	2560		–	–	–	–	
BPV		–	–	–	–	–			–	–	–	–		256	–	–	–	–	
PPV		–	–	–	–	–			–	–	–	–	–	–	1280	320	–	–	–
	KBSH	–	–	–	–	–			–	–	–				1280	1280	–	–	–
LuIII		–	–	–	–	–			–	–	–	–	–		–	–	2560	–	–
TVX		–	–	–	–	–			–	–	–	–	–		–	–	–	1280	–
LPV		–	–	–	–	–			–	–	–	–			–	–	–	–	640

[a] This table is based on Table 9 in Siegl, G. The Parvoviruses, in: *Virology monographs* (S. Gard and C. Hallauer, eds.), Vol. 15, 1976; it is reproduced and extended with permission of Springer-Verlag, Berlin.

[b] ≦20.

In general, virus strains belonging to a certain antigenic group cannot be distinguished by HI. The validity of this observation was also claimed for the host-range variants of FPV, i.e., for FPLV, MEV, and CPV (Carmichael *et al.*, 1980; Lenghaus and Studdert, 1980). Following preadsorption of anti-CPV serum with FPLV antigen, however, Flower *et al.*, (1980) have provided evidence for the existence of a specific extra antigenic determinant in CPV, Tratschin *et al.* (1982) also could show that, in agreement with the results of neutralization tests, antisera prepared against FPLV or MEV do not distinguish between the three viruses in HI-tests. Antisera raised against CPV, on the other hand, react with the homologous virus at considerabley (8 to 64 times) higher dilutions than the two heterologous agents.

To the extent tested, antisera against the two non hemagglutinating autonomous parvoviruses ADV and CPV did not interfere with hemagglutination by some of the viruses listed in Table III.

C. Serum Neutralization

As far as the serologic relationship between the viruses grouped in Table III is concerned, cross neutralization brought about no new points of view. No antigen in common to the various viruses could be revealed by such tests (Moore, 1962b; Portella, 1963; Lum and Schreiner, 1963; Toolan, 1964; Crawford, 1966; Gorham *et al.*, 1966; Johnson, 1967b; Johnson *et al.*, 1967; Scott *et al.*, 1970b; Cross and Parker, 1972; Storz *et al.*, 1972; Inaba *et al.*, 1973b; Joo *et al.*, 1975; Siegl, 1976; Tratschin *et al.*, 1982). Slight antigenic differences between individual isolates of one and the same virus, however, apparently are more readily detected than by means of HI. This has been observed in studies within RV, H-3, and L-S virus (Moore, 1962b; Portella, 1963; Lum and Schreiner, 1963) where immune sera showed considerably higher titers for the homologous than for the heterologous virus. Similarly, Lenghaus and Studdert (1980) reported that, by serum neutralization, CPV proved closely related but clearly distinguishable from wild-type and vaccine strains of FPLV. The results of Tratschin *et al.* (1982) are in agreement with these findings.

Of the two nonhemagglutinating parvoviruses GPV and ADV the goose parvovirus was shown by neutralization tests to be unrelated to MVM, PPV, BPV, and H-1 virus (J. Kisary, personal communication). Moreover, cross neutralization with the French isolate, the isolate from Hungary, from the Netherlands, and one from the USSR suggested that probably all GPV strains isolated in Europe are antigenically identical (Dannacher *et al.*, 1974; Kisary, 1974). Neutralization tests with ADV, on the other hand, present a special and so far unresolved problem. As will be discussed elsewhere in more detail, ADV infection of mink results in the formation of large amounts of circulating antiviral antibodies. These antibodies could be shown by electron microscopy to react with

ADV. The resulting virus–antibody complexes, however, remain infectious for both susceptible animals and sensitive cell cultures. Such results were obtained with sera from chronically infected mink, with sera of mink with a nonpersistent ADV infection, as well as with sera from ferrets with ADV antibody and no disease (Larsen and Porter, 1975; Porter and Cho, 1980).

Neutralization tests have been reported to be more sensitive than HI tests in the detection of antiviral antibody (Johnson and Collings, 1971; Joo *et al.*, 1975; Siegl and Kronauer, 1980). This is especially true for plague-reduction tests where sera nonreactive in HI can frequently be diluted several-hundred-fold before end titers in neutralization are reached (Siegl and Kronauer, 1980). On the basis of assay systems which detect viral infectivity via the induction of cytopathologic changes in the infected monolayer or by screening for intranuclear inclusion bodies, however, neutralizing serum antibodies frequently are found to be of the same magnitude as antibody titers read by HI (Goto, 1975; Konishi *et al.*, 1975; Mochizuki *et al.*, 1978).

D. Immunofluorescent Studies

Immunofluorescent staining has successfully been used to detect parvovirus antigens in tissues of infected animals (cf. Csiza *et al.*, 1971b; Storz and Bates, 1973; Mengeling *et al.*, 1975; Carmichael and Binn, 1981) to study the replication of parvoviruses in cell cultures (Mayor and Ito, 1968; Cole and Nathanson, 1969; Siegl and Gautschi, 1973; Bachmann and Danner, 1976; Hahn *et al.*, 1977a,b), and to follow synthesis and accumulation of viral polypeptides in infected cells (Croft *et al.*, 1974). IF-staining is very helpful in determining infectivity titers of parvoviruses. It is usually much easier and much less time consuming to detect single infected cells by this technique than by screening cultures infected with end point dilutions for, e.g., FPV-specific intranuclear inclusion bodies (King and Croghan, 1965; Mochizuki *et al.*, 1978; Siegl and Kronauer, 1980). In consequence, much higher virus titers are usually recorded and the apparent infectious unit consists of less virus particles. This, in return, affects the results of neutralization tests (Scott *et al.*, 1970b). Moreover, under conditions where virus infections neither lead to cytopathologic alterations in the unstained monolayer nor to formation of inclusion bodies, IF-staining is the only reasonable means to detect viral replication. At the present time, this does apply to the propagation of ADV *in vitro*. Infectivity of the latter virus is titrated by a fluorescent focus assay (Hahn *et al.*, 1977a,b; Porter and Cho, 1980).

IF-staining has also been used to investigate the antigenic relationship of several parvoviruses. Cross and Parker (1972) infected rat embryo cell cultures with RV, H-1, and MVM and, by cross-IF staining with the respective antisera, found these viruses to share a common antigen (s).

Similarly, Black *et al.* (1979) and Eugster (1980) reported on an antigen in common to CPV and PPV which can be revealed by fluorescent antibody staining. Most authors, however, agree that the latter two viruses are antigenically unrelated (for reference see Chapter 9). In our laboratory repeated attempts to reveal a common antigen for PPV and CPV by indirect IF-staining have failed (G. Siegl and G. Kronauer, unpublished).

On the other hand, fluorescent antibody staining indicated that GPV is serologically unrelated to RV, H-1, MVM, RTV, PPV, BPV, FPV, TVX, and LuIII virus (P.A. Bachmann, personal communication).

E. Complement Fixation and Further Serologic Techniques

Complement fixation (CF) has been repeatedly used to identify and to quantitate parvovirus antigens (Kääriäinen *et al.*, 1966; Cross and Parker, 1972, Schettler, 1973; Ruckerbauer *et al.*, 1978; Mochizuki *et al.*, 1978). In such studies it was found that CF reactivity is associated with the well-known types of virus particles detected also by infectivity assay and/or hemagglutination. The time course the appearance of CF-antibody subsequent to the contact with parvovirus antigens has only been studied in rabbits infected with FPLV. In this animal the CF antibody response paralleled the behavior of HI and SN titers but was distinctly lower. Finally, in agreement with HI and SN tests, attempts to detect a common CF-antigen for some of the viruses listed in Table III failed (Cross and Parker, 1972). It proved also impossible to discriminate between individual strains of FPLV by this technique (Mochizuki *et al.*, 1978).

Of further serologic techniques applied to the study of parvoviruses, counter-immune electrophoresis (CIEP) was found to be the most useful approach to detect and to quantitate ADV antigen in the serum of infected mink (Cho and Ingram, 1972, 1973, 1974; Bloom *et al.*, 1975; Notani *et al.*, 1976; Porter and Cho, 1980). Immunodiffusion finally was used by Joo *et al.* (1978) in the study of PPV antigenicity, and, very recently, Flower *et al.* (1980) have applied the latter technique to reveal an extra antigenic determinant in CPV by means of which it can be distinguished from FPLV and MEV.

III. PROPAGATION AND ASSAY SYSTEMS

A. Dependence of Viral Replication on Cell Physiologic State

Although the viruses dealt with in this review are referred to as nondefective or autonomous parvoviruses, they depend for replication on cellular helper functions. These functions are expressed only transiently in late S- or early G-2 phase during the life cycle of a susceptible cell (Tennant *et al.*, 1969; Hampton, 1970; Tattersall, 1972; Rhode, 1973;

Siegl and Gautschi, 1973). Our present state of knowledge concerning the characteristics of the cellular functions as well as the steps in virus replication governed by them will not be discussed in this context. Here I shall only summarize some basic *in vitro* observations and shall place emphasis on the implications of this specific virus–cell relationship for the propagation and assay of parvoviruses.

In 1965 Margolis and Kilham related the pronounced destructive action of viruses like RV and H-1 in rapidly proliferating tissues of newborn rodents to their affinity for cells undergoing mitosis. At about the same time Johnson (1965, 1967a) attempted *in vitro* propagation of feline panleukopenia virus. He then listed a number of critical parameters either stimulating or hampering replication of the virus in primary feline kidney cells. All those parameters which affected mitotic activity within the cell cultures (e.g., density of the cell monolayer, the quality of sera used to supplement growth medium, or the temperature of incubation) apparently favored virus replication whenever they increased the mitotic activity. Since these early observations, comparable cultural prerequisites have been established for *in vitro* propagation of every autonomous parvovirus studied in greater detail, i.e., for RV, H-1, FPV, PPV, BPV, ADV, GPV, LPV, RTV, TVX, and LuIII virus (Cole and Nathanson, 1969; Ledinko *et al.*, 1969; Tennant *et al.*, 1969; Hallauer *et al.*, 1972; Tattersall, 1972; Storz and Bates, 1973; Lubeck and Johnson, 1976; Hahn *et al.*, 1977a,b; O'Shea and Studdert, 1978; Kisary, 1979; Carmichael and Binn, 1981; Johnson and Spradbrow, 1979). In particular it was found essential to either infect cultures "not yet completely confluent" or to add the virus already at the time of seeding the cells. At rather low multiplicity of infection, for example, in cultures inoculated with end point dilutions of virus suspensions, even this technique may fail to stimulate viral replication to a degree inducing CPE before the monolayer reaches confluence. In such cultures, in which mitotic activity has slowed down due to density-dependent inhibition of cellular growth, viral replication can be activated again by trypsinization and splitting of the cells at a reasonable rate (Hallauer *et al.*, 1972; Johnson *et al.*, 1974).

In susceptible culture systems the ability of the cells to proliferate and to pass actively through the division cycle thus is the predominant prerequisite of successful virus replication. If this is taken into account, some of the observations of Cole and Nathanson (1969) and Ledinko *et al.* (1969) concerning the replication of RV and H-1 virus in L-cells and in the diploid cell line Wi-26, respectively, can probably be explained. Judged on the production of infectious virus particles, replication of these viruses in the cell systems originally appeared to be incomplete or even abortive. Fluorescent antibody staining, however, revealed that viral replication proceeded successfully in a small proportion of cells apparently undergoing active growth.

Finally, it has to be mentioned that a cell's ability to support replication of a parvovirus—in other words, its susceptibility or resistance to infection—seems to be a function of its metabolic capacity as expressed under the control of differentiation and development. This conclusion can be drawn from *in vitro* experiments conducted by Tattersall (1978b). Using murine teratoma cells he could show that, within the pathway these cells can differentiate, only one type of cell evolved which supported replication of MVM.

B. Host Cell Range and Cultivation

It is obvious that the dependence of parvovirus replication on helper functions expressed under the control of cellular differentiation has a direct influence on the *in vitro* host range of the viruses. This host range therefore is both of basic theoretical and of practical laboratory significance. In most of the studies available to date, however, mere practical points of view prevailed and investigations were mainly centered at the question whether a certain parvovirus can be propagated to high titers in a readily accessible and easily maintainable culture system. The respective published laboratory experience is condensed to Table IV. To the best of our knowledge, this table lists the most appropriate culture systems for the various viruses.

As is evident, most of the viruses can be propagated in primary and secondary cultures derived from tissues of embryos or neonates of their natural animal host. Moreover, permanent cell lines originating from such sources in general proved to be very useful. For some of the parvoviruses, e.g., RTV, MVM, BPV, MVC, LPV, and GPV, primary and permanent cell cultures derived from tissues of their natural host apparently are the only susceptible ones (Crawford, 1966; Binn *et al.*, 1970; Parker *et al.*, 1970a; Hallauer *et al.*, 1971, 1972; Storz and Bates, 1973; Kisary, 1974; Matsunaga *et al.*, 1977). These culture systems frequently also have to be used to achieve isolation of RV, PPV, and FPLV under routine diagnostic conditions (Kilham and Olivier, 1959; Lum and Schreiner, 1963; Payne *et al.*, 1964; Mayr *et al.*, 1968; Johnson, 1964). Of these viruses, however, variants are known which can be readily propagated in cells of widely unrelated hosts. Thus, H-3 virus—antigenically closely related to RV—can be grown in cell lines of human origin (Toolan and Ledinko, 1965; Hallauer *et al.*, 1972) and related viruses were recovered from a cloned line of Detroit-6 cells (neoplastic human bone marrow) (Melnik *et al.*, 1971) as well as from human embryonic lung (HEL) cells (Berquist *et al.*, 1972). Likewise, Hallauer *et al.* (1971, 1972) found the great majority of parvoviruses isolated as contaminants of permanent human cell lines (e.g., KB, HeLa, HEp-2, and FL-Amnion) to be serologically indistinguishable from PPV. Attempts to isolate PPV from tissues of infected pigs in cells of human, monkey, and rat origin originally had failed (Mayr *et al.*, 1968), but Cartwright *et al.* (1969) succeeded in adapting PPV to growth in a variety of human cell lines. Finally, the determination of the

TABLE IV. The Most Convenient *in Vitro* Systems for the Propagation of Parvoviruses[a]

Virus: Strain		Primary cultures	Permanent cell lines
RV		Rat embryonic	Rat nephroma (Babcock and Southam)
	H-3	Rat embryonic	NB (human transformed, Shein & Enders)
	X-14	Rat embryonic	Rat nephroma
	L-S	Rat embryonic	Rat nephroma
	HER	Rat embryonic	Rat nephroma
RT		n.t.[b]	AT (rat; Tournier)
H-1		Rat embryonic	NB (human transformed, Shein & Enders)
MVM		Mouse and rat embryonic	A-9 (mouse L-cells)
FPV:	FPLV	Feline kidney	CRFK (feline kidney)
	MEV	Feline kidney	CRFK
	CPV	Feline and canine kidney	CRFK; MDCK (canine kidney), A 72 (canine neoplastic)
PPV		Fetal or neonatal porcine kidney and testicle	PK-14 (porcine kidney) SK (swine testicle)
	KBSH	?	HeLa, KBS (human neoplastic)
BPV		Bovine fetal lung, spleen and kidney	Buffalo lung fibroblast
MVC		?	WRCC (canine)
ADV		Mink, ferret and feline kidney	CRFK (feline kidney)
LPV		Rabbit kidney	RK (rabbit kidney)
GPV		Goose embryo, duck embryo	?
TVX		?	HeLa, KB (human neoplastic)
LuIII		?	HeLa, NB (human transformed, Shein & Enders) Chimpanzee liver (Douglas)

[a] For references the reader is referred to the text as well as to the specific reviews (Toolan, 1968; Siegl, 1976; Porter and Cho, 1980; Mengeling, 1981; Siegl, Chapter 9).
[b] Not tested.

in vitro host cell range is one of the very few possibilities to distinguish CPV from FPLV and MEV. The latter two variants of the feline parvovirus have a rather limited host range and are routinely propagated in feline cells only. CPV, on the other hand and well in accordance with *in vivo* observations, was found to replicate exceptionally well in primary and permanent cultures of both feline and canine origin (Johnson and Spradbrow, 1979; Eugster, 1980; Lenghaus and Studdert, 1980; Carmichael *et al.*, 1980; Tratschin *et al.*, 1982).

Serological surveys and epidemiological observations (cf. Mengeling, 1981) clearly indicate that the natural host of PPV is the pig. Viruses like KBSH and comparable isolates (Hallauer *et al.*, 1971, 1972) therefore may be regarded as host cell variants selected by cultivation in the laboratory. Similar considerations were drawn into account to explain the generation of CPV from FPV (see Chapter 9). In this special case the new host-range variant also established itself as pathogen in a so far unaffected species. In contrast to these two examples, however, the situation with H-1 virus is not conclusive. This virus was found to replicate readily in human cells *in vivo* and *in vitro* (Toolan *et al.*, 1965; Toolan and Ledinko, 1965; Toolan, 1968); yet, primary cultures of rat embryo cells support virus replication to the same extent (Moore, 1962a; Nicoletti *et al.*, 1969) and there is evidence for a widespread dissemination of the agent in rat populations (Kilham and Margolis, 1969). Neither the *in vivo* nor the *in vitro* host range of H-1 therefore provides helpful hints in the search for its natural host.

Similar difficulties are encountered with two of the parvoviruses isolated by Hallauer *et al.* (1971) from contaminated human cell lines. To date, LuIII virus could only be propagated in permanent cell lines of human and chimpanzee origin (Hallauer *et al.*, 1972; Soike *et al.*, 1976; Siegl and Gautschi, 1978) but antibodies to the virus could never be detected in sera of these species. TVX also was shown to replicate readily in various human cell lines. As with LuIII virus, no antibodies could be detected in human sera. Screening of rat sera, however, revealed HI-antibody titers as high as 1:1280 in 16 out of 50 sera (G. Siegl and G. Kronauer, unpublished). Whether the virus can replicate in rat cells *in vitro* still awaits testing.

There is sound reason to regard the viruses dealt with in the preceding paragraphs as easily cultivable parvoviruses. For each of the agents several susceptible culture systems have been described and, given incubation conditions stimulating cellular proliferation and mitosis, all viruses can easily be propagated. In contrast, at least two of the better characterized parvoviruses as well as a series of possible members of the autonomous parvovirus group are rather delicate with respect to cultivation. MVC, the canine virus isolated by Binn *et al.*, in 1970, could only be propagated in the WRCC (Walter Reed canine cell) line. All parallel attempts to adapt the virus to growth in primary cells of dog, cat, mouse, rat, hamster, guinea pig, rabbit, pig, bovine, monkey, and human origin failed. Permanent lines of cells derived from the same species also proved insusceptible. Since that time many laboratories have tried to propagate the original isolate of MVC in canine cells or to recover the virus again in culture from fecal canine specimens. According to various personal communications all these attempts, including those undertaken in our own laboratory, were in vain. Impressed by these negative results R. H. Johnson (personal communication) assumes that MVC does not belong to the autonomous but rather, is a member of the helper-dependent parvoviruses. On the basis of such a hypothesis, propagation in the WRCC line could then have been possible due to a preexisting latent infection with, or the unrecognized presence in the inoculum of an adenovirus helper. The necessary critical experiment to test for this possibility consists in the inoculation of MVC into canine cells together with an adequate helper virus. This has not yet been attempted. It is also quite questionable whether the results of such an experiment would be conclusive. Brailovsky and Chany (1965), Chany and Brailovsky (1965), as well as Ledinko and Toolan (1968) have shown that, comparable to the situation with the dependoviruses, multiplication of autonomous parvoviruses like RV and H-1 in inappropriate culture systems can be stimulated by coinfection with an adenovirus helper. The necessary critical helper function was found to be expressed late in adenovirus replication (Ledinko *et al.*, 1969). Under its influence H-1 virus replicated successfully both in human embryonic lung cells as well as in cells of the diploid human cell line Wi-26 which otherwise did not support progressive virus growth.

In spite of the difficulties in interpretation apparently associated with the experiments suggested by R.H. Johnson, the idea might be worthwhile trying with the various noncultivable, parvoviruslike particles detected repeatedly during the past decade in human feces and sera (Cossart *et al.*, 1975; Paver *et al.*, 1973, 1974, 1975a,b; Paver and Clarke, 1976; Pattison *et al.*, 1981; Serjeant *et al.*, 1981). Samples containing these viruses are in general very rare and even a low degree of multiplication of the viruses—whether completely helper virus dependent or not—would be helpful in determining their characteristics and in facilitating the search for a susceptible culture system.

Besides the "priming" of an otherwise unsuitable cell culture system by coinfection with a helper adenovirus, an additional factor stimulating or inducing replication of poorly or even noncultivable parvoviruses could consist in a proper selection of the temperature of incubation. This procedure proved to be rewarding in studies with ADV. In spite of the fact that the body temperature of mink is 39°C, frequent attempts to propagate ADV at 37°C failed or could not be reproduced (for reference see Hahn *et al.*, 1977a). In 1977, however, Porter *et al.* succeeded in cultivating the virus in primary or permanent feline kidney cells at 31.8°C. After 10 passages under these conditions of incubation, the temperature optimum shifted to 37°C. In parallel, Hahn and co-workers (1977a) could adapt the highly virulent Utah-1 strain of ADV directly to growth in feline kidney cells at 37°C whereas virus strains of low virulence underwent abortive replication under the same conditions of cultivation. Porter and Cho (1980) recently reported that such less virulent ADV strains are even difficult to propagate at 31.8°C; yet, they could be serially passaged when the temperature of incubation was further reduced to 28°C. In all the *in vitro* experiments conduced so far with ADV, the CRFK permanent line of feline kidney cells was found to be the best system for the propagation of the virus (Hahn *et al.*, 1977a; Porter and Cho, 1980; Bloom *et al.*, 1980). It is noteworthy, however, that primary mink and ferret cells were also quite susceptible. Finally, expression of ADV antigen has been observed by means of fluorescent antibody staining in guinea pig peritoneal macrophages and even in NB (simian virus 40-transformed newborn human kidney) cells (Hahn *et al.*, 1977a). To what extent, if at all, accumulation of this antigen is indicative of synthesis of progeny infectious virus has not yet been tested.

There are two additional points which have to be mentioned in context with the propagation of parvoviruses *in vitro*. The first concerns the transiency of virus replication as observed in several culture systems. It is closely related to the dependence of virus synthesis on actively growing cells. As lined out already under Section III.A and as is referred to again in the following chapter, virus replication is halted whenever cellular proliferation is arrested. In culture systems prone to reach a state of density-dependent inhibition of cellular growth, virus synthesis then might become completely undetectable. In general, this phenomenon is less evident during propagation of the viruses in permanent line cells than in primary or secondary cell cultures.

The final facet of parvovirus replication which has to be referred to concerns the strict cell association frequently observed with parvoviruses. Strict cell association, however, is not a stable characteristic of all parvovirus–cell systems. It can vary considerably both with the virus strain and the cell type used. Therefore, no generally valid rule can be established; rather, both the supernatant culture medium and the cellular extracts have to be assayed for presence of progeny virus at regular intervals to obtain reliable information on the distribution of the virus in the culture system of interest. For that purpose, cellular extracts can either be prepared by treatment of monolayers with alkaline glycine buffer according to Hallauer and Kronauer (1965) or by repeated freezing and thawing of cells followed by digestion with RDE or treatment with Freon.

C. Cytopathology

With only the exception of ADV, replication of the autonomous parvoviruses in one of the *in vitro* culture systems listed in Table IV is accompanied by cytopathologic changes. These virus induced cellular changes can be observed both in the unstained monolayer and after histologic staining. In the unstained monolayer they are in general characterized by diffuse granulation, rounding up of cells, detachment of single or groups of pyknotic cells and, finally, by complete destruction of the culture (Moore, 1962a; Toolan and Ledinko, 1965; Johnson, 1967a; Mayr *et al.*, 1968; Parker *et al.*, 1970a Binn *et al.*, 1970; Hallauer *et al.*, 1972; Storz and Bates, 1973; Matsunaga *et al.*, 1977). Complete CPE, however, will only develop if a mitotically active cell sheet is inoculated at a sufficiently high multiplicity of infection. If this requirement is not met, i.e., if the infected monolayer reaches a state of density-dependent inhibition of cellular growth before infection has spread to the greater majority of a culture's cells, the cytopathologic changes are transient in nature (Johnson, 1965, 1967a; Lubeck and Johnson, 1967). Under the latter circumstances the virus-affected, pyknotic cells are lost from the cell sheet which then resumes an uninfected appearance and can be distinguished from control cultures only on the basis of a more or less characteristic thinning. Detection of parvovirus replication by CPE, therefore, may require a careful, sometimes lengthy observation period and repeated subculturing may be necessary to continuously provide a source of actively proliferating cells.

In many of the early studies centered at the *in vitro* replication of parvoviruses, the time necessary for the development of clear CPE was reported to vary from two to 15 days. From the more recent studies in synchronized culture systems, however, we know that the shortest possible replication cycle of an autonomous parvovirus, that is to say, the time elapsing between infection of a susceptible cell in mid S-phase and the height of maturation of progeny virus, is about 8 to 10 hr (Hampton, 1970; Rhode, 1973; Siegl and Gautschi, 1973a,b). After that period of time, phase contrast microscopy may already reveal early intranuclear changes consisting of a more or less prominent granulation. The full spectrum of cytopathology accompanying cellular death, however, will not develop before 24 to 48 hr after infection.

In contrast to the delayed morphologic reaction of the parvovirus-infected cell as observed by light microscopy, histologic staining reveals virus-induced cytopathology at a much earlier stage. The changes are detectable by H & E, May–Gruenwald–Giesma, or Feulgen staining. They consist in the generation of intranuclear Cowdry type-A inclusion bodies and have been described for H-1, as well as for RV and related isolates (Dawe *et al.*, 1961; Moore, 1962; Bernhard *et al.*, 1963; Portella, 1963; Hampton, 1964; Mayor and Jordan, 1966; Nicoletti *et al.*, 1969), for FPLV (Johnson, 1965; Goto and Ishiko, 1976; O'Shea and Studdert, 1978), PPV (Mayr *et al.*, 1968; Cartwright *et al.*, 1969), MVC (Binn *et al.*, 1970), CPV (Johnson and Spradbrow, 1979; Gagnon and Povey, 1979), BPV (Storz and Warren, 1970), GPV (Kisary and Derzsy, 1974), and for LPV (Matsunaga *et al.*, 1977). In an unsynchronized, randomly proliferating cell culture the first visible histologic changes become evident already between 8 and 12 hr after infection (O'Shea and Studdert, 1978). In general, they consist in an enlargement of the nucleoli, which is followed by rarefaction of normal chromatin around the nucleoli and the replacement of normal chromatin by granular, mostly eosinophilic material. With progressing time, continued accumulation of granular material leads to margination of the cellular chromatin, to thickening of the nuclear membrane, and, finally, to the development of a homogeneous, mostly basophilic inclusion. At this stage the nucleoli are in general dislocated to the nuclear membrane.

An additional, different type of inclusion has been described for BPV (Inaba *et al.*, 1973; Hinaidy *et al.*, 1979; Leary and Storz, 1980). This type either consists of multiple round, eosinophilic to basophilic foci spread throughout the nucleus or of similar structures arranged along the nuclear membrane. The factors predisposing the development of the multifocal or the classical Cowdry type A inclusion in a BPV infected cell are unknown. On the basis of preliminary experiments with plague purified BPV, however, Leary and Storz (1980) have suggested that the genotype of the virus may play a significant role. Yet, additional parameters such as the actual cellular phenotype or the number of virus particles entering a cell may also influence the result of infection. In this context it has to be mentioned that the multifocal type of inclusion body has not yet been observed in tissues of infected animals.

According to fluorescent antibody staining, both the classical Cowdry type A inclusion and the multifocal inclusion body consist of an accumulation of parvovirus antigen. The structure of this antigen as well as its intranuclear distribution has been repeatedly analyzed by electron microscopy (Bernhard *et al.*, 1963; Portella, 1964; Mayor and Jordan, 1966; Mayr *et al.*, 1968; Al-Lami *et al.*, 1969; Vincent, 1971; Siegl *et al.*, 1972; Bates and Storz, 1974, O'Shea and Studdert, 1978; Leary and Storz, 1980). The most thorough ultrahistological study was conducted with H-1 virus (Singer, 1975, 1976; Singer and Toolan, 1975; Singer and Rhode, 1977a,b, 1978). Singer and co-workers took into consideration the influence of the physiologic state of the infected cell as well as the genetic characteristics of the infecting virus. They also related ultrastructural observations to biochemical and immunocytochemical findings. Under the limitations of this review it is impossible to discuss the multitude of these investigators' ultrastructural observations in detail. Due to significant differences in the experimental design—parasynchronously rather than randomly growing cells were infected and analyzed—it is also difficult to relate the sequence of events recorded for H-1 to the findings with other parvoviruses. Nevertheless, a general picture of the morphologic aspects of parvovirus replication can be deduced from the available studies:

The first specific ultrastructural changes are observed as early as 8 to 12 hr after infection. In the case of H-1 virus they consist of vacuolization of nucleolar fibrillar centers and are accompanied by the appearance of "full" and "empty" virus particles within the nucleolus. Involvement of the nucleolus early in the replication of parvoviruses has also been claimed for RV (Nicolletti *et al.*, 1969) and for BPV (Bates and Storz, 1974). The subsequent, prominent accumulation of viral antigen, however, occurs outside the nucleolus. Most of this antigen in the form of assembled virus capsids is localized on nuclear euchromatin fibers. Immunocytochemical studies of H-1 infected cells, however, also pointed to the presence of significant concentrations of H-1 proteins not assembled into capsids, in condensing and marginating nucleolus-associated chromatin and in the extranucleolar heterochromatin. Whether predominantly "full" or "empty" particles finally accumulate in the nucleus seems to depend on the host cell type (Singer and Toolan, 1965). The host cell type apparently also determines whether the virus particles are randomly distributed throughout the nucleoplasm or whether cristalline arrays are formed (Al-Lami *et al.*, 1969). Experiments with temperature-sensitive mutants of H-1 virus, however, seem to indicate that the genotype of the infecting virus plays an additional role (Singer and Rhode, 1977a).

Fluorescent antibody staining of infected cells revealed that, in terminal stages of parvovirus replication, virus antigen is released from the nucleus into the cytoplasm. This observation is corroborated by ultrahistological studies where virus particles were found within the cytoplasm in the vicinity of nuclear pores at later phases of infection. Since at that time the nuclear membrane frequently shows signs of fragmentation, it is not clear whether the presence of virions in the cytoplasm is the result of an active transport process or reflects a passive release of progeny virions from a degenerating nucleus into a similarly degenerating cytoplasm.

D. Persistently Infected Cell Cultures

Latency and persistent infection are common features of the interaction of parvovirus with host cells. This applies both for the situation *in vivo* (cf. Section IV.D) and for propagation *in vitro*. In fact, many of the autonomous parvoviruses have been recovered when tissues latently infected *in vivo* were trypsinized and the cells were explanted (Kilham and Olivier, 1959; Mayr *et al.*, 1968). Others apparently were introduced by chance into permanent cell lines in the laboratory and remained undetected throughout a great number of passages (Wozniak and Hetrick, 1969; Lum, 1970a,c:, Hallauer *et al.*, 1971; Bonnard *et al.*, 1976; Mengeling, 1975; Nettleton and Rweyemamu, 1980). In both cases the infected monolayers mostly showed a quite normal aspect. The contaminating virus was produced in rather low quantity and remained highly cell associated. Presence of infection therefore was indicated by not more than a slight degree of cellular degeneration similar to that characteristic for an aging, yet uninfected monolayer. Occasionally, however, virus replication in such cultures can be activated and then leads to the production of high concentrations of progeny virus in parallel to complete destruction of the cell sheet. It is under these circumstances only that a persistent infection of cell cultures with parvoviruses is finally recognized.

We have at present no detailed insight into the factors controlling, on the one side, the development and maintenance of persistent infection and, on the other side, confering latent infection into fulminant, frank virus replication. In primary and low-passage cell cultures established from latently infected tissues a virus–cell relationship could exist which might be described as "supraoptimal steady state infection." In such cultures the equilibrium between virus replication, virus-specific cytopathology, and cell proliferation is very delicate and may be controlled by two parameters. The first of these is assumed to consist in the percentage of cells infected in the original tissue. The second then concerns the number of cell doublings possible before density-dependent inhibition of cellular proliferation in the developing monolayer ensues. These two parameters could confine infection to a limited, but nevertheless increasing number of cells and, hence, would determine the time and the number of passages after which virus replication culminates in complete destruction of the cell sheet.

The virus/cell relationship in persistently infected, long-term passaged cell strains cannot be explained in the same way. As described by Lum (1970a), Hallauer *et al.* (1971), and by Bass and Hetrick (1978b), such cultures can continuously produce low concentrations of parvoviruses over a period of months or even years. In most cases, however, a sequence of alternating "virus positive" and "virus negative" passage cycles is observed. The terms "negative" and "positive" thereby refer to the detectability of infectious or hemagglutinating virus under the prevailing laboratory conditions. They do not imply complete absence of virus-synthesizing cells. Unfortunately, cell cultures exhibiting such fluctuations in virus replication have not been examined by fluorescent antibody staining. For persistently infected cultures which continuously released RV over a period of 212 days, Bass and Hetrick (1978b), however, have shown that, at any given time, not more than 1–5% of the cells were engaged in virus synthesis.

It is not very likely that in long-term cultures the low percentage of susceptible cells is equivalent to the percentage of cells undergoing mitosis. If this were true, virus infection should sooner or later involve the whole monolayer. Similarly, it is difficult to ascribe the restriction of virus replication to a limited number of cells to the influence of antiviral factors in spite of the fact that at least RV is very sensitive to inhibition by interferon (Kilham *et al.*, 1968). Rather, it may be hypothesized that, at any given time, only a limited percentage of cells of an otherwise virus-resistent culture become susceptible to infection by derepression of certain genetic functions. Whether the result of this derepression consists in the appearance of specific receptors at the cell surface or whether distinct intracellular processes are turned on, remains to be shown. In any case, the frequency of the generation of susceptible cells at every passage level argues against mutational events playing an important role. Some observations of Hallauer *et al.* (1971) may be interpreted in favor of the above theory. These authors reported that an unrecognized latent infection can be activated if the cells are subjected to "stress" situations, such as incubation at the wrong temperature, by freeze-storing, or by addition of culture medium either lacking calf serum or having the wrong osmolarity.

At the level of the virus, persistent infection may also be the result of the presence of defective interfering (DI) particles or the generation of temperature-sensitive mutants. So far, these aspects of the parvovirus–cell relationship have not been analyzed in detail. It is however, well known that autonomous as well as the helper-dependent parvoviruses readily give rise to the formation of incomplete and DI particles (Müller *et al.*, 1978; Rhode, 1978; Carter *et al.*, 1979; Siegl *et al.*, unpublished) and that presence of this type of particles in the inoculum greatly affects the outcome of virus replication. On the other hand, ADV strains of low virulence apparently can be adapted to growth in CRFK cells only at temperatures in the range of 28°C (Porter and Cho, 1980). Under this condition persistent infections are readily established whereas the viruses apparently are lost from the cultures as soon as the temperature of incubation is shifted to 37°C.

E. Assay Systems

Infectivity of the autonomous parvoviruses can be assayed in cell cultures by reading of unstained CPE, by counting intranuclear inclusion bodies after histologic staining, by staining with fluorescent antibody, or on the basis of plaque formation. Each of these approaches has its advantages and disadvantages; all, however, yield suitable results only if the already repeatedly mentioned main prerequisite of parvovirus replication—the availability of cycling cells—is fulfilled. Even then unstained CPE may be weak, uncharacteristic, and transient in parallel with the disappearance of cycling cells as soon as the infected monolayer reaches confluence. In cultures inoculated at a low multiplicity of infection virus replication therefore frequently will pass undetected unless attempts are made to reactivate virus replication by repeated subculturing of the cell sheet. Moreover, extraction and assay of viral hemagglutinin may prove to be rewarding. Hallauer *et al.* (1972) have shown that by skillful combination of subcultivation and extraction of hemagglutinin up to three log higher infectivity titers can be recorded for one and the same virus suspension than are observed by reading of end points on the basis of virus-induced cytopathology.

It is obvious that titration of parvoviruses by this technique is laborious and time consuming. Moreover, the reliability of the results will depend on the patience of the investigator. This is also true for the assay of parvovirus infectivity by screening for specific intranuclear inclusions in stained monolayers. In this respect detection of one single characteristic inclusion is sufficient to prove the presence of infectious virus in the inoculum; yet, formation of inclusions was found to be time dependent (Johnson, 1965b, 1967a; Scott *et al.*, 1970a,b). In consequence, multiple cultures of each dilution have to be screened at various times after infection to assure maximum sensitivity and reliability.

Under properly controlled experimental conditions staining of coverslip cultures by the direct or indirect fluorescent antibody technique probably provides the most feasible means to detect parvovirus infected cells. It is in general much easier to localize a single fluorescing cell against the background of an unstained monolayer than to discriminate a cell with more or less pronounced intranuclear changes from a multitude of only slightly different neighbors. Besides, fluorescent antibody staining is able to detect those early stages in the intranuclear accumulation of parvovirus antigen which do not show up by H&E or May–Gruenwald–Giemsa staining. On the basis of a comparable period of time allowed for screening of cultures the assay of parvovirus infectivity by the fluorescent antibody technique, therefore, can be of ten-fold to a hundredfold higher sensitivity than the reading of cytopathologic changes (Siegl and Kronauer, 1980). In addition, replication of ADV in CRFK cells can only be detected by this means. Hahn *et al.* (1977a,b) have developed a fluorescent focus assay for this parvovirus which, under proper conditions, can be as effective as a plaque assay.

Plaque titration of parvovirus infectivity overcomes most of the problems associated with the methods listed in the previous paragraphs. In general, it yields the highest infectivity titers and, because its sensitivity is primarily not a function of the user's patience, it also yields the most reproducible results. So far, plaque tests have been developed for RV (Brailovsky, 1966), H-1 (Ledinko, 1967:, Rhode, 1973), MVM (Tattersall, 1972; Kajioka *et al.*, 1974), BPV (Bates and Storz, 1973; Lubeck and Johnson, 1976), FPV (Siegl and Kronauer, 1980), and LuIII virus (Siegl and Kronauer, unpublished). The basis of each of these tests is a viable monolayer composed of cells which continue to proliferate under the overlay of the solidified medium. To meet this requirement, both viability and number of cells used to establish the monolayers as well as the time elapsing between seeding of cells and the inoculation of virus suspensions have to be carefully controlled. Moreover, solidification of the culture medium by addition of agar frequently is not suitable and electrophoresis grade agarose or methylcellulose have to be used to allow formation of plaques. There is some circumstantial evidence suggesting that unpurified agar and agarose contain some type of inhibitor(s) interfering with the spread of parvovirus from cell to cell. Substances active at the same level were also detected at a high frequency in fetal calf sera (Siegl and Kronauer, 1980; Tokuhisa *et al.*, 1981). In consequence, pretesting of the ingredients of the overlay is an absolute necessity to guarantee reproducible results.

For most viruses plaque size and plaque morphology are assumed to reflect the genetic composition of the infecting particle. This is not necessarily true for parvoviruses. For the various members of this virus group it could be shown that the diameter of a plaque apparently is a function of the density of the infected monolayer rather than an expression of the virus genetic information. The use of plaque size as a genetic marker for parvovirus strains therefore should be treated with caution.

IV. PATHOGENESIS

Parvoviruses have been frequently referred to as "viruses in search of a disease." This proved to be true especially in the early years following the discovery of these agents by Kilham and Olivier (1959) and by Toolan *et al.* (1960). At that time serologic surveys as well as the frequency of accidental isolations indicated that RV and H-1 were wide spread in populations of both wild and laboratory rats and, yet, specific syndromes could only be observed after experimental infection of laboratory rodents. During the past two decades thorough studies have now related parvoviruses to a broad variety of syndromes developing both upon natural and upon experimental infection. These studies also shed some light on the conditions predisposing the course of parvovirus infection. In most instances, however, we are still far from being able to predict the course of events let alone to understand the interactions between viral and cellular function, the reaction of the organism, and the influence of adventitious environmental factors. It is hoped that, by presenting the pathogenic properties of parvoviruses in concise chapters, at least the outlines of some of the mechanisms active in parvovirus disease might show up.

A. Natural and Experimental Host Range

At first sight the results of various serological surveys and of transmission experiments suggest that parvoviruses can be divided into viruses easily crossing the species barrier and those replicating and inducing disease in their natural host species only. In fact, however, it is dangerous to make such a statement because the information available on the host range of individual members of the parvovirus group differs considerably with respect to both quantity and quality.

The most extended studies have been conducted with RV and antigenically related isolates. In accordance with the frequence of isolation of the viruses from rat tissues serologic surveys revealed antibodies in up to 85% of rat sera collected at various occasions (Moore and Nicastry, 1965; Kilham, 1966; Nathanson *et al.*, 1968; Robey *et al.*, 1968). Therefore, it appears to be safe to assume that the natural host of these viruses is the rat. Whereas, however, experimental injection of RV into neonates of the syrian hamster almost regularly resulted in the development of characteristic syndromes and frequently also in death of the animal (Kilham, 1961b), inoculation of RV into newborn rats usually led to inapparent infection rather than to acute lethal disease (Kilham, 1961a; Kilham and Moloney, 1964; Kilham and Margolis, 1966; El Dadah *et al.*, 1967; Novotny and Hetrick, 1970). Due to pretesting of experimental animals in at least some of the studies, this failure to induce clinical disease in the most likely natural host cannot be attributed to the presence of anti-RV antibodies in the infected animals. Rather, genetic aspects of the infecting virus may play a significant role. In support of this view, Moore and Nicastri (1965) have observed that an initially apathogenic virus will gain in pathogenicity in the course of repeated successive animal-to-animal passages and, finally, may be able to cause acute lethal disease in newborn rats.

Without previous adaptation, however, RV was found to replicate and to induce disease in newborn mastomys (*Rattus natalensis*) (Rabson *et al.*, 1961). Moreover, El Dadah *et al.* (1967) and Matsuo and Spencer (1969) recorded fatal paralysis and the development of acute lethal disease after intracerebral injection of the HER-virus and of RV, respectively, into less than day-old baby mice. Previous attempts to transmit RV to mice had failed (Kilham and Olivier, 1959; Lum and Schreiner, 1963). This failure very likely has been due to inappropriate experimental conditions because El Dadah *et al.* as well as Matsuo and Spencer could provide evidence that the route of infection (intracerebral instead of intraperitoneal) and the virus dose administered (at least 10^4 $TCID_{50}$) were necessary prerequisites for successful induction of disease.

In contrast to the relatively clear situation with RV, the designation of the main natural host of H-1, HT, and HB viruses is still controversal. These viruses were isolated from cell-free fractions of human tumors and other fast-growing human tissues by injection into hamster neonates (Toolan, 1961a, 1964, 1968; Toolan *et al.*, 1961). Because thousands of uninfected hamsters tested by Toolan were seronegative for the viruses (Toolan, personal communication) and because blind passages of hamster tumors or tissues under identical conditions never yielded similar viruses, Toolan has concluded that man is the natural host of these viruses. Some additional support in favor of this view was derived from serological surveys which revealed antibodies to H-1 virus in two out of 130 sera of woman with a history of abortions or congenital defects in their infants (Monif *et al.*, 1965) and in "very few" of the 500 serum samples collected from cancer patients (Toolan, 1968). A considerably higher incidence of anti H-1 positive sera (13.71%) has recently been reported in a survey of 350 sera collected from woman with a history of repeated abortions and still birth (Guglielmino *et al.*, 1978). 9.14% of the same sera also contained HI-antibodies to RV. The prevalence of antibodies to H-1 and RV in 300 control samples collected from people in the same region but with no history of reproductive problems were 2.66 and 1.66%, respectively. Finally, it has been repeatedly observed, that laboratory workers handling H-1 may develop high titer HI-antibody which can persist over years (Toolan, personal communication).

It is beyond question that H-1 virus can proliferate in humans. This was shown by intramuscular injection of the virus into human volunteers (Toolan *et al.*, 1965) as well as by direct isolation of the virus from both the tumor and the embryo of a pregnant woman with metastatic breast carcinoma (Toolan, 1964). The latter patient had developed HI-antibody to the virus at a titer of 1 : 640.

Data which are not in agreement with the above results and contradict the view taken by Toolan have been reported by Newman *et al.* (1970) and by Kilham and Margolis (1969). Newman *et al.* tried to isolate H-1 or similar viruses from as many as 50 tissue specimens collected from cases of spontaneous human abortions by means of injection of tissue homogenates into newborn hamsters. In contrast to the frequency of isolates reported by Toolan (about 20%) not a single virus could be recovered in these studies. Moreover, Kilham and Margolis (1969) isolated H-1 virus from tissues of rats and HI-antibodies to the virus were detected in as many as 80% of sera collected from both wild and laboratory rats. The wild rats bled in the latter surveys had been caught in the surroundings of human garbage dumps near Hanover, New Hampshire.

At the present time, it is thus difficult if not impossible to decide whether man *or* rat or man *and* rat represent the natural reservoir for the H-1 viruses. In any case, both species apparently are susceptible to natural and experimental infection. The idea that H-1 virus infection has the characteristics of a zoonosis is tempting and might explain the differences in antibody prevalence and in frequency of isolations reported by the various investigators on a social and environmental basis. On the other hand, however, it is also conceivable that rats could have contracted infections by feeding on virus-contaminated human garbage.

On the basis of an increase in antibody titers it was also assumed that, after experimental injection, H-1 virus can proliferate in newborn rhesus monkeys or in adult, tumor-bearing animals of this species (Toolan, 1966). Attempts to induce antibodies to RV, H-3, and HB-virus in monkeys were unsuccessful.

In respect to natural and experimental host range MVM, RTV, TVX, and perhaps also LuIII virus seem to be related to the "hamster osteolytic agents" RV and H-1. Thus, according to the results of serologic surveys (Parker *et al.*, 1970b), MVM is highly prevalent in colonies of wild and laboratory mice. Clinically apparent disease due to MVM infection, however, has not been observed in naturally infected animals. After experimental infection of susceptible baby mice nothing but a retardation in growth of the inoculated animals has been recorded (Kilham and Margolis, 1970). Low titer HI-antibody to MVM was also detected at a high incidence in sera of rats (Kilham and Margolis, 1970). The reliability of this finding has been questioned (Parker *et al.*, 1970); yet, MVM apparently proliferates in various organs of newborn rats both after intraperitoneal and intracerebral injection. Clinical disease, however, is only observed after experimental injection of hamster neonates. In such animals the syndromes induced resemble those resulting from infection with RV and H-1 (Kilham and Margolis, 1970, 1971).

RTV, TVX, and LuIII virus have only been isolated from contaminated permanent cell lines (Hallauer *et al.*, 1971) and no direct line to a potential natural host can be drawn. Screening of sera from many species for HI-antibody also yielded not the slightest indication in respect to the natural host of LuIII virus. For RTV and TVX, however, HI-antibodies were detected in 9/50 and 16/50 sera of laboratory rats, respectively (Siegl and Kronauer, unpublished), thus suggesting that the latter two viruses are at least able to proliferate in this rodent. Finally, RTV and LuIII virus were found to induce acute lethal disease after injection into day-old hamsters (Siegl, 1976; Soike *et al.*, 1976). Soike *et al.* (1976) also could show that, after infection of pregnant hamsters, LuIII virus crossed the placenta and caused abortions at a high frequency.

The origin and natural hosts of parvoviruses like BPV, PPV, GPV, and LPV have been established beyond doubt. Numerous serological surveys provided evidence for the worldwide occurence of the viruses in populations of the respective natural host, i.e., in cattle, swine, and geese. Besides, BPV, PPV, and GPV are well known to cause characteristic syndromes in these animals both after natural and experimental infection (cf. Derzsy, 1967; Siegl, 1976; Storz and Leary, 1979; Matsunaga and Chino, 1981; Mengeling, 1981). Transmission to other animals has been attempted in some limited experiments only. Mayr *et al.* (1968), Cartwright *et al.* (1969), and Hallauer *et al.* (1971) reported that neither suckling mice and hamsters nor embryonated hen's eggs could be infected with PPV. GPV, on the other hand, replicated in embryonated eggs of geese, Muskovy ducks, and, after adaption by serial blind passages, also in embryonated eggs of white Peking ducks (Schettler, 1971b; Hoeckstra *et al.*, 1977).

The minute virus of canines (MVC) was isolated from the feces of dogs and antibody to MVC was detected at an incidence of 70% in dog sera (Binn *et al.*, 1970). The pathogenic potency of the virus as well as its experimental host range, however, are still unresolved.

Of the remaining two autonomous parvoviruses listed in Table I—ADV and FPV—Aleutian disease virus of mink apparently has a rather broad natural and experimental host range. Under natural conditions antibodies to ADV were found in sera of mink, ferrets, feral skunks, raccoons, and foxes (Ingram and Cho, 1974; Porter *et al.*, 1980); yet, clinical disease was only observed to occur in mink and ferrets (Kenyon *et al.*, 1966, 1967; Ohshima *et al.*, 1978). After experimental infection of as many as 14 species of Mustelidae, Kenyon *et al.* (1978) detected antibodies in the mink, ferret, weasel, fisher, marten and striped skunk. In accordance with observations under field conditions, characteristic lesions were only present in experimentally infected mink and ferrets. Lesions observed in the striped skunk and in martens could not be unambiguously related to the injected virus.

The host range of feline parvovirus finally presents a complicated yet interesting problem. As described elsewhere in more detail (cf. Siegl, Chapter 9 of this volume), huge epidemics of "feline enteritis" have swept through Europe North America, and South America during the first decades of this century. The virus also has been shown to persist as endemic infection in feline populations all over the world (for references see Siegl, 1976). Up to the end of the 1930s, however, neither circumstantial nor experimental observations pointed to the existence of a susceptible species other than the cat (Urbain, 1933; Kikuth, *et al.*, 1940; Syverton *et al.*, 1943). It therefore aroused considerable interest when, beginning with 1947 in Canada (Shofield, 1949), an infectious agent indistinguishable from the virus incriminated in feline enteritis was found responsible for the outbreaks of severe enteritis in ranch mink. Subsequent studies revealed that the same or a very similar virus can proliferate and cause disease in raccoons and in coatimundi (Gorham *et al.*, 1966; Johnson, 1967b; Nettles *et al.*, 1980). The virus also could be transmitted to neonatal but not to adult ferrets (Schofield, 1949; Johnson, 1965; Johnson *et al.*, 1967; Kilham *et al.*, 1967).

In more recent years pandemics of enteritis have been observed in canines. All available evidence indicates that this highly contagious disease results from infection by CPV, a previously nonexisting host-range mutant of FPV. For example, antibody to FPV could not be found in canine sera collected before 1976 and it is still a matter of discussion whether FPLV or MEV can proliferate in dogs after experimental infection. Felines, however, are susceptible to infection with CPV (for references see Chapter 9).

The generation of host-range mutants of FPV which are able to spread and to cause disease in populations of a previously unaffected and, consequently, fully unsusceptible species necessarily raises the question for the factors predisposing the evolution of such viruses. At the present time there is at least some evidence that, in the case of CPV, handling of FPLV or MEV in the laboratory might have played a significant role (Johnson and Spradbrow, 1978; McMaster *et al.*, 1981; Tratschin *et al.*, 1982). If definite proof for this assumption can be obtained, experiments centered at the *in vivo* host range and the pathogenicity of parvoviruses will have to be conducted in the future under the strictest precautions only.

B. Transmission

As outlined in the previous sections, parvovirus infections are highly endemic in populations of their natural hosts. This observation points to the existence of a potent reservoir as well as to a very efficient mode of transmission of the viruses. With respect to reservoir there is reliable evidence that both infected animals succumbing to acute disease as well as those experiencing clinically inapparent infection shed the virus in their feces, in urine, saliva, and nasal secretions (Urbain, 1933; Lawrence *et al.*, 1943; Schofield, 1949; Kenyon *et al.*, 1963; Gorham *et al.*, 1964; Derzsy, 1967; Eklund *et al.*, 1968; Toolan, 1968; Parker *et al.*, 1970b; Binn *et al.*, 1970; Lipton *et al.*, 1972; Storz and Leary, 1979; Mengeling, 1981). Information concerning the duration of shedding, however, is scarce. In the course of acute enteritis CPV has been recovered from the feces of dogs for only 10 days after the onset of disease (M.J.G. Appel, personal communication), and Johnson *et al.* (1976) also did not observe shedding of PPV beyond the interval of acute infection. The situation is different, however in the case of clinically inapparent infections (cf. Section IV.D). Circumstantial evidence suggests that persistently infected animals may harbor parvoviruses in the kidneys throughout their lives. In accordance with Kilham and Margolis (1966b) and Csiza *et al.* (1971c) such animals therefore can be suspected to shed the viruses into the environment by way of urine. Nothing is known about whether this shedding occurs in all infected animals at a constant low level, whether it takes place more or less periodically and is triggered by hormonal conditions, superinfections, and other unknown factors, or whether only some few individual excreters exist.

Due to their outstanding resistance to environmental factors such as temperature, pH, detergents, and enzymes, excreted parvoviruses remain infectious over a prolonged period of time and constitute a continuous and frequently undetected source of infection. This has been recognized already in the early studies with FPLV where it proved almost impossible to protect susceptible control cats from contracting disease when kept in the usual animal quarters with infected animals (Hammon and Enders, 1939a,b; Kikuth *et al.*, 1940; Lawrence *et al.*, 1943). Similarly, susceptible weanling rats readily developed antibodies to RV when placed in clean cages in a rat colony room where this infection was endemic (Toolan, 1968; Robinson *et al.*, 1971). The ability of PPV to survive under farm conditions was demonstrated by the seroconversion of pigs placed in a stable that had been routinely cleaned and then kept empty for 135 days after infected pigs were removed (Johnson *et al.*, 1976). Finally, Pollock *et al.*(L.E. Carmichael, personal communication) could show that CPV survived in canine feces at room temperature for as long as 6 months. It may be concluded from all these observations that the main routes of transmission of parvoviruses are fecal–oral, saliva–aerosol–respiratory, or dust–respiratory. Besides, mechanical transport of the viruses via contaminated food (Lawrence *et al.*, 1943), contaminated gloves (Larson, 1969, cited by Porter and Cho, 1980), or clothing (Carmichael and Binn, 1981) had been suspected to play a role in the spread from animal quarter to animal quarter as well as over long distances.

The isolation of PPV from semen and from testicles, as well as from scrotal lymph nodes of infected boars (Cartwright and Huck, 1967; Cartwright *et al.*, 1969; McAdaragh and Anderson, 1975; Lucas *et al.*, 1974) suggests that parvoviruses can also be sexually transmitted. However, vertical transmission by transplacental infection may be more important in the disemination of parvoviruses. This has been observed by Toolan *et al.* (1960) already in the early experimental studies with H-1 virus. Since then this mode of transmission was reported for RV (Ferm and Kilham, 1964; Kilham and Ferm, 1964), LuIII (Soike *et al.*, 1976), MVM (Kilham and Margolis, 1971), FPV (Kilham and Margolis, 1966; Kilham *et al.*, 1967; Csiza *et al.*, 1971), PPV (for reference see Mengeling *et al.*, 1980; Mengeling, 1981), BPV (Storz *et al.*, 1972; Inaba *et al.*, 1973b; Storz *et al.*, 1978b), as well as for ADV (Padgett *et al.*, 1967; Porter *et al.*, 1977a;

Porter and Cho, 1980). Factors such as the time of gestation at which the animals acquire infection, their state of immunity, and last but not least, the genetic composition of the infecting virus apparently determine the outcome of *in utero* infection. The main prerequisite for transplacental infection, however, seems to be the presence of the virus in the blood stream independent of whether this is brought about by *intraveneous* injection of the virus or by viremia as result of the replication of virus in maternal tissues (Kilham and Margolis, 1969; Csiza *et al.*, 1971a,b; Joo *et al.*, 1976a; Mengeling and Cutlip, 1977; Storz *et al.*, 1978b).

Several attempts have also been made to elucidate the transmission of parvoviruses at the level of ova. Kilham and Margolis (1968) observed that ova, whether mitotically inert or arrested in meiosis, were refractory to infection by RV. Vegetative cells of the ovary, however, apparently supported virus replication. In contrast, proliferation of MVM was reported to occur after *in vitro* infection of fertilized, two-cell mouse eggs deprived of their zona pellucida (Mohanty and Bachmann, 1974); yet, the development of the cultivated embryo was not affected. Under conditions where the zona pellucida was not removed, PPV did not penetrate into fertilized pig eggs (Wrathall and Mengeling, 1979a). The antigen remained associated with the zona pellucida and infectious virus could be transferred to seronegative gilts by transplantation of contaminated ova (Wrathall and Mengeling, 1979b). Early embronic loss and pregnancy failure then was assumed to result from infection of the uterine epithelium and, at later times, from infection and damage of the developing fetus. In context with the increasing importance of egg transplantations in animal live stock the potential risk of spreading parvovirus infections by this route therefore should be seriously considered.

Transmission of goose parvovirus via ova has also been reported by Derzsy (1967).

C. The Spectrum of Disease

1. *In Utero* Infection, Fetal and Neonatal Death, and Congenital Malformations

In their review of intrauterine and neonatal infections Kilham and Margolis (1975) have stated that the ability to induce transplacental infections is more widespread among parvoviruses than among any other group of viruses. In fact, the greater part of the spectrum of disease caused by these agents is related to intrauterine infection. It results either in death of the embryo and fetus, in obvious malformations, or in specific defects which sooner or later become evident during the animal's postnatal life. There is good evidence that the spectrum of syndromes reflects the state of histogenesis and organogenesis of the fetus at the time of infection. However, maternal and fetal immune response as well as hormonal changes accompanying pregnancy also seem to modulate the effect of parvoviruses during *in utero* infection.

Direct *in utero* inoculation of fetuses as well as transplacental infection after injection of pregnant animals suggest that the embryos and fetuses of rats, hamsters, mice, cats, dogs, ferrets, pigs, and of cattle are susceptible to infection with the respective parvoviruses throughout the time of gestation (Toolan *et al.*, 1960; Ferm and Kilham, 1963, 1964; Kilham *et al.*, 1967, 1971; Kilham and Margolis, 1969, 1971; Joo *et al.*, 1977; Storz *et al.*, 1978b; Lenghaus *et al.*, 1980; Mengeling, 1981). The question whether parvoviruses also play a role in transplacental infections in man cannot be answered conclusively on the basis of the available data.

The reaction of the embryo and fetus to infection with parvoviruses varies considerably with the time of gestation. As already outlined in the section on transmission, the very early stages of the developing embryo apparently resist infection with parvoviruses. At what time this state of apparent insusceptibility is terminated has not been investigated to a reliable degree. In the hamster, implantation occurs around day six of gestation. It was reported that, up to that time, infection of pregnant animals with H-1, RV, and LuIII virus has no deleterious effect on the developing embryo (Ferm and Kilham, 1964, 1965; Soike *et al.*, 1976; Toolan, 1978) Nevertheless, animals born to mothers infected during the preimplantation period may harbor the virus at the time of birth.

The hamster embryo seems to be most susceptible to infection by H-1 and LuIII virus at around eight days or, in other words, at about the middle of the short gestation period of this species (Ferm and Kilham, 1964, 1965; Soike *et al.*, 1976; Toolan, 1978). A high percentage of fetuses infected at that time die and are resorbed or aborted. Littermates surviving to parturition are frequently born severely ill and usually die shortly after birth. The effects of BPV and PPV on bovine and porcine embryos or fetuses, respectively, are quite similar. In both species the most severe consequences of *in utero* infection were recorded subsequent to infection in the first and second trimester of gestation. They consisted in death of the embryo or fetus, followed either by resorption, mummification, or abortion (Storz *et al.*, 1978b; Joo *et al.*, 1976a; Bachmann *et al.*, 1975; Mengeling and Cutlip, 1977; Mengeling, 1981). In the pig mummified fetuses may be farrowed at term. Their presence in the uterus, however, can also prolong gestation and lead to stillbirth of apparently normal littermates (Narita *et al.*, 1975). These immediate consequences of transplacental infection in the pig and a probably related, reduced neonatal vitality of affected litters are summarized under the term "reproductive failure." This syndrome is assumed to represent the major and usually only clinical response of pigs to infection with PPV.

Fetuses infected in the third trimester towards the end of gestation run a good chance of overcoming infection without obvious clinical manifestations. This has been observed after injection of the rodent parvoviruses as well as of LuIII virus into pregnant rats and hamsters (Ferm and Kilham, 1964, 1965; Kilham and Margolis, 1966b; Soike *et al.*, 1976). Toolan (1978), however, reported that injection of H-1 virus into pregnant hamsters on days 12 to 14 of gestation, i.e., shortly before delivery, may cause as high a mortality and morbidity in the fetus and in newborns as does transplacental infection at midgestation. Because the progesterone levels in the gravid hamster closely parallel the observed bimodal mortality pattern, it was hypothesized that this hormone plays a distinct role in fetal mortality due to parvovirus infection. The most likely mode of action of the hormone was assumed to consist in its role as a T-cell suppressor, by means of which it could interfere with the development of an immunological response to the infecting agent. In this context it is also of interest that the resistance of the bovine and procine fetuses to infection with their specific parvoviruses late in gestation has been attributed to the beginning ability of the fetuses to generate an immune response (Redman *et al.*, 1974; Bachmann *et al.*, 1975; Cutlip and Mengeling, 1975; Storz *et al.*, 1978b); The presence of antiviral antibody in considerable quantity and a continuously reduced ability of fetal tissues to support virus replication could be the reason for a limited proliferation and spread of the infecting agents. At the same time, however, restriction of virus replication in still differentiating tissues, partially inaccessible to antiviral antibody, could favor the development of distinctly located congenital malformations such as cerebellar hypoplasia (Kilham and Ferm, 1964; Kilham and Margolis, 1966a,b, 1971), hydrocephalus (Csiza *et al.*, 1971), and perhaps also myocarditis (Lenghaus *et al.*, 1980).

Neither the varying susceptibility of embryonal and fetal tissues nor the maternal and fetal immune response to the infecting parvovirus are sufficient criteria to explain the full and changing spectrum of disease resulting from *in utero* infection. As emphasized by Kilham and co-workers, genetically determined properties of the infecting virus are at least of similar importance. Thus, it was found (Ferm and Kilham, 1963) that a strain of RV which had been carried for 42 passages in neonatal hamsters readily crossed the placenta of pregnant rats. The virus was recovered from tissues of the uterus, the placenta, and the embryos of infected animals at significant titers. In contrast to the evident pathogenicity of this strain when injected directly into fetal rats or neonatal hamsters, however, none of the rat fetuses acquiring infection transplacentally showed any sign of clinical disease. Quite similar results were obtained with another strain of RV passaged many times in tissue culture (Kilham and Ferm, 1964). From these and further observations (Kilham and Margolis, 1969) it was concluded that the fewer laboratory passages a strain of H-1 and RV has been subjected to, the more readily transplacental infections in hamsters and rats are accompanied by overt disease. In agreement with these observations Paul and Mengeling (1980) were recently able to select a strain of PPV for the production of a live virus vaccine which did not pass the placenta of pregnant gilts. However, the virus replicated in the tissues of the gilt as well as in the tissue of fetuses after direct *in utero* inoculation.

The results available to date allow no conclusion with respect to the nature of the virus-specific function(s) responsible for transplacental passage and *in utero* pathogenicity. No virus strain has been isolated so far which exhibited a specific affinity for a distinct type of embryonal or fetal tissue; rather, both embryos severely affected and those surviving *in utero* infection without clinical disease frequently were found to harbor the virus at comparable concentrations in a quite similar broad spectrum of tissues. Moreover, pathogenic as well as apathogenic strains caused viremia in pregnant animals and, as far as rodent parvoviruses are concerned, were able to pass the placenta. It is therefore quite unlikely that pathogenic strains are distinguished by an exceptional capability to proliferate and to spread in maternal and fetal tissues. On the other hand it has been repeatedly reported that the concentration of virus injected into pregnant hamsters or rats determines the severity of *in utero* disease. Injection of high concentrations of H-1 virus into pregnant animals regularly resulted in death of embryos *in utero* or at birth. Inoculation of smaller quantities of the same virus, however, frequently induced distinct malformations or an overall deformation of the developing animal (Ferm and Kilham, 1964, 1965b; Toolan, 1978).

As regards the pathologic features of *in utero* disease, numerous macroscopic changes and microscopic lesions in the embroys and fetuses infected by the different parvoviruses have been recorded in great detail. Compilation of the vast information would exceed the scope of this review. In summary, however, it was observed that virus replication and the resulting lesions were most severe in tissues with the most pronounced cellular proliferation. In early gestation mesendermal tissues were heavily involved. With beginning differentiation the various organs represented the main goal of attack, and during late gestation the lamina propria and muscularis are preferentially affected. For detailed information on this subject the interested reader is referred to the reviews of Toolan (1968), Margolis and Kilham (1975), Siegl (1976), and Mengeling, (1981).

2. The Osteolytic Syndrome

Both after transplacental infection and after direct inoculation into newborn hamsters RV and related isolates, MVM, as well as the H-viruses frequently give rise to the development of the so-called osteolytic syndrome. Affected animals develop either dwarfism and/or mongoloidlike features characterized by small flat face, microcephalic domed head, protruding eyes and tongue, missing or abnormal teeth, and fragile bones (Toolan, 1960; Dalldorf, 1960; Kilham, 1961b; Baer and Kilham, 1962a,b; Ferm and Kilham, 1965a; Baer and Kilham, 1965; Lipton and Johnson, 1972). Dwarfism as the result of parvovirus infection is also occasionally observed in litters of infected rats and mice (Moore, 1963; Moore and Nicastri, 1965; Matsuo and Spencer, 1969) and retardation in growth was reported to result from infection of calves with BPV, of piglets with PPV, and of goslings with GPV (Derzsy, 1967; Storz and Bates, 1973; Mengeling, 1981). Only in hamsters and rats, however, is the overall picture of dwarfism conserved throughout an animal's life, whereas affected mice, calves, piglets, and geese may overcome the difference in size to uninfected controls with progressive age.

The full spectrum of the osteolytic syndrome is most consistently observed in hamsters inoculated within 48 hr after birth (Toolan, 1960). Inoculation after that time produces fewer malformations the older the animals are at the time of infection. This observation has been taken as indication of a restriction of parvovirus pathogenicity to proliferating and differentiating tissues. Moreover, the various facets of the osteolytic syndrome have greatly helped to formulate the hypothesis concerning the dependence of parvovirus replication on "dividing" cells (Margolis and Kilham, 1965). Finally, comparative studies with H-1 virus and various strains of RV also suggested that, as is the case with transplacental infection, the genetic characteristics of the virus strain determine the spectrum and the severity of clinical disease in newborn rodents (Baer and Kilham, 1965; Ferm and Kilham, 1965a; Toolan, 1968).

As the term "osteolytic" already implies, the characteristic features of this syndrome can be related to the cytolytic attack of parvoviruses on osteogenic tissues (Ferm and Kilham, 1965b). Virus replication interferes with the growth of bones and teeth and, consequently, affects the development of the skeleton of infected hamsters. Dalldorf (1960), for example, observed virus-induced degeneration of osteoblasts at the growing costochondral junctions of long bones, and Ferm and Kilham (1965a) noticed severe effects of two strains of RV on the proper differentiation of the cranial sutures. The viruses used in the latter studies had a pronounced affinity for the membranous bones of the skull and the mandible. Interference of virus replication with the growth of the skull apparently leads to the characteristic microcephalic domed head of the "mongoloid" hamster with protruding eyes due to lack of skull space. Interestingly, however, the skeleton of infected animals in general is affected in all its parts and its proportionality usually is conserved.

The other facet of the osteolytic syndrome, i.e., the effect of rodent parvoviruses on the development of the denture of infected hamsters, has been thoroughly investigated in a series of studies by Baer and Kilham (1962a,b, 1964a,b, 1965, 1974) and by Cohen and Shklar (1964). In the main part of these experiments five-day-old suckling hamsters were infected with H-3 virus and the development of lesions was recorded over a period of 18 months. Because the molar teeth of the suckling hamster develop on different days, the resulting abnormalities of individual teeth could be directly related to their stage of development at the time of infection. Within only 24 hr after infection intranuclear inclusions were detected in cells of the most susceptible periodontal membrane. Virus replication then led to destruction of the membrane as well as of the supporting alveolar bone and, within 7 to 12 days after inoculation, resorption of roots and bony ankylosis could be observed. Shortened and deformed roots, resorption of roots and bones, as well as anomalous deposition of osteodentin and cementum were characteristic features of the teeth of the infected, grown-up hamster. Moreover, odontogenic tumors were detected within the body of the mandible and about the apical portion of the incisors in the maxilla in animals kept for 12 to 18 months after infection.

Inoculation of rodent parvoviruses into healthy adult hamsters and rats has never been reported to result in destruction of bones and teeth (Toolan, 1965) Engler *et al.* (1969) and Baer *et al.* (1971a), however, have shown that these viruses may interfere with the healing process of osseous wounds. Engler injected H-3 virus intravenously into an adult hamster one or three days after fracture of the forearm or after extraction of a maxillary second molar tooth. The virus replicated at the fracture site and replication apparently interfered with the formation of a fibrocartilagenous callus and subperiosteal osteogenesis along the diaphysis. It also caused retardation of osteogenesis in the alveoli. In this context it should be emphasized that replication of the virus in these regions cannot be attributed to a specific affinity of the virus for osteoblasts. Rather, the affinity of parvoviruses for cells undergoing mitosis triggers its accumulation at the fracture site. In consequence, the interference of parvoviruses with the healing of osseous wounds is just another aspect of the hazardous association of these agents with regenerating tissues (cf. Section IV.C.5).

The overall appearance of hamsters surviving the osteolytic attack of rodent parvoviruses has many features in common with the mongoloid phenotype of humans suffering from Down's syndrome. The similarities include the dental defects (Cohen *et al.*, 1961), the conservation of an infantile skull with abnormal proportions (Baer *et al.*, 1971b), as well as behavioral changes and reduced learning capacities (Landauer *et al.*, 1967). Whereas the human deformities, however, are considered as being due to trisomy of chromosome 21, no indication for a significant chromosomal anomaly could be found in the cells of deformed hamsters (Galton and Kilham, 1966). Moreover, in contrast to the genetic fixation of the syndrome in man, mongoloid animals gave birth to normal offspring (Toolan, 1960; Ferm and Kilham, 1965b). It has therefore been repeatedly speculated whether the physiologic conditions resulting from trisomy 21 might provide the medium for the cytolytic replication of one of the known or of an unknown parvovirus (Toolan, 1968; Kilham and Margolis, 1975). Neither serologic surveys nor isolation attempts, however, have furnished any reliable results in favor of this hypothesis (Monif *et al.*, 1965; G. Siegl and G. Kronauer, unpublished).

3. Acute Lethal Disease in Newborn Animals

Newborn animals are highly susceptible to generalized and often lethal infections with parvoviruses. This has been established in particular for the rodent parvoviruses. Infection of hamsters and rats with H-1, RTV, LuIII, and RV-type viruses shortly after birth usually terminated fatally within 5 to 14 days (Toolan, 1961b; Kilham, 1961a,b; Moore and Nicastri, 1965; Siegl, 1976; Soike *et al.*, 1976). Clinical disease thereby consisted mostly in apathetic behavior, a sudden slugishness, and, occasionally, a tendency of the animals to gasp for breath. These symptoms usually became evident only some few hours before death ensued. At necropsy, macroscopic examinations frequently revealed a sanguineous exudate in the intestines, hemorrhages in the stomach and the kidneys, as well as congestion of the liver. Microscopic and virologic examinations, on the other hand, pointed to a systemic involvement of almost all organs and tissues in virus replication.

Generalized disease as a consequence of infection with parvoviruses has also been reported for FPLV in newborn kittens (Csiza *et al.*, 1971a,b) and for CPV in a litter of puppies (Lenghaus *et al.*, 1980). In cats and dogs, however, acute generalized disease in neonates seems to be the exception and more specific syndromes like enteritis, panleukopenia, and myocarditis apparently prevail. On the contrary, acute disease accompanied by a sometimes extraordinarily high incidence of mortality is characteristic for the infection of goslings with GPV (Kraus, 1965; Derzsy, 1967; Schettler, 1971a). In some outbreaks up to 95% of goslings between 1 and 30 days of age succumbed to infection. The course of clinical disease as well as the tissues preferentially involved in virus replication were found to vary with the age of the animals. Young animals between 4 and 8 days of age are most severely affected. They develop diarrhea, nasal catarrh, and incoordination of movements. After a rather short period of slugishness they may die within one to two days after the appearance of the first clinical signs. The most typical lesions found upon necropsy of animals dead after such a short episode of disease consisted of edema and diffuse degeneration of the myocardium as well as an exudative hepatitis (Derzsy, 1967; Kraus, 1965; Palya and Kisary, 1978; Schettler, 1971a). In older animals the liver, thyroids, and pancreas probably are the main target organs of viral attack. Animals older than two weeks at the time of infection have a good chance to survive (Coudart *et al.*, 1974); yet, they may show retardation in growth for a certain period of time.

4. Enteritis

Enteritis is one of the most prominent features of parvovirus-induced disease. It frequently develops after infection of newborn animals but may also constitute a main syndrome after infection of adults. As outlined in the preceding section, newborn hamsters and rats inoculated with the rodent parvoviruses or with LuIII virus may show signs of hemorrhagic enteritis shortly before death. Moreover, diarrhea and vomiting are prominent clinical symptoms in feline enteritis (feline panleukopenia) (Urbain, 1933; Lawrence and Syverton, 1938; Lucas and Riser, 1945), in mink enteritis (Schofield, 1949; Wills, 1952; Myers *et al.*, 1959; Reynolds, 1969), in canine parvovirus enteritis (Appel *et al.*, 1978; Kelly, 1978; Robinson *et al.*, 1980b), in goslings infected with GPV (Kraus, 1965; Derzsy, 1967), and in neonatal calves affected by BPV (Abinanti and Warfield, 1961; Storz *et al.*, 1978a). Whether those viruses like MVC and LPV which have been recovered from the feces of apparently healthy dogs (Binn *et al.*, 1970) and rabbits (Matsunaga *et al.*, 1977), respectively, are also able to cause acute diarrhea, remains to be shown.

The enteric syndrome related to parvovirus infection represents an enigma. It is readily observed during epidemics and, thereby, may take an acute and drastic course. Discharge of watery, sometimes blood-stained stools can be accompanied by vomiting, pyrexia, anorexia, depression, and dehydration in varying extent. However, in all instances where naturally and experimentally infected animals were thoroughly examined both at the clinical and at the pathological level, diarrhea was found to be only one out of several and, in addition, the most inconsistent manifestation of disease. Nevertheless, the same studies have revealed that replication of the infecting virus always occurred in the cryptal epithelium of the small intestine (Csiza *et al.*, 1971b; Robinson *et al.*, 1980b; Storz *et al.*, 1978a). Specific lesions were also observed in the cranial and mid-jejunum (Carlson *et al.*, 1977), and, under extraordinary circumstances, the ileum and the colon of cats were found to be affected (Langheinrich and Nielsen, 1971; Larsen *et al.*, 1976). In spite of the regular involvement of the intestinal tract in virus replication, attempts to reproduce the full spectrum of enteric disease under laboratory conditions have always met with difficulties (Bittle *et al.*, 1961; O'Reilly, 1970; Carlson *et al.*, 1977; Carlson and Scott, 1977; Robinson *et al.*, 1980b). It was therefore concluded that circumstantial factors such as the physical condition of the animal at the time of infection, the presence of adventitious viruses, or coinfection with bacteria may be responsible for the induction of the enteric syndrome. Overt enteritis also was more frequently observed after infection of animals by the intraveneous and respiratory route than after oral exposure to the virus. Finally, there is evidence that feline parvovirus rapidly loses its enteropathogenicity during serial propagation in tissue culture (Carlson *et al.*, 1977). Hence, freshly isolated or *in vivo* passaged virus strains seem to be necessary for experimental induction of disease.

5. Hepatitis

Involvement of the liver in parvovirus disease was first observed after spontaneous *in utero* infection of rat fetuses with RV (Kilham and Margolis, 1966a). Some of the offspring in such litters were jaundiced and large quantities of RV could be isolated from their livers. Subsequent studies indicated that severe hepatitis could be induced by inoculation of the virus into susceptible weanling rats and hamsters as late as 12 days postnatally (Margolis *et al.*, 1968). At that age, the liver of rodents exhibits a considerable and continuing mitotic activity (Steiner *et al.*, 1966) which apparently provides the medium for progressive replication of RV and the manifestation of disease in this tissue. Margolis *et al.* (1968) and Bergs and Scotti (1967) described the histopathologic changes resulting from replication of RV in the liver of neonatal rats as cryptic lesions of the parenchymal peliosis type.

The mitotic activity in the liver of rodents remains high for as long as 4–6 weeks after birth. Cessation of this activity then coincides with an apparent insusceptibility of the liver to parvovirus infection. Nevertheless, Henry and Diario (1975) have observed that H-1 virus can induce hepatitis in adult, 4–6-month-old hamsters. No clinical symptoms could be recorded, but focal degeneration of hepatic cells as well as elevated liver enzymes were believed to be indicative of a continuing susceptibility of the differentiated liver to infection with parvoviruses. Because the liver has an extraordinary regenerative potential, susceptibility of the adult differentiated organ can be stimulated or even restored in the course of regeneration subsequent to hepatic injury. In this respect partial hepatectomy (Ruffalo *et al.*,1966), infection of rates with *Cysticercus fasciolaris* (the larval form of the feline tapeworm *Taenia taeniaformis*, Kilham *Kilham et al.*, 1970), as well as carbon tetrachloride poisoning (Margolis and Kilham, 1975) were found to result in a vigorous reparative activity which provided the medium for massive replication of a simultaneously injected parvovirus.

Hepatitis is also one of the prominent manifestations of GPV infections in goslings. In the acute phase of disease the main pathologic lesions were reported to consist in an exudative hepatitis with serous parenchymal damage whereas at later stages subacute interstitial hepatitis developed (Palya and Kisary, 1978). In each stage replication of GPV in hepatocytes was indicated by the presence of Cowdry type A intranuclear inclusions.

6. Myocarditis

Post mortem examination of newborn animals dead from acute, generalized parvovirus infection revealed virus-specific intranuclear inclusion bodies in almost all tissues and also in the myocardium. Particularly widespread replication of parvoviruses in this tissue has been observed in newborn kittens infected with FPLV (Csiza *et al.*, 1971a,b) and in calves infected with BPV (Storz and Leary, 1979). In none of these animals, however, was the involvement of the heart muscle in virus replication paralleled by a distinct clinical symptomatology.

Lesions exceeding the mere presence of virus-infected cells in the myocardium are a consistent finding in goslings succumbing to infection with GPV. According to Derzsy (1967) the most typical changes are distension of the heart, rounding off of the cardiac apex, and diffuse degeneration of the myocardium. At later times the spectrum of lesions may be completed by reparative inflammation. Cowdry type A inclusion bodies were always present in the affected tissue. Such typical intranuclear inclusions in the myocardium also were the first indication for the involvement of a parvovirus in the etiology of the recently noticed acute epidemic heart failure in dogs. As has been shown by various investigators (cf. Hayes *et al.*, 1979; Kelly and Atwell, 1979; Robinson *et al.*, 1979a,b 1980a; Lenghaus *et al.*, 1980), this exceptional syndrome is caused by infection with CPV. However, myocarditis in dogs usually occurs without concurrent parvoviral enteritis. It is limited to puppies and, in general, onset and progression of clinical disease are rapid. Two clinical presentations have been described. Ths most common one in very young animals is characterized by sudden death occasionally preceded by a rather short period of dullness and dyspnoe. In dogs over eight weeks of age death frequently intervenes within 24 hr after acute respiratory distress, depression, and weakness were observed. Survivers may develop congestive heart failure. According to the many detailed histologic examination of field cases (cf. Siegl, Chapter 9), the pathology of the disease presents as a marked interstitial myocarditis with loss of myofibers, multifocal myofiber necrosis, and mononuclear cell infiltrate. So far, attempts to reproduce myocarditis in dogs were successful only after *in utero* inoculation of puppies eight days before parturition (Lenghaus *et al.*, 1980). Analysis of surviving animals in these studies also suggested that congestive heart disease in chronic myocarditis may be due to extensive focal fibrosis within the myocardium.

7. Cerebellar Ataxia

About a century ago, Herringham and Andrewes (1888) described the clinical and gross pathological features of cerebellar ataxia in kittens. Under field conditions usually the whole litter is affected and ataxic kittens in general show no other signs of disease. The syndrome therefore has been regarded as a congenital anomaly of genetic origin up to recent times. A first lead pointing to an infective etiology of this disease was obtained in the course of experiments with RV in newborn hamsters. After intracerebral injection of a specific strain of this parvovirus into baby hamsters less than four days of age (Kilham and Margolis, 1964), the typical clinical picture of ataxia became slowly evident after four to five weeks. The affected animals developed an unsteadiness in gait as well as an instability of balance. On sacrifice, it was noted that infection had resulted in severe hypoplasia of the cerebellum. Histologic examinations finally suggested this pathologic condition as being due to the selective attack of virus replication on the outer germinal layer of the differentiating cerebellum. Clinical ataxia and cerebellar hypoplasia were also observed subsequent to transplacental infection of rat fetuses with RV (Kilham and Margolis, 1966).

Attempts to induce ataxia in kittens by intracerebral injection of RV failed; yet, as was indicated by the appearance of intranuclear inclusion bodies and by partial destruction of the tissue, the virus was found to proliferate in the external germinal layer of the cerebellum of these animals (Kilham and Margolis, 1965). Clinical ataxia as well as the underlaying cerebellar hypoplasia finally could be reproduced by intracerebral inoculation of feline panleukopenia virus into newborn kittens (Kilham and Margolis, 1966; Johnson *et al.*, 1967). Both in these kittens and in neonatal ferrets—which proved to be of exceptional value in the study of feline ataxia—histopathologic changes were indistinguishable from those observed in the ataxic hamster and rat. Characteristically, virus replication in the outer germinal layer led to the absence of a definitive granular cortex, and, sometimes, destruction of Purkinje cells could be observed.

Susceptibility of the cerebellum to parvovirus infection is limited to those developmental stages of the animals during which the differentiating outer germinal layer provides a maximum number of dividing, susceptible cells for virus replication. In the cat this stage is terminated at about nine days after birth. It may therefore be expected that, depending on the time of infection, infection with FPLV up to that age of the animal may induce lesions of verying severity. In fact, according to Csiza *et al.* (1971b) specific intranuclear fluorescence could be detected in the cerebellum, the cerebrum, and the spinal cord within three days after oral exposure or intranasal infection of newborn kittens. Virus replication subsequently resulted in a marked reduction of the cerebellar size. Clinical signs of ataxia, however, were absent in the animals. These observations plus additional experimental data (Kilham and Margolis, 1966a; Kilham *et al.*, 1967; Duenwald *et al.*, 1971), as well as circumstantial evidence suggest that most of the observed cases of spontaneous ataxia in kittens result from *in utero* infections.

Finally, very recent experiments with BPV seem to indicate that bovine fetuses exposed to the virus around 130 days of fetal development experience brain lesions resembling closely the findings with feline and rat parvoviruses. According to Storz (personal communication) they consisted of a depopulation of the germinative external granular cell layer of the cerebellum with granuloprival cortical atrophy.

8. Hemorrhagic Encephalopathy

After treatment of adult, apparently healthy rats with cyclophosphamide, El Dadah *et al.* (1967) recovered a strain of RV from the brain and spinal cord of animals exhibiting paralysis of the hind limbs. Intracerebral inoculation of this virus into suckling rats again resulted in acute fatal paralysis associated with hemorrhages and necrosis in the brain and spinal cord. This entity—hemorrhagic encephalopathy—could be subsequently reproduced with various strains of RV in newborn rats, mice, and hamsters (Nathanson *et al.*, 1970; Cole *et al.*, 1970; Margolis and Kilham, 1970, 1972). Although less common, it was also observed after infection of kittens and ferrets *in utero* or postnatally with FPLV. Margolis and Kilham (1970, 1975) have suggested that the basis of this disease is the attack of parvoviruses on the proliferating, supporting structure of the neonatal CNS. However, the contribution of various other factors to the development of the syndrome, such as, for example, the attack of the virus upon the pool of developing megacaryocytes also have to be considered (Margolis and Kilham, 1972).

9. Panleukopenia

Between 1938 and 1943 experimental work in several laboratories was centered on the clinicopathological features of feline infectious enteritis (Hammon and Enders, 1939a,b; Lawrence and Syverton, 1938; Kikuth *et al.*, 1940; Lawrence *et al.*, 1940, 1943). These studies revealed that, besides a more or less constant involvement of the intestine, a prominent characteristic of this devastating disease consisted in severe hematologic changes. In particular, a profound leukopenia was reported to occur. According to these investigations as well as to more recent comparative experiments in conventionally raised cats, in specific pathogen-free, and in germ-free animals leukopenia is the most constantly found facet of FPLV infection (Rohovsky and Fowler, 1971; Langheinrich and Nielsen, 1971; Larsen *et al.*, 1976; Carlson and Scott, 1977; Carlson *et al.*, 1978).

Characteristically, the fall in total white blood cell counts starts in parallel with the initial slow rise in temperature at about two to three days after infection. Lowest cell counts—as few as 100 leucocytes per mm^3 have been reported in severe cases—are usually observed at the height of clinical disease. Recovery, on the other hand, is signaled by a detectable increase in leucocyte counts. Detailed studies have revealed that lymphocytes and neutrophiles are especially involved in the variation of the blood picture. There is also convincing evidence that the changes result from the primary and direct attack of FPLV upon the lymphoid tissues and the blood-forming organs. By three to four days after infection cytocidal replication of FPLV in lymphoid cells of the thymic cortex, in both follicular and paracortical areas of lymph nodes, as well as in the follicular and periarteriolar areas of the spleen can be observed. Virus specific inclusion bodies are also found in the bone marrow, and destruction of the primitive blood cells is believed to result in neutropenia.

Although not studied in such detail, hematologic changes comparable to those sketched in the preceding paragraph were also found to constitute a rather constant aspect of disease induced by the two other variants of FPV—MEV and CPV—in mink and dogs, respectively (Reynolds, 1969; Jacobs *et al.*, 1980; Robinson *et al.*, 1980b; Woods *et al.*, 1980). Leukopenia resulting from infection with these viruses, however, in general is less pronounced, transient and, consequently, is frequently missed if no serial white blood cell counts are performed. Likewise, many pigs, irrespective of age, experience a transient, usually mild leukopenia within about the first ten days after initial exposure to PPV (Johnson and Collings, 1969, 1971; Mengeling and Cutlip, 1977; Joo *et al.*, 1976a). Moreover, infection of lymphatic organs of calves by BPV also parallels the pathogenic events described for panleukopenia in newborn kittens (Storz *et al.*, 1978a).

10. The Impact of Parvoviruses on the Immune System

It has been hypothesized that destruction by parvoviruses of cells responsible for the immune mechanism should have a considerable impact on the immunological responsiveness of the infected organism. Rather unexpectedly, first evidence for such an effect of parvovirus replication has come from *in vitro* studies. Bonnard *et al.* (1976) reported that filtered culture fluids of a subline of mouse EL-4 lymphoma inhibited the generation of cytolytic effect in allogenic mixed lymphocytic cultures. Immunosuppression could be attributed to a contaminating virus, a lymphotropic variant of MVM. In subsequent experiments this virus, but not the prototype strain of MVM, was shown to inhibit a variety of T-cell functions, e.g., antigen-induced proliferation, generation of cytolytic T-cell activity, and T-helper function. It did not, however, interfere with B-cell specific plaque-forming cell generation *in vitro* (McMaster *et al.*, 1981; Engers *et al.*, 1982). According to Bonnard *et al.* (1976), immunosuppressive effects can also be observed with RV in rat lymphocyte cultures.

As concerns the impact of parvovirus replication on the immunologic responsiveness of the infected organism, Schultz *et al.* (1976) showed that, in the course of feline panleukopenia, the lymphocyte responsiveness to phytohemagglutinin and to concanavalin A were depressed. Responses to B lymphocyte mitogens as well as antibody responses to rabbit red blood cells and to the infecting virus, however, were not impaired. In the course of clinically mild FPLV infections the humoral immune mechanisms may even be stimulated rather than inhibited (Carlson *et al.*, 1978). In contrast to the latter observations, however, the immune responsiveness of mink suffering from Aleutian disease appears to be considerably reduced. Suppression of responsiveness to a variety of antigens thereby starts at about the time the infected animal develops the most outstanding feature of this syndrome, i.e., hypergammaglobulinemia (Porter *et al.*, 1965, 1969; Kenyon, 1966; Lodmell *et al.*, 1970; Trautwein *et al.*, 1974).

11. Aleutian Disease of Mink

Aleutian disease of mink is a rather complex entity which has several facets in common with disease induced by well-known and better understood parvoviruses; yet, it also includes a number of aspects resisting interpretation on the basis of today's knowledge. The most distinguishing features of ADV infection consist of a chronic, slowly progressive course and the involvement of immunological mechanisms in the production of lesions. The disease is characterized by widespread plasmacytosis, glomerulonephritis, arteritis, and focal hepatitis. Death may occur any time from 2 to 24 months after infection (Henson *et al.*, 1966; Pan *et al.*, 1970; Porter and Larsen, 1969; Drommer and Trautwein, 1975; Barcley, 1976; Porter and Cho, 1980).

Replication of ADV takes place after infection of fetal, newborn, and adult mink. It gives rise to chronic viremia; however, no disease seems to be caused by virus replication itself. It is assumed that the constant presence of huge amounts of viral antigen induces a maximal immune response as reflected by rapid proliferation of antibody-producing cells and the development of hypergammaglobulinemia. Porter (1981) has recently provided evidence that most of the gammaglobulin produced in the course of ADV infection is specific antiviral antibody. This antibody, however, is not capable of neutralizing viral infectivity and, hence, the circulating antigen–antibody complexes found in persistently infected mink are infectious.

The majority of lesions observed in the course of Aleutian disease are very likely mediated by deposition of immune complexes. Glomerular and arterial cell wall lesions were shown to contain deposits of IgG, C3, as well as of ADV antigen, and anti-ADV antibody could be eluted from affected kidneys (Henson *et al.*, 1969; Porter *et al.*, 1969, 1973). Moreover, immunosuppressive therapy can suppress the development of lesions (Cheema *et al.*, 1972). Immunization of animals with a killed virus vaccine prior to challenge with live virus, on the other hand, increases drastically the severity of disease (Porter and Cho, 1980).

The severity and the spectrum of disease observed after infection of mink with ADV apparently is influenced by the ADV strain used, the host genotype, and the age at which the animal is infected. The relative contribution of these factors to the disease process has been discussed in detail by Porter and Cho (1980). This aspect is also referred to in Section IV. F of this review.

D. Persistent, Clinically Inapparent Infection

As indicated by the high incidence of antiviral antibodies, parvovirus infections are rather common in populations of their natural hosts. This observation suggests that for every virus a potent reservoir must exist from which infection of susceptible animals and reinfection of already immune animals takes place. It is not very likely that this reservoir consists exclusively in those animals experiencing acute clinical disease. With the exception of Aleutian disease of mink, the number of such animals in populations with endemic parvovirus infections is rather small. However, carriers shedding the virus in feces, urine, saliva, and seminal fluid over a long period of time after recovering from acute disease have been noticed. Csiza *et al.* (1971c), for example, reported that after *in utero* infection a kitten was found to disseminate FPLV in urine until it was sacrificed at 12 days of age. Other animals shed the virus in their feces for at least 43 days. Likewise, PPV could be recovered from the seminal fluid of pigs infected *in utero* and sacrificed at the age of 8 months (Johnson and Collings, 1971). Characteristically, such animals are found to have high titers of antibodies to the respective virus. A similar coexistence of antiviral humoral immunity and the presence of virus in the feces has also been reported to occur with RV, MVC, BPV, LPV, and GPV in apparently healthy hosts. In addition, almost all these parvoviruses were repeatedly recovered from kidneys, testicles, and lung tissue of such animals. There is sound evidence that the higher the titer of circulating antiviral antibody, the higher the probability of isolating infectious virus from an animal's tissue. It may be concluded, therefore, that persistent, clinically inapparent infection is the rule rather than the exception with parvoviruses.

The situation is different with ADV infections in mink and ferrets. Here, persistent infection is equivalent to acute clinical disease. In general it takes a progressive, fatal course. This is in particular true for mink of the Aleutian genotype. In non-Aleutian mink, however, up to one quarter of animals may develop a clinically inapparent, or nonprogressive type of ADV infection (Larsen and Porter, 1975; An and Ingram, 1977, 1978). Nevertheless, the animals contain low amounts of ADV in blood and tissues, and constant shedding of the virus seems to occur. Mink with inapparent infections therefore may constitute a continuing unrecognized source of virus for both horizontal and vertical transmission (An and Ingram, 1978).

We still do not know by which mechanisms the persistence of parvoviruses in the presence of a sometimes impressive immunity is maintained. Integration of the nucleic acid of autonomous parvoviruses into the genome of an infected cell has not been reported; yet, it seems to be possible that the virus can survive in contact-inhibited or chemically arrested, noncycling cells to the time this block for virus replication is released (Siegl and Kronauer, unpublished observation). In addition, direct cell-to-cell transfer of newly synthesized virus in tissue of low but continuing mitotic activity could perpetuate infection. Such reduced spread of virus may prevent the development of lesions and, at the same time would be barely accessible to the interruptive activity of humoral immunity.

Because the presence of antibodies to parvoviruses seems to indicate a carrier state rather than to guarantee the absence of the virus, such immune animals may disseminate the virus if introduced into "clean" antibody-negative animal colonies. Moreover, persistent inapparent infections in "immune" animals could provide an explanation of the syndrome "reproductive failure" in pig herds. In this case, latent infection could be activated by the presentation to the virus of rapidly proliferating tissues as a consequence of pregnancy (Cartwright and Huck, 1967; Johnson and Collings, 1969; Mengeling, 1981). It is conceivable that a similar mechanism is responsible for the appearance of parvoviruses in tissues of healing wounds as well as for the repeatedly observed association of parvoviruses with tumors. In this context, derangement of the apparently delicate equilibrium between latent infection and the state of immunity by drugs or other treatment with immunosuppressive effect may play an additional role (El Dadah *et al.*, 1967; Zhdanov and Merekalova, 1962; Payne *et al.*, 1963; Lum and Schreiner, 1963).

E. Parvovirus Infections and Cancer

Much of the interest virologists have taken in parvoviruses at the time of their first isolation and identification around 1960 came from the frequent association of these agents with tumors. As has been already outlined in detail (cf. Introduction), the H-viruses were discovered in the course of serial transplantation of human tumors in rodents (Toolan *et al.*, 1960), RV and many of its isolates frequently appeared in rats bearing tumors or subjected to treatment with cancerogeneous agents (Kilham and Olivier, 1959; Zhdanov and Merekalova, 1962; Payne *et al.*, 1963), and Hallauer *et al.* (1971) recovered parvoviruses at a high frequency from cell cultures derived from tumor tissues. This association of the virus with neoplastic tissues has led to speculations concerning their possible involvement in oncogenesis. In this respect, however, only negative evidence has been obtained (Kilham, 1966; Toolan, 1967a). Accumulation of parvoviruses in tumors can be explained on the basis of their affinity for dividing, constantly proliferating cells, and just contrary to the previously suspected role as oncogenic agents, they are now considered to harbor an antineoplastic potential.

Based on the known osteolytic activity of H-1, Toolan *et al.* (1965) injected the virus intramuscularly into two young patients who suffered from osteosarcoma, the rapid spread of which could not be influenced by conventional treatment. H-1 virus proliferated readily in the patients and multiplication led to cyclic viremia. Unfortunately, however, virus replication proved to be without influence on the growth of the sarcoma. The "oncolytic" potential of parvoviruses has subsequently been studied in tumor-bearing hamsters and rats. Toolan (1967a) observed that hamsters infected soon after birth with H-1 and H-3 virus have an unusually low incidence of spontaneous tumors throughout their lives. Likewise, infection with parvoviruses interfered in a significant way with the induction of leukemias and solid tumors by Moloney leukemia virus and adenovirus 12, respectively (Toolan and Ledinko, 1968; Bergs, 1969). Additional and more detailed experiments are in progress. The respective preliminary results seem to conform to previous observations and indicate that, at least in hamsters, the "oncolytic" effect of H-1 virus infections is demonstrable against the induction of tumors by adenovirus, Simian virus 40, and dimethylbenzanthrene (Soike, personal communication; Rhode, personal communication). It is quite conceivable that the reduced incidence of spontaneous or experimentally induced tumors in infected organisms results from the selective infection and rapid destruction of *de novo* transformed cells. Thereby, both latent and direct infection seem to excert the same effect of antineoplastic surveillance. Already developed and metastasing tumors, however, apparently provide an excellent medium for virus replication but cannot be removed before the immunologic response of the organism prevents further replication and spread of virus.

F. Factors Influencing Pathogenicity

In context with the presentation of the various aspects of disease resulting from infection with parvoviruses it proved continuously necessary to point to factors and conditions favoring or mitigating the course of clinicopathologic events. As more than twenty years of thorough experimental studies have shown, there is a series of such factors. Some control pathogenicity at the level of cellular susceptibility, some at the level of the infected organism, and epidemiologic conditions apparently control susceptibility of whole animal populations. The picture is completed by genetic variation of the infecting virus itself.

The dependence of the replication of autonomous parvoviruses on cells undergoing mitosis or, correctly speaking, on those passing actively through their division cycle, has been established beyond doubt. It is therefore quite understandable that parvovirus replication preferentially is observed in differentiating tissues and also in tissues which, though fully differentiated, have retained a high rate of cellular proliferation. Besides, replication of parvoviruses in general is cytocidal. In consequence, it is to be expected that fulminant replication is reflected by the generation of lesions in affected organs. Potent examples illustrating this relationship between cell proliferation, virus replication, and the development of disease are found in the embryo, the fetus, and in newborn animals (Margolis *et al.*, 1971). Affinity of parvoviruses for "dividing" cells is also reflected by the accumulation of the agents in the callus of healing wounds as well as by their effect on lymphopoietic and hematopoietic tissues in the adult host. However, dependence of virus replication on cycling cells explains only a small facet of parvovirus pathogenicity.

Generalized disease observed in various host species after infection of fetuses or neonates supports the view of Kilham and Margolis (1966b) that parvoviruses are pantropic. In spite of this fact and despite the presence of rapidly proliferating cells in almost every part of the immature organism, generalized, acute disease is a relatively rare event. Rather, both transplacental infections of fetuses and infection of neonates frequently culminate in distinct, well circumscribed syndromes such as cerebellar ataxia, myocarditis, and hepatitis. It is, however, quite unlikely, that, for instance, cerebellar ataxia or myocarditis ensues due to an unique susceptibility of the outer germinal layer of the immature cerebellum or of the differentiating myocardium to infection at the stage of development concerned. These tissues, however, lack the evident regenerative ability of, e.g., the liver. Therefore, it appears to be much more probable that the degree of definite differentiation at the time of infection determines the extent of virus-specific lesions and, hence, simulates a more or less pronounced pathogenic potential of the infecting virus. It should nevertheless be borne in mind that there is convincing experimental evidence for a developmentally controlled, cell-cycle-independent change in a cell's ability to support synthesis of a parvovirus (Tattersall, 1978b).

Another type of host "differentiation" interfering with the pathogenic potential of parvoviruses is illustrated by the ability of the bovine and porcine fetuses to muster a considerable immunologic response against viruses already during the third trimester of gestation. This capability apparently prevents systemic infection of fetuses as well as the induction of distinct lesions. Humoral immunity and differentiation-dependent susceptibility of cells and tissues finally may influence the picture of parvovirus pathogenicity at the level of the newborn and maturing animal. Serologic surveys have shown that parvoviruses in general are endemic in populations of their natural hosts. For example, HI-antibodies to RV were found in rat colonies at an incidence of more than 85% (Robey *et al.*, 1968). In pigs and calves the respective figures are 78% (Bachmann, 1969) and 86% (Abinati and Warfield, 1961; Storz *et al.*, 1972). This high prevalence of passive immunity will prevent infection of the offspring and thus, suppresses manifestation of disease characteristic for this developmental stage of the animal. The efficiency of this mechanism is well illustrated by the changing epidemiologic situation since the recent appearance of CPV in canine populations. The first rush of the virus in a then virgin, antiviral antibody-free canine population was accompanied by manifestation of the enteric form of disease in a high percentage of dogs of all ages. Similarly, the myocardial form was observed in an epidemic extent. Within the past two to three years, however, more than 60–80% of dogs in all parts of the world have developed antibodies to the virus (Walker *et al.*, 1980; Carmichael and Binn, 1981). In parallel to this increase in herd immunity myocarditis has become a rare finding and enteric disease is only observed in puppies exposed to infection at the time maternally transmitted, passive immunity vanishes.

Canine parvovirus enteritis, feline panleukopenia, mink enteritis, and goose hepatitis also may serve to illustrate the importance of copathogens—viruses, bacteria, fungi, and parasites—in the course of parvovirus-induced disease. In every instance, field cases in general take a severe clinical course associated with the development of massive histopathologic lesions. Under controlled experimental conditions and, particularly, in specific pathogen-free or germ-free animals, however, the disease is usually rather mild. Signs of enteric disease are rarely observed; yet, virus replication can be demonstrated in almost all parts of the intestine (cf. Section IV.C.4).

Finally, pathogenicity of parvoviruses clearly is influenced by genetic factors of the infecting virus as well as by the genetic background of the infected host. As far as the virus is concerned, there are many experimental observations indicating that the spectrum of disease may change due to "adaptation" of a virus in the course of *in vitro* or *in vivo* passages. *In vitro* passages thereby tend to yield parvovirus strains of reduced pathogenicity (Kilham and Ferm, 1964; Kilham and Margolis, 1966b, 1969; Paul and Mengeling, 1980) whereas during *in vivo* passages the pathogenicity of the viruses in general is maintained or even enhanced (cf. Kilham and Margolis, 1966b; Toolan, 1968; Carlson *et al.*, 1977). Moreover, recent comparison of the genomes of virus strains distinguished either by low pathogenicity—as in the case of FPLV vaccine viruses (Tratschin *et al.*, 1982)—or by affinity for a distinct type of cell—as in the case of wild-type virus and the lymphotropic variant of MVM (McMaster *et al.*, 1981; Tattersall, 1981; Ron and Tal, 1981)—have shown that such differences are paralleled by changes at the level of the viral genome. The viral functions affected by these changes and, hence, responsible for the variation in pathogenicity are widely unknown. It may be speculated that they control affinity of the virus for a certain type of cell, ability and efficiency of virus replication, or the type of virus particles (e.g., defective interfering particles) produced. If, for example, all these functions result in slow spread of the virus in the organism, its immunologic system of surveillance as well as the regenerative activity of the affected tissues may be able to keep virus-specific damage at a minimum.

Virus strains of greatly varying virulence have also been described for ADV (cf. Porter and Cho, 1980). In the case of this syndrome, however, the genetic background of the infected animal seems to play an extraordinary role with respect to the severity and course of clinical disease. Classic Aleutian disease was first described in mink homozygous for the recessive Aleutian coat color (Hartsough and Gorham, 1956). This genetic constellation apparently determines a lysosomal abnormality similar to that of the Chediak-Higashi syndrome. It was speculated that the same gene may increase susceptibility of mink to infection with ADV (Padgett *et al.*, 1967; Johnson *et al.*, 1975). Whether "susceptibility" in the sense used throughout this chapter in fact is responsible for the differences in disease observed in Aleutian-type and non-Aleutian-type mink remains to be shown. Nevertheless, mink with the said genetic background typically die of renal failure three or five months after infection. On the other hand, mink of non-Aleutian type rarely die earlier than five months after infection and most survive for more than one year (Porter and Cho, 1980).

In summary, the most evident factors controlling parvovirus pathogenicity are the genetic structure of the infecting virus, the genetic background of the infected host, the developmental stage of the host, the presence of viral, bacterial, fungal, or parasitic copathogens and, last but not least, the overall prevalence of the virus.

ACKNOWLEDGMENTS. I am greatly indebted to many colleagues who provided information and papers prior to publication. Thanks are also due to B. Thomson, who typed the manuscript.

REFERENCES

Abinanti, F.R., and Warfield, M.S., 1961, Recovery of a hemadsorbing virus (HADEN) from gastrointestinal tract of calves, *Virology* **14:**288.

Al-Lami, F., Ledinko, N., and Toolan, H., 1969, Electron microscope study of human NB and SMH cells infected with the parvovirus H-1: Involvement of the nucleus, *J. Gen. Virol.* **5:**485.

An, S.H., and Ingram, D.G., 1977, Detection of inapparent Aleutian disease virus infection in mink, *Am. J. Vet. Res.* **38:**1619.

An, S.H., and Ingram, D.G., 1978, Transmission of Aleutian disease from mink with inapparent infections, *Am. J. Vet. Res.* **39:**309.

Appel, M.J.G., Cooper, B.J., Greisen, H., and Carmichael, L.E., 1978, Status report: Canine viral enteritis, *J. Am. Vet. Assoc.* **173:**1516.

Bachmann, P.A., 1969, Vorkommen und Verbreitung von Picodna (Parvo)-Virus beim Schwein, *Zentralbl. Veterinaermed. Reihe B* **16:**341.

Bachmann, P.A., 1971, Properties of a bovine parvovirus, *Zentralbl. Veterinaermed. Reihe B.* **18:**80.

Bachmann, P.A., and Danner, K., 1976, Porcine parvovirus infection in vitro: A study model for the replication of parvoviruses. II. Kinetics of virus and antigen production, *Zentralbl. Veterinaermed. Reihe B.* **23:**355.

Bachmann, P.A., Sheffy, B.E., and Vaughan, J.T., 1975, Experimental in utero infection of fetal pigs with a porcine parvovirus, *Infect. Immun.* **12:**455.

Baer, P.N., and Kilham, L., 1962a, Rat virus and peridontal disease. I. The peridontum in mongoloid hamsters, *Oral Surg. Oral Med. Oral Pathol.* **15:**756.

Baer, P.N., and Kilham, L., 1962b, Rat virus and peridontal disease, II. Onset and effect of age and time of inoculation, *Oral Surg. Oral Med. Oral Pathol.* **15:**1302.

Baer, P.N., and Kilham, L., 1964a, Rat virus and peridontal disease. III. The histopathology of early lesions in the first molar, *Oral Surg. Oral Med. Oral Pathol.* **17:**116.

Baer, P.N., and Kilham, L., 1964b, Rat virus and periodontal disease. IV. The aged hamster, *Oral Surg. Oral Med. Oral Pathol.* **18:**803.

Baer, P.N., and Kilham, L., 1965, A comparison of the effects of four viruses on the periodontium of the Syrian hamster, *Periodontol. Res.* **36:**127.

Baer, P.N., and Kilham, L., 1974, Dental defects in hamsters infected with minute virus of mice, *Oral Surg.* **37**:385.

Baer, P.N., Garrington, G.E., and Kilham, L., 1971a, Effect of age and H-1 virus on healing fractures in hamsters, *J. Geront.* **26**:373.

Baer, P.N., Coccaro, P.J., Baer, M.J., and Kilham, L., 1971b, Craniofacial manifestations of virus induced mongolism in the hamster and Down's syndrome in man, *Am. J. Orthod.* **60**:221.

Bass, R., and Hetrick, F.M., 1978, Persistent infection of a human lymphocyte cell line (Molt-4) with the Kilham rat virus, *J. Infect. Dis.* **137**:210.

Bates, R.C., and Storz, J., 1973, Host cell range and growth characteristics of bovine parvoviruses, *Infect. Immun.* **7**:398.

Bates, R.C., and Storz, J., 1974, Morphogenesis of bovine parvoviruses and associated cellular changes, *Exp. Mol. Pathol.* **20**:208.

Bates, R.C., Storz, J., and Reed, D.E., 1972, Isolation and comparison of bovine parvoviruses, *J. Infect. Dis.* **126**:531.

Bazeley, P.O., 1976, The nature of Aleutian disease in mink. I. Two forms of hypergammaglobulinemia as related to method of disease transmission and type of lesion, *J. Infect. Dis.* **134**:252.

Bergs, V.V., 1969, Rat virus-mediated suppression of leukemia induction by Moloney virus in rats, *Cancer Res.* **29**:1669.

Bergs, V.V., and Scotti, T.M., 1967, Virus-induced peliosis hepatitis in rats, *Science* **128**:377.

Bernhard, W., Kasten, F.H., and Chany, Th., 1963, Etude cytochimique et ultrastructurale de cellules infectées par le virus K du rat et le virus H-1, *C.R. Acad. Sci.* **257**:1566.

Berquist, K.R., Maynard, J.E., Sheller, M., and Schable, C.A., 1972, Comparative studies of hepatitis candidate agents and parvovirus in Detroit 6 cell cultures, *J. Infect. Dis.* **126**:203.

Binn, L.N., Lazar, E.C., Eddy, G.A., and Kajima, M., 1970, Recovery and characterization of a minute virus of canines, *Infect. Immun.* **1**:503.

Bittle, J.L., Emery, J.B., York, C.J., and McMillen, J.K., 1961, Comparative study of feline cytopathogenic viruses and feline panleukopenia virus, *Am. J. Vet. Res.* **22**:374.

Black, J.W., Holscher, M.A., and Povell, H.S., 1979, Parvoviral Enteritis and panleukopenia in dogs, *Vet. Med. Sm. An. Clin.* **74**:47.

Bloom, M.E., Race, R.E., Hadlow, W.J., and Chesebro, R., 1975, Aleutian disease of mink: the antibody response of sapphire and pastel mink to Aleutian disease virus, *J. Immunol.* **115**:1034.

Bloom, M.E., Race, R.E., and Wolfinbarger, J.B., 1980, Characterization of Aleutian disease virus as a parvovirus, *J. Virol.* **35**:836.

Bonnard, G.D., Manders, E.K., Campbell, D.A., Herberman, R.B., and Collins, M.J., 1976, Immunosuppressive activity of a subline of the mouse EL-4 lymphoma. Evidence for minute virus of mice causing the inhibition, *J. Exp. Med.* **143**:187.

Bouillant, A., and Hanson, R.P., 1965, Epizootiology of mink enteritis. III. Carrier state in mink, *Can. J. Comp. Med.* **29**:183.

Brailowsky, C., and Chany, C., 1965, Un facteur produit par l'adenovirus 12 en culture cellulaire, stimulant la multiplication du virus K du rat, *C.R. Acad. Sci.* **260**:2634.

Carlson, J.H., and Scott, F.W., 1977, Feline panleukopenia. II. The relationship of intestinal mucosal cell proliferation rates to viral infection and development of lesions, *Vet. Pathol.* **14**:173.

Carlson, J.H., Scott, F.W., and Duncan, J.R., 1977, Feline panleukopenia. I. Pathogenesis in germfree and specific pathogen free cats, *Vet. Pathol.* **14**:79.

Carlson, J.H., Scott, F.W., and Duncan, J.R., 1978, Feline panleukopenia. III. Development of lesions in lymphoid tissues, *Vet. Pathol.* **15**:383.

Carmichael, L.E., and Binn, L.N., 1981, New enteric viruses in the dog, in: *Advances in Veterinary Science and Comparative Medicine* (C.E. Cornelius and C.F. Simpson, eds.), Vol. 25, p. 1–37, Academic Press, New York.

Carmichael, L.E., Jaubert, J.C., and Pollock, R.V.H., 1980, Hemagglutination by canine parvovirus: Serologic studies and diagnostic applications, *Am. J. Vet. Res.* **41**:784.

Carter, B.J., Laughlin, C.A., de la Maza, L.M., and Myers, M., 1979, Adeno-associated virus autointerference, *Virology* **92**:449.

Cartwright, S.F., and Huck, R.A., 1967, Viruses isolated in association with herd infectivity, abortions and still births in pigs, *Vet. Rec.* **81**:196.

Cartwright, S.F., Lucas, M., and Huck, R.A., 1969, A small hemagglutinating porcine DNA virus. I. Isolation and properties, *J. Comp. Pathol.* **79**:371.

Castro, A., Cambria, M., and Russo, G., 1971a, Chemical characteristics of the erythrocyte receptors for picodnaviruses RV, H-1, and X 14, *Boll. Ist. Sieroter. (Milan)* **50**:107.

Castro, A., Nicoletti, G., and Cocuzza, G., 1971b, Characteristics of hemagglutinins of the picodnaviruses (RV, H-1, X 14), *Arch. Gesamte Virusforsch.* **34**:261.

Chany, C.H., and Brailovsky, C., 1965, Les stimulons, facteurs antagonistes de l'interféron favorisant la multiplication intracellulaire des virus, *C.R. Acad. Sci.* **261**:4282.

Cheema, A., Henson, J.B., and Gorham, J.R., 1972, Aleutian disease of mink. Prevention of lesions by immunosuppression, *Am. J. Pathol.* **66**:543.

Cho, H.J., and Ingram, D.G., 1972, Antigen and antibody in Aleutian disease in mink. I. Precipitation reaction by agar-gel electrophoresis, *J. Immunol.* **108**:555.

Cho, H.J., and Ingram, D.G., 1973, Antigen and antibody in Aleutian disease in mink. II. The reaction of antibody with the Aleutian disease agent using immunodiffusion and immunoelectrophoresis, *Can. J. Comp. Med.* **37**:217.

Clarke, D.H., and Casals, J., 1958, Techniques for hemagglutination and hemagglutination-inhibition with arthropod-borne viruses, *Am. J. Trop. Med. Hyg.* **7**:561.

van Cleef, S.A.M., and Miltenburg, J.T., 1966, En erustige acuut verlopende virusziekte bij ganzekuikens gepaarde met grote sterfte, *Tijdschr. Diergenesk.* **91**:6.

Clinton, M., and Hayashi, M., 1976, The parvovirus MVM: A comparison of heavy and light particle infectivity and their density conversion in vitro, *Virology* **74**:57.

Cocuzza, G., and Russo, G., 1969, Erythrozyte receptors in hemagglutination by picodnaviruses, *Boll. Ist. Sieroter. Milan.* **48**:204.

Cohen, M.M., and Shklar, G., 1964, Deformities of the cranofacial and dental complex in the H-1 virus modified hamster, *Oral Surg., Oral Med. Oral Pathol.* **17**:533.

Cohen, M.M., Winer, R.A., Schwartz, S., and Shklar, G., 1961, Oral aspects of mongolism. Part I. Periodontal disease in mongolism, *Oral Surg.* **14**:92.

Cole, G.A., and Nathanson, N., 1969, Immunofluorescent studies of the replication of rat virus (HER strain) in tissue culture, *Acta Virol.* **13**:515.

Cole, G.A., Nathanson, N., and Rivet, H., 1970, Viral hemorrhagic encephalopathy of rats, II. Pathogenesis of central nervous system lesions, *Am. J. Epidemiol.* **91**:339.

Cossart, Y.E., Field, A.M., Cant, B., and Widdows, D., 1975, Parvovirus like particles in human sera, *Lancet* **I**:72.

Coudert, M., Fedida, M., Dannacher, G., and Peillon, M., 1974, The late form of parvovirus disease of gosling, *Rec. Med. Vet.* **150**:899.

Crawford, L.V., 1966, A minute virus of mice, *Virology* **29**:605.

Croft, G.F., Hoggan, M.D., and Johnson, F.B., 1974, Production and reactivity of immune sera specific for HADEN virus polypeptide antigens, *J. Virol.* **13**:608.

Croghan, D.L., Matchett, A., and Koski, T.A., 1973, Isolation of porcine parvovirus from commercial trypsin, *Appl. Microbiol.* **26**:431.

Cross, S.S., and Parker, J.C., 1972, Some antigenic relationships of the murine parvoviruses: Minute virus of mice, rat virus and H-1 virus, *Proc. Soc. Exp. Biol. Med.* **139**:105.

Csiza, C.K., Scott, F.W., de Lahunta, A., and Gillespie, J.H., 1971a, Pathogenesis of feline panleukopenia virus in susceptible newborn kittens. I. Clinical signs, hematology, serology, and virology, *Infect. Immun.* **3**:833.

Csiza, C.K., de Lahunta, A., Scott, F.W., and Gillespie, J.H., 1971b, Pathogenesis of feline panleukopenia virus in susceptible newborn kittens. II. Pathology and immunofluorescence, *Infect. Immun.* **3**:838.

Csiza, C.K., Scott, F.W., de Lahunta, A., and Gillespie, J.H., 1971c, Immune carrier state of feline panleukopenia virus-infected cats, *Am. J. Vet. Res.* **32**:419.

Cutlip, R.C., and Mengeling, W.L., 1975, Pathogenesis of in utero infection of eight and ten-week-old porcine fetuses with porcine parvovirus, *Am. J. Vet. Res.* **36**:1751.

Dalldorf, G., 1960, Viruses and human cancer, *Bull. N.Y. Acad. Med.* **36**:795.

Dannacher, G., Fouillet, X., Coudert, M., Fedida, M., and Peillon, M., 1974, Etiologie de la maladie a virus de l'oison: le virus β. *Rec. Med. Vet.* **150**:49.

Dawe, C.J., Kilham, L., and Morgan, W.D., 1961, Intranuclear inclusions in tissue cultures infected with rat virus, *J. Natl. Cancer Inst.* **27**:221.

Derzsy, D., 1967, A viral disease of goslings. I. Epidemiological, clinical, pathological and aetiological studies, *Acta Vet. Acad. Sci. Hung.* **17**:443.

Drommer, W., and Trautwein, G., 1975, Die Pathogenese der Aleutenkrankheit der Nerze. VII. Chronische Hepatitis mit Gallengangproliferation, *Vet. Pathol.* **12**:77.

Duenwald, J.C., Holland, J.M., Gorham, J.R., and Ott, R.L., 1971, Feline panleukopenia: Experimental cerebellar hypoplasia produced in neonatal ferrets with live virus vaccine, *Res. Vet. Sci.* **12**:394.

Eklund, C.M., Hadlow, W.J., Kennedy, R.C., Boyle, C.C., and Jackson, T.A., 1968, Aleutian disease of mink. Properties of the etiologic agent and the host responses, *J. Infect. Dis.* **118**:510.

El Dadah, A.N., Nathanson, N., Smith, K.O., Squire, R.A., Santos, G.W., and Melby, E.C., 1967, Viral hemorrhagic encephalopathy of rats, *Science* **156**:392.

Engers, H.D., Louis, J.A., Zubler, R.H., and Hirt, B., 1981, Inhibition of T-cell mediated functions by MVM (i), a parvovirus closely related to minute virus of mice, *J. Immunol.* **127**:2280.

Engler, W.O., Baer, P.N., and Kilham, L., 1966, Effects of rat virus on healing osseous wounds, *Arch. Pathol.* **82**:93.

Eugster, A.K., 1980, Studies of canine parvovirus infections: Development of an inactivated vaccine, *Am. J. Vet. Res.* **41**:2020.

Ferm, V.H., and Kilham, L., 1963, Rat virus (RV) infection in fetal and pregnant hamsters, *Proc. Soc. Exp. Biol. Med.* **112**:623.

Ferm, V.H., and Kilham, L., 1964, Congenital anomalies induced in hamster embryos with H-1 virus, *Science* **145**:510.

Ferm, V.H., and Kilham, L., 1965a, Skeletal studies of virus-induced dwarfism, *Growth* **29**:7.

Ferm, V.H., and Kilham, L., 1965b, Histopathologic basis of the teratogenic effects of H-1 virus on hamster embryos, *J. Embryol. Exp. Morphol.* **13**:151.

Flower, R.L.P., Wilcox, G.E., and Robinson, W.F., 1980, Antigenic differences between canine parvovirus and feline panleucopenia virus, *Vet Rec.* **107**:254.

Fowler, E.H., and Rohovsky, M.W., 1970, Enzyme histochemistry of the small intestine in germfree and specific-pathogen-free cats inoculated with feline infectious enteritis virus. Enzyme histochemistry of lymphoid tissues in germfree cats inoculated with feline enteritis, *Am. J. Vet. Res.* **31**:2055.

Gagnon, A.N., and Povey, R.C., 1979, A possible parvovirus associated with an epidemic gastroenteritis of dogs in Canada, *Vet. Rec.* **104**:263.

Galton, M., and Kilham, L., 1966, Chromosomes of "mongoloid" hamsters, *Proc. Soc. Exp. Biol. Med.* **122**:18.

Gierthy, J.F., Ellem, K.A.O., and Singer, I.I., 1974, Environmental pH and the recovery of H-1 parvovirus during single cycle infection, *Virology* **60**:548.

Gorham, J.R., Leader, R.W., and Henson, J.B., 1964, The experimental transmission of a virus causing hypergammaglobulinemia in mink: Sources and modes of infection, *J. Infect. Dis.* **114**:341.

Gorham, J.R., Hartsough, G.R., Sato, N., and Lust, S., 1966, Studies on cell culture adapted feline panleukopenia virus–virus neutralization and antigenic extinction, *Vet. Med.* **61**:35.

Gorham, J.R., Leader, R.W., Crawford, T.B., and Padgett, G.A., 1976, The epizootiology of Aleutian disease, in: *Slow Virus Diseases of Animals and Man* (R.H. Kimberlin, ed.), pp. 135–158, North-Holland, Amsterdam.

Goto, H., 1975, Feline panleukopenia in Japan. II. Hemagglutinability of the isolated virus, *Jpn. J. Vet. Sci.* **37**:431.

Goto, H., and Ishiko, H., 1976, Feline panleukopenia in Japan. III. Propagation of the virus in cat kidney cell cultures, *Jpn. J. Vet. Sci.* **38**:65.

Greene, E.L., 1964, Ph.D. thesis, Cornell University, Ithaca, N.Y.

Greene, E.L., 1965, Physical and chemical properties of H-1 virus. I. pH and heat stability of the hemagglutinating property, *Proc. Soc. Exp. Biol. Med.* **118**:973.

Guglielmino, S., Tempera, G., Pappalardo, G., and Castro, A., 1978, H-1 and X14 parvovirus antibodies in women with abortions or still-births, *Acta Virol.* **22**:426.

Hahn, E.C., Ramos, L., and Kenyon, A.J., 1977a, Expression of Aleutian mink disease antigen in cell culture, *Infect. Immun.* **15**:204.

Hahn, E.C., Ramos, L., and Kenyon, A.J., 1977b, Properties of Aleutian disease virus assayed with feline kidney cells, *Arch. Virol.* **55**:315.

Hallauer, C., and Kronauer, G., 1960, Nachweis von Gelbfieber virus-Hemagglutinin in menschlichen Explantaten, *Arch. Gesamte Virusforsch.* **10**:267.

Hallauer, C., and Kronauer, G., 1962, Nachweis eines nicht identifizierten Hámagglutinins in meschlichen Tumorzellstämmen, *Arch. Gesamte Virusforsch.* **11**:754.

Hallauer, C., and Kronauer, G., 1965, Extraction of cell-associated virus without damage of the culture, *Arch. Gesamte Virusforsch.* **15**:433.

Hallauer, C., Kronauer, G., and Siegl, G., 1971, Parvoviruses as contaminants of permanent human cell lines. I. Virus isolations from 1960–1970, *Arch. Gesamte Virusforsch.* **35**:80.

Hallauer, C., Siegl, G., and Kronauer, G., 1972, Parvoviruses as contaminants of permanent human cell lines. III. The biologic properties of the isolated viruses, *Arch. Gesamte Virusforsch.* **38**:366.

Hammon, W.D., and Enders, J.F., 1939a, A virus disease of cats, principally characterized by aleucocytosis, enteric lesions and the presence of intranuclear inclusion bodies, *J. Exp. Med.* **69**:327.

Hammon, W.D., and Enders, J.F., 1939b, Further studies on the blood and the hematopoietic tissues in malignant panleukopenia of cats, *J. Exp. Med.* **70**:557.

Hampton, E.G., 1970, H-1 virus growth in synchronized rat embryo cells, *Can. J. Microbiol.* **16**:266.

Hartsough, C.F., and Gorham, J.R., 1956, Aleutian disease of mink, *Natl. Fur. News* **28**:10.

Hayes, M.A., Russell, R.G., and Babiuk, L.A., 1979, Sudden death in young dogs with myocarditis caused by parvovirus, *J. Am. Vet. Med. Assoc.* **174**:1197.

Henry, C.J., and Diorio, A.F., 1975, Induction of hepatitis in adult Syrian hamsters by H–1 virus, *Proc. Soc. Exp. Biol. Med.* **149**:23.

Henson, J.B., Gorham, J.R., Leader, R.W., and Wagner, B.M., 1962, Experimental hypergammaglobulinemia in mink, *J. Exp. Med.* **116**:357.

Henson, J.B., Leader, R.W., Gorham, J.R., and Padgett, G.A., 1966, The sequential development of lesions in spontaneous Aleutian disease of mink, *Pathol. Vet.* **3**:289.

Henson, J.B., Gorham, J.R., and Padgett, G.A., 1969, Pathogenesis of the glomerular lesions in Aleutian disease of mink: Immunofluorescent studies, *Arch. Pathol.* **87**:21.

Herringham, W.P., and Andrewes, F.W., 1888, Two cases of cerebellar disease in cats with staggering, *St. Barth's Hosp.* **24**:112.

Hinaidy, B., Messner, A., and Bürki, F., 1979, Bovine Parvoviren—Isolierung in Zellkulturen, Zytopathologie und Kulturausbeute, *Wien. Tieraerztl. Mschr.* **66**:359.

Hoekstra, J., Smit, Th., and van Brakel, C., 1973, Observations on host range and control of goose virus hepatitis, *Avian Pathol.* **2**:169.

Horzinek, M., Mussgay, M., Maess, J., and Petzoldt, K., 1967, Nachweis dreier Virusarten (Schweinepest-, Adeno-, Picodna-Virus) in einem als cytopathogen bezeichneten Schweinepest-Virusstamm, *Arch. Gesamte Virusforsch.* **21**:98.

Inaba, Y., Omori, T., Kono. M., Ishii, S., and Matumoto, M., 1973a, A new serotype of bovine parvovirus, *Jpn, J. Microbiol.* **17**:85.

Inaba, Y., Kurogi, H., Takahashi, E., Sato, K., Tanaky, Y., Goto, Y., Omori, T., and Matumoto, M., 1973b, Isolation and properties of bovine parvovirus type I from Japanese calves, *Arch. Gesamte Virusforsch.* **42**:54.

Ingram, D.G., and Cho, H.J., 1974, Aleutian disease in mink: Virology, immunology and pathogenesis, *J. Rheumatol.* **1**:74.

Jacobs, R., Weiser, M.G., Hall, R.L., and Kowalski, J.J., 1980, Clinicopathologic features of canine parvoviral enteritis, *J. Am. Anim. Hosp. Assoc.* **16**:809.

Johnson, M.I., Henson, J.B., and Gorham, J.R., 1975, The influence of genotype on the development of glomerular lesions in mink with Aleutian disease virus. A correlated light, fluorescent, and electron microscopic study, *Am. J. Pathol.* **81**:321.

Johnson, R.H., 1964, Isolation of a virus from a condition stimulating feline panleukopenia in a leopard, *Vet. Rec.* **76**:1008.

Johnson, R.H., 1965a, Feline panleukopenia. I. Identification of a virus associated with the syndrome, *Res. Vet. Sci.* **6**:466.

Johnson, R.H., 1965b, Feline panleukopenia virus. II. Some features of the cytopathic effects in feline kidney monolayers, *Res. Vet. Sci.* **6**:472.

Johnson, R.H., 1967a, Feline panleukopenia virus. I.V. Methods for obtaining reproducible in vitro results, *Res. Vet. Sci.* **8**:256.

Johnson, R.H., 1967b, Feline panleukopenia virus—In vitro comparison of strains with a mink enteritis virus, *J. Small Anim. Pract.* **8**:319.

Johnson, R.H., and Collings, D.F., 1969, Experimental infection of piglets and pregnant gilts with a parvovirus, *Vet. Rec.* **85**:446.

Johnson, R.H., and Collings, D.F., 1971, Tranplacental infection of piglets with a porcine parvovirus, *Res. Vet. Sci.* **12**:570.

Johnson, R.H., and Cruickshank, J.G., 1966, Problems in classification of feline panleukopenia virus, *Nature (London)* **212**:622.

Johnson, R.H., and Spradbrow, P.B., 1979, Isolation from dogs with severe enteritis of a parvovirus related to feline panleukopenia virus, *Aust. Vet. J.* **55**:151.

Johnson, R.H., Margolis, G., and Kilham, L., 1967, Identity of feline ataxia virus with feline panleukopenia virus, *Nature, (London)* **214**:175.

Johnson, R.H., Siegl, G., and Gautschi, M., 1974, Characteristics of feline panleukopenia virus strains enabling definitive classification as parvoviruses, *Arch. Gesamte Virusforsch.* **46**:315.

Johnson, R.H., Donaldson-Wood, C.R., and Joo, H.J., 1976, Observations on the epidemiology of porcine parvovirus, *Aust. Vet. J.* **52**:80.

Joo, H.S., Donaldson-Wood, C.R., and Johnson, R.H., 1975, A microneutralization test for the assay of porcine parvovirus antibody, *Arch. Virol.* **47**:337.

Joo, H.S., Donaldson-Wood, C.R., and Johnson, R.H., 1976a, Observations on the pathogenesis of porcine parvovirus infection, *Arch. Virol.* **51**:123.

Joo, H.S., Donaldson-Wood, C.R., and Johnson, R.H., 1976b, A standardised, haemagglutination inhibition test for porcine parvovirus antibody, *Aust. Vet. J.* **52**:422.

Joo, H.S., Donaldson-Wood, C.R., Johnson, R.H., and Campbell, R.R.F., 1977, Pathogenesis of porcine parvovirus infection: Pathology and immunofluorescence in the fetus, *J. Comp. Pathol.* **87**:383.

Joo, H.S., Johnson, R.H., and Watson, D.L., 1978, Serological procedures to determine time of infection of pigs with porcine parvovirus, *Aust. Vet. J.* **54**:125.

Kääriäinen, L., Kangas, J., Keränen, S., Nyholm, M., and Weckström, P., 1966, Studies on mink enteritis virus, *Arch. Gesamte Virusforsch.* **19**:197.

Kajioka, J., Taguchi, F., and Nagaki, D., 1974, Plaque formation and replication of minute virus of mice, *Arch. Exp. Med.* **47**:33.

Karasaki, S., Toolan, H.W., and Usategui-Gomez, M., 1965, A human placental fluid inhibitor to hemagglutination by H-1 and HB-viruses. II. Electron microscopic studies, *Proc. Soc. Exp. Biol. Med.* **120**:391.

Karstad, L., and Pridham, T.J., 1962, Aleutian disease of mink: Evidence of its viral etiology, *Can. J. Comp. Med.* **26**:97.

Kelly, W.R., 1978, An enteric disease of dogs resembling feline panleukopenia, *Aust. Vet. J.* **54**:593.

Kelly, W.R., and Atwell, R.B., 1979, Diffuse subacute myocarditis of possible viral aetiology—A cause of sudden death in pups, *Aust. Vet. J.* **55**:36.

Kenyon, A.J., 1966, Immunologic deficiency in Aleutian disease of mink, *Am. J. Vet. Res.* **27**:1780.

Kenyon, A.J., Helmboldt, C.F., and Nielsen, S.W., 1963, Experimental transmission of Aleutian disease with urine, *Am. J. Vet. Res.* **24**:1066.

Kenyon, A.J., Magnano, T., Helmboldt, C.F., and Buko, L., 1966, Aleutian disease in the ferret, *J. Am. Vet. Med. Assoc.* **149**:920.

Kenyon, A.J., Howard, E., and Buko, L., 1967, Hypergammaglobulinemia in ferrets with lymphoproliferative lesions (Aleutian disease), *Am. J. Vet. Res.* **28**:1167.

Kenyon, A.J., Kenyon, B.J., and Hahn, E.C., 1978, Protides of the mustelidae: Immunoresponse of mustelids to Aleutian mink disease virus, *Am. J. Vet. Res.* **39**:1011.

Kikuth, W., Gönnert, R., and Schweickert, M., 1940, Infektiöse Aleukozytose der Katzen, *Zentralbl. Bakt. I. Orig.* **146**:1.

Kilham, L., 1961a, Rat virus (RV) infections in hamsters, *Proc. Soc. Exp. Biol. Med.* **106**:825.

Kilham, L., 1961b, Mongolism associated with rat virus (RV) infection in hamsters, *Virology* **13**:141.

Kilham, L., 1966, Viruses of laboratory and wild rats, *Natl. Cancer Inst. Monogr.* **20**:117.

Kilham, L., and Ferm, V.H., 1964, Rat virus (RV) infections of pregnant, fetal and newborn rats, *Proc. Soc. Exp. Biol. Med.* **117**:874.

Kilham, L., and Margolis, G., 1964, Cerebellar ataxia in hamsters inoculated with rat virus, *Science* **143**:1047.

Kilham, L., and Margolis, G., 1965, Cerebellar disease in cats induced by inoculation of rat virus, *Science* **148**:244.

Kilham, L., and Margolis, G., 1966a, Viral etiology of spontaneous ataxia of cats, *Am. J. Pathol.* **48**:991.

Kilham, L., and Margolis, G., 1966b, Spontaneous hepatitis and cerebellar hypoplasia in suckling rats due to congenital infection with rat virus, *Am. J. Pathol.* **49**:457.

Kilham, L., and Margolis, G., 1969, Transplacental infection of rats and hamsters induced by oral and parenteral inoculations of H-1 and rat viruses (RV), *Teratology* **2**:111.

Kilham, L., and Margolis, G., 1970, Pathogenicity of minute virus of mice (MVM) for rats, mice, and hamsters, *Proc. Soc. Exp. Biol. Med.* **133**:1447.

Kilham, L., and Margolis, G., 1971, Fetal infections of hamsters, rats and mice induced with the minute virus of mice (MVM), *Teratology* **4**:43.

Kilham, L., and Margolis, G., 1970, Pathogenicity of minute virus of mice (MVM) for rats, mice, and hamsters, *Proc. Soc. Exp. Biol. Med.* **133**:1447.

Kilham, L., and Margolis, G., 1971, Fetal infections of hamsters, rats and mice induced with the minute virus of mice (MVM), *Teratology* **4**:43.

Kilham, L., and Margolis, G., 1975, Association of human concern arising from animal models of intrauterine and neonatal infections due to viruses: A review I. Introduction and virologic studies, *Progr. Med. Virol.* **20**:113.

Kilham, L., and Molony, V.B., 1964, Association of rat virus and Moloney leukemia virus in tissues of inoculated rats, *J. Natl. Cancer Inst.* **32**:523.

Kilham, L., and Olivier, L.J., 1959, A latent virus of rats isolated in tissue culture, *Virology* **7**:428.

Kilham, L., Margolis, G., and Colby, E.D., 1967, Congenital infections of cats and ferrets by feline panleukopenia virus manifested by cerebellar hypoplasia, *Lab. Invest.* **17**:465.

Kilham, L., Buchler, C.E., Ferm, V.H., and Baron, S., 1968, Production of interferon during rat virus infection, *Proc. Soc. Exp. Biol. Med.* **129**:274.

Kilham, L., Margolis, G., and Colby, E.D., 1970, Enhanced proliferation of H-1 virus in livers of rats infected with Cysticercus fasciolaris, *J. Infect. Dis.* **121**:648.

Kilham, L., Margolis, G., and Colby, E.D., 1971, Cerebellar Ataxia and its congenital transmission in cats by feline panleukopenia virus, *J. Am. Vet. Med. Ass.* **158**:888.

King, D.A., and Crogham, D.L., 1965, Immunofluorescence of feline panleukopenia virus in cell culture: Determination of immunological status of felines by serum neutralization, *Can. J. Comp. Med.* **29**:85.

Kisary, J., 1974a, Cross-neutralization tests on parvoviruses isolated from goslings, *Avian Pathol.* **3**:293.

Kisary, J., 1974b, Some growth characteristics of goose parvovirus strain B, *Acta Vet. Acad. Sci. Hung.* **24**:329.

Kisary, J., 1979, Interaction in replication between the goose parvovirus strain B and duck plaque herpesvirus, *Arch. Virol.* **59**:81.

Kisary, J., and Derzsy, D., 1974, A viral disease of goslings. IV. Characterization of the causal agent in tissue culture system, *Acta Vet. Acad. Sci. Hung.* **24**:287.

Kongsvik, J.R., and Toolan, H.W., 1972a, Capsid components of the parvovirus H-1, *Proc. Soc. Exp. Biol. Med.* **139**:1202.

Kongsvik, J.R., and Toolan, H.W., 1972b, Effect of proteolytic enzymes on the hemagglutinating property of the parvoviruses H-1, H-3, and RV, *Proc. Soc. Exp. Biol. Med.* *140*:140.

Kongsvik, J.R., Singer, I.I., and Toolan, H.W., 1974, Studies on the red cell and antibody reactive sites of the parvovirus H-1: effect of fixatives, *Proc. Soc. Exp. Biol. Med.* **145**:763.

Kongsvik, J.R., Hopkins, M.S., and Ellem, K.A.O., 1978, Two populations of infectious virus produced during H-1 infection of synchronized transformed cells, in: *Replication of Mammalian Parvoviruses* (D.C. Ward and P. Tattersall, eds.), pp. 505–520, Cold Spring Harbor Laboratory, Cold Spring Harbor, New York.

Konishi, S., Mochizuki, M., and Ogata, M., 1975, Studies on feline panleukopenia. I. Isolation and properties of virus strains, *Jpn. J. Vet. Sci.* **37**:439.

Krauss, H., 1965, Eine verlustreiche Aufzuchtkrankheit bei Gänseküken (Erste Mitteilung), *Berl. Meunch. Tieraerztl. Wochenschr.* **78**:372.

Landauer, K., Kilham, L., and Buchtel, H.A., 1967, Behavioral characteristics associated with the rat-virus-induced "hamster mongolism" syndrome, *J. Psychiat. Res.* **5**:95.

Langheinrich, K.A., and Nielsen, S.W., 1971, Histopathology of feline panleukopenia: Report of 65 cases, *J. Am. Vet. Med. Assoc.* **158**:863.

Larsen, A.E., and Porter, D.D., 1975, Pathogenesis of Aleutian disease in mink: Identification of non-persistent infections, *Infect. Immun.* **11**:92.

Larsen, S.A., Flagstad, A., and Aalbaek, B., 1976, Experimental feline panleukopenia in the conventional cat, *Vet. Pathol.* **13**:216.

Lawrence, J.S., and Syverton, J.T., 1938, Spontaneous agranulocytosis in the cat, *Proc. Soc. Exp. Biol. Med.* **38**:914.

Lawrence, J.S., Syverton, J.T., Shaw, J.S. Jr., and Smith, F.P., 1940, Infectious feline agranulocytosis, *Am. J. Pathol.* **16**:333.

Lawrence, J.S., Syverton, J.T., Ackart, R.J., Adams, W.S., Ervin, D.M., Haskins, A.L., Jr., Sounders, R.H., Jr., Stringfellow, M.B., and Wetrick, R.M., 1943, The viruses of infectious feline agranulocytosis. II. Immunological relations to other viruses, *J. Exp. Med.* **77**:57.

Leary, J.J., and Storz, J., 1980, Nucleopathic changes in parvovirus-infected, cultured cells, *Exp. Mol. Pathol.* **32**:188.

Ledinko, N., 1967, Plaque assay of the effects of cytosine arabinoside and 5-iodo-deoxiuridine on the synthesis of H-1 virus particles, *Nature* (*London*) **214**:1346.

Ledinko, N., and Toolan, H.W., 1968, Human adenovirus type 12 as a "helper" for growth of H-1 virus, *J. Virol.* **2**:155.

Ledinko, N., Hopkins, S., and Toolan, H., 1969, Relationship between potentiation of H-1 growth by human adenovirus 12 and inhibition of the "helper" adenovirus by H-1, *J. Gen. Virol.* **5**:19.

Lenghaus, C., and Studdert, M.J., 1980, Relationships of canine panleukopenia (enteritis) and myocarditis parvoviruses to feline panleukopenia virus, *Aust. Vet. J.* **56**:152.

Lenghaus, M.J., Studdert, M.J., and Finnie, J.W., 1980, Acute and chronic canine parvovirus myocarditis following intrauterine inoculation, *Aust. Vet. J.* **56**:465.

Lipton, H.L., and Johnson, R.T., 1972, The pathogenesis of rat virus infections in the newborn hamster, *Lab. Invest.* **27**:508.

Lipton, H., Nathanson, N., and Hodous, J., 1972, Enteric transmission of parvoviruses: Pathogenesis of rat virus infection in adult rats, *Am. J. Epidemiol.* **96**:443.

Lodmell, D.L., Hadlow, W.J., Munoz, J.J., and Whitford, H.W., 1970, Hemagglutinin antibody response of normal and Aleutian disease-affected mink to key hole limpet hemocyanin, *J. Immunol.* **104**:878.

Lubeck, M.D., and Johnson, F.B., 1976, Multiplication of bovine parvovirus in two cell strains, *Infec. Immunity* **13**:1289.

Lucas, A.M., and Riser, W.H., 1945, Intranuclear inclusions in panleucopenia of cats. A correlation with the pathogenesis of the disease and comparison with inclusions of herpes, B-virus, yellow-fever, and burns, *Am. J. Pathol.* **21**:435.

Lucas, M.H., Cartwright, S.F., and Wrathall, A.E., 1974, Genital infection of pigs with porcine parvovirus, *J. Comp. Pathol.* **84**:347.

Lum, G.S., 1970a, In-vitro studies of rat viruses. I. Effects of long-term culture, *Oncology* **24**:401.

Lum, G.S., 1970b, In-vitro studies of rat viruses. II. Effects of heat, X-irradiation and carcinogenic drugs, *Oncology* **24**:416.

Lum, G.S., 1970c, Serological studies of rat viruses in relation to tumors, *Oncology* **24**:335.

Lum, G.S., and Schreiner, A.W., 1963, Study of a virus isolated from a chloroleukemic Wistar rat, *Cancer Res.* **23**:1742.

Majaniemi, I., Tratschin, J.D., and Siegl, G., 1981, A reassessment of the nucleic acid and protein components of parvovirus LuIII, *Proceedings 5th International Congress Virology,* Strasbourg, p. 366.

Margolis, G., and Kilham, L., 1965, Rat virus, an agent with an affinity for the dividing cell, in: *Slow, Latent and Temperate Virus Infections* (D.C. Gajdusek, C.J. Gibbs, and M. Alpers, eds.), pp 361–367, NINDB Monograph No. 2, U.S. Government Printing Office, Washington, D.C.

Margolis, G., and Kilham, L., 1970, Parvovirus infections, vascular endothelium, and hemorrhagic encephalopathy, *Lab. Invest.* **22**:478.

Margolis, G., and Kilham, L., 1972, Rat virus infection of megakariocytes: A factor in hemorrhagic encephalopathy? *Exp. Mol. Pathol.* **16**:326.

Margolis, G., and Kilham, L., 1975, Problems of human concern arising from animal models of intrauterine and neonatal infections due to viruses: A review. II. Pathologic studies, *Progr. Med. Virol.* **20**:144.

Margolis, G., Kilham, L., and Ruffolo, P.R., 1968, Rat virus disease, an experimental model of neonatal hepatitis, *Exp. Mol. Pathol.* **8**:1.

Margolis, G., Kilham, L., and Johnson, R.H., 1971, The parvoviruses and replicating cells: Insights into the pathogenesis of cerebellar hypoplasia, *Progr. Neuropathol.* **1**:168.

Matsunaga, Y., and Chino, F., 1981, Experimental infection of young rabbits with rabbit parvovirus, *Arch. Virol.* **68**:257.

Matsunaga, Y., Matsuno, S., and Mukoyama, J., 1977, Isolation and characterization of a parvovirus of rabbits, *Infect. Immunity* **18**:495.

Matsuo, Y., and Spencer, H.J., 1969, Studies on the infectivity of rat virus (RV) in BALB/c mice, *Proc. Soc. Exp. Biol. Med.* **130**:294.

Mayor, H.D., and Ito, M., 1968, The early detection of picodnavirus X14 by immunofluorescence, *Proc. Soc. Exp. Biol. Med.* **129**:684.

Mayor, H.D., and Jordan, E.L., 1966, Electron microscopic study of the rodent "Picodnavirus" X14, *Exp. Mol. Pathol.* **5**:580.

Mayr, A. Bachmann, P.A., Siegl, G., Mahnel, H., and Sheffy, B.E., 1968, Characterization of a small porcine DNA virus, *Arch. Gesamte Virusforsch.* **25**:38.

McAdaragh, J.P., and Anderson, G.A., 1975, Transmission of viruses through boar semen, *Proc. 18th Am. Assoc. Vet. Lab. Diag.* pp. 69–76.

McMaster, G.K., Beard, P., Engers, H.D., and Hirt, B., 1981a, Characterization of an immunosuppressive parvovirus related to the minute virus of mice, *J. Virol.* **38**:317.

McMaster, G.K., Tratschin, J.D., and Siegl, G., 1981b, Comparison of canine parvovirus with mink enteritis virus by restriction site mapping, *J. Virol.* **38**:368.

McPherson, J.W., 1956, Feline enteritis virus—Its transmission to mink under natural and experimental conditions, *Can. J. Comp. Med.* **20:**197.

Melnick, J.L., Boucher, D.W., Clarke, J., and Boggs, J., 1971, Properties of a virus isolated from patients with MS-1 infectious hepatitis, *J. Infect. Dis.* **124:**76.

Mengeling, W.L., 1972, Porcine parvovirus: Properties and prevalence of a strain isolated in the United States, *Am. J. Vet. Res.* **33:**2239.

Mengeling, W.L., 1975, Porcine parvovirus: Frequency of naturally occuring transplacental infection and viral contamination of fetal porcine kidney cell cultures, *Am. J. Vet. Res.* **36:**41.

Mengeling, W.L., 1981, Porcine parvovirus infection, in: *Diseases of Swine* (A.D. Leman, ed.), pp. 352–365, Iowa State Press, Ames, Iowa.

Mengeling, W.L., and Cutlip, R.C., 1977, Reproductive disease experimentally induced by exposing pregnant gilts to porcine parvovirus, *Am. J. Vet. Res.* **37:**1393.

Mengeling, W.L., Cutlip, R.C., Wilson, R.A., and Parks, J.B., 1975, Fetal mumification associated with porcine parvovirus infection, *J. Am. Vet. Med. Assoc.* **166:**993.

Mengeling, W.L., Paul, P.S., and Brown, T.T., 1980, Transplacental infection and embryonic death following maternal exposure to porcine parvovirus near the time of conception, *Arch. Virol.* **65:**55.

Mirkovic, R.R., Adamova, V., Boucher, D.W., and Melnick, J.L., 1971, Identification of the Kirk hepatitis virus as a member of the parvovirus (picodnavirus) group, *Proc. Soc. Exp. Biol. Med.* **138:**626.

Mochizuki, M., Konishi, S., and Ogata, M., 1978, Studies on feline panleukopenia. II. Antigenicities of the virus, *Jpn. J. Vet. Sci.* **40:**375.

Mohanty, S.B., and Bachmann, P.A., 1974, Susceptibility of fertilized mouse eggs to minute virus of mice, *Infect. Immunity* **9:**762.

Monif, G.R.G., Sever, J.L., and Cochran, W.D., 1965, The H-1 and the RV viruses and pregnancy: Serological study of certain groups of pregnant women, *J. Pediatrics* **67:**253.

Moore, A.E., 1962a, Characteristics of certain viruses isolated from transplantable tumors, *Virology* **18:**182.

Moore, A.E., 1962b, Relationship between H-1, H-3, and RV viruses, *Proc. Am. Assoc. Cancer. Res.* **3:**345.

Moore, A.E., 1963, Fatal infection of A × C rats with H-viruses, *Proc. Am. Assoc. Cancer Res. 4:*45.

Moore, A.E., and Nicastri, A.D., 1965, Lethal infection and pathological findings in A × C rats inoculated with H-virus and RV, *J. Natl. Cancer Inst.* **35:**937.

Müller, H.P., Gautschi, M., and Siegl, G., 1978, Defective particles of parvovirus LuIII, in: *Replication of Mammalian Parvoviruses* (D.C. Ward and P. Tattersall, eds.), pp. 231–240, Cold Spring Harbor Laboratory, Cold Spring Harbor, New York.

Myers, W.L., Alberts, J.O., and Brandly, C.A., 1959, Certain characteristics of the virus of infectious enteritis of mink and observations on pathogenesis of the disease—Preliminary report, *Can. J. Comp. Med.* **23:**283.

Narita, M., Inui, S., Kawakami, Y., Kitamura, K., and Maeda, A., 1975, Histopathological changes of the brain in swine fetuses naturally infected with porcine parvovirus, *Natl. Inst. Anim. Health Qt.* **15:**24.

Nathanson, N., Cole, G.A., Santos, G.W., Squire, R.A., and Smith, K.O., 1970, Viral hemorrhagic encephalopathy of rats. I. Isolation, identification, and properties of the HER strains of rat virus, *Am. J. Epidemiol.* **91:**328.

Newman, S.J., McCallin, P.F., and Sever, J.L., 1970, Attempts to isolate H-1 virus from spontaneous human abortions: A negative report, *Teratology* **3:**279.

Nettles, V.F., Pearson, J.E., Gustafson, G.A., and Blue, J.L., 1980, Parvovirus infection in translocated raccoons, *J. Am. Vet. Med. Assoc.* **177:**787.

Nettleton, P.F., and Rweyemamu, M.M., 1980, The association of calf serum with the contamination of BHK 21 clone 13 suspension cells by a parvovirus serologically related to the minute virus of mice, *Arch. Virol.* **64:**359.

Nicoletti, G., Castro, A., Russo, G., and Cocuzza, G., 1969, Citopatologia da picodnavirus (RV, H-1, X14) in cellule di embrione di ratto, *Ann. Sclavo* **11**:188.

Notani, G.W., Hahn, E.C., Sarkar, N.H., and Kenyon, A.J., 1976, Characterization of Aleutian disease antigens, *Nature (London)* **261**:56.

Novotny, J.F., and Hetrick, F.M., 1970, Pathogenesis and transmission of Kilham rat virus infection in rats, *Infect. Immunity* **2**:298.

O'Reilly, K.J., and Whitaker, A.M., 1969, The development of feline cell lines for the growth of feline infectious enteritis (panleukopenia) virus, *J. Hyg.* **67**:115.

Ohshima, K., Shen, D.T., Henson, J.B., and Gorham, J.R., 1978, Comparison of lesions of Aleutian disease in mink and hypergammaglobulinemia in ferrets, *Am. J. Vet. Res.* **39**:653.

O'Shea, J.D., and Studdert, M.J., 1978, Growth of an autonomously replicating parvovirus (feline panleukopenia): Kinetics and morphogenesis, *Arch. Virol.* **57**:107.

Padgett, G.A., Gorham, J.R., and Henson, J.B., 1967, Epizootiologic studies of Aleutian disease. I. Transplacental transmission of the virus, *J. Infect. Dis.* **177**:35.

Palya, V., and Kisary, J., 1978, Histopathological studies on goslings infected experimentally with the wild-type and attenuated virus of Derzsy's disease, *Magy. Allatorv. Lapja* **12**:830.

Pan, I.C., Tsai, K.S., and Karstad, L., 1970, Glomerulonephritis in Aleutian disease of mink: Histological and immunofluorescence studies, *J. Pathol.* **101**:119.

Parker, J.C., Cross, S.S., Collins, Jr., M.J., and Rowe, W.P., 1970a, Minute virus of mice. I. Procedures for quantitation and detection, *J. Natl. Cancer Inst.* **45**:297.

Parker, J.C., Cross, S.S., Collins, Jr., M.J., and Rowe, W.P., 1970b, Minute virus of mice. II. Prevalence, epidemiology and occurrence as a contaminant of transplanted tumors, *J. Natl. Cancer Inst.* **45**:305.

Pattison, J.R., Jones, S.E., Hodgson, J., Davis, L.R., White, J.M., Stroud, C.E., and Murtaza, L., 1981, Parvovirus infections and hypoplastic crisis in sickle-cell anaemia, *Lancet* **8221**:664.

Paul, P.S., and Mengeling, W.L., 1980, Evaluation of a live-virus vaccine for the prevention of porcine parvovirus-induced reproductive disease in swine, *Am. J. Vet. Res.* **41**:2007.

Paver, W.K., and Clarke, S.K.R., 1976, Comparison of human fecal and serum parvo-like viruses, *J. Clin. Microbiol.* **4**:67.

Paver, W.K., Caul, E.O., Ashley, C.R., and Clarke, S.K.R., 1973, A small virus in human feces, *Lancet* **1**:237.

Paver, W.K., Caul, E.O., and Clarke, S.K.R., 1974, Comparison of a 22 nm virus from human faeces with animal parvoviruses, *J. Gen. Virol.* **22**:447.

Paver, W.K., Caul, E.O., and Clarke, S.K.R., 1975a, Parvovirus-like particles in human sera, *Lancet* **1**:232.

Paver, W.K., Caul, E.O., and Clarke, S.K.R., 1975b, Parvovirus-like particles in human faeces, *Lancet* **1**:691.

Payne, F.E., Shellabarger, C.J., and Schmidt, R.W., 1963, A virus from mammary tissue of rats treated with X-ray or methylcholanthrene, *Proc. Am. Assoc. Cancer Res.* **4**:51.

Payne, F.E., Beals, T.F., and Prestor, R.E., 1964, Morphology of a small DNA-virus, *Virology* **23**:109.

Portella, O.B., 1964, Hemadsorption and related studies on the hamster-osteolytic viruses, *Arch. Gesamte Virusforsch.* **14**:277.

Porter, D.D., 1981, Much of the increased IgG in Aleutian disease of mink is viral antibody, *Proceedings 5th International Congress Virology*, p. 363, Strasbourg.

Porter, D.D., and Larsen, A.E., 1968, Virus-host interactions in Aleutian disease of mink, *Perspect. Virol.* **6**:173.

Porter, D.D., and Cho, H.J., 1980, Aleutian disease of mink: A model for persistent infection, in: *Comprehensive Virology* (H. Fraenkel-Conrat and R.R. Wagner, eds.), Vol. 16, Plenum Press, New York.

Porter, D.D., Dixon, F.J., and Larsen, A.E., 1965, Metabolism and function of gamma globulin in Aleutian disease of mink, *J. Exp. Med.* **121**:889.

Porter, D.D., Larsen, A.E., and Porter, H.G., 1969, The pathogenesis of Aleutian disease of mink. I. In vivo viral replication and the host antibody responses to viral antigen, *J. Exp. Med.* **130**:575.

Porter, D.D., Larsen, A.E., and Porter, H.G., 1977a, Reduced severity of lesions in mink infected transplacentally with Aleutian disease virus, *J. Immunol.* **119**:872.

Porter, D.D., Larsen, A.E., Cox, N.A., Porter, H.G., and Suffin, S.C., 1977b, Isolation of Aleutian disease virus of mink in cell culture, *Intervirology* **8**:129.

Porter, D.D., Larsen, A.E., and Porter, H.G., 1980, Aleutian disease of mink, *Adv. Immunol.* **29**:261.

Rabson, A.S., Kilham, L., and Kirschstein, R.L., 1961, Intranuclear inclusions in Rattus (Mastomys) natalensis injected with rat virus, *J. Natl. Cancer. Inst.* **27**:1217.

Redman, D.R., Bohl, E.H., and Ferguson, L.C., 1974, Porcine parvovirus; natural and experimental infections of the porcine fetus and prevalence in mature swine, *Infect. Immunity* **10**:718.

Reynolds, H.A., 1969, Some clinical and hematological features of virus enteritis of mink, *Can. J. Comp. Med.* **33**:155.

Rhode, S.L., 1973, Replication process of the parvovirus H-1. I. Kinetics in a parasynchronous cell system, *J. Virol.* **11**:856.

Rhode, S.L., 1978, Defective interfering particles of parvovirus H-1, *J. Virol.* **27**:347.

Robey, R.E., Woodman, D.R., and Hetrick, F.M., 1968, Studies on the natural infection of rats with the Kilham rat virus, *Am. J. Epidemiol.* **88**:139.

Robinson, G.W., Nathanson, N., and Hodous, J., 1971, Seroepidemiological study of rat virus infection in a closed laboratory colony, *Am. J. Epidemiol.* **94**:91.

Robinson, W.F., Wilcox, G.E., Flower, R.L.P., and Smith, J., 1979a, Evidence for a parvovirus as the aetiologic agent in myocarditis of puppies, *Aust. Vet. J.* **55**:294.

Robinson, W.F., Huxtable, C.R.R., Pass, D.A., and McHowell, J., 1979b, Clinical and electrocardiographic findings in suspected viral myocarditis of pups, *Aust. Vet. J.* **55**:351.

Robinson, W.F., Huxtable, C.R., and Pass, D.A., 1980a, Canine parvoviral myocarditis: A morphologic description of the natural disease, *Vet. Pathol.* **17**:282.

Robinson, W.F., Wilcox, G.E., and Flower, R.L.P., 1980b, Canine parvoviral disease: Experimental reproduction of the enteric form with a parvovirus isolated from a case of myocarditis, *Vet. Pathol.* **17**:589.

Rohovsky, M.W., and Fowler, E.H., 1971, Lesions of experimental feline panleukopenia, *J. Am. Vet. Med. Assoc.* **158**:872.

Ron, D., and Tal, J., 1981, Expanded host range and altered genomic DNA of minute virus of mice "B" formed in restrictive A9 cells, in: *Proceedings 5th Internal Congress Virology*, p. 368, Strasbourg.

Ruckerbauer, G.M., Dulac, G.C., and Boulanger, P., 1978, Demonstration of parvovirus in Canadian swine and antigenic relationships with isolates from other countries, *Can. J. Comp. Med.* **42**:278.

Ruffalo, P.R., Margolis, G., and Kilham, L., 1966, The induction of hepatitis by prior partial hepatectomy in resistent adult rats injected with H-1 virus, *Am. J. Pathol.* **49**:795.

Salzman, L.A., and Jori, L.A., 1970, Characterization of the Kilham rat virus, *J. Virol.* **5**:114.

Schettler, C.H., 1971a, Isolation of a highly pathogenic virus from geese with hepatitis, *Avian Dis.* **15**:323.

Schettler, C.H., 1971b, Virus hepatitis of geese. II. Host range of goose hepatitis virus, *Avian. Dis.* **15**:809.

Schettler, C.H., 1973, Virus hepatitis of geese. 3. Properties of the causal agent, *Avian Pathol.* **2**:179.

Schofield, F.W., 1949, Virus enteritis in mink, *N. Am. Vet.* **30**:651.

Schultz, R.D., Mendel, H., and Scott, F.W., 1976, Effect of feline panleukopenia virus infection on development of humoral and cellular immunity, *Cornell Vet.* **66**:324.

Scott, F.W., 1980, Virucidal desinfectants and the feline viruses, *Am. J. Vet. Res.* **41**:410.

Scott, F.W., Csiza, C.K., and Gillespie, J.H., 1970a, Feline viruses. IV. Isolation and characterization of feline virus in tissue culture and comparison of cytopathogenicity with feline picornavirus, herpesvirus, and reovirus, *Cornell Vet.* **60**:165.

Scott, F.W., Csiza, C.K., and Gillespie, J.H., 1970b, Feline viruses. V. Serum-neutralization test for feline panleukopenia, *Cornell Vet.* **60:**183.

Serjeant, G.R., Topley, J.M., Mason, K., Serjeant, B.E., Pattison, J.R., Jones, S.E., and Mohamed, R., 1981, Outbreak of aplastic crises in sickle cell anemia associated with parvovirus-like agent, *Lancet* **8247:**595.

Siegl, G., 1976, The Parvoviruses, in *Virology Monographs* (S. Gard and C. Hallauer, eds.), Vol. 15, Springer Verlag, Berlin.

Siegl, G., and Gautschi, M., 1973a, The multiplication of parvovirus LuIII in a synchronized culture system. I. Optimum conditions for virus replication, *Arch. Gesamte Virusforsch.* **40:**105.

Siegl, G., and Gautschi, M., 1973b, The multiplication of parvovirus LuIII in a synchronized culture system. II. Biochemical characteristics of virus replication, *Arch. Gesamte Virusforsch.* **40:**119.

Siegl, G., and Gautschi, M., 1978, Purification and properties of replicative-form and replicative-intermediate DNA molecules of parvovirus LuIII, in: *Replication of Mammalian Parvoviruses* (D.C. Ward and P. Tattersall, eds.), pp 315–325, Cold Spring Harbor Laboratory, Cold Spring Harbor, New York.

Siegl, G., and Kronauer, G., 1980, A plaque assay for feline panleukopenia virus, *J. Gen. Virol.* **46:**211.

Siegl, G., Hallauer, C., Novak, A., and Kronauer, G., 1971, Parvoviruses as contaminants of permanent human cell lines. II. Physicochemical properties of the isolated viruses, *Arch. Gesamte Virusforsch.* **35:**80.

Siegl, G., Hallauer, C., and Novak, A., 1972, Parvoviruses as contaminants of permanent human cell lines. IV. Multiplication of KBSH-virus in KB cells, *Arch. Gesamte Virusforsch.* **36:**351.

Singer, I.I., 1975, Ultrastructural studies of H-1 parvovirus replication. II. Induced changes in the deoxyribonucleoprotein and ribonucleoprotein components of human NB cell nuclei, *Exp. Cell Res.* **95:**205.

Singer, I.I., 1976, Ultrastructural studies of H-1 parvovirus replication. III. Intracellular localization of viral antigens with immuno-cytochrome C, *Exp. Cell. Res.* **99:**346.

Singer, I.I., and Rhode, S.L., 1977a, Ultrastructural studies of H-1 parvovirus replication. IV. Crystal development and structure with the temperature-sensitive mutant *ts*1, *J. Virol.* **24:**343.

Singer, I.I., and Rhode, S.L., 1977b, Ultrastructural studies of H-1 parvovirus replication. V. Immunocytochemical demonstration of separate chromatin-associated and inclusion-associated antigens, *J. Virol.* **24:**353.

Singer, I.I., and Rhode, S.L., 1978, Ultrastructural studies of H-1 parvovirus replication. VI. Simultaneous autoradiographic and immunochemical intranuclear localization of viral DNA synthesis and protein accumulation, *J. Virol.* **25:**349.

Singer, I.I., and Toolan, H.W., 1975, Ultrastructural studies of H-1 parvovirus replication. I. Cytopathology produced in human NB epithelial cells and hamster embryo fibroblasts, *Virology* **65:**40.

Soike, K.F., Iatropoulis, M., and Siegl, G., 1976, Infection of newborn and fetal hamster induced by inoculation of LuIII parvovirus, *Arch. Virol.* **51:**235.

Spencer, H.J., 1967, Recovery of rat virus from a series of chemically induced rat leukemia, *Proc. Am. Assoc. Cancer Res.* **8:**62.

Steiner, J.W., Perez, Z.M., and Taichman, L.B., 1966, Cell population dynamics in the liver. A review of quantitative morphological techniques applied to the study of physiological and pathological growth, *Exp. Mol. Pathol.* **5:**146.

Storz, J., and Bates, R.C., 1973, Parvovirus infections in calves, *J. Am. Vet. Med. Assoc.* **163:**884.

Storz, J., and Leary, J.J., 1979, Bovine parvoviruses: Epidemiology and host-virus relationships, in: *Viral enteritis*, INSERM, Vol. 90, pp 63–80.

Storz, J., and Warren, G.S., 1970, Effect of antimetabolites and actinomycin D on the replication of HADEN, a bovine parvovirus, *Arch. Gesamte Virusforsch.* **30:**271.

Storz, J., Bates, R.C., Warren, G.S., and Howard, T.H., 1972, Distribution of antibodies against bovine parvovirus 1 in cattle and other animal species, *Am. J. Vet. Res.* **33**:269.

Storz, J., Leary, J.J., Carlson, J.H., and Bates, R.C., 1978a, Parvoviruses associated with diarrhea in calves, *J. Am. Med. Assoc.* **173**:624.

Storz, J., Young, S., Carroll, E.J., Bates, R.C., Bowen, R.A., and Keney, D.A., 1978b, Parvovirus infection of the bovine fetus: Distribution of infection, antibody response, and age-related susceptibility, *Am. Vet. Res.* **39**:1099.

Syverton, J.T., Lawrence, J.S., Ackart, R.J., Adams, W.S., Ervin, D.M., Haskins, A.L., Sounders, R.H., Stringfellow, M.B., and Wetrick, R.M., 1943, The virus of infectious feline agranulocytosis. I. Characters of the virus: Pathogenicity, *J. Exp. Med.* **77**:41.

Tabel, H., Ingram, D.G., and Fletch, S.M., 1970, Natural antibodies in sera of mink before and after the development of Aleutian disease (viral plasmocytosis), *Canad. J. Comp. Med.* **34**:320.

Tattersall, P., 1972, Replication of the parvovirus MVM. I. Dependence of virus multiplication and plaque formation on cell growth, *J. Virol.* **10**:586.

Tattersall, P., 1978a, Parvovirus protein structure and virion maturation, in: *Replication of Mammalian Parvoviruses* (D.C. Ward and P. Tattersall, eds.), pp. 53–72. Cold Spring Harbor Laboratory, Cold Spring Harbor, New York.

Tattersall, P., 1978b, Susceptibility to minute virus of mice as a function of host-cell differentiation, in: *Replication of Mammalian Parvoviruses* (D.C. Ward and P. Tattersall, eds.), pp 131–149, Cold Spring Harbor Laboratory, Cold Spring Harbor, New York.

Tattersall, P.J., 1981, Analysis of a sequence determining specific tissue-tropism within the genome of the minute virus of mice (MVM). in: *Proceedings 5th International Congress Virology*, p 369, Strasbourg.

Tattersall, P., Cawte, J., Shatkin, J., and Ward, D.C., 1976, Three structural polypeptides coded for by minute virus of mice, a parvovirus, *J. Virol.* **20**:273.

Tattersall, P., Shatkin, A.J., and Ward, D.C., 1977, Sequence homology between the structural polypeptides of minute virus of mice, *J. Mol. Biol.* **111**:375.

Tennant, R.W., Layman, K.R., and Hand, E.R., 1969, Effect of cell physiological state on infection by rat virus, *J. Virol.* **4**:872.

Tokuhisa, S., Inaba, Y., Sato, K., Miura, Y., Akashi, H., Satoda, K., and Matumoto, M., 1981, Inhibitors of bovine parvovirus, coronavirus and rotavirus in precolostral and fetal bovine sera, *Vet. Microbiol.* **6**:143.

Toolan, H.W., 1960, Experimental production of mongoloid hamsters, *Science* **131**:1446.

Toolan, H.W., 1961a, A virus associated with transplantable human tumors, *Bull. N.Y. Acad. Med.* **37**:305.

Toolan, H.W., 1961b, Studies on a viral agent associated with human tissues, *Proc. Am. Assoc. Cancer Res.* **3**:273.

Toolan, H.W., 1964, Studies on H-viruses, *Proc. Am. Assoc. Cancer Res.* **5**:64.

Toolan, H.W., 1965, H-1 virus in the adult hamster, *Proc. Soc. Exp. Biol. Med.* **119**:715.

Toolan, H.W., 1966, Susceptibility of the rhesus monkey (Macaca mulatta) to H-1 virus, *Nature (London)* **209**:833.

Toolan, H.W., 1967a, Lack of oncogenic effect of the H-viruses for hamsters, *Nature (London)* **214**:1036.

Toolan, H.W., 1967b, Agglutination of the H-viruses with various types of red blood cells, *Proc. Soc. Exp. Biol. Med.* **124**:144.

Toolan, H.W., 1968, The picodna viruses: H, RV, and AAV, *Int. Rev. Exp. Pathol.* **6**:135.

Toolan, H.W., 1972, The parvoviruses, *Prog. Exp. Tumor Res.* **16**:410.

Toolan, H.W., 1978, Maternal role in susceptibility of embryonic and newborn hamsters to H-1 parvovirus, in: *Replication of Mammalian Parvoviruses* (D.C. Ward and P. Tattersall, eds.), pp. 161–176, Cold Spring Harbor Laboratory, Cold Spring Harbor, New York.

Toolan, H., and Ledinko, N., 1965, Growth and cytopathogenicity of H-viruses in human and simian cell cultures, *Nature (London)* **208**:812.

Toolan, H.W., and Ledinko, N., 1968, Inhibition by H-1 virus of the incidence of tumors produced by adenovirus 12 in hamsters, *Virology* **35**:475.

Toolan, H.W., Dalldorf, G., Barclay, M., Chandra, S., and Moore, A.E., 1960, An unidentified filtrable agent isolated from transplanted human tumors, *Proc. Natl. Acad. Sci. USA.* **46**:1256.
Toolan, H.W., Buttle, G.A.H., and Kay, H.E.M., 1962, Isolation of the H-1 and H-3 viruses directly from human embryos, *Proc. Am. Assoc. Cancer Res.* **3**:368.
Toolan, H.W., Saunders, E.L., Southam, C.M., Moore, A.E., and Levin, A.G., 1965, H-1 virus viremia in human, *Proc. Soc. Exp. Biol. Med.* **119**:711.
Tratschin, J.D., McMaster, G.K., Kronauer, G., and Siegl, G., 1982, Canine parvovirus: Relationship to wild-type and vaccine strains of feline panleukopenia virus and mink enteritis virus, *J. Gen. Virol.* **61**:33.
Trautwein, G., Schneider, P., and Ernst, E., 1974, Untersuchungen über die Pathogenese der Aleutenkrankheit. VIII. Depression der Antikörperbildung gegen Meerrettich-Peroxydase, *Zentralbl. Vet. Med. Reihe B* **21**:467.
Urbain, A., 1933, Contribution à l'étude de la gastro-enterite infectieuse des chats, *Ann. Inst. Pasteur* **51**:202.
Usategui-Gomez, M., 1965, A human placental fluid inhibitor to hemagglutination by H-1 and H-B viruses: I. Purification, *Proc. Soc. Exp. Biol. Med.* **120**:385.
Usategui-Gomez, M., 1968, Further purification of a human placental inhibitor to hemagglutination by H-1 virus, *Proc. Soc. Exp. Biol. Med.* **127**:244.
Vincent, J., 1971, Isolement en Algérie de quatre souches de parvovirus bovis, *Ann. Inst. Pasteur* **121**:811.
Wachnik, Z., and Nowaki, J., 1962, Wirosowje zapalenie watroby u gesiat, *Medicyny Weterynaryjnej* **6**:344.
Walker, S.T., Feilen, C.P., and Sabine, M., 1980, A serological survey of canine parvovirus infection in New South Wales, Australia, *Vet. Rec.* **106**:324.
Wills, C.G., 1952, Notes on infectious enteritis of mink and its relationship to feline enteritis, *Can. J. Comp. Med.* **16**:419.
Woods, C.B., Pollock, R.V.H., and Carmichael, L.E., 1980, Canine parvoviral enteritis, *J. Am. Anim. Hosp. Assoc.* **16**:171.
Wozniak, J., and Hetrick, F., 1969, Persistent infection of a rat nephroma cell line with Kilham rat virus, *J. Virol.* **4**:313.
Wrathall, A.E., and Mengeling, W.L., 1979a, Effect of porcine parvovirus on development of fertilized pig eggs in vitro, *Br. Vet. J.* **135**:249.
Wrathall, A.E., and Mengeling, W.L., 1979b, Effect of transferring parvovirus-infected fertilized pig eggs into seronegative gilts, *Br. Vet. J.* **135**:255.
Zhdanov, V.M., and Merekalova, Z.J., 1962, Isolation of a virus from connective tissue of carcinogen-treated rats, *Vopr. Virusol.* **7**:339.
Zschokke, E., 1900, Ueber coli-bacilläre Infektionen, *Schweiz. Arch. Tierheilk.* **42**:20.

CHAPTER 9

Canine Parvovirus
Origin and Significance of a "New" Pathogen

GÜNTER SIEGL

I. INTRODUCTION

Until very recently individual autonomous parvoviruses were assumed to have a characteristic, rather narrow natural host range. Moreover, parvoviruses pathogenic for a certain species were found to exhibit a distinct and unique antigenicity. Up to 1978 this "law" also applied to the parvoviruses infecting dogs. The autonomous minute virus of canines (MVC) was isolated from the feces of asymptomatic dogs by Binn in 1970. It proved to be serologically unrelated to any other member of the genus Parvovirus as well as to the helper-dependent canine adeno-associated virus (cAAV) isolated by Sugimura and Yanagawa (1968) and characterized by Onuma and Yanagawa (1972a,b). Both agents apparently are widespread in canine populations; however, no direct relationship between infection and a distinct syndrome could be conclusively established. Therefore, considerable interest was aroused when, starting in the spring of 1978, the appearance of a large number of articles, both in the lay press and in scientific journals, signaled the rapid spread of a parvovirus in dogs almost simultaneously in widely separated geographical regions. Infection was manifested either as a fulminant enteritis of high morbidity and mortality in dogs of all ages or as a subacute myocarditis in puppies three to eight weeks of age. The causative virus, rapidly known by concerned dog owners as "Killer Virus," "The Parvovirus," or merely "The Virus"

GÜNTER SIEGL • Institute for Hygiene and Medical Microbiology, University of Bern, CH 3010 Bern, Switzerland.

generally is referred to in scientific articles as canine parvovirus (CPV) or, to distinguish between MVC and CPV, as CPV-2 (Carman and Povey, 1980). It has several quite unusual features. The most outstanding is the fact that this virus not only causes a panleukopenia–enteritis syndrome in dogs similar to that known to occur in felines and mink (Siegl, 1976), but has also proved to be serologically and genetically closely related to the feline parvovirus (FPV). In consequence, "CPV" is now considered to be a host range variant of the latter virus rather than a true canine virus. Throughout this review, however, it will be referred to as canine parvovirus (CPV).

Before 1978 all experimental attempts to transmit the feline parvovirus to young or adult dogs had failed (Siegl, 1976). Retrospective analysis of serum samples also indicated that, up to that time, dogs apparently were naturally refractory to infection with both wild type and attenuated strains of FPV. The sudden appearance and rapid dissemination of FPV in dogs therefore raises several unpleasant, yet very important questions. For example:

1. Can the canine variant of FPV be readily distinguished from the feline and mink variants for diagnostic and scientific purposes?
2. Is canine parvovirus infection a naturally occurring or a man-made disease? In other words, did the switch in host range from felines or mink to canines occur under field conditions or has it been brought about in a laboratory where strains of FPV—accidentally or deliberately—were adapted to growth in canine cells?
3. Which factors contributed to the worldwide, rapid spread of the "new" pathogen, thereby circumventing all legal and quarantine barriers established to prevent importation of foreign diseases?
4. Should we anticipate a further extension of the host range of FPV to other animal species or even to man? If so, what are the measures to be taken to prevent such an event?

It is with consideration of these points that the data contained in the many papers (more than 110 have come to this author's notice at the time this chapter was completed in summer, 1981!) published already on "canine parvovirus" have been compiled, evaluated, and compared to the characteristics of the two well-known variants of feline parvovirus—feline panleukopenia (FPLV) and mink eneteritis virus (MEV). The features of clinical disease, pathologic findings, and diagnostic procedures, are only briefly summarized and the reader interested in details is referred to the specific articles.

II. THE VIRUS

The first correlation between canine enteritis and parvoviruses was based on the detection by electron microscopy of small spherical, virus-like particles in the feces of puppies with nonfatal diarrhea (Eugster and Nairn, 1977). To what extent, and if at all, these particles were related to the virus later implicated in the worldwide outbreaks of canine enteritis is not known. The virus could not be propagated in cell cultures and neither serological nor physiochemical data other than a diameter of 18–20 nm were reported. In contrast, the virus involved in the outbreaks of enteritis and myocarditis starting in 1978 has been the subject of numerous reports. The many individual pieces of data collected in various laboratories now add up to a fairly complete picture.

A. Physicochemical Characteristics

According to various observations, the spherical, nonenveloped virus particle has a diameter of 21 ± 3 nm (e.g., Appel *et al.*, 1978; Cooper *et al.*, 1979; Eugster *et al.*, 1978; Gagnon and Povey, 1979; Fletcher *et al.*, 1979; Johnson and Spradbrow, 1979; Osterhaus *et al.*, 1980a). The infectious virus bands at densities between 1.38 and 1.43 g/ml in CsCl (Osterhaus *et al.*, 1979; Bourtonboy *et al.*, 1979b; Williams, 1980) and sediments with 110 S (Siegl *et al.*, unpublished). It resists heating to 60°C for at least 1 hr and is also stable at pH 3 (Johnson and Spradbrow, 1979). The viral genome is a single-stranded DNA 4,900 ± 100 nucleotides long. This is the same size that has been determined for the genomes of both MEV and FPLV (McMaster *et al.*, 1981; Tratschin *et al.*, 1982). Nothing is known with respect to the physical properties of the structural polypeptide(s) of the virus.

B. Biological Properties

1. Hemagglutination

The known physicochemical parameters do not allow discrimination between the canine virus and the feline and mink variants of FPV. Such a distinction is possible, however, on the basis of biological properties. Until recently, FPLV has been assumed to agglutinate pig erythrocytes only at 4°C and to elute quickly from the red blood cells at 20°C (Johnson and Cruickshank, 1966; Siegl, 1976). Carmichael *et al.* (1980), however, have found that both FPLV and MEV also agglutinate rhesus monkey erthrocytes and CPV has been reported to react with pig erythrocytes, rhesus monkey, African green monkey, cynomolgous monkey, crab-eating macaque, and cat red blood cells (Burtonboy *et al.*, 1979b; Carmichael *et al.*, 1980; Gagnon and Povey, 1979; Johnson and Spradbrow, 1979, Osterhaus *et al.*, 1980). Hemagglutination by the canine virus apparently is relatively insensitive to temperatures above 4°C. It has also been reported to be unaffected by changes in pH between pH 5.8 and 7.2 (Carmichael *et al.*, 1980; Moraillon *et al.*, 1980; Tratschin *et al.*, 1982). In fact Carmichael *et al.* (1980) suggest that it is possible to distinguish between FPLV and CPV by comparing the ability to agglutinate pig erythrocytes at pH 6.5 and 7.2. Whereas FPLV agglutinates this type of red blood cells preferentially at pH <6.5, the canine variant also shows considerable hemagglutination activity at pH 7.2. We have been able to corroborate the observation of Carmichael *et al.* (1980), and, in agreement with Moraillon *et al.* (1980), have found that CPV more closely resembles MEV than FPLV (Tratschin *et al.*, 1982). The use of this characteristic to distinguish among the three variants under routine conditions, however, is dangerous, because pig erythrocytes tend to agglutinate spontaneously below pH 6.5.

Inconsistent hemagglutination results with the canine virus have been observed with horse (Carmichael *et al.*, 1980; Gagnon and Povey, 1979) and rat red blood cells (Appel *et al.*, 1979a; Osterhaus *et al.*, 1980) and no reaction was reported with human O, dog, cattle, goat, sheep, guinea pig, mouse, hamster, turkey, goose, and chicken erythrocytes (Carmichael *et al.*, 1980; Osterhaus *et al.*, 1980).

2. Antigenicity and Serology

The most important and most intriguing finding in the early attempts to characterize the causative agent of canine epidemic enteritis and myocarditis was that this virus proved to be antigenically closely related to, if not identical with, feline parvovirus. This was demonstrated repeatedly by hemagglutination-inhibition (HI) tests, immunofluorescent (IF) studies, and serum neutralization (SN) assays in which anti-FPLV sera were as effective in revealing the respective antigenic determinants of the newly isolated virus as homologous anti-CPV antibodies (Appel *et al.*, 1979a; Johnson and Spradbrow, 1979; Tratschin *et al.*, 1981). Most authors therefore concluded that differentiation between FPLV, MEV, and CPV is not possible on the basis of serological reactivity. In contrast, Lenghaus and Studdert (1980) observed that, although two strains of virus isolated from cases of canine parvovirus enteritis and canine parvovirus myocarditis, respectively, were indistinguishable, they proved closely related to but clearly distinguishable from both wild-type and vaccine strains of FPLV by serum neutralization. Neutralization titers were always higher with the homologous than with the heterologous antisera. Hemagglutination-inhibition, however, could not distinguish between the viruses.

Flower *et al.* (1980) have provided direct evidence for the existence of a specific extra antigenic determinant on the canine virus particle. It was demonstrable both by agar gel precipitation and, following preadsorption with FPLV, by hemagglutination inhibition with anti-CPV serum. Our results (Tratschin *et al.*, 1982) are well in agreement with these findings. Using both dog sera collected in natural outbreaks of the disease and sera prepared in laboratory dogs against the three variant viruses, it was shown that antisera against FPLV and MEV do not distinguish between the viruses in HI and plaque reduction tests. Antisera prepared against CPV, however, react with the homologous virus at considerably higher dilutions than with the two heterologous agents, which again cannot be distinguished from each other.

With regard to the antigenic relationship of CPV to other members of the genus Parvovirus, Black *et al.* (1979) and Eugster (1980) reported a serologic relatedness between the canine parvovirus and porcine parvovirus (PPV) which could be observed by fluorescent antibody staining. Most authors, however, agree that CPV is unrelated to PPV, the minute virus of canines (MVC), and to bovine parvovirus (BPV) (Appel *et al.*, 1980, Carmichael *et al.*, 1980). Moreover, CPV did not cross react with the Norwalk agent, a human parvovirus-like agent incriminated in human enteric disease (Appel *et al.*, 1980a,b). Tratschin *et al.* (1982) recently performed HI tests with antisera raised in rabbits against partially purified antigens of MVC, PPV, BPV, MVM, RV, Hl virus, RT virus, TVX virus, and LuIII virus. All these sera failed to inhibit agglutination of pig red blood cells by CPV. Attempts to localize CPV antigen in infected feline kidney cell cultures with anti-PPV serum in an indirect immunofluorescent test were also negative. Therefore, the possibility that the new canine virus is serologically related to any member of the genus parvovirus other than FPV is remote.

3. Host Cell Spectrum and Replication

The *in vitro* host cell range of CPV parallels the extended red blood cell spectrum. Whereas FPLV and MEV were reported to replicate in feline, mink, and ferret cells only (Siegl, 1976), CPV could be propagated—although with variable success—in canine, feline, mink, racoon, and bovine cells (Appel *et al.*, 1979a; Black *et al.*, 1979; Carmichael *et al.*, 1980; Gagnon and Povey, 1979; Hitchcock and Scarnell, 1979; Johnson and Spradbrow, 1979; Osterhaus *et al.*, 1980). Eugster (1980) has reported replication in Vero (African green monkey kidney) cells, and according to Appel *et al.* (1980b), human cells are also susceptible.

Of the primary cell cultures tested, canine fetal lung, kidney, and intestine (Johnson and Spradbrow, 1979; Eugster, 1980), as well as feline fetal lung and kidney (Eugster, 1980) gave the most consistent results. The permanent cell lines found most suitable for the isolation and propagation of the virus were the canine cell line A72 established recently by Binn *et al.* (1980) as well as the feline cell lines NLFK (Siegl *et al.*, unpublished), CRFK (Eugster, 1980, Robinson *et al.*, 1980b; Hinaidy, 1981), and FEmb (Lenghaus and Studdert, 1980). Results ranging from completely insusceptible (Carmichael and Binn, 1981) to excellent replication (Eugster, 1980, Osterhaus *et al.*, 1980, Hinaidy, 1981; Siegl and Kronauer, unpublished) were observed with the Madin Darby canine kidney (MDCK) cell line.

In all cell culture studies viral replication has been found to be dependent on cellular proliferation. Resting cell sheets and aged cultures support virus replication only to a very low degree. Cytopathologic changes observable by light microscopy in the unstained monolayer do occur; yet, depending on the type of cell, the age of the culture, and the multiplicity of infection, such changes may be highly variable and sometimes indistinguishable from the alterations associated with aging of the monolayer. Likewise, the basophilic intranuclear inclusion bodies detected by May-Grünwald/Giesma or H&E staining in canine kidney (Gagnon and Povey, 1979), canine fetal lung (Johnson and Spradbrow, 1979), MDCK (Osterhaus *et al.*, 1980; Hinaidy, 1981), and NLFK (Siegl and Kronauer, unpublished), may be ill defined or completely missing in other cell types (e.g., Johnson and Spradbrow, 1979). The most reliable way to monitor viral replication therefore is either by the extraction and quantification of viral hemagglutinin or by fluorescent antibody staining of infected monolayers (Johnson and Spradbrow, 1979; Carmichael and Binn, 1981; Eugster, 1980; Hinaidy, 1981; G. Siegl and G. Kronauer, unpublished). Results of studies based on the use of these techniques suggest that the kinetics of replication of the three variants of FPV do not differ significantly. Finally, plaque assays and plaque neutralization tests with CPV can be performed in NLFK, CRFK, and A72 cells (G. Siegl and G. Kronauer, unpublished; L.E. Carmichael, personal communication) according to the method used by Siegl and Kronauer (1980) to plaque FPLV and MEV.

4. *In Vivo* Host Range

The *in vivo* host range of CPV cannot be discussed without making detailed reference to the *in vivo* host range of feline panleukopenia virus. For the latter virus direct evidence and circumstantial observations suggest that all members of the family Felidae—including domestic cats, leopards, panthers, tigers and lions—are naturally susceptible.

On the other hand, all laboratory animals, including young and adult dogs, tested in the extended studies performed between 1930 and 1945 were found to be refractory to infection (for reference see Siegl, 1976). However, in 1947 huge outbreaks of a feline panleukopenia–enteritis-like disease which started in Canada (Schofield, 1949) and spread rapidly to farms in other regions of the world, occurred in ranch mink. The virus incriminated in these epidemics—mink enteritis virus (MEV)—was repeatedly shown to be serologically indistinguishable from the already well-known feline panleukopenia virus (McPherson, 1956; Gorham *et al.*, 1965; Johnson, 1967; Scott *et al.*, 1970). An exception of these results was the report by Myers *et al.* (1959) that the antigenic relationship between MEV and FPV is only partial. Whether the disease in mink was started by the introduction of wild-type feline panleukopenia virus into the farm animals or whether a mutant or variant virus with an expanded host range suddenly manifested in a new species is difficult to determine by retrospective analysis. It must be mentioned, however, that feline panleukopenia virus could be transmitted experimentally without previous adaptation to another member of the family Mustelidae, the ferret (Johnson *et al.*, 1967).

The 1947 outbreaks of mink enteritis now are referred to frequently as the first practical example of the potential of feline parvovirus to expand its host range to other species. In fact, however, there is reliable evidence that, as early as 1938, the enteric syndrome had already been noticed in wild raccoons roaming in the vicinity of cat farms with a high reported incidence of feline panleukopenia–enteritis. In 1940, the disease also occurred in raccoons kept in fur farms. It could be prevented by prophylactic treatment of exposed animals with anti-feline-enteritis serum (Waller, 1940). More recently, experimental infection and natural disease in raccoons has been reported by Gorham *et al.* (1966) and Nettles *et al.* (1980). A virus indistinguishable from MEV has also been isolated from a coatimundi (Johnson, 1969).

Neither young nor adult dogs could be infected with wild-type feline parvovirus (Urban, 1933; Kikuth *et al.*, 1940). This finding is in agreement with the more recent observation that no antibodies to this virus can be detected in canine sera collected before 1976 (Black *et al.*, 1979; Johnson and Spradbrow, 1979; Schwers *et al.*, 1979; Carmichael *et al.*, 1980; Osterhaus *et al.*, 1980; Walker *et al.*, 1980) in spite of the documented, continued presence of the virus in feline populations since the end of the last century (Zschokke, 1900; Siegl, 1976). The insusceptibility of dogs to FPV infections apparently also extends to the modified FPV strains contained in commercial attenuated vaccines. Lloyd-Evans (1980) recently reported that such strains of FPV did not replicate in vaccinated dogs. Injected virus could neither be recovered from nor detected by flourescent antibody staining in any internal organ of vaccinated pups. In contrast, Appel *et al.* (1980a) and Carmichael and Binn (1981) feel that attenuated FPV strains replicate, yet, to a very limited extent, in the dog. The generally better antibody response observed after injection of attenuated living virus than after the use of killed vaccines is thought to support this point of view. With both types of vaccine, however, the success of vaccination is highly dose dependent (Appel *et al.*, 1980a,b). At least 10^4 $TCID_{50}$ of live FPV virus have to be injected to successfully trigger an immunological reaction in dogs (Carmichael and Binn, 1981) and it may be argued that the quality of the antigen (which in the case of attenuated vaccines is unaffected by inactivating agents) rather than the replication of the virus is responsible for the better vaccination results.

Whether or not the variant virus incriminated in mink enteritis is able to replicate in dogs has never been the aim of extended experimental studies. However, Moraillon *et al.* (1980) recently reported on the successful induction of acute fatal enteritis in dogs after feeding them homogenates prepared from the intestines of naturally diseased mink (Moraillon *et al.*, 1979b). The virus—strain V22—could be reisolated from infected dogs in a feline lung cell line. It showed the hemagglutination characteristics of MEV and closely resembled the latter virus in its restriction site pattern (see Section II.B.5). However, neither Tratschin *et al.* (1982) nor Carmichael (personal communication) was able to propagate the virus in the canine cell lines A72 and MDCK which otherwise readily supported replication of all CPV strains. According to Appel (personal communication), the parvovirus isolated from raccoons with parvovirus enteritis proved to be not infectious for dogs by oral–nasal inoculation.

The canine variant of FPV is distinguished from the feline and the mink virus by its undisputed ability to infect probably all members of the family Canidae. Dogs of all breeds were found to be equally susceptible. The enteric form of canine parvovirus disease was observed in a captive coyote population (Everman *et al.*, 1980), in maned wolves, and in a crab-eating fox (Fletcher *et al.*, 1979; Mann *et al.*, 1980). Besides these closely related species, domestic cats without antibodies were found to be susceptible to experimental infection with the virus (Osterhaus *et al.*, 1980; Carmichael, personal communication). Apart from a mild leukopenia and rise in temperature, however, disease symptoms were absent in these animals. In contrast, raccoons were not susceptible to CPV infection (Appel, personal communication). After oral–nasal exposure, they did not shed virus in feces and they did not seroconvert.

There have been frequent reports of human vomiting and diarrhea in persons closely associated with dogs suffering from parvovirus disease. Significant attention was paid to these observations. However, neither Carmichael (personal communication), who searched more than 300 sera of such individuals for hemagglutination-inhibiting antibodies, nor Lenghaus and Studdert (1980), who tested more than 200 sera for both HI and/or SN antibody to CPV, found any serologic evidence for transmission of the virus from dog to man. Lenghaus and Studdert also were unable to demonstrate the presence of parvoviruses in samples of human diarrheal feces and vomitus. We have analyzed only a few episodes of human enteric disease observed in parallel to canine parvovirus-like enteritis (Siegl and Kronauer, unpublished), yet, our results are identical to those reported by the American and the Australian groups. In this context, it has to be mentioned that infections with *Campylobacter* jejuni can mimic parvovirus enteritis or may be present in dogs together with CPV (Fleming, 1980). *Campylobacter* jejuni is known to be readily transmitted from dogs to man and vice versa (Bruce *et al.*, 1980). Therefore, it may be advisable to test cases of human vomiting and diarrhea which are observed concomitantly with parvovirus-like disease in dogs for the bacterial pathogen.

C. Restriction Endonuclease Cleavage Site Analysis of the Viral Genome

Restriction endonucleases cleave double-stranded DNA at specific nucleotide sequences. They thus can be used to produce a set of fragments of the DNA molecule which, when separated by electrophoresis, give a characteristic restriction site pattern. This technique of characterizing a viral genome has been repeatedly used with great success to reveal the genetic relationship between different serotypes, strains, or isolates of DNA viruses (Mulder *et al.*, 1974; Osborn *et al.*, 1974; Buchman *et al.*, 1978; Rymo *et al.*, 1979). For viruses containing a single-stranded DNA genome such an analysis is also possible if the double-stranded replicative form (RF) DNA can be isolated in sufficient quantity and purity from infected cells. This has proved possible for various parvoviruses, and McMaster *et al.* (1981) and Tratschin *et al.* (1982) have used the technique to characterize the genetic relatedness between the canine, feline, and mink variants of feline parvovirus, between individual strains of CPV, as well as several attenuated vaccine virus strains. In the initial set of experiments (McMaster *et al.*, 1981) the relationship between CPV and MEV was determined using 25 restriction enzymes. Seventeen of these enzymes cleaved the RF DNA of both viruses at least once and a total 79 restriction sites were mapped. Of these sites, 68, or 86%, were found to be common for both types of DNA, indicating that the canine and mink variants are indeed closely related viruses. In additional studies only those seven enzymes were used which previously were shown to distinguish between CPV and MEV (Tratschin *et al.*, 1981). They were responsible for the generation of 60, or 76% of all restriction sites mapped. Under these conditions, reference strains of FPLV and MEV were distinguished by only one mapped site and MEV strains V1 and V22 isolated recently by Moraillon *et al.* (1979, 1980) also showed only minor differences. On the other hand, the CPV strains isolated in the U.S.A. and in Belgium in 1978, in Switzerland and in Germany in 1980, yielded a relatively constant restriction site pattern which allowed their undisputed differentiation from both the wild-type and the attenuated vaccine strains of FPLV and MEV. An example is given in Figs. 1a and 1b. The restriction site pattern derived from the genomes of the vaccine viruses resembled more closely those of the wild-type FPLV and MEV viruses than the one characteristic for the genome of the canine variant. However, at least in the case of one of the vaccine viruses the process of attenuation has led to the generation of those three HinfI restriction sites which are characteristic for CPV (See Fig. 2). It then could be speculated that this virus represents an ancestor of CPV from which the genome of the canine virus was derived by introduction of five additional specific mutations.

All differences in genome structure revealed by these studies are located in a region of CPV-nucleic acid which, by analogy with data reported for other parvoviruses (Green *et al.*, 1979; Tal *et al.*, 1979), has to be assumed to code for viral capsid proteins. Moreover, variations in genome length observed with the attenuated vaccine strains at least partially appear to be due to insertions and/or deletions within the same region of the nucleic acid close to the 5′-end of the virion DNA molecule. Therefore, it may be hypothesized that the genomic differences revealed by restriction site mapping are directly related to structural and func-

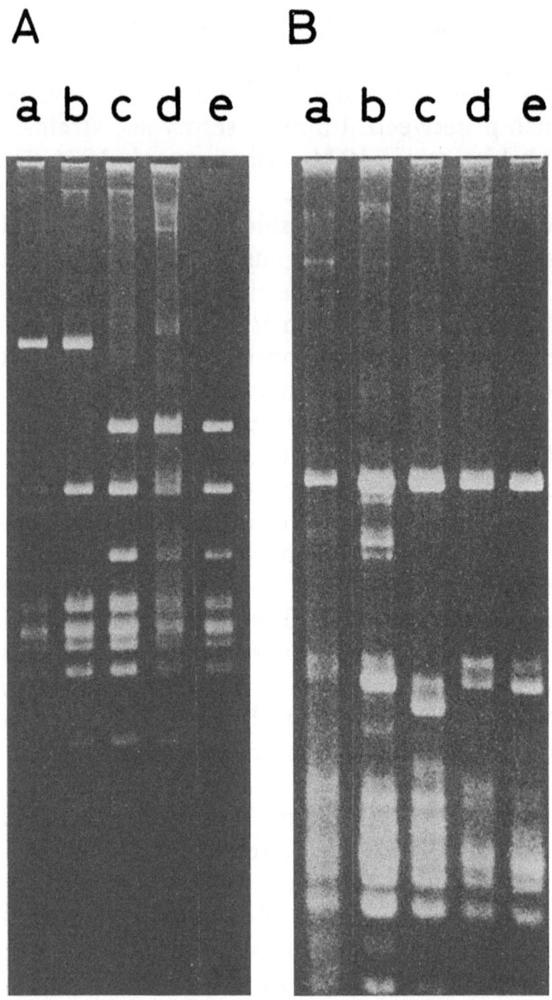

FIGURE 1. Gel electrophoresis patterns of viral RF DNAs restricted with the enzyme MboII. (A) Comparison of MEV (a) and FPLV (b) with CPV isolated in Switzerland (c), Belgium (d) and the United States (e). (B) Comparison of FPLV (a) with four live attenuated vaccine strains (b–e).

tional differences of the viral capsid proteins as reflected by the differences in antigenicity, hemagglutination pattern, and host cell range.

III. THE DISEASE

In contrast to the early observations which suggested that canine parvovirus infections are highly lethal—up to 80% of affected dogs were reported to die in certain outbreaks—it is now understood that there is a wide range in the severity of infections from the completely asymptomatic case to the peracute dysenteric animal which dies within 24 hr.

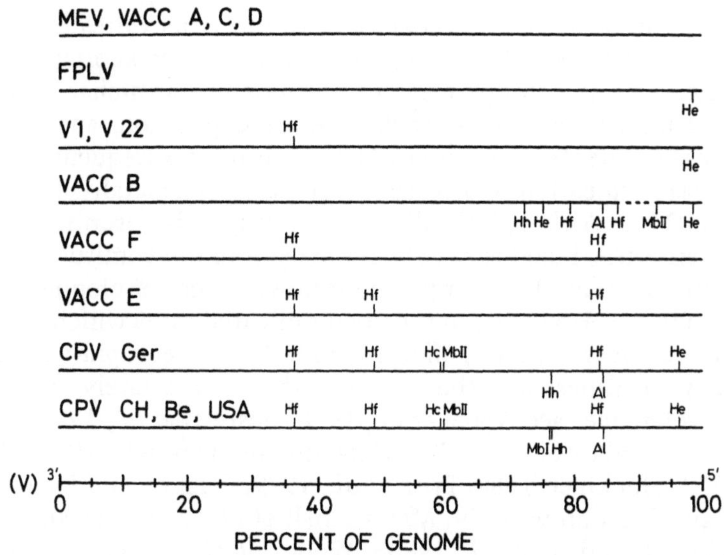

FIGURE 2. Comparative restriction enzyme maps of the DNAs of CPV, FPLV, MEV, and six live attenuated vaccine strains. Each horizontal line represents one single-stranded virus genome with its 3′-end at the left. The scale is given in map units (percentage of the genome). The map of MEV is used as a basis for comparison. Restriction enzyme sites in common to all viruses are not shown; additional sites are drawn above the line, whereas those missing are drawn below the line. The dotted line at the right-hand end of the map of vaccine strain B indicates a deletion of about 0.2 kb. V1 and V22 are two MEV strains isolated recently in France. CPV strains were derived from the feces of naturally infected dogs in Switzerland (CH), Belgium (Be), Germany (Ger) and the U.S.A. The vaccine strains are commercially available and originate from FPLV (strains A,B,C,E, and F) and MEV (strain D). Restriction enzymes used were HinfI (Hf); HaeIII (He); HphI (Hh); AluI (Al); MboII (MbII); HincII (Hc); MboI (MbI). [This figure has been reproduced from Tratschin *et al.* (1982) with permission of the *Journal of General Virology*.]

Moreover, it is clearly evident that with increasing distribution of the virus in dog population, both the age at which animals preferentially succumb to infection and the predominant clinicopathological features of the infection have changed. However, it is still true that two distinct clinical forms—the enteric–panleukopenic form and the cardiac form—can be observed.

A. Enteritis and Panleukopenia

During the original outbreaks of canine parvovirus disease in 1978 acute enteritis occurred in dogs of almost any age; yet, young dogs represented a significant proportion of all affected animals. At the present time, enteritis has become the characteristic manifestation of infection in weaned pups at between the ages of seven and fourteen weeks.

Retrospective analysis of naturally occurring infections, as well as experimental studies have indicated that the incubation period of the enteric form after oral or nasal exposure to the virus varies from five to ten days. In experimental infections this time may be shortened to four days if the virus is administered intravenously (Carmichael and Binn, 1981). The initial signs of overt disease are depression, anorexia, and, more or less frequently, fever up to 41°C. Vomiting is frequently the first characteristic indication of disease. Diarrhea then generally commences within 6 to 24 hr. Although the disease was originally described as "hemorrhagic" enteritis, blood is not always present in the vomitus and/or the stools. Animals may die at any time following the development of the enteric syndrome. However, in the uncomplicated cases which now seem to outnumber the fatal ones, affected animals recover rapidly after three to four days of illness and the course of the disease rarely exceeds one week (for references see Appel *et al.*, 1978; Bestetti *et al.*, 1979; Black *et al.*, 1979; Burtonboy *et al.*, 1979a; Coignoul and Dewaele, 1979; Fletcher *et al.*, 1979; Flückiger, 1980; Fritz *et al.*, 1979; Gagnon and Povey, 1979; Johnson and Spradbrow, 1979; McCandlish *et al.*, 1979; Moraillon *et al.*, 1979a, 1980; Robinson *et al.*, 1980b; Woods *et al.*, 1980).

In parallel to the development of the enteritis syndrome, hematological examinations frequently revealed the development of a frank leukopenia. Various observations suggest that this leukopenia generally is due to an absolute neutropenia, a moderate shift of band neutrophiles, absolute lymphopenia, and an absolute monocytosis. During the first two to five days of illness cell counts as low as 100 cells per mm^3 have been reported but counts of 500 to 4000 per mm^3 are much more common at the peak of illness. In uncomplicated cases as well as after experimental infection of laboratory dogs an early transient leukopenia may be the only clinical parameter indicative of successful infection. Recovery of dogs is always signaled by a rise of white blood cell counts. During convalescence the majority of dogs then experience a compensatory leukocytosis with cell counts as high as 80,000 per mm^3 (Fritz, 1979; Flückiger, 1980).

The pathological changes associated with the enteric form of canine parvovirus disease are very similar to those found in feline panleukopenia. They have been described repeatedly in great detail (Bestetti *et al.*, 1979; Carpenter *et al.*, 1980; Cooper *et al.*, 1979; Fletcher *et al.*, 1979; Fritz, 1979; Hayes *et al.*, 1979; Kelly, 1978; Meunier *et al.*, 1981; Nelson *et al.*, 1979; Pletcher *et al.*, 1979; Robinson *et al.*, 1980b), therefore, only a brief summary will be given. Macroscopically, the changes may vary from a mild sequential enteritis to a diffuse enteritis with dark intestinal contents. Microscopically, the changes mostly consist of an acute necrosis of epithelial cells of the small intestine with occasional eosinophilic inclusion bodies in the nuclei of adjacent, undestroyed epithelial cells. Necrosis of lymphoid tissues can be extensive and may involve Peyer's patches, lymphnodes, spleen, and thymus. Changes in the bone marrow include destruction of the blast cells and a massive appearance of immature neutrophiles (Appel *et al.*, 1980b; Black *et al.*, 1979; Cooper *et al.*, 1979; Fritz, 1979; Thomson and Gagnon, 1978).

In attempts to reproduce the enteric disease under experimental conditions, Carmichael and co-workers (cited by Carmichael and Binn, 1981) have shown that, upon infection by the oral route, the virus apparently replicated first in the lymphoid tissues of the oropharynx. This initial replication was followed by a brief viremia which could be demonstrated three to five days after infection. In parallel with and subsequent to the spread of the virus, viral antigen was detected by immunofluorescent staining in intestinal crypt epithelial cells as well as in the thymus, spleen, lymph nodes, and bone marrow. Necrosis of the cells of these tissues therefore seems to be due to the direct attack of the virus. Viral shedding in the feces usually commenced three to four days after infection, i.e., again in parallel with the viremic phase. Maximal concentrations of up to 10^9 $TCID_{50}$ of virus per g of stool have been reported at the height of clinical disease. Excretion of virus generally is limited to the period of acute enteritis and rarely extends for more than 10–12 days (M. Appel, personal communication).

As has been mentioned in the introduction to this chapter, the outcome of parvoviral infection in young and adult dogs can vary from the clinically inapparent case to acute dysenteric disease. The factors which determine the severity of the disease are not completely understood. Circumstantial evidence, however, suggests that among the factors responsible are the route and dose of infection, the general health of the animal and, last but not least, the presence of intestinal parasites as well as the influence of secondary bacterial infections. Thus, it is a rather constant finding that experimentally infected, specific-pathogen-free dogs rarely experience the enteric form of the disease. Most frequently they react with nothing but a very brief episode of leukopenia (Appel *et al.*, 1979a; Osterhaus *et al.*, 1980). On the other hand, conventionally raised dogs, dogs with a history of parasitic infections, or dogs starved for 24 to 48 hr before infection develop the symptoms of enteritis much more readily (Carman and Povey, 1980; Carmichael and Binn, 1981; Eugster, 1980; Moraillon *et al.*, 1980; Robinson *et al.*, 1980b). Finally, the age of the animal at the time of infection apparently is important with respect to the susceptibility of the cells of the different organs. This will be discussed in the following chapter in context with the cardiac form of canine parvovirus infection.

B. Myocarditis

Myocarditis due to infection with CPV occurs in two clinical presentations. The most common one is seen in puppies three to eight weeks of age (McCandlish *et al.*, 1980; Lenghaus *et al.*, 1980). It is characterized by sudden death of the affected animals, occasionally preceded by a rather short period of dullness and dyspnea. In puppies eight weeks and older premonitory signs of the disease consist of acute severe respiratory distress, depression, and weakness. In almost 90% of cases the affected animal then dies within 1 to 24 hr. The mortality rate in affected litters was seen to vary between 50 and 100%. Some of the survivers subsequently died at up to one year of age (McCandlish *et al.*, 1980).

Upon necropsy, the findings at both the macroscopic and the microscopic level are variable. In severe cases pale areas of myocardium may be recognized. Edematous changes are found predominantly in the lung; yet, the spleen, liver, and gallbladder also can be affected. Histologically the disease presents as a marked interstitial myocarditis with loss of myofibers, multifocal myofiber necrosis, and mononuclear cell infiltrate. Most typical are large basophilic and strongly Feulgen-positive intranuclear inclusion bodies. According to immunofluorescent staining, the inclusions contain viral antigen and electron microscopy reveals masses of roughly spherical particles with a diameter of 15–22 nm. Such inclusions are not found in the lung, spleen, or liver of animals dying from acute myocarditis (Bestetti *et al.*, 1979; Hayes *et al.*, 1979; Huxtable *et al.*, 1979; Jeffries and Blakemore, 1979; Jezyk *et al.*, 1979; Kelly and Atwell, 1979; Lenghaus *et al.*, 1980; Lenihan *et al.*, 1980; McCandlish *et al.*, 1979; Robinson *et al.*, 1979a,b, 1980a; Thompson *et al.*, 1979; Van den Ingh, 1980).

Attempts to reproduce experimentally the cardiac syndrome by inoculation of puppies at the age of four to seven weeks either with homogenates from affected myocardium or with virus isolated in cell culture from such homogenates failed (Hayes *et al.*, 1979; Lenghaus *et al.*, 1980; Robinson *et al.*, 1980b). In most instances the infected animals experienced nothing but a mild diarrhea; yet, Robinson *et al.* (1980b) observed the development of the complete enteric syndrome in two litters at three to five days after inoculation. On the other hand, Lenghaus *et al.* (1980) succeeded in inducing both acute and chronic myocarditis by inoculation of puppies *in utero* eight days before parturition. Two puppies—clinically normal at birth—developed acute disease 23 and 27 days after inoculation, respectively. Two other puppies were euthanized at 87 and 131 days after inoculation and showed extensive focal fibrosis within the myocardium. All four animals had high antibody titers to the virus one day after birth.

As in most other syndromes related to infection with parvoviruses, attempts have been made to explain the genesis of canine parvovirus enteritis and especially of canine parvovirus myocarditis on the basis of the evident predilection of the agent for actively dividing cells in rapidly proliferating tissues. Such a concept apparently fits excellently with the interaction of the virus with the intestinal epithelial cells as well as with the cells of the lymphopoietic and hematopoietic tissues. Adaptation of the concept to the genesis of myocarditis, however, meets with difficulties. Bishop (1972) has shown that in newborn puppies, 2–4% of myocardial cells undergo division until approximately day 15 postparturition. As pointed out by Robinson *et al.* (1980a), this period of relatively rapid cell division correlates well with the strict age incidence of the myocardial syndrome. However, both in the pre- and in the postnatal phase of life the heart is only one out of many differentiating and proliferating tissues. Therefore, if dividing cells represent the only parameter controlling the susceptibility of tissues, generalized rather than localized viral replication should occur. Such generalized disease including pneumonitis, hepatitis, gastritis, nephritis and enteritis has been observed by Lenghaus *et al.* (1980) in a litter of puppies within ten days after birth. However, a generalized course of CPV infection seems to be the exception and Robinson *et al.* (1980a) never found histopathological evidence of myocarditis in parallel with necrosis in theoretically vulnerable tissues such as intestinal epithelium, bone marrow, and lymphoid organs.

There are most likely three main factors which determine the outcome of canine parvovirus infection in newborn puppies: the age of the animal at the time of infection, the immune status of the bitch, and the ability of individual tissues to support viral replication. It is self-evident that congenital infection leading to myocarditis and neonatal infection resulting in generalized disease can only occur in offsprings of nonimmune bitches. Puppies born to immune bitches, however, obviously become susceptible only as passively transmitted maternal antibodies disappear; i.e., generally between 7 and 14 weeks of age. On the other hand, it is tempting to speculate whether upon infection of fetuses *in utero*, the apparently rapid immunologic reaction of the bitch (Lenghaus *et al.*, 1980) is capable of restricting viral replication to the myocytes or whether both pre- and postnatally the myocytes are the only cell type able to support viral replication. In this context, Lenghaus *et al.* (1980) pointed to the fact that, according to the observations of Bishop and Hine (1975), myocytes of dogs pass from a mononuclear state at birth to a largely binuclear state at the age of about eight weeks. Such a transition could provide a special cellular environment for the replication of the virus. The idea is well in line with the most recent trend in experimental parvovirus research which has been initiated by Tattersall (1978) and clearly demonstrates that susceptibility to parvovirus infection is in fact controlled by cellular differentiation.

A final explanation could be that congenital infection results in myocarditis because all the other organs and tissues are able to undergo repair whereas the myocardium—comparable to the cerebellum in feline ataxia—underwent a definite and uncorrectable defect.

C. Immunity, Antibody Prevalence, and Manifestation of Infection

Both by retrospective serological analysis of naturally occurring infections and in serological follow-up studies of experimentally induced disease, the immunological response of dogs to infections with the canine parvovirus was found to be rather fast. HI-antibodies can be detected as early as three or four days after infection (Appel *et al.*, 1980b; Carmichael *et al.*, 1980) and may be present in considerable concentration at the height of clinical disease. Intestinal antibodies apparently are present in the feces at later stages of infection (Eugster, 1980). They give rise to clumping of virus particles in the fecal extracts and have to be assumed to interfere with the isolation of virus in cell cultures unless special precautions are taken.

In early convalescent sera extremely high HI-antibody titers (sometimes > 1:80,000) can be recorded. Such titers may decline within days and weeks to a considerably lower level but then they persist in parallel with SN antibodies at only slowly diminishing titers for months or even years. Appel *et al.* (1980a) and Carmichael *et al.* (1980) have stated that HI titers > 1:80 are equivalent to solid immunity. Circumstantial evidence, however, suggests that even barely detectable antibody titers are sufficient to prevent infection or, cautiously speaking, at least induction of clinically overt disease. Whether such low humoral antibody titers can prevent replication and excretion of the virus in the feces remains a matter of discussion. Eugster (1980) reported intermittent shedding of the virus up to 8 days after challenge infection of dogs immunized with an experimental inactivated canine parvovirus vaccine.

Only relatively little information is available concerning the significance and development of antiviral immunity in the course of infections resulting in myocarditis. Lenghaus *et al.* (1980) have shown that, upon experimental *in utero* infection of fetuses, the bitch developed high levels of HI-antibodies within five days after surgery. Comparable high HI-antibody titers were detected in the sera of the offspring one day after parturition. In spite of this immunity, acute myocarditis could be diagnosed in two puppies at 15 and 19 days after parturition, respectively. In two others infection resulted in a clinically inapparent, chronic myocarditis. It may be assumed, however, that congenital infection of fetuses which could terminate in myocarditis is prevented by humoral immunity of the bitch. The increasing prevalence of antibodies in canine populations due to the rapid spread of natural infection as well as to the measures of vaccination taken by concerned dog owners therefore can be expected to reduce drastically the incidence of myocarditis. The same factors already have changed the age prevalence of the enteric form of the disease. Pups born to immune dams passively acquire antibody to CPV via the placenta and the colostrum (Appel *et al.*, 1980b; Meunier *et al.*, 1981). Depending on the bitch's titer, these antibodies either disappear within some few weeks or persist as long as 12 to 16 weeks. They interfere both with natural infection and prophylactic vaccination (Appel *et al.*, 1980a; McCandlish *et al.*, 1980). In a virus-contaminated surrounding the enteric disease then will occur preferentially in weaned pups 7 to 14 weeks of age which succumb to infection as soon as passively transmitted immunity vanishes.

IV. THE ORIGIN OF CPV

During the first half of this century feline panleukopenia virus has repeatedly given rise to huge epidemics in various parts of the world (Siegl, 1976). Since then effective vaccines have been developed, but the virus has remained endemic and is still a major pathogen of felines. Exposure of dogs to the virus therefore is continuously possible. However, neither circumstantial nor experimental evidence point to a natural susceptibility of this species to feline panleukopenia virus before or after the recent outbreaks of canine disease. In consequence, the agent incriminated in canine disease must represent a viral entity with altered biologic properties in spite of the difficulty in distinguishing it serologically from feline panleukopenia virus under routine diagnostic conditions.

As discussed in Section II in more detail, the differences between CPV and both wild-type and attenuated strains of FPLV/MEV consist of the undisputed ability of CPV to replicate in canine cells, in subtle antigenic differences, and in the red blood cell spectrum as well as the pH and temperature sensitivity of the viral hemagglutinin. With respect to the latter characteristics, however, CPV more closely resembles MEV than FPLV. Finally, unambiguous identification of CPV is possible by restriction enzyme digestion of the viral replicative form (RF) DNA. The multiple structural differences revealed by this technique are located in a region of the viral genome which, by analogy with other parvoviruses, has to be assumed to code for the viral structural proteins.

There is convincing evidence that all CPV strains characterized to date have identical biological properties. This observation conforms to the extraordinarily close genetic relationship revealed by restriction enzyme analysis of four CPV strains which were isolated in widely separated geographical regions within a period of two years. It is quite unlikely that these viruses have evolved in parallel and separately from different FPLV/MEV ancestor strains. Rather, the possible evolution of a descendant of FPLV or MEV into a prototype canine parvovirus and its efficient and rapid spread all over the world has to be considered.

The factors predisposing the evolution of a prototype virus with the biological characteristics of CPV have been a matter of much speculation. We can put aside all theories based upon a hypothetical recombination between the minute virus of canines (MVC) and canine adenovirus (Evermann, 1980) or between several other nondefective parvoviruses such as MVM, PPV, and BPV (Danson, 1980). The serologic and genetic data strongly argue against them. On the other hand, the possibility that CPV derives from MEV must be considered more seriously. These variants of the feline parvovirus have largely identical hemagglutination characteristics; yet, they are distinguished by both host cell range and restriction enzyme pattern. Nevertheless, Moraillon *et al.* (1980) reported on the ability to induce the full spectrum of enteric disease in dogs by feeding them the intestines of mink which died from natural MEV infection. The genome of the virus reisolated from diseased dogs showed a restriction enzyme pattern consistent with the one recorded for wild-type MEV. In contrast to all isolates of viruses recovered so far from infected dogs, however, it also failed to grow in canine cells (Tratschin *et al.*, 1981; L.E. Carmichael *et al.*, personal communication). Therefore, further transmission experiments with additional strains of MEV seem to be necessary to allow for a thorough evaluation of the intriguing results reported by the French investigators.

Considering the multiple, yet constant structural differences between the genomes of CPV and FPLV/MEV, the possibility that CPV has arisen by mutation under field conditions without selective pressure and within a relatively short period of time seems remote. It is much more likely that wild-type FPLV or MEV has been adapted to growth in canine cells under laboratory conditions. Such an adaption could have occurred accidentally. Contamination of cell cultures with parvoviruses takes place quite frequently and can be prevented only by the most stringent precautions (Hallauer *et al.*, 1971; Siegl, 1976; Bonnard *et al.*, 1976; Nettleton and Rweyemamu, 1980). Thereby, the source of virus could have been cells experimentally infected with FPLV/MEV and handled together with canine cells in one and the same laboratory as well as primary cell cultures prepared from latently infected tissues such as feline kidneys (O'Reilly and Whitaker, 1969) and passaged in parallel with canine cells in a cell culture laboratory. Cross contamination of cell cultures may remain undetected for a rather long time (Hallauer *et al.*, 1971; Nettleton and Rweyemamu, 1980). However, the contaminating virus could be spread efficiently in every product derived from such cultures.

On the other hand, the process of adaptation could have been initiated deliberately in attempts to achieve rapid attenuation of a wild-type FPLV/MEV strain for vaccine purposes by continuous passage in heterologous cells. The resulting variant virus then would have been spread in the final commercial product and introduced into dogs via contact with vaccinated cats or mink. Whereas it will prove impossible to resolve the origin of CPV if the prototype virus has arisen accidentally and was distributed in a cell culture-derived product unrelated to FPLV/MEV, samples of modified living FPLV/MEV vaccines can be tested for the presence of virus with characteristics similar to CPV. This has been done with six different vaccines produced in various countries and commercially available in Switzerland (Tratschin *et al.*, 1982). None of these samples contained virus that replicated in canine cells. Changes in the restriction enzyme pattern of the respective genomes obviously had occurred in the course of attenuation; however, the patterns of all vaccine strains proved much more closely related to those of the wild-type viruses than to the canine variant of FPV. The latter study, based on the analysis of a rather limited number of vaccine strains, allows no straight conclusion as to whether or not FPLV/MEV vaccines were involved in the generation and distribution of the new canine virus. Yet, further strains and even individual lots of vaccines can easily be tested and the experimental design may prove useful in preventing similar unpleasant events in the future (see Conclusions).

It would be very helpful, in coming to a reliable decision in favor of one or the other theories, if the epidemiological data related to the spreading of CPV disease would give a more coherent picture. The first well-documented cases of canine enteritis and myocarditis resulting from infection with a virus unambiguously identified as CPV were observed between spring and early fall 1978. This is true for the U.S.A., Canada, Australia, New Zealand, South Africa, and Europe. With the exception of Europe, the time of appearance of clinical diseases conforms excellently to the appearance of antibodies to CPV in canine sera (Appel *et al.*, 1980b; Burrows, 1978; Carmichael *et al.*, 1980; Flückiger, 1980; Helfer-Baker *et al.*, 1980; Hoffmann *et al.*, 1980; Kelly, 1978; McCandlish *et al.*, 1979; Meunier *et al.*, 1981; Mulvey *et al.*, 1980; Osterhaus *et al.*, 1980; Smith *et al.*, 1980; Walker *et al.*, 1980). Schwers *et al.* (1979), however, reported that 3 out of 56 sera collected in Belgium between June 1976 and June 1977 contained HI-antibody ($\geq$ 1280) to the virus. In France, Chappuis (1980, Recueil du Syndicat des Véterinaires Praticien) initially was unable to detect antibodies in a "large number" of diagnostic sera collected before 1978. More recent studies (Petermann and Chappuis, 1981), however, suggest an extraordinary high incidence of CPV infections already in 1977, as 20% of sera collected at this time proved positive.

Most veterinarians involved in the study of canine parvovirus disease agree that the introduction of virus into a virgin, fully susceptible population of canines is unlikely to pass unnoticed. Under these conditions infection is signaled by a rapidly increasing incidence of acute, frequently fatal enteritis. Therefore, it remains to be explained how the virus could have given rise to a considerable proportion of inapparently infected animals about two years before epidemic disease was recorded in Belgium (in 1978) as well as in France (around October 1979). Moreover, it is barely understandable that the virus should have been introduced almost simultaneously into the U.S.A., Australia, Great Britain, and South Africa in 1978 whereas countries in the vicinity of France and Belgium like Switzerland, Germany, and Austria, experienced the first documented outbreaks of canine enteritis in 1979, i.e., about one year later and in parallel with the overt outbreak of disease in France.

There is a final enigma associated with the epidemiology of the canine parvovirus. Most countries have established strict legal regulations to prevent importation of foreign pathogens. The known stability of parvoviruses that allow CPV to survive even under unfavorable conditions for long periods of time may account for a relative inefficiency of quarantine regulations. Besides, dog owners may have transported virus passively on contaminated clothing and shoes thus giving rise to localized outbreaks of canine enteritis subsequent to dog shows (L.E. Carmichael, personal communication). However, it is unlikely that these were the main mechanisms by which infection spread almost simultaneously to North America, South Africa, Australia, and in Europe. Therefore, the idea that the virus was spread in a commercial product designed for veterinary use (Johnson and Spradbrow, 1979) is appealing. This would in fact explain the close genetic relationship of the CPV isolates characterized to date, the efficient and almost simultaneous spread of the virus to far distant locations, as well as the readiness by which all quarantine regulations were circumvented.

V. CONCLUSIONS

All available data concerning the characteristics of the pathogen responsible for the worldwide outbreaks of canine epidemic enteritis and myocarditis indicate that this virus is a variant of feline parvovirus. It can be distinguished unambiguously from the other two variants—feline panleukopenia virus and mink enteritis virus—on the basis of an extended *in vitro* host cell range and a characteristic restriction enzyme pattern. Antigenic differences have also been reported; their significance and usefulness for diagnostic purposes, however, are still a matter of discussion. There is reliable evidence that all CPV strains isolated to date are identical.

The first well-documented and virologically proven episodes of canine parvovirus disease were observed within a rather brief period of time in widely separated geographical locations. In most of these regions canine sera collected before that event did not contain antibodies to the virus. These epidemiological data and the evident complex biological and genetic differences between CPV and FPLV/MEV argue against the possibility that CPV evolved under field conditions by stepwise mutation and acquired its pathogenicity in the course of frequent dog-to-dog passages. Rather, the generation of the new pathogen under highly selective pressures provided in a biological laboratory must be seriously considered.

Transmission of parvoviruses is favored by their extraordinary ability to survive even under the most unfavorable conditions for a rather long period of time. Therefore, dog owners may have transported the virus during international flights on contaminated clothing. However, it is open to question whether these factors alone can account for the almost simultaneous appearance of canine parvovirus disease on four continents as well as for the rapidity with which all quarantine regulations were circumvented. In consequence, dissemination of the virus in a commercial product for veterinary use such as, for example, in a viral vaccine still cannot be excluded. Independent of the vehicle involved in the spread of CPV, however, it has become evident that the existing legal regulations apparently are rather ineffective in preventing the spread of a parvovirus.

Epidemic canine parvovirus enteritis and myocarditis is the second serious episode in which a variant of feline parvovirus has established itself as a threatening pathogen in a previously unaffected and even unsusceptible species. The origin and genesis of mink enteritis virus, observed for the first time in ranch mink in Canada in 1947, has never been analyzed in detail. There are also no suggestive experimental data as to whether the potential to cross species barriers is inherent to all parvoviruses or whether it is limited to feline parvovirus. In any case, the potential for spread to further, unrelated species seem to be signaled for CPV by its ability to replicate in cultures of bovine and human cells.

The scenario of an epidemic disease with the features of feline panleukopenia, mink enteritis, or canine parvovirus disease in valuable animal livestock or even in man is highly worrying. Therefore, every possible measure should be taken to prevent such an event. In this respect, the most obvious possibility is to avoid any selective pressure on the virus which would favor mutation in the direction of variants with new biological properties. Selective pressure evidently is created by the repeated and uncritical injection of living modified FPLV or MEV vaccine virus into dogs or further animals as well as by the use of heterologous cell culture systems to speed up attenuation of viruses for vaccine production. It would also seem highly desirable to avoid the use of attenuated parvovirus vaccines altogether. Injection of such vaccines obviously adds to the dissemination of virus mutants of largely unknown biological potential. As long as this is not feasible, safety tests should be requested for every attenuated parvovirus vaccine proving at least the inability of the virus to replicate in cell cultures derived from human tissues as well as in cells of the most important domestic animals. Additionally, a restriction enzyme pattern should be established and deposited, thus allowing the unambiguous identification of the virus if necessary. Finally, experimental analysis of the factors and mechanisms controlling the replication and pathogenicity of parvoviruses should be given the highest priority.

ACKNOWLEDGMENTS. The photographs of the restriction enzyme patterns of the viral RF DNAs were kindly provided by Dr. J-D. Tratschin. Preprints and personal communications of results were unselfishly contributed by Drs. M.J.G. Appel, L.E. Carmichael, and M.J. Studdert.

REFERENCES

Appel, M.J.G., Cooper, B.J., Greisen, H., and Carmichael, L.E., 1978, Status report: Canine viral enteritis, *J. Amer. Vet. Med. Assoc.* **173**:1516.

Appel, M.J.G., Scott, F.W., and Carmichael, L.E., 1979a, Isolation and immunisation studies of a canine parvo-like virus from dogs with haemorrhagic enteritis, *Vet. Rec.* **105**:156.

Appel, M.J.G., Cooper, B.J., Greisen, H., and Carmichael, L.E., 1979b, Canine viral enteritis, I. Status report on corona and parvo-like viral enteritis, *Cornell Vet.* **69**:123.

Appel, M.J.G., Carmichael, L.E., McGregor, D.D., and Pollack, R.V.H., 1980a, Canine parvovirus vaccination, *Mod. Vet. Pract.* **61**:983.

Appel, M., Meunier, P., Pollock, R., Greisen, H., Carmichael, L., and Glickman, L., 1980b, Canine viral enteritis, *Canine Pract.* **7**:22.

Bestetti, G., Hãni, H., Dudan, F., Meister, V., Waber, S., and Luginbühl, H., 1979, Panelukopenie-ähnliche Enteritis und plötzlich Todesfälle bei Welpen infolge Myocarditis, wahrscheinlich verursacht durch Parvoviren, *Schweiz. Arch. Tierheilk.* **121**:663.

Binn, L.N., Lazer, E.C., Eddy, G.A., and Kajima, M., 1970, Recovery and characterization of minute virus of canines, *Infect. Immun.* **1**:503.

Binn, L.N., Marchwicki, R.H., and Stephenson, E.H., 1980, Establishment of a canine cell line: Derivation, characterization, and viral spectrum, *Am. J. Vet. Res.* **41**:855.

Bishop, S.P., 1972, Effect of aortic stenosis on myocardial cell growth, hyperplasia and ultrastructure in neonatal dogs, *Recent Adv. Stud. Cardiac Struc. Metab.* **3**:637.

Bishop, S.P., and Hine, P., 1975, Cardiac muscle cytoplasmic and nuclear development during canine neonatal growth, *Recent Adv. Stud. Cardiac Struc. Metab.* **8**:77.

Black, J.W., Holscher, M.A., Povell, H.S., and Byerly, C.S., 1979, Parvoviral enteritis and panleukopenia in dogs, *Vet. Med. Small Anim. Clin.* **74**:47.

Blaser, M., Powers, B.W., Cravens, J., and Wang, W.L., 1978, Campylobacter enteritis associated with canine infection, *Lancet* **2**:979.

Bonnard, G.D., Manders, E.K., Campbell, D.A., Herberman, R.B., and Collins, M.J., 1976, Immunosuppressive activity of a subline of the mouse EL 4 lymphoma. Evidence for minute virus of mice causing the inhibition, *J. Exp. Med.* **143**:187.

Bruce, D., Zochowski, W., and Fleming, G.A., 1980, Campylobacter infections in cats and dogs, *Vet. Rec.* **107**:200.

Buchman, T.G., Roizman, B., Adams, G., and Stover, B.H., 1978, Restriction endonuclease fingerprinting of herpes simplex virus DNA. A novel epidemiological tool applied to a nosocomial outbreak, *J. Infect. Dis.* **138**:488.

Burrows, C.F., 1978, Canine hemorrhagic gastroenteritis, *J. Am. Anim. Hosp. Assoc.* **12**:492.

Burtonboy, G., Coignoul, F., and Pastoret, P., 1979a, L'énterite a parvovirus du chien, *Ann. Méd. Vét.* **123**:123.

Burtonboy, G., Coignoul, F., Delferriere, N., and Pastoret, P.P., 1979b, Canine hemorrhagic enteritis: Detection of viral particles by electron microscopy, *Arch. Virol.* **61**:1.

Carman, P.S., and Povey, R.C., 1980, Experimental challenge of dogs with canine parvovirus-2, *Vet. Rec.* **107**:447.

Carmichael, L.E., and Binn, L.N., 1981, New enteric viruses in the dog, in: *Advances in Veterinary Science and Comparative Medicine* (C.E. Cornelius and C.F. Simpson, eds.), Vol. 25, pp. 1–37, Academic Press, New York.

Carmichael, L.E., Jaubert, J.C., and Pollock, R.V.H., 1980, Hemagglutination by canine parvovirus: Serologic studies and diagnostic applications, *Am. J. Vet. Res.* **41**:784.

Carpenter, J.L., Roberts, R.M., Harpster, N.K., and King, N.W., 1980, Intestinal and cardiopulmonary forms of parvovirus infection in a litter of pups, *J. Am. Vet. Med. Assoc.* **176**:1269.

Coignoul, F., and Dewaele, A., 1979, Canine haemorrhagic enteritis—pathology of a syndrome, *Ann. Méd. Vét.* **123**:47.

Cooper, B.J., Carmichael, L.E., Appel, M.J.G., and Greisen, H., 1979, Canine viral enteritis. II: Morphologic lesions in naturally occurring parvovirus infection, *Cornell Vet.* **69**:134.

Danson, D.L.G., 1981, Origins of parvoviral infections, *J. Am. Vet. Med. Assoc.* **178**:373.

Eugster, A.K., 1980, Studies on canine parvovirus infections: Development of an inactivated vaccine, *Am. J. Vet. Res.* **41**:2020.

Eugster, A.K., and Mairn, C., 1977, Diarrhea in Puppies: Parvovirus-like particles demonstrated in their feces, *Southwest Vet.* **30**:59.

Eugster, A.K., Bendele, A., and Jones, C.P., 1978, Parvoviral infection in dogs, *J. Am. Vet. Med. Assoc.* **173**:1340.

Evermann, J.F., 1981, Origin of canine parvovirus, *J. Am. Vet. Med. Assoc.* **178**:340.

Evermann, J.F., Foreyt, W., Maag-Miller, L., Leathers, C.W., McKeirnan, A.J., and Master, B.L., 1980, Acute hemorrhagic enteritis associated with canine coronavirus and parvovirus infections in a captive coyote population, *J. Am. Vet. Med. Assoc.* **177**:784.

Fleming, M.P., 1980, Incidence of campylobacter infection in dogs, *Vet. Rec.* **107**:202.

Fletcher, K.C., Eugster, A.K., Schmidt, R.E., and Hubbard, G.B., 1979, Parvovirus infection in maned wolves, *J. Am. Vet. Med. Assoc.* **175**:897.

Flower, R.L.P., Wilcox, G.E., and Robinson, W.F., 1980, Antigenic differences between canine parvovirus and feline panleukopenia virus, *Vet. Rec.* **107**:254.

Flückiger, M., 1980, Die Parvovirus-Enteritis des Hundes, eine Analyse von 50 Fällen, *Schweiz. Arch. Tierheilk.* **122**:573.

Fritz, T.E., 1979, Canine enteritis caused by a parvovirus-Illinois, *J. Am. Vet. Med. Assoc.* **174**:3.

Gagnon, A.N., and Povey, R.C., 1979, A possible parvovirus associated with an epidemic gastroenteritis of dogs in Canada, *Vet. Rec.* **104**:263.

Gorham, J.R., Hartsough, G.R., Burger, D., Lust, S., and Sato, N., 1965, The preliminary use of attenuated feline panleukopenia virus to protect cats against panleukopenia and mink against virus enteritis, *Cornell Vet.* **55**:559.

Gorham, J.R., Hartsough, G.R., Sato, N., and Lust, S., 1966, Studies on cell culture adapted feline panleukopenia virus—virus neutralization and antigenic extinction, *Vet. Med.* **61**:35.

Green, M.R., Lebowitz, R.M., and Roeder, R.G., 1979, Expression of the autonomous parvovirus H1 genome: Evidence for a single transcriptional unit and multiple spliced polyadenylated transcripts, *Cell* **17**:967.

Hallauer, C., Kronauer, G., and Siegl, G., 1971, Parvoviruses as contaminants of permanent human cell lines. I. Virus isolation from 1960–1970, *Arch. Gesamte Virusforsch.* **35**:80.

Hayes, M.A., Russell, R.G., and Babiuk, L.A., 1979, Sudden death in young dogs with myocarditis caused by parvovirus, *J. Am. Vet. Med. Assoc.* **174**:1197.

Helfer-Baker, C., Evermann, J.F., McKeirnan, A.J., Morrison, W.B., Slack, R.L., and Miller, C.W., 1980, Serological studies on the incidence of canine enteritis viruses, *Canine Pract.* **7**:37.

Hinaidy, B., 1981, Aetiologische Diagnose durch canines Parvovirus bedingter Enteritisfälle in Oesterreich, *Wien. tierärztl. Mschr.* **68**:145.

Hitchcock, L.M., and Scarnell, J., 1979, Canine parvovirus isolated in UK., *Vet. Rec.* **105**:172.

Hoffmann, R., Frese, K., Reinacher, M., and Kraus, H., 1980, Parvovirusinfektionen bei akuten Magen- und Darmerkrankungen des Hundes, *Berl. Münch. Tierärztl. Wochenschr.* **93**:121.

Huxtable, C.R., Howell, J.McC., Robinson, W.F., Wilcox, G.E., and Pass, D.A., 1979, Sudden death in puppies associated with a suspected viral myocarditis, *Aust. Vet. J.* **55**:37.

van den Ingh, T.S.G.A.M., van den Linde-Sipman, J.S., and Wester, P.W., 1980, Parvovirus-like particles in myocarditis in pups, *J. Small Anim. Pract.* **21**:81.

Jefferies, A.R., and Blakemore, W.F., 1979, Myocarditis of probable viral origin in pups of weaning age, *J. Am. Vet. Med. Assoc.* **174**:1204.

Jezyk, P.F., Haskins, M.E., and Jones C.L., 1979, Myocarditis of probable viral origin in pups of weaning age, *J. Am. Vet. Med. Assoc.* **174**:1204.

Johnson, R.H., 1967, Feline panleukopenia virus—In vitro comparison of strains with a mink enteritis virus, *J. Small Anim. Pract.* **8**:319.

Johnson, R.H., 1969, Feline panleukopenia, *Vet. Rec.* **84**:338.

Johnson, R.H., and Cruickshank, J.G., 1966, Problems in classification of feline panleukopenia virus, *Nature (London)* **212**:622.

Johnson, R.H., and Spradbrow, P.B., 1979, Isolation from dogs with severe enteritis of a parvovirus related to feline panleukopenia virus, *Aust. Vet. J.* **55**:151.

Johnson, R.H., Margolis, G., and Kilham, L., 1967, Identity of feline ataxia virus with feline panleukopenia virus, *Nature (London)* **214**:175.

Kelly, W.R., 1978, An enteric disease of dogs resembling feline panleukopenia, *Aust. Vet. J.* **54**:593.

Kelly, W.R., and Atwell, R.B., 1979, Diffuse subacute myocarditis of possible viral aetiology—A cause of sudden death in pups, *Aust. Vet. J.* **55**:36.

Kikuth, W., Gönnert, R., and Schweickert, M., 1940, Infektiöse Aleukozytose der Katzen, *Zentralbl. Bakteriol. I. Orig.* **146**:1.

Lenghaus, C., and Studdert, M.J., 1980, Relationships of canine panleukopenia (enteritis) and myocarditis parvoviruses to feline panleukopenia virus, *Aust. Vet. J.* **56**:152.

Lenghaus, C., Studdert, M.J., and Finnie, J.W., 1980, Acute and chronic canine parvovirus myocarditis following intrauterine inoculation, *Aust. Vet. J.* **56**:465.

Lenihan, P., Bassett, H.F., and Weavers, E.D., 1980, Demonstration by electron microscopy of parvovirus-like particles in canine parvovirus myocarditis, *Vet. Rec.* **107**:201.

Lloyd-Evans, L.P., 1980, Vaccine and canine parvovirus, *Vet. Rec.* **106**:445.

Mann, P.C., Bush, M., Appel, M.J.G., Beehler, B.A., and Montali, R.J., 1980, Canine parvovirus infection in South American canids, *J. Am. Vet. Med. Assoc.* **177**:779.

McCandlish, I.A.P., Thompson, H., Cornwell, H.J.C., Laird, H., and Wright, N.G., 1979, Isolation of a parvovirus from dogs in Britain, *Vet. Rec.* **105**:167.

McCandlish, I.A.P., Thompson, H., Cornwell, H.J., and McCartney, L., Canine parvovirus infection, *Vet. Rec.* **107**:204.

McMaster, G.K., Tratschin, J-D., and Siegl, G., 1981, Comparison of canine parvovirus with mink enteritis virus by restriction site mapping, *J. Virol.* **38**:368.

McPherson, I.W., 1956, Feline enteritis virus—Its transmission to mink under natural and experimental conditions, *Can. J. Comp. Med.* **20**:197.

Meunier, P.C., Glickman, L.T., Appel, M.J.G., and Shin, S.J., 1981, Canine parvovirus in a commercial kennel: Epidemiologic and pathologic findings, *Cornell Vet.* **71**:96.

Moraillon, A., Moraillon, R., Delisle, F., Cotard, J.P., Pouchelon, J.L., and Clerc, B., 1979a, Identification en France et étude générale de la gastro entérite canine à parovivurs, *Le Point Vét.* **9**:13.

Moraillon, A., Parodi, A., Moraillon, R., and Gras, B., 1979b, Une enzootie d'entérite à virus dans un élavage de visons des Charentes, *Rec. Méd. Vét.* **155**:29.

Moraillon, A., Moraillon, R., Person, J.M., and Parodi, A.L., 1980, Parvovirose canine: L'ingestion d'organes de vision atteint d'entérite à virus décleuche chez le chien une maladie identique à la maladie spontanée, Rec. Méd. Vét. **156**:539.

Mulder, C., Sharp, P.A., Delius, H., and Pettersson, U., 1974, Specific fragmentation of DNA of adenovirus serotypes 3,5,7, and 12 and adeno-simian virus 40 hybrid virus Ad2$^+$ NDI by restriction endonuclease R.EcoRI, *J. Virol.* **14**:68.

Mulvey, J.J., Bech-Nielsen, S., Haskins, M.E., Jezyk, P.F., Taylor, H.W., and Eugster, A.K., 1980, Myocarditis induced by parvoviral infection in weanling pups in the United States, *J. Am. Vet. Med. Assoc.* **177**:695.

Myers, W.L., Alberts, J.O., and Brandly, C.A., 1959, Certain characteristics of the virus of infectious enteritis of mink and observations in pathogenesis of the disease—Preliminary report, *Can. J. Comp. Med.* **23**:283.

Nettles, V.F., Pearson, J.E., Gustafson, G.A., and Blue, J.L., 1980, Parvovirus infection in translocated raccoons, *J. Am. Vet. Med. Assoc.* **177**:787.

Nettleton, P.F., and Rweyemamu, M.M., 1980, The association of calf serum with the contamination of BHK 21 clone 13 suspension cells by a parvovirus serologically related to the minute virus of mice, *Arch. Virol.* **64**:359.

Onuma, M., and Yanagawa, R., 1972a, Additional characterization of the adenosatellite virus associated with the infectious canine hepatitis virus, *Jpn. J. Vet. Res.* **20**:13.

Onuma, M., and Yanagawa, R., 1972b, Distribution of antibodies in dogs against adenosatellite viruses, *Jpn. J. Microbiol.* **16**:438.

O'Reilly, K.J., and Whitaker, A.M., 1969, The development of feline cell lines for the growth of feline infectious enteritis (panleukopenia) virus, *J. Hyg.* **67**:115.

Osborn, J.E., Robertson, S.M., Padgett, B.L., zu Rhein, G.M., Walker, D.L., and Weisblum, B., 1974, Comparison of JC and BK human papovaviruses with simian virus 40: Restriction endonuclease digestion and gel electrophoresis of resultant fragments, *J. Virol.* **13**:614.

Osterhaus, A.D.M.E., van Steenis, G., and de Kreek, P., 1980, Isolation of virus closely related to feline panleukopenia virus from dogs with diarrhea, *Zentralbl. Vet. Med. B.* **27**:11.

Petermann, H.G., and Chappuis, G., 1981, Immunprophylaxe der Parvovirusinfektion beim Hund, *Prakt. Tierarzt* **62**:52.

Pletcher, J.M., Toft, J.D., Frey, R.M., and Casey, H.W., 1979, Histopathologic evidence for parvovirus infection in dogs, *J. Am. Vet. Med. Assoc.* **175**:825.

Robinson, W.F., Wilcox, G.E., Flower, R.L.P., and Smith, J., 1979a, Evidence for a parvovirus as the aetiologic agent in myocarditis of puppies, *Aust. Vet. J.* **55**:294.

Robinson, W.F., Huxtable, C.R.R., Pass, D.A., and McHowell, J., 1979b, Clinical and electrocardiographic findings in sucpected viral myocarditis of pups, *Aust. Vet. J.* **55**:351.

Robinson, W.F., Huxtable, C.R., and Pass, D.A., 1980a, Canine parvoviral myocarditis: A morphologic description of the natural disease, *Vet. Pathol.* **17**:282.

Robinson, W.F., Wilcox, G.E., and Flower, R.L.P., 1980b, Canine parvoviral disease: Experimental reproduction of the enteric form with a parvovirus isolated from a case of myocarditis, *Vet. Pathol.* **17**:589.

Rymo, L., Lindahl, T., and Adams, A., 1979, Sites of sequence variability in Epstein–Barr Virus DNA from different sources, *Proc. Natl. Acad. Sci. USA* **76**:2794.

Schofield, F.W., 1949, Virus enteritis in mink, *North. Am. Vet.* **30**:651.

Schwers, A., Pastoret, P.P., Burtonboy, G., and Thiry, E., 1979, Fréquence en Belgique de l'infection à Parvovirus chez le chien, avant et aprés l'observation des premiers cas cliniques, *Ann. Méd. Vét.* **123**.561.

Scott, F.W., Csiza, C.K., and Gillespie, J.H., 1970, Feline viruses. V. Serum-neutralization test for feline panleukopenia, *Cornell Vet.* **60**:183.

Siegl, G., 1976, The Parvoviruses, in: *Virology Monographs* (S. Gard and C. Hallauer, eds.), Vol. 15, Springer Verlag, Berlin.

Siegl, G., and Kronauer, G., 1980, A plaque assay for feline panleukopenia virus, *J. Gen. Virol.* **46**:211.

Smith, J.R., Farmer, T.S., and Johnson, R.H., 1980, Serological observations on the epidemiology of parvovirus enteritis, *Aust. Vet. J.* **56**:149.

Sugimura, T., and Yanagawa, R., R., 1968, Effect of sodium deoxycholate on infectious canine hepatitis virus, and an additional description of a small associated virus, *Jpn. J. Vet. Res.* **16**:1.

Tal, J., Ron, D., Tattersall, P., Bratosin, S., and Aloni, Y., 1979, About 30% of minute virus of mice RNA is spliced out following polyadenylation, *Nature (London)* **279**:649.

Tattersall, P., 1978, Susceptibility to minute virus of mice as a function of host-cell differentiation, in: *Replication of Mammalian Parvoviruses* (D.C. Ward and P. Tattersall, eds.), pp 131–149, Cold Spring Harbor Laboratory, Cold Spring Harbor, New York.

Thompson, H., McCandlish, I.A.P., Cornwell, H.I.C., Wright, N.G., and Rogerson, P., 1979, Myocarditis in puppies, *Vet. Rec.* **104**:107.

Thomson, G.W., and Gagnon, A.N., 1978, Canine gastroenteritis associated with a parvovirus-like agent, *Can. Vet. J.* **19**:346.

Tratschin, J.D., McMaster, G., Kronauer, G., and Siegl, G., 1982, Canine Parvovirus: Relationship to wild-type and vaccine strains of feline panleukopenia virus and mink enteritis virus *J. Gen. Virol.* **61**:33.

Urbain, A., 1933, Contribution à l'étude de la gastroenterite infectieuse des chats, *Ann. Inst. Pasteur* **51**:202.

Verge, J., and Christoforoni, N., 1928, La gastroentérite infectieuse des chats est-ella due à un virus filtrable?, *C. R. Soc. Biol. Paris* **99**:312.

Walker, S.T., Feilen, C.P., and Savine, M., 1980, A serological survey of canine parvovirus infection in New South Wales, Australia, *Vet. Rec.* **106**:324.

Waller, E.F., 1940, Infectious gastroenteritis in raccoons (=Procyon Lotor), *J. Am. Vet. Med. Assoc.* **96**:266.

Williams F.P., Jr., 1980, Astrovirus-like, coronavirus-like, and parvovirus-like particles detected in the diarrheal stools of beagle pups, *Arch. Virol.* **66**:215.

Wilson, N.D., 1980, Origin of canine parvorirus, *Vet. Rec.* **106**:392.

Woods, C.B., Pollock, R.V.H., and Carmichael, L.E., 1980, Canine parvoviral enteritis, *J. Am. Anim. Hosp. Assoc.* **16**:171.

Zschokke, E., 1900, Ueber coli-bacilläre Infektionen, *Schweiz. Arch. Tierheilk.* **42**:20.

ADDENDUM

The Human Parvovirus

GÜNTER SIEGL

I. A UBIQUITOUS VIRUS ASSOCIATED WITH ERYTHEMA INFECTIOSUM?

Since parvoviruses have come to our attention almost 25 years ago, repeated attempts have been made to relate these agents to specific human disease. H-1, HT, HB, H-3, TVX, and Lu III viruses have thus on occasion been thought to represent candidates for the role of a genuine human parvovirus. In most instances, however, such conclusions had to be put aside because of rapidly evolving contradictory evidence and in a few cases (e.g., H-1 and HB viruses), the true role of the virus in human infections is still under dispute (cf. Chapter 8). Several candidate viruses, however, were unquestionably recovered from human clinical specimens. Among them are the parvovirus-like particles identified by Paver *et al.* (1973) in human feces, by Cossart *et al.* (1975) in sera of asymptomatic blood donors and, more recently, by Schneerson *et al.* (1980) in sera of patients suffering from febrile illness. In contrast to the agents first listed, the virus-like particles resisted progagation *in vitro*. They could only be detected by either counter immuno-electrophoresis (CIE) or immune electron microscopy (IEM), and their tentative classification as parvoviruses was based on morphologic and physical parameters (i.e., isometric, probably icosahedral particles with a diameter of 20–25 nm which banded at a buoyant density of 1.36–1.40 g/ml in CsCl). Serologic surveys conducted by CIE and IEM also revealed that infections with these particles were rather common. Nevertheless, it proved difficult to link particles and infections to a specific clinical syndrome.

At the present time, the human parvovirus-like particles, designated

---

GÜNTER SIEGL • Institute for Hygiene and Medical Microbiology, University of Bern, CH 3010 Bern, Switzerland.

B19 by Cossart *et al.* (1975) and frequently referred to as serum parvovirus-like virus (SPLV), can not yet be propagated *in vitro*. The evident association of B19 with aplastic crisis in sickle cell anemia (Pattison *et al.*, 1981; Serjeant *et al.*, 1981), however, restimulated interest in this virus and has furthered investigations into the structure and biochemistry of the particle as well as into the epidemiologic background of infections. The results of these studies, among them a great number of unpublished data, were put up for discussion at a recent workshop in London. They shall be summarized below. Because the data evidently support classification of the serum parvovirus-like particles as the first autonomous parvovirus of man, the virus will consequently be referred to as human parvovirus (HPV).

The physicochemical properties of HPV have been determined on the basis of particles purified from sera of asymptomatic blood donors or from sera of patients suffering from aplastic crisis in hemolytic anemias. Such particles occasionally revealed a hexagonal outline in the electron microscope and measured 19–25 nm in diameter. They banded in CsCl at a density around 1.4 g/ml. Moreover, they cosedimented with the minute virus of mice (MVM), for which a sedimentation coefficient of 110S has been recorded (Cossart *et al.*, 1975; Paver and Clark, 1976; Schneerson *et al.*, 1980; Anderson *et al.*, 1982a; Summers *et al.*, 1983; J. P. Clawley, unpublished; P. Tattersall, unpublished).

Judged by the known outstanding resistance of parvoviruses to environmental factors, the repeatedly observed lability of HPV in the presence of CsCl (J. P. Clawley, unpublished; P. Tattersall, unpublished; N. S. Young, unpublished) and the apparent inactivation of the virus during heating at 56°C for 30 min (Mortimer *et al.*, 1983) represent unexpected findings.

The data of three laboratories are well in agreement that the genome of HPV consists of DNA (J. P. Clawley, unpublished; Summers *et al.*, 1983; P. Tattersall, unpublished). Extraction of this DNA under routine conditions and electrophoresis in agarose gels revealed double-stranded, linear molecules 5.3–5.5 kilo base pairs (kbp). When, however, the nucleic acid was extracted under conditions of low ionic strength, the vast majority of the molecules migrated in the gels like a 5.5 kb single-stranded DNA. Therefore, HPV apparently incorporates equal numbers of complementary DNA strands into separate virus particles. Upon extraction and under appropriate reannealing conditions they come together and form double-stranded molecules. This behavior is known to be characteristic for the genome of the genera *Dependovirus* and *Densovirus* within the family Parvoviridae; it has also recently been reported for Lu III virus within the genus *Parvovirus* (Majaniemi *et al.*, 1981; Muller and Siegl, 1983). Whereas the 3′ and 5′ ends of the single-stranded genome of those viruses belonging to the genus *Parvovirus* show no sequence homology, the terminal palindromic structures of the *Dependovirus* genome consist of inverted repeats and can be hybridized to each other. Hybridization could not be achieved with the ends of the HPV genome (P. Tattersall, unpublished). Therefore, HPV is indeed likely to be classified with the autonomous parvoviruses.

With regard to the genetic relatedness between HPV and other parvoviruses, no homology could be detected between the genome of HPV and the nucleic acids of bovine or canine parvovirus under conditions of high stringency of hybridization (<15% mismatch of bases). Weak homology, however, was recorded with the genomes of Lu III virus, MVM, and porcine parvovirus. Finally, and apparently in line with the absence of inverted repeats at the genome termini, the genomes of HPV and of AAV-2 showed no homology even under conditions of low stringiency of hybridization (<25% mismatch) (P. Tattersall, unpublished).

The polypeptide spectrum of HPV is characterized by the presence of three structural proteins with molecular weights of 83000 (VP1), 62000–64000 (VP2), and 60000 (VP3). These polypeptides constitute 10, 20, and 70% of the total protein mass of the viral capsid (P. Tattersall, unpublished).

Little is known of the biologic properties of HPV. The virus was found to lack the ability to agglutinate red blood cells of rat, mouse, hamster, ferret, fowl, rhesus monkey, and human (type O) origin (Cossart *et al.*, 1975). In contrast to most other parvoviruses, serologic studies aimed at the antigenic relatedness of HPV thus can not be performed by a convenient hemagglutination inhibition test. Of the tests available, CIE failed to reveal an antigenic relationship to rat virus and to AAV types 1–4 (Cossart *et al.*, 1975). Finally, the fragments of information concerning replication of HPV point to a strict dependence of virus multiplication on cellular helper functions which can only be supplied by a distinct developmental stage of the differentiating erythrocyte. Suggestive evidence for such a conclusion will be discussed in context with HPV pathogenicity.

The involvement of HPV in aplastic crisis in sickle cell anemia was first signaled by the almost constant presence of viral antigen in acute-phase sera of affected patients. In addition, seroconversion could be recorded in the great majority of convalescent sera and, more recently, appearance of specific IgM was regularly observed concomitant with, or shortly after, clearance of antigen from the patient's blood (Pattison *et al.*, 1981; Serjeant *et al.*, 1981; Anderson 1982; Anderson *et al.*, 1982 a,b). In this context the development of a radioimmunoassay for anti-HPV-IgM has greatly facilitated diagnosis of HPV infections (Anderson *et al.*, 1982b; B. Cohen, unpublished). In contrast to earlier observations (Paver and Clarke, 1976), which seemed to indicate that the parvovirus-like particles are present in peripheral blood for a period of weeks, there is now sound evidence pointing to a rather short-lived viremia extending over about 2–7 days only (Anderson, 1982; Anderson *et al.*, 1982b; P. P. Mortimer, unpublished). Moreover, the maximum of this viremia is evidently located well before the onset of clinical disease with viral antigen clearing rapidly during the first few days of illness.

To the best of our present knowledge, viremia results from replication of HPV in the bone marrow, specifically from multiplication of the virus in erythrocyte precursor cells. Young *et al.* (1982) and Mortimer *et al.* (1983) have provided suggestive evidence that human sera containing HPV antigen, as well as purified virus particles isolated from such a source, are able to interfere with hematopoietic colony formation *in vitro*. According to electron optical studies (N. S. Young, unpublished) inhibition is the direct result of cytolytic replication of HPV in a distinct cellular intermediate in the differentiation of red blood cells. In patients with homozygous sickle-cell disease this process is reflected by depletion of the bone marrow from erythrocyte precursors as well as by the complete disappearance of circulating reticulocytes. Such changes, however, are transient in nature and the cell spectra tend to return to normal within a period of 1–2 weeks as soon as the organism musters an immune response against HPV. In a normal individual, the transient stop in erythrocyte maturation will hardly become noticeable against the background of long-lived (~120 days) virus-insusceptible peripheral red blood cells. In patients with sickle cell anemia, on the other hand, where the life span of erythrocytes is in the order of only 10–15 days, the virus-induced halt of erythrocyte maturation rapidly leads to a drastic reduction of peripheral hemoglobin levels and, hence, to distress of the infected individual. The situation appears to be identical in HPV infections of patients suffering from other types of chronic hemolytic disease such as hereditary spherocytosis (Kelleher *et al.*, 1983), thalassemia intermedia (Rao *et al.*, 1983), or pyruvate kinase deficiency (Duncan and Kurtz, 1982).

In individuals lacking the factors predisposing to aplastic crisis, infection with HPV usually resulted in minor febrile illness with malaise, headache and, occasionally, vomiting. A considerable proportion of these patients had leukopenia, and in a few reticulocytopenia, could be observed (Cant and Widdows, 1975; Schneerson *et al.*, 1980; Kelleher *et al.*, 1983). Finally, one of the blood donors from whom Cossart *et al.* (1975) originally isolated B19 virus subsequently developed a rash (Cant and Widdows, 1975). In light of the most recent findings, this might be taken as the first indication for a possible link between HPV infections and erythema infectiosum.

Erythema infectiosum or "fifth disease" was first described almost a century ago (Tschamer, 1889). At that time the disease was considered to represent an atypical form of rubella. Herrick (1926) finally established erythema infectiosum as a separate entity. It is now known as an epidemic disease of world-wide distribution which tends to occur in late winter and spring. Children of the age 4–14 years are most frequently affected; yet, adults may also contract the infection. The clinical course of erythema infectiosum is usually rather mild. It starts off with the sudden appearance of an erythema of the cheeks. This is followed by a reticular or lacy rash at the extremities. The rash usually fades within a few days. Characteristically, however, it may reoccur over a period of several weeks. Complications are uncommon in children. Adults, on the other hand, are known to develop arthritis and adenopathy (Gershon, 1979).

Erythema infectiosum was always believed to be caused by a virus. Throughout the past decades, however, none of the known viruses could convincingly be related to this disease. Evidence for an involvement of HPV was obtained only a few months ago when, between March and late May, 1983, an outbreak of erythema infectiosum was observed among pupils of two primary schools in northern London (Anderson *et al.*, 1983). Serum samples from 31 children aged 4–12, from 2 adolescents (15 and 17 yr), and from 8 adults, were subsequently screened for the presence of anti-HPV-IgM by radioimmunoassay. All but two serum samples drawn from adults proved positive. Anti-HPV-IgM concentrations were also consistently higher in sera drawn shortly after the onset of disease than in samples collected during convalescence. In spite of the fact that HPV remains to be demonstrated in clinical specimens collected during the described epidemic, these serological data strongly suggest an etiologic role of the virus in erythema infectiosum.

The observation that HPV infection is identical with an epidemic disease of childhood closely matches the fragments of epidemiological data collected since the first demonstration of HPV antigen in sera of asymptomatic blood donors. Thus, infections with HPV evidently are quite common. According to Cossart *et al.* (1975; unpublished observations) 30–45% of adults in the United Kingdom and Australia have antibodies to HPV as demonstrable by CIE. P. P. Mortimer (unpublished) even recorded antibodies in 59% of sera of English blood donors. His results, as well as the observations of Paver and Clarke (1976), also suggest that infection with the virus preferentially occurs in early life because slightly more than 30% of the normal population are antibody-positive by the age of 16.

We do not know how HPV is normally transmitted. In comparison with other parvoviruses, the fecal–oral route may be expected to play an important role. So far, however, this could not be ascertained by identification of HPV antigen in feces of diseased persons in spite of the fact that diarrhea is not uncommon during acute infection. Clearly, however, HPV can be spread in blood and blood products. This is not only suggested by the well-documented isolation of particles from donated blood but also by the increased proportion of antibody-positive sera from multiple-transfused patients and hemophiliacs, as well as from drug adicts. In the latter group, 81% of the sera were found to contain antibodies to HPV (P. P. Mortimer, unpublished).

The data listed before evidently justify the conclusion that the virus-like particles originally detected in the sera of asymptomatic blood donors represent a human parvovirus and are identical with the etiologic virus of erythema infectiosum. Further studies concerning this virus and the disease it causes will profit from the vast body of knowledge accumulated so far for animal parvoviruses. It will then be interesting to see whether syndromes described as a consequence of infection with the latter viruses, and, known to have their counterparts in human disease, can in fact be related to the presence of HPV.

REFERENCES

Anderson, M.J., 1982, The emerging story of a human parvovirus-like agent, *J. Hyg. Camb.* **89**:1.

Anderson, M.J., Davis, L.R., Hodgson, J., Jones, S.E., Murtaza, L., Pattison, J.R., Stroud, C.E., and White, J.M., 1982a, Occurrence of infection with a parvovirus-like agent in children with sickle cell anaemia during a two-year period, *J. Clin. Pathol.* **35**:744.

Anderson, M.J., Davis, L.R., Jones, S.E., Pattison, J.R., and Serjeant, G.R., 1982b, The development and use of an antibody capture radioimmunoassay for specific IgM to a human parvovirus-like agent, *J. Hyg. Camb.* **88**:309.

Anderson, M.J., Jones, S.E., Fischer-Hoch, S.P., Lewis, E., Hall, S.M., Bartlett, C.L.R., Cohen, B.J., Mortimer, P.P., and Pereira, M.S., 1983, Human parvovirus, the cause of erythema infectiosum (fifth disease)? *Lancet* **i**:1378.

Cant, B., and Widdows, D., 1975, *Abstr. Conf. Int. Soc. Blood Transfusion*, p. 73.

Cossart, Y.E., Field, A.M., Cant, B., and Widdows, D., 1975, Parvovirus-like particles in human sera, *Lancet* **i**:72.

Duncan, J., and Kurtz, J., 1982, Parvovirus-like particles in blood, *Comm. Dis. Rep.* **24**:3.

Gershon, A.A., 1979, Erythema infectiosum (fifth disease), in: *Principles and Practice of Infectious Diseases*, (G.L. Mandel, R.H. Douglas, and J.E. Bennet, eds.) pp. 1453–1456, J. Wiley and Son, New York.

Herrick, T.P., 1926, Erythema infectiosum; A clinical report of 74 cases, *Am. J. Dis. Child.* **31**:486.

Kelleher, J.F., Luban, N.L.C., Mortimer, P.P., and Kamimura, T., 1983, Human serum "parvovirus": A specific cause of aplastic crisis in children with hereditary sperocytosis, *J. Pediat.* **102**:720.

Majaniemi, I.J., Tratschin, J.D., and Siegl, G., 1981, A reassessment of the nucleic acid and protein components of parvovirus LuIII, *Proc. 5th Internat. Congr. Virol*, Strasbourg, p. 366.

Mortimer, P.P., Humphries, R.K., Moore, J.G., Purcell, R.H., and Young, N.S., 1983, A human parvovirus-like virus inhibits haematopoietic colony formation *in vitro*, *Nature*, (*London*) **302**:426.

Muller, D.E., and Siegl, G., 1983, Maturation of parvovirus LuIII in a subcellular system. Optimal conditions for *in vitro* synthesis and encapsidation of viral DNA, *J. Gen. Virol.* **64**:1043.

Pattison, J.R., Jones, S.E., Hodgson, J., Davis, L.R., White, J.M., Stroud, C.E., and L. Murtaza, 1981, Parvovirus infections and hypoplastic crisis in sickle cell anaemia, *Lancet*, **i**:664.

Paver, W.K., and Clarke, S.K.R., 1976, Comparison of human fecal and serum parvo-like viruses, *J. Clin. Microbiol.* **4**:1.

Paver, W.K., Caul, E.O., Ashley, C.R., and Clarke, S.K.R., 1973, A small virus in human faeces, *Lancet* **i**:237.

Rao, K.P.R., Patel, A.R., Anderson, M.J., Hodgson, J., and Jones, S.E., 1983, Infection with parvovirus-like virus and aplastic crisis in chronic hemolytic anemia, *Ann. Int. Med.*, **98**:930.

Schneerson, J.M., Mortimer, P.P., and Vandervelde, E.M., 1980, Febrile illness due to a parvovirus, *Brit. Med. J.*, **ii**:1580.

Serjeant, G.R., Toplay, J.M., Mason, K., Serjeant, B.E., Pattison, J.R., Jones, S.E., and Mohamed, R., 1981, Outbreak of aplastic crisis in sickle cell anaemia associated with parvovirus-like agent, *Lancet* **ii**:595.

Summers, J., Jones, S.E., and Anderson, M.J., 1983, Characterisation of the genome of the agent of erythrocyte aplasia permits its classification as a human parvovirus, *J. Gen. Virol.* (in press).

Tschamer, A., 1889, Ueber örtliche Rötheln, *Jahrb. Kinderheilk.* **29**:372.

Young, N.S., Humphries, R.K., Moore, J.G., Purcell, R.H., and Mortimer, P.P., 1982, A human virus inhibits haematopoiesis *in vitro*, *Blood* **60**(Suppl. I):42a.

Appendix

During 1983 the complete genome sequences of AAV-2 and of the autonomous parvoviruses H-1 and MVM were published. Knowledge of the sequences constitutes a major advance in efforts to understand the molecular biology of the parvoviruses. In the following pages the three sequences are reproduced with permission.

Kenneth I. Berns

```
→ Inverted terminal repeat                                          'Flip' orientation
5'-TTGGCCACTC CCTCTCTGCG CGCTCGCTCG CTCACTGAGG CCGGGCGACC AAAGGTCGCC CGACGCCCGG GCTTTGCCCG GGCGGCCTCA GTGAGCGAGC 100
              A                                   B          B'          C          C'          A'
                        Inverted terminal repeat ←
GAGCGCGCAG AGAGGGAGTG GCCAACTCCA TCACTAGGGG TTCCTGGAGG GGTGGAGTCG TGACGTGAAT TACGTCATAG GGTTAGGGAG GTCCTGTATT
                                                            Promoter 1                  5' terminus of RNA 1
AGAGGTCACG TGAGTGTTTT GCGACATTTT GCGACACCAT GTGGTCACGC TGGGTATTTA AGCCCGAGTG AGCACGCAGG GTCTCCATTT TGAAGCGGGA
                    Coding region 1
GGTTTGAACG CGCAGCCGCC ATGCCGGGGT TTTACGAGAT TGTGATTAAG GTCCCCAGCG ACCTTGACGG GCATCTGCCC GGCATTTCTG ACAGCTTTGT

GAACTGGGTG GCCGAGAAGG AATGGGAGTT GCCGCCAGAT TCTGACATGG ATCTGAATCT GATTGAGCAG GCACCCCTGA CCGTGGCCGA GAAGCTGCAG 500

CGCGACTTTC TGACGGAATG GCGCCGTGTG AGTAAGGCCC CGGAGGCCCT TTTCTTTGTG CAATTTGAGA AGGGAGAGAG CTACTTCCAC ATGCACGTGC

TCGTGGAAAC CACCGGGGTG AAATCCATGG TTTTGGGACG TTTCCTGAGT CAGATTCGCG AAAAACTGAT TCAGAGAATT TACCGCGGGA TCGAGCCGAC

TTTGCCAAAC TGGTTCGCGG TCACAAAGAC CAGAAATGGC GCCGGAGGCG GGAACAAGGT GGTGGATGAG TGCTACATCC CCAATTACTT GCTCCCCAAA
                                              Promoter 2                      5' terminus of RNA 2
ACCCAGCCTG AGCTCCAGTG GGCGTGGACT AATATGGAAC AGTATTTAAG CGCCTGTTTG AATCTCACGG AGCGTAAACG GTTGGTGGCG CAGCATCTGA
                                                                                                 Coding region 2
CGCACGTGTC GCAGACGCAG GAGCAGAACA AAGAGAATCA GAATCCCAAT TCTGATGCGC CGGTGATCAG ATCAAAAACT TCAGCCAGGT ACATGGAGCT 1000

GGTCGGGTGG CTCGTGGACA AGGGGATTAC CTCGGAGAAG CAGTGGATCC AGGAGGACCA GGCCTCATAC ATCTCCTTCA ATGCGGCCTC CAACTCGCGG
```

FIGURE 1. AAV-2 DNA sequence (Srivastava, A., Lusby, E., and Berns, K.I., 1983, *J. Virol.* 45:555–564, reprinted with permission).

(*continued*)

TCCCAAATCA AGGCTGCCTT GGACAATGCG GGAAAGATTA TGAGCCTGAC TAAAACCGCC CCCGACTACC TGGTGGGCCA GCAGCCCGTG GAGGACATTT

CCAGCAATCG GATTTATAAA ATTTTGGAAC TAAACGGGTA CGATCCCCAA TATGCGGCTT CCGTCTTTCT GGGATGGGCC ACGAAAAAGT TCGGCAAGAG

GAACACCATC TGGCTGTTTG GGCCTGCAAC TACCGGGAAG ACCAACATCG CGGAGGCCAT AGCCCACACT GTGCCCTTCT ACGGGTGCGT AAACTGGACC

AATGAGAACT TTCCCTTCAA CGACTGTGTC GACAAGATGG TGATCTGGTG GGAGGAGGGG AAGATGACCG CCAAGGTCGT GGAGTCGGCC AAAGCCATTC 1500

TCGGAGGAAG CAAGGTGCGC GTGGACCAGA AATGCAAGTC CTCGGCCCAG ATAGACCCGA CTCCCGTGAT CGTCACCTCC AACACCAACA TGTGCGCCGT

GATTGACGGG AACTCAACGA CCTTCGAACA CCAGCAGCCG TTGCAAGACC GGATGTTCAA ATTTGAACTC ACCCGCCGTC TGGATCATGA CTTTGGGAAG

GTCACCAAGC AGGAAGTCAA AGACTTTTTC CGGTGGGCAA AGGATCACGT GGTTGAGGTG GAGCATGAAT TCTACGTCAA AAAGGGTGGA GCCAAGAAAA

Promoter 3 → 5' terminus of RNA 3
GACCCGCCCC CAGTGACGCA GATATAAGTG AGCCCAAACG GGTGCGCGAG TCAGTTGCGC AGCCATCGAC GTCAGACGCG GAAGCTTCGA TCAACTACGC

→ Splice
AGACAGGTAC CAAAACAAAT GTTCTCGTCA CGTGGGCATG AATCTGATGC TGTTTCCCTG CAGACAATGC GAGAGAATGA ATCAGAATTC AAATATCTGC 2000

TTCACTCACG GACAGAAAGA CTGTTTAGAG TGCTTTCCCG TGTCAGAATC TCAACCCGTT TCTGTCGTCA AAAAGGCGTA TCAGAAACTG TGCTACATTC

Polyadenylation signal
ATCATATCAT GGGAAAGGTG CCAGACGCTT GCACTGCCTG CGATCTGGTC AATGTGGATT TGGATGACTG CATCTTTGAA CAATAAATGA TTTAAATCAG

Splice ← Terminator
GTATGGCTGC CGATGGTTAT CTTCCAGATT GGCTCGAGGA CACTCTCTCT GAAGGAATAA GACAGTGGTG GAAGCTCAAA CCTGGCCCAC CACCACCAAA

GCCCGCAGAG CGGCATAAGG ACGACAGCAG GGGTCTTGTG CTTCCTGGGT ACAAGTACCT CGGACCCTTC AACGGACTCG ACAAGGGAGA GCCGGTCAAC

GAGGCAGACG CCGCGGCCCT CGAGCACGTA CAAAGCCTAC GACCGGCAGC TCGACAGCGG AGACAACCCG TACCTCAAGT ACAACCACGC CGACGCGGAG 2500

TTTCAGGAGC GCCTTAAAGA AGATACGTCT TTTGGGGGCA ACCTCGGACG AGCAGTCTTC CAGGCGAAAA AGAGGGTTCT TGAACCTCTG GGCCTGGTTG

AGGAACCTGT TAAGACGGCT CCGGGAAAAA AGAGGCCGGT AGAGCACTCT CCTGTGGAGC CAGACTCCTC CTCGGGAACC GGAAAGGCGG GCCAGCAGCC

TGCAAGAAAA AGATTGAATT TTGGTCAGAC TGGAGACGCA GACTCAGTAC CTGACCCCCA GCCTCTCGGA CAGCCACCAG CAGCCCCCTC TGGTCTGGGA

Coding region 3 →
ACTAATACGA TGGCTACAGG CAGTGGCGCA CCAATGGCAG ACAATAACGA GGGCGCCGAC GGAGTGGGTA ATTCCTCCGG AAATTGGCAT TGCGATTCCA

FIGURE 1 (*continued*)

CATGGATGGG CGACAGAGTC ATCACCACCA GCACCCGAAC CTGGGCCCTG CCCACCTACA ACAACCACCT CTACAAACAA ATTTCCAGCC AATCAGGAGC 3000

CTCGAACGAC AATCACTACT TTGGCTACAG CACCCCTTGG GGGTATTTTG ACTTCAACAG ATTCCACTGC CACTTTTCAC CACGTGACTG GCAAAGACTC

ATCAACAACA ACTGGGGATT CCGACCCAAG AGACTCAACT TCAAGCTCTT TAACATTCAA GTCAAAGAGG TCACGCAGAA TGACGGTACG ACGACGATTG

CCAATAACCT TACCAGCACG GTTCAGGTGT TTACTGACTC GGAGTACCAG CTCCCGTACG TCCTCGGCTC GGCGCATCAA GGATGCCTCC CGCCGTTCCC

AGCAGACGTC TTCATGGTGC CACAGTATGG ATACCTCACC CTGAACAACG GGAGTCAGGC AGTAGGACGC TCTTCATTTT ACTGCCTGGA GTACTTTCCT

TCTCAGATGC TGCGTACCGG AAACAACTTT ACCTTCAGCT ACACTTTTGA GGACGTTCCT TTCCACAGCA GCTACGCTCA CAGCCAGAGT CTGGACCGTC 3500

TCATGAATCC TCTCATCGAC CAGTACCTGT ATTACTTGAG CAGAACAAAC ACTCCAAGTG GAACCACCAC GCAGTCAAGG CTTCAGTTTT CTCAGGCCGG

AGCGAGTGAC ATTCGGGACC AGTCTAGGAA CTGGCTTCCT GGACCCTGTT ACCGCCAGCA GCGAGTATCA AAGACATCTG CGGATAACAA CAACAGTGAA

TACTCGTGGA CTGGAGCTAC CAAGTACCAC CTCAATGGCA GAGACTCTCT GGTGAATCCG GCCATGGCAA GCCACAAGGA CGATGAAGAA AAGTTTTTTC

CTCAGAGCGG GGTTCTCATC TTTGGGAAGC AAGGCTCAGA GAAAACAAAT GTGAACATTG AAAAGGTCAT GATTACAGAC GAAGAGGAAA TCGGAACAAC

CAATCCCGTG GCTACGGAGC AGTATGGTTC TGTATCTACC AACCTCCAGA GAGGCAACAG ACAAGCAGCT ACCGCAGATG TCAACACACA AGGCGTTCTT 4000

CCAGGCATGG TCTGGCAGGA CAGAGATGTG TACCTTCAGG GGCCCATCTG GGCAAAGATT CCACACACGG ACGGACATTT TCACCCCTCT CCCCTCATGG

GTGGATTCGG ACTTAAACAC CCTCCTCCAC AGATTCTCAT CAAGAACACC CCGGTACCTG CGAATCCTTC GACCACCTTC AGTGCGGCAA AGTTTGCTTC

CTTCATCACA CAGTACTCCA CGGGACACGG TCAGCGTGGA GATCGAGTGG GAGCTGCAGA AGGAAAACAG CAAACGCTGG AATCCCGAAA TTCAGTACAC

BglI ↓ Hinfl ↓

Terminator

TTCCAACTAC AACAAGTCTG TTAATCGTGG ACTTACCGTG GATACTAATG GCGTGTATTC AGAGCCTCGC CCCATTGGCA CCAGATACCT GACTCGTAAT

3' termini of RNAs

Polyadenylation signal

CTGTAATTGC TTGTTAATCA ATAAACCGTT TAATTCGTTT CAGTTGAACT TTGGTCTCTG CGTATTTCTT TCTTATCTAG TTTCCATGGC TACGTAGATA 4500

→Inverted terminal repeat

AGTAGCATGG CGGGTTAATC ATTAACTACA AGGAACCCCT AGTGATGGAG TTGGCCACTC CCTCTCTGCG CGCTCGCTCG CTCACTGAGG CCGGGCGACC

A B

'Flop' orientation Inverted terminal repeat←

AAAGGTCGCC CGACGCCCGG GCTTTGCCCG GGCGGCCTCA GTGAGCGAGC GAGCGCGCAG AGAGGGAGTG GCCAA-3'

B' C C' A'

FIGURE 1 (*continued*)

0001 CATTTTTAGA ACTGACCAAC CATGTTCACG CAAGTGACGT GATGACGCGC
0051 GCTGCGCGCG CTGCCTTCGG CAGTCACACG TCACTAGCGT TTCACATGGT
0101 TGGTCAGTTC TAAAAATGAT AAGCGGTTCA GAGAGTTTGA AACCAAGGCG
0151 GGAAACGGAA GTGGGCGTGG CTAACTGTAT ATAAGCAGTC ACTCTGGTCG
0201 GTTACTCACT CTGCTTTCAT TTCTGAGTTT GTGAGACACA GGAGCGAGAC
0251 TAACCAACTA ACC ATG GCT GGA AAC GCT TAC TCC GAT GAG GTT TTG GGA
0300 GTA ACA AAC TGG CTG AAG GAC AAA AGT AGC CAG GAG GTG TTC TCA TTT
0348 GTT TTT AAA AAT GAA AAC GTC CAA CTA AAT GGA AAG GAC ATC GGT TGG
0396 AAT AGT TAC AGA AAG GAG CTA CAA GAT GAC GAG CTG AAG TCT CTA CAA
0444 CGA GGG GCG GAG ACC ACT TGG GAC CAA AGC GAG GAC ATG GAA TGG GAG
0492 AGC GCA GTG GAT GAC ATG ACC AAA AAG CAA GTA TTT ATT TTT GAT TCT
0540 TTG GTT AAG AAG TGT TTG TTT GAA GTG CTC AGC ACA AAG AAC ATA GCT
0588 CCT AGT AAT GTT ACT TGG TTC GTG CAG CAT GAA TGG GGA AAG GAC CCA
0636 GGC TGG CAC TGT CAT GTG CTG ATT GGA GGC AAG GAC TTT AGT CAA CCT
0684 CAA GGA AAA TGG TGG AGA AGG CAG CTA AAT GTG TAC TGG AGT AGA TGG
0732 TTG GTG ACT GCC TGT AAT GTT CAA CTA ACA CCA GCT GAA AGA ATT AAA
0780 CTG AGA GAA ATA GCA GAG GAC AGT GAA TGG GTC ACT TTG CTT ACC TAT
0828 AAG CAT AAG CAC ACC AAG AAG GAC TAT ACC AAG TGT GTT CTT TTT GGA
0876 AAC ATG ATT GCT TAT TAC TTT TTA AGC AAA AAG AAA ATA TGT ACC AGT
0924 CCA CCA AGG GAC GGA GGC TAT TTT CTT AGC AGT GAC TCT GGC TGG AAA
0972 ACT AAC TTT TTG AAA GAG GGC GAG CGC CAT CTA GTG AGC AAA CTG TAT
1020 ACT GAT GAG ATG AAA CCA GAA ACG GTC GAG ACC ACA GTG ACC ACT GCA
1068 CAG GAA GCT AAG CGC GGC AGA ATT CAA ACT AGA GAG GAG GTC TCG ATT
1116 AAA ACC ACA CTC AAA GAG TTG GTA CAT AAA AGA GTA ACC TCA CCA GAA
1164 GAC TGG ATG ATG ATG CAG CCA GAC AGT TAC ATT GAA ATG ATG GCT CAA
1212 CCA GGT GGA GAA AAC TTG CTT AAA AAT ACA CTA GAG ATC TGT ACA CTG

1260 ACT CTA GCA AGA ACC AAA ACA GCC TTT GAC TTG ATT CTG GAA AAA GCT
1308 GAA ACC AGC AAA CTA GCC AAC TTT TCC ATG GCT AGC ACC AGA ACC TGT
1356 AGA ATC TTT GCT GAG CAT GGC TGG AAC TAT ATT AAA GTC TGC CAT GCC
1404 ATC TGT TGT GTG CTG AAT AGA CAA GGA GGC AAA AGG AAC ACT GTG CTC
1452 TTT CAC GGA CCA GCC AGC ACA GGC AAA TCT ATT ATT GCA CAA GCC ATA
1500 GCA CAA GCA GTT GGT AAT GTT GGT TGT TAC AAT GCT GCC AAT GTG AAC
1548 TTT CCA TTT AAT GAC TGT ACC AAC AAA AAC TTG ATT TGG GTG GAA GAA
1596 GCT GGT AAC TTT GGC CAG CAA GTA AAC CAA TTC AAA GCT ATT TGT TCT
1644 GGC CAA ACC ATA CGC ATT GAT CAA AAA GGA AAA GGC AGC AAA CAG ATT
1692 GAA CCA ACA CCA GTT ATT ATG ACC ACC AAC GAG AAC ATT ACC GTG GTT
1740 AGA ATA GGC TGT GAG GAA AGA CCA GAA CAC ACT CAA CCA ATC AGA GAC
1788 AGA ATG CTC AAC ATT CAC CTG ACA CGT ACA CTA CCT GGT GAC TTT GGT
1836 TTG GTG GAT AAG CAC GAA TGG CCT CTG ATC TGT GCT TGG TTG GTG AAG
1884 AAT GGT TAC CAA TCT ACC ATG GCT TGT TAC TGT GCT AAA TGG GGC AAA
1932 GTT CCT GAT TGG TCA GAG GAC TGG GCG GAG CCG AAG CTA GAC ACT CCT
1980 ATA AAT TCG CTA GGT TCA ATG CGC TCA CCA TCT CTG ACT CCG AGA AGT
2028 ACG CCT CTC AGC CAA AAC TAC GCT CTT ACT CCA CTT GCA TCG GAC CTT
2076 GCG GAC CTA GCT CTA GAG CCT TGG AGC ACA CCA AAT ACT CCT GTT GCG
2124 GGC ACT GCA GCA AGC CAA AAC ACT GGG GAG GCT GGT TCC ACA GCC TGC
2172 CAA GGT GCT CAA CGG AGC CCA ACC TGG TCC GAG ATC GAG GCG GAT TTG
2220 AGA GCT TGC TTC AGT CAA GAA CAG TTG GAG AGC GAC TTC AAC GAG GAG
2268 CTG ACC TTG GAC TAA GGTACA ATG GCA CCT CCA GCT AAA AGA GCT AAA
2316 AGA GGT AAG GGG CTA AGG GAT GGT TGG TTG GTG GGG
2352 TAC TAA TGTATGACTACCTGTTTTACAGGCCTGA frame 2: AAT
2389 CAC TTG GTT CTA GGT TGG GTG CCT CCT GGC TAC AAG TAC CTG GGA CCA
2437 GGG AAC AGC CTT GAC CAA GGA GAA CCA ACC AAC CCT TCT GAC GCC GCT
2485 GCC AAA GAA CAC GAC GAA GCC TAC GAC CAA TAC ATC AAA TCT GGA AAA
2533 AAT CCT TAC CTG TAC TTC TCT CCT GCT GAT CAA CGC TTC ATT GAC CAA

2581 ACC AAA GAC GCC AAG GAC TGG GGC GGC AAG GTT GGT CAC TAC TTT TTT
2629 AGA ACC AAG CGA GCT TTT GCA CCT AAG CTT TCT ACT GAC TCT GAA CCT
2677 GGC ACT TCT GGT GTG AGC AGA CCT GGT AAA CGA ACT AAA CCA CCT GCT
2725 CAC ATT TTT GTA AAT CAA GCC AGA GCT AAA AAA AAA CGC GCT TCT CTT
2773 GCT GCA CAG CAG AGG ACT CTG ACA ATG AGT GAT GGC ACC GAA ACA AAC
2821 CAA CCA GAC ACT GGA ATC GCT AAT GCT AGA GTT GAG CGA TCA GCT GAC
2869 GGA GGT GGA AGC TCT GGG GGT GGG GGC TCT GGC GGG GGT GGG ATT GGT
2917 GTT TCT ACT GGG ACT TAT GAT AAT CAA ACG ACT TAT AAG TTT TTG GGA
2965 GAT GGA TGG GTA GAA ATA ACT GCA CAT GCT TCT AGA CTT TTG CAC TTG
3013 GGA ATG CCT CCT TCA GAA AAC TAC TGC CGC GTC ACC GTT CAC AAT AAT
3061 CAA ACA ACA GGA CAC GGA ACT AAG GTA AAG GGA AAC ATG GCC TAT GAC
3109 ACA CAT CAA CAA ATT TGG ACA CCA TGG AGC TTG GTA GAT GCT AAT GCT
3157 TGG GGA GTT TGG TTC CAA CCA AGT GAC TGG CAG TTC ATT CAA AAC AGC
3205 ATG GAA TCG CTG AAT CTT GAC TCA TTG AGC CAA GAA CTA TTT AAT GTA
3253 GTA GTC AAA ACA GTC ACT GAA CAA CAA GGA GCT GGC CAA GAT GCC ATT
3301 AAA GTC TAT AAT AAT GAC TTG ACG GCC TGT ATG ATG GTT GCT CTG GAT
3349 AGT AAC AAC ATA CTG CCT TAC ACA CCT GCA GCT CAA ACA TCA GAA ACA
3397 CTT GGT TTC TAC CCA TGG AAA CCA ACC GCA CCA GCT CCT TAC AGA TAC
3445 TAC TTT TTC ATG CCT AGA CAA CTC AGT GTA ACC TCT AGC AAC TCT GCT
3493 GAA GGA ACT CAA ATC ACA GAC ACC ATT GGA GAG CCA CAG GCA CTA AAC
3541 TCT CAA TTT TTT ACT ATT GAG AAC ACC TTG CCT ATT ACT CTC CTG CGC
3589 ACA GGT GAT GAG TTT ACA ACT GGC ACC TAC ATC TTT AAC ACT GAC CCA
3637 CTT AAA CTT ACT CAC ACA TGG CAA ACC AAC AGA CAC TTG GCA TGC CTC
3685 CAA GGA ATA ACT GAC CTA CCA ACA TCA GAT ACA GCA ACA GCA TCA CTA
3733 ACT GCA AAT GGA GAC AGA TTT GGA TCA ACA CAA ACA CAG AAT GTG AAC
3781 TAT GTC ACA GAG GCT TTG CGC ACC AGG CCT GCT CAG ATT GGC TTC ATG
3829 CAA CCT CAT GAC AAC TTT GAA GCA AAC AGA GGT GGC CCA TTT AAG GTT

3878 CCA GTG GTA CCG CTA GAC ATA ACA GCT GGC GAG GAC CAT GAT GCA AAC
3925 GGA GCC ATA CGA TTT AAC TAT GGC AAA CAA CAT GGC GAA GAT TGG GCC
3973 AAA CAA GGA GCA GCA CCA GAA AGG TAC ACA TGG GAT GCA ATT GAT AGT
4021 GCA GCT GGG AGG GAC ACA GCT AGA TGC TTT GTA CAA AGT GCA CCA ATA
4069 TCT ATT CCA CCA AAC CAA AAC CAG ATC TTG CAG CGA GAA GAC GCC ATA
4118 GCT GGC AGA ACT AAC ATG CAT TAT ACT AAT GTT TTT AAC AGC TAT GGT
4165 CCA CTT AGT GCA TTT CCT CAT CCA GAT CCC ATT TAT CCA AAT GGA CAA
4213 ATT TGG GAC AAA GAA TTG GAC CTG GAA CAC AAA CCT AGA CTA CAC GTA
4261 ACT GCA CCA TTT GTT TGT AAA AAC AAC CCA CCA GGT CAA CTA TTT GTT
4309 CAC TTG GGG CCT AAT CTG ACT GAC CAA TTT GAC CCA AAC AGC ACA ACT
4357 GTT TCT CGC ATT GTT ACA TAT AGC ACT TTT TAC TGG AAG GGT ATT TTG
4405 AAA TTC AAA GCC AAA CTA AGA CCA AAT CTG ACC TGG AAT CCT GTA TAC
4453 CAA GCA ACC ACA GAC TCT GTT GCC AAT TCT TAC ATG AAT GTT AAG AAA
4501 TGG CTC CCA TCT GCA ACT GGC AAC ATG CAC TCT GAT CCA TTG ATT TGT
4549 AGA CCT GTG CCT CAC ATG ACA TAC TAA CCAAC CAACTATGTT TCTCTGTTTG
4601 CTTCACATAA TACTTAAACT AACTAGACTA CAACATAAAA ATATACACTT
4651 AATAATAGAT TATTAAAAAT AACATAATAT GGTAGGTTAA CTGTAAAAAA
4701 TAATAGAACT TTTGGAATAA ATATAGTTAG TTGGTTAATG TTAGATAGAA
4751 TATAAAAAGA TTTTGTATTT TAAAATAAAT ATAGTTAGTT GGTTAATGTT
4801 AGATAGAATA TAAAAAGATT TTGTATTTGG GAAATAAAAA GGGTGGTTGG
4851 GTGGTTGGTT GGTACTCCCT TAGACTGAAT GTTAGGGACC AAAAAAATAA
4901 TAAAATAATT AAAATGAACA AGGACTACTG TCTATTCAGT TGACCAACTG
4951 AACCTATAGT ATCACTATGT TTTTAGGGTG GGGGGGTGGG AGATACATAC
5001 GTTCGCTATG GACCAAGTGG TACCGGTTGG TTGCTAAGCT CGAACAAGAC
5051 GGCTAAGCCG GTCCGGTTGG TTGAGCGCAA CCAACCGGTA CCACTTGGTC
5101 CATAGCGAAC GTATGTATCT CCCACCCCCC CACCCTAAAA ACATAGTGAT
5151 ACTATAGGTT CAGTTGGTCA ACTGAA(T)-5176-3' TERMINUS

FIGURE 2. H-1 DNA sequences (Rhode, S.L., III and Paradiso, P., 1983, J. Virol. 45:173–184, reprinted with permission).

```
              15                  30                  45                  60                  75                  9C
ATTTTTAGAACTGACCAACCATGTTCACGTAAGTGACGTGATGACGCGCGCTGCGCGCGCGCCTTCGGACGTCACACGTCACTTACGTTT
TAAAAATCTTGACTGGTTGGTACAAGTGCATTCACTGCACTACTGCGCGCGACGCGCGCGCGGAAGCCTGCAGTGTGCAGTGAATGCAAA
                          SacIII    SacIII    HgaIFnu4HIHhaI        Acy1SacIII SacIII   SacIII

                3 (m  i  s  g  s  g  s  l  n  q  g  a  k  r  k  w  a  w  f  k  v  y
             105                 120                 135                 150                 165                 180
CACATGGTTGGTCAGTTCTAAAAATGATAAGCGGTTCAGGGAGTTTAAACCAAGGCGCGAAAAGGAAGTGGGCGTGGTTTAAAGTATATA
GTGTACCAACCAGTCAAGATTTTTACTATTCGCCAAGTCCCTCAAATTTGGTTCCGCGCTTTTCCTTCACCCGCACCAAATTTCATATAT
                                                      FnuDIIHhaI

k  q  l  l  k  s  v  t  y  l  f  f  h  s  v  s  r  d  a  q  k  e  s  n  q  l  t) M  A  G
             195                 210                 225                 240                 255                 270
AGCAACTACTGAAGTCAGTTACTTATCTTTTCTTTCATTCTGTGAGTCGAGACGCACAGAAAGAGAGTAACCAACTAACCATGGCTGGAA
TCGTTGATGACTTCAGTCAATGAATAGAAAAGAAAGTAAGACACTCAGCTCTGCGTGTCTTTCTCTCATTGGTTGATTGGTACCGACCTT
                                            HinFITaqIHgaI                        NcoI

N  A  Y  S  D  E  V  L  G  A  T  N  W  L  K  E  K  S  N  Q  E  V  F  S  F  V  F  K  N  E
             285                 300                 315                 330                 345                 360
ATGCTTACTCTGATGAAGTTTTGGGAGCAACCAACTGGTTAAAGGAAAAAAGTAACCAGGAAGTGTTCTCATTTGTTTTTAAAAATGAAA
TACGAATGAGACTACTTCAAAACCCTCGTTGGTTGACCAATTTCCTTTTTTCATTGGTCCTTCACAAGAGTAAACAAAAATTTTTACTTT
                                                       BstNI

N  V  Q  L  N  G  K  D  I  G  W  N  S  Y  K  K  E  L  Q  E  D  E  L  K  S  L  Q  R  G  A
             375                 390                 405                 420                 435                 450
ATGTTCAACTGAATGGAAAAGATATCGGATGGAATAGTTACAAAAAAGAGCTGCAGGAGGACGAGCTGAAATCTTTACAACGAGGAGCGG
TACAAGTTGACTTACCTTTTCTATAGCCTACCTTATCAATGTTTTTTCTCGACGTCCTCCTGCTCGACTTTAGAAATGTTGCTCCTCGCC
                    EcoRV                        AluIBbvIPstIMnlI AluI              MnlI

E  T  T  W  D  Q  S  E  D  M  E  W  E  T  T  V  D  E  M  T  K  K  Q  V  F  I  F  D  S  L
             465                 480                 495                 510                 525                 540
AAACTACTTGGGACCAAAGCGAGGACATGGAATGGGAAACCACAGTGGATGAAATGACCAAAAAGCAAGTATTCATTTTTGATTCTTTGG
TTTGATGAACCCTGGTTTCGCTCCTGTACCTTACCCTTTGGTGTCACCTACTTTACTGGTTTTTCGTTCATAAGTAAAAACTAAGAAACC
          AvaII    MnlI                                                           HinFI

V  K  K  C  L  F  E  V  L  N  T  K  N  I  F  P  G  D  V  N  W  F  V  Q  H  E  W  G  K  D
             555                 570                 585                 600                 615                 630
TTAAAAAATGTTTATTTGAAGTGCTTAACACAAAGAATATATTTCCTGGTGATGTTAATTGGTTTGTGCAACATGAATGGGGAAAAGACC
AATTTTTTACAAATAAACTTCACGAATTGTGTTTCTTATATAAAGGACCACTACAATTAACCAAACACGTTGTACTTACCCCTTTTCTGG
                                             BstNIHphI   EcoRI

Q  G  W  H  C  H  V  L  I. G  G  K  D  F  S  Q  A  Q  G  K  W  W  R  R  Q  L  N  V  Y  W
             645                 660                 675                 690                 705                 720
AAGGCTGGCACTGCCATGTACTAATTGGAGGAAAGGACTTTAGTCAAGCTCAAGGGAAATGGTGGAGAAGGCAACTAAATGTTTACTGGA
TTCCGACCGTGACGGTACATGATTAACCTCCTTTCCTGAAATCAGTTCGAGTTCCCTTTACCACCTCTTCCGTTGATTTACAAATGACCT
                  RsaI EcoRI*MnlI              AluI

S  R  W  L  V  T  A  C  N  V  Q  L  T ·P  A  E  R  I  K  L  R  E  I  A  E  D  N  E  W  V
             735                 750                 765                 780                 795                 810
GCAGATGGTTGGTAACAGCCTGTAATGTGCAACTAACACCAGCTGAAAGAATTAAACTAAGAGAAATAGCAGAAGACAATGAGTGGGTTA
CGTCTACCAACCATTGTCGGACATTACACGTTGATTGTGGTCGACTTTCTTAATTTGATTCTCTTTATCGTCTTCTGTTACTCACCCAAT
                                         AluIPvuII  EcoRI* DdeI         MboII

T  L  L  T  Y  K  H  K  Q  T  K  K  D  Y  T  K  C  V  L  F  G  N  M  I  A  Y  Y  F  L  T
             825                 840                 855                 870                 885                 900
CTCTACTTACTTATAAGCATAAGCAAACCAAAAAAGACTATACCAAGTGTGTTCTTTTTGGAAACATGATTGCTTACTATTTTTTAACTA
GAGATGAATGAATATTCGTATTCGTTTGGTTTTTTCTGATATGGTTCACACAAGAAAAACCTTTGTACTAACGAATGATAAAAAATTGAT

K  K  K  I  S  T  S  P  P  R  D  G  G  Y  F  L  S  S  D  S  G  W  K  T  N  F  L  K  E  G
             915                 930                 945                 960                 975                 990
AAAAGAAAATAAGCACTAGTCCACCAAGAGACGGAGGCTATTTTCTTAGCAGTGACTCTGGCTGGAAAACTAACTTTTTAAAAGAAGGCG
TTTTCTTTTATTCGTGATCAGGTGGTTCTCTGCCTCCGATAAAAGAATCGTCACTGAGACCGACCTTTTGATTGAAAAATTTTCTTCCGC
                                    MnlI      DdeI     HinFI

E  R  H  L  V  S  K  L  Y  T  D  D  M  R  P  E  T  V  E  T  T  V  T  T  A  Q  E  T  K  R
             1005                1020                1035                1050                1065                1080
AGCGCCATCTAGTGAGCAAACTATACACTGATGACATGCGGCCAGAAACGGTTGAAACCACAGTAACCACTGCGCAGGAAACTAAGCGCG
TCGCGGTAGATCACTCGTTTGATATGTGACTACTGTACGCCGGTCTTTGCCAACTTTGGTGTCATTGGTGACGCGTCCTTTGATTCGCGC
HaeIIHhaI                                 Fnu4HIHaeIII                       MstIHhaI DdeIHhaIFnuDII
```

FIGURE 3. MVM DNA sequence (Astell, C.R., Thomson, M., Merchlinsky, M., and Ward, D.C., 1983, Nucl. Acids Res. 11:999–1014, reprinted with permission).

```
G R I Q T K K E V S I K T T L K E L V H K R V T S P E D W M
        1095         1110         1125         1140         1155         1170
GCAGAATTCAAACTAAAAAAGAAGTTTCTATTAAAACTACACTTAAAGAGCTGGTGCATAAAAGAGTAACCTCACCAGAGGACTGGATGA
CGTCTTAAGTTTGATTTTTTCTTCAAAGATAATTTTGATGTGAATTTCTCGACCACGTATTTTCTCATTGGAGTGGTCTCCTGACCTACT
Fnu4HI EcoRI                                  AluI                 MnlIHphIMnlI

M M Q P D S Y I E M M A Q P G G E N L L K N T L E I C T L T
        1185         1200         1215         1230         1245         1260
TGATGCAGCCAGACAGTTACATTGAAATGATGGCTCAACCAGGTGGAGAAAACCTGCTGAAAAATACGCTAGAGATTTGTACACTAACTC
ACTACGTCGGTCTGTCAATGTAACTTTACTACCGAGTTGGTCCACCTCTTTTGGACGACTTTTTATGCGATCTCTAAACATGTGATTGAG
BbvISfaNI                             BstNI                                  RsaI

L A R T K T A F D L I L E K A E T S K L T N F S L P D T R T
        1275         1290         1305         1320         1335         1350
TAGCCAGAACCAAAACAGCATTTGACTTAATTTTAGAAAAAGCTGAAACCAGCAAACTAACCAACTTTTCACTGCCTGACACAAGAACCT
ATCGGTCTTGGTTTTGTCGTAAACTGAATTAAAATCTTTTTCGACTTTGGTCGTTTGATTGGTTGAAAAGTGACGGACTGTGTTCTTGGA
                           EcoRI*      AluI

C R I F A F H G W N Y V K V C H A I C C V L N R Q G G K R N
        1365         1380         1395         1410         1425         1440
GCAGAATTTTTGCTTTTCATGGCTGGAACTATGTTAAAGTTTGCCATGCTATTTGCTGTGTTTTAAACAGACAAGGAGGCAAAAGAAATA
CGTCTTAAAAACGAAAAGTACCGACCTTGATACAATTTCAAACGGTACGATAAACGACACAAAATTTGTCTGTTCCTCCGTTTTCTTTAT
PstIEcoRI*                                                                MnlI

T V L F H G P A S T G K S I I A Q A I A Q A V G N V G C Y N
        1455         1470         1485         1500         1515         1530
CTGTTTTATTTCATGGACCAGCCAGCACAGGCAAATCTATTATTGCACAAGCCATAGCACAAGCAGTTGGCAATGTTGGTTGCTATAATG
GACAAAATAAAGTACCTGGTCGGTCGTGTCCGTTTAGATAATAACGTGTTCGGTATCGTGTTCGTCAACCGTTACAACCAACGATATTAC
              Sau96IAvaII

A A N V N F P F N D C T N K N L I W V E E A G N F G Q Q V N
        1545         1560         1575         1590         1605         1620
CAGCCAATGTAAACTTTCCATTTAATGACTGTACCAACAAGAACTTGATTTGGGTAGAAGAAGCTGGTAACTTTGGACAGCAAGTAAACC
GTCGGTTACATTTGAAAGGTAAATTACTGACATGGTTGTTCTTGAACTAAACCCATCTTCTTCGACCATTGAAACCTGTCGTTCATTTGG
                                RsaI                       MboIIAluI

Q F K A I C S G Q T I R I D Q K G K G S K Q I E P T P V I M
        1635         1650         1665         1680         1695         1710
AGTTTAAAGCCATTTGCTCTGGTCAAACTATTCGCATTGATCAAAAAGGAAAAGGCAGCAAACAGATTGAACCAACACCAGTCATCATGA
TCAAATTTCGGTAAACGAGACCAGTTTGATAAGCGTAACTAGTTTTTCCTTTTCCGTCGTTTGTCTAACTTGGTTGTGGTCAGTAGTACT
                                        BclISau3A   BbvIFnu4HI

T T N E N I T V V R I G C E E R P E H T Q P I R D R M L N I
        1725         1740         1755         1770         1785         1800
CCACAAATGAGAACATTACAGTGGTCAGAATAGGCTGCGAAGAAAGACCAGAACACACTCAACCAATCAGAGACAGAATGCTTAACATTC
GGTGTTTACTCTTGTAATGTCACCAGTCTTATCCGACGCTTCTTTCTGGTCTTGTGTGAGTTGGTTAGTCTCTGTCTTACGAATTGTAAG
                                    BbvIFnu4HIMboII

H L T H T L P G D F G L V D K N E W P M I C A W L V K N G Y
        1815         1830         1845         1860         1875         1890
ATCTAACACATACCTTGCCTGGTGACTTTGGTTTGGTTGACAAAAATGAATGGCCCATGATTTGTGCTTGGTTGGTAAAGAATGGTTACC
TAGATTGTGTATGGAACGGACCACTGAAACCAAACCAACTGTTTTTACTTACCGGGTACTAAACACGAACCAACCATTTCTTACCAATGG
                 BstNIHphI          HinDII           Sau96IHaeIII                   BstEII

Q S T M A S Y C A K W G K V P D W S E N W A E P K V P T P I
        1905         1920         1935         1950         1965         1980
AATCTACCATGGCAAGCTACTGTGCTAAATGGGGCAAAGTTCCTGATTGGTCAGAAAACTGGGCGGAGCCAAAGGTGCCAACTCCTATAA
TTAGATGGTACCGTTCGATGACACGATTTACCCCGTTTCAAGGACTAACCAGTCTTTTGACCCGCCTCGGTTTCCACGGTTGAGGATATT
      NcoI   AluI                                                                   EcoRI*

N L L G S A R S P F T T P K S T P L S Q N Y A L T P L A S D
        1995         2010         2025         2040         2055         2070
ATTTACTAGGTTCGGCACGCTCACCATTCACGACACCGAAAAGTACGCCTCTCAGCCAGAACTATGCACTAACTCCACTTGCATCGGATC
TAAATGATCCAAGCCGTGCGAGTGGTAAGTGCTGTGGCTTTTCATGCGGAGAGTCGGTCTTGATACGTGATTGAGGTGAACGTAGCCTAG
                    HphI                  RsaI MnlIDdeI                       Sau3ASfaNIXhoII

L E D L A L E P W S T P N T P V A G T A E T Q N T G E A G S
        2085         2100         2115         2130         2145         2160
TCGAGGACCTGGCTTTAGAGCCTTGGAGCACACCAAATACTCCTGTTGCGGGCACTGCAGAAACCCAGAACACTGGGGAAGCTGGTTCCA
AGCTCCTGGACCGAAATCTCGGAACCTCGTGTGGTTTATGAGGACAACGCCCGTGACGTCTTTGGGTCTTGTGACCCCTTCGACCAAGGT
TaqIMnlIBstNIAvaII                                    PstI                   AluI
```

FIGURE 3 (*continued*)

(*continued*)

```
K   A   C   Q   D   G   Q   L   S   P   T   W   S   E   I   E   E   D   L   R   A   C   F   G   A   E   P   L   K   K
          2175           2190           2205           2220           2235           2250
AAGCCTGCCAAGATGGTCAACTGAGCCCAACTTGGTCAGAGATCGAGGAGGATTTGAGAGCGTGCTTCGGTGCGGAACCGTTGAAGAAAG
TTCGGACGGTTCTACCAGTTGACTCGGGTTGAACCAGTCTCTAGCTCCTCCTAAACTCTCGCACGAAGCCACGCCTTGGCAACTTCTTTC
                HinDIIDdeI                 Sau3ATaqIMnlIMnlI                         MboII

                                                                                  1 M   V   G
D   F   S   E   P   L   N   L   D   *   3 M   A   P   P   A   K   R   A   K   R   G   K   G   L   R   D   G   W
          2265           2280           2295           2310           2325           2340
ACTTCAGCGAGCCGCTGAACTTGGACTAAGGTACGATGGCGCCTCCAGCTAAAAGAGCTAAAAGAGGTAAGGGTTTAAGGGATGGTTGGT
TGAAGTCGCTCGGCGACTTGAACCTGATTCCATGCTACCGCGGAGGTCGATTTTCTCGATTTTCTCCATTCCCAAATTCCCTACCAACCA
          Fnu4HI        DdeI RsaI HaeIINarIMnlI AluI  AluI    MnlI

 W   W   G   I   N   V   *              1 (N   H   L   V   L   G   W   V   P   P   G   Y   K   Y   L   G
            2 M   F   N   Y   L   F   Y   R   P   E   I   T   W   F   *
L   V   G   Y   *
          2355           2370           2385           2400           2415           2430
TGGTGGGGTATTAATGTTTAATTACCTGTTTTACAGGCCTGAAATCACTTGGTTTTAGGTTGGGTGCCTCCTGGCTACAAGTACCTGGGA
ACCACCCCATAATTACAAATTAATGGACAAAATGTCCGGACTTTAGTGAACCAAAATCCAACCCACGGAGGACCGATGTTCATGGACCCT
                   EcoRI*      HaeIHaeIII                      MnlIBstNI    RsaIBstNI

 P   G   N   S   L   D   Q   G   E   P   T   N   P   S   D   A   A   A   K   E   H   D   E   A   Y   D   Q   Y   I   K
          2445           2460           2475           2490           2505           2520
CCAGGGAACAGCCTTGACCAAGGAGAACCAACCAATCCATCTGACGCCGCTGCCAAAGAGCACGACGAGGCCTATGATCAATACATCAAA
GGTCCCTTGTCGGAACTGGTTCCTCTTGGTTGGTTAGGTAGACTGCGGCGACGGTTTCTCGTGCTGCTCCGGATACTAGTTATGTAGTTT
AvaIIBstNI                            AcyIBbvIHgaIFnu4HI    HaeIMnlI   BclISau3A

 S   G   K   N   P   Y   L   Y   F   S   A   A   D   Q   R   F   I   D   Q   T   K   D   A   K   D   W   G   G   K   V
          2535           2550           2565           2580           2595           2610
TCTGGAAAAAATCCTTACCTGTACTTCTCTGCTGCTGATCAACGCTTTATTGACCAAACCAAGGACGCCAAAGACTGGGGAGGCAAGGTT
AGACCTTTTTTAGGAATGGACATGAAGAGACGACGACTAGTTGCGAAATAACTGGTTTGGTTCCTGCGGTTTCTGACCCCTCCGTTCCAA
                   RsaI   BbvIBclIFnu4HISau3A              AcyIHgaI        MnlI

 G   H   Y   F   F   R   T   K   R   A   F   A   P   K   L   A   T   D   S   E   P   G   T   S   G   V   S   R   A   G
          2625           2640           2655           2670           2685           2700
GGTCACTACTTTTTTAGAACCAAGCGCGCTTTTGCACCTAAGCTTGCTACTGACTCTGAACCTGGAACTTCTGGTGTAAGCAGAGCTGGT
CCAGTGATGAAAAAATCTTGGTTCGCGCGAAAACGTGGATTCGAACGATGACTGAGACTTGGACCTTGAAGACCACATTCGTCTCGACCA
                     FnuDIIHhaI  AluIDdeIHinDIII HinFI   BstNI                   AluI

 K   R   T   R   P   P   A   Y   I   F   I   N   Q   A   R   A   K   K   K   L   T   S   S   A   A   Q   Q   S   S   Q
          2715           2730           2745           2760           2775           2790
AAACGCACTAGACCACCTGCTTACATTTTTATTAACCAAGCCAGAGCTAAAAAAAAACTTACTTCTTCTGCTGCACAGCAAAGCAGTCAA
TTTGCGTGATCTGGTGGACGAATGTAAAAATAATTGGTTCGGTCTCGATTTTTTTTTGAATGAAGAAGACGACGTGTCGTTTCGTCAGTT
                                          AluI                MboIIBbvIFnu4HI

 T)M   S   D   G   T   S   Q   P   D   S   G   N   A   V   H   S   A   A   R   V   E   R   A   A   D   G   P   G   G
          2805           2820           2835           2850           2865           2880
ACCATGAGTGATGGCACCAGCCAACCTGACAGCGGAAACGCTGTCCACTCAGCTGCAAGAGTTGAACGAGCAGCTGACGGCCCTGGAGGC
TGGTACTCACTACCGTGGTCGGTTGGACTGTCGCCTTTGCGACAGGTGAGTCGACGTTCTCAACTTGCTCGTCGACTGCCGGGACCTCCG
                                             DdeIPvuIIBbvI    PvuIIBbvI BglIHaeIIIMnlIBstNI

 S   G   G   G   G   S   G   G   G   G'  V   G   V   S   T   G   S   Y   D   N   Q   T   H   Y   R   F   L   G   D   G
          2895           2910           2925           2940           2955           2970
TCTGGGGGTGGGGGCTCTGGCGGGGGTGGGGTTGGTGTTTCTACTGGGTCTTATGATAATCAAACGCATTATAGATTCTTGGGTGACGGC
AGACCCCCACCCCCGAGACCGCCCCCACCCCAACCACAAAGATGACCCAGAATACTATTAGTTTGCGTAATATCTAAGAACCCACTGCCG
                                                                           HinFI   HphI

 W   V   E   I   T   A   L   A   T   R   L   V   H   L   N   M   P   K   S   E   N   Y   C   R   I   R   V   H   N   T
          2985           3000           3015           3030           3045           3060
TGGGTAGAAATTACTGCACTAGCAACTAGACTAGTACATTTAAACATGCCTAAATCAGAAAACTATTGCAGAATCAGAGTTCACAATACA
ACCCATCTTTAATGACGTGATCGTTGATCTGATCATGTAAATTTGTACGGATTTAGTCTTTTGATAACGTCTTAGTCTCAAGTGTTATGT
         EcoRI*                     RsaI                                 HinFI

 T   D   T   S   V   K   G   N   M   A   K   D   D   A   H   E   Q   I   W   T   P   W   S   L   V   D   A   N   A   W
          3075           3090           3105           3120           3135           3150
ACAGACACATCAGTCAAAGGCAACATGGCAAAAGATGATGCTCATGAGCAAATTTGGACACCATGGAGCTTGGTGGATGCTAATGCTTGG
TGTCTGTGTAGTCAGTTTCCGTTGTACCGTTTTCTACTACGAGTACTCGTTTAAACCTGTGGTACCTCGAACCACCTACGATTACGAACC
                                     SfaNI        EcoRI*   NcoI  AluI     SfaNI
```

FIGURE 3 (*continued*)

```
G V W L Q P S D W Q Y I C N T M S Q L N L V S L D Q E I F N
3165 3180 3195 3210 3225 3240
GGAGTTTGGCTCCAGCCAAGTGACTGGCAATACATTTGCAACACCATGAGCCAGCTTAACTTGGTATCACTTGATCAAGAAATATTCAAT
CCTCAAACCGAGGTCGGTTCACTGACCGTTATGTAAACGTTGTGGTACTCGGTCGAATTGAACCATAGTGAACTAGTTCTTTATAAGTTA
AluI BclISau3A

V V L K T V T E Q D L G G Q A I K I Y N N D L T A C M M V A
3255 3270 3285 3300 3315 3330
GTAGTGCTGAAAACTGTTACAGAGCAAGACTTAGGAGGTCAAGCTATAAAAATATACAACAATGACCTTACAGCTTGCATGATGGTTGCA
CATCACGACTTTTGACAATGTCTCGTTCTGAATCCTCCAGTTCGATATTTTTATATGTTGTTACTGGAATGTCGAACGTACTACCAACGT
DdeI MnlI AluI AluI

V D S N N I L P Y T P A A N S M E T L G F Y P W K P T I A S
3345 3360 3375 3390 3405 3420
GTAGACTCAAACAACATTTTGCCATACACACCTGCAGCAAACTCAATGGAAACACTTGGTTTCTACCCCTGGAAACCAACCATAGCATCA
CATCTGAGTTTGTTGTAAAACGGTATGTGTGGACGTCGTTTGAGTTACCTTTGTGAACCAAAGATGGGGACCTTTGGTTGGTATCGTAGT
AccIHinFI BbvIPstIFnu4HI BstNI HphISfaNI

P Y R Y Y F C V D R D L S V T Y E N Q E G T V E H N V M G T
3435 3450 3465 3480 3495 3510
CCATACAGGTACTATTTTTGCGTTGACAGAGATCTTTCAGTGACCTACGAAAATCAAGAAGGCACAGTTGAACATAATGTGATGGGAACA
GGTATGTCCATGATAAAAACGCAACTGTCTCTAGAAAGTCACTGGATGCTTTTAGTTCTTCCGTGTCAACTTGTATTACACTACCCTTGT
RsaI HinDII BglII

P K G I P Q F F T I E N T Q Q I T L L R T G D E F A T G T Y
3525 3540 3555 3570 3585 3600
CCAAAAGGAATTCCTCAATTTTTTACCATTGAGAACACACAACAAATCACATTGCTCAGAACAGGGGACGAATTTGCCACAGGTACTTAC
GGTTTTCCTTAAGGAGTTAAAAAATGGTAACTCTTGTGTGTTGTTTAGTGTAACGAGTCTTGTCCCCTGCTTAAACGGTGTCCATGAATG
EcoRIMnlIEcoRI* DdeI EcoRI* RsaI

Y F D T N S V K L T H T W Q T N R Q L G Q P P L L S T F P E
3615 3630 3645 3660 3675 3690
TACTTTGACACAAATTCAGTTAAACTCACACACACGTGGCAAACCAACCGTCAACTTGGACAGCCTCCACTGCTGTCAACCTTTCCTGAA
ATGAAACTGTGTTTAAGTCAATTTGAGTGTGTGTGCACCGTTTGGTTGGCAGTTGAACCTGTCGGAGGTGACGACAGTTGGAAAGGACTT
EcoRI* SacIII HinDII MnlI HinDII AluI

A D T D A G T L T A Q G S R H G T T Q M G V N W V S E A I R
3705 3720 3735 3750 3765 3780
GCTGACACTGATGCAGGTACACTTACTGCTCAAGGGAGCAGACATGGAACAACACAAATGGGGGTTAACTGGGTGAGTGAAGCAATCAGA
CGACTGTGACTACGTCCATGTGAATGACGAGTTCCCTCGTCTGTACCTTGTTGTGTTTACCCCCAATTGACCCACTCACTTCGTTAGTCT
SfaNI RsaI HinDIIHpaIHphI

T R P A Q V G F C Q P H N D F E A S R A G P F A A P K V P A
3795 3810 3825 3840 3855 3870
ACCAGACCTGCTCAAGTAGGATTTTGTCAACCACACAATGACTTTGAAGCCAGCAGAGCTGGACCATTTGCTGCCCCAAAAGTTCCAGCA
TGGTCTGGACGAGTTCATCCTAAAACAGTTGGTGTGTTACTGAAACTTCGGTCGTCTCGACCTGGTAAACGACGGGGTTTTCAAGGTCGT
HinDII AluIAvaII BbvIFnu4HI

D I T Q G V D K E A N G S V R Y S Y G K Q H G E N W A S H G
3885 3900 3915 3930 3945 3960
GATATTACTCAAGGAGTAGACAAAGAAGCCAATGGCAGTGTTAGATACAGTTATGGCAAACAGCATGGTGAAAATTGGGCTTCACATGGA
CTATAATGAGTTCCTCATCTGTTTCTTCGGTTACCGTCACAATCTATGTCAATACCGTTTGTCGTACCACTTTTAACCCGAAGTGTACCT
AccI HphI EcoRI* AvaII

P A P E R Y T W D E T S F G S G R D T K D G F I Q S A P L V
3975 3990 4005 4020 4035 4050
CCAGCACCAGAGCGCTACACATGGGATGAAACAAGCTTTGGTTCAGGTAGAGACACCAAAGATGGTTTTATTCAATCAGCACCACTAGTT
GGTCGTGGTCTCGCGATGTGTACCCTACTTTGTTCGAAACCAAGTCCATCTCTGTGGTTTCTACCAAAATAAGTTAGTCGTGGTGATCAA
HaeIIHhaI AluIHinDIII

V P P P L N G I L T N A N P I G T K N D I H F S N V F N S Y
4065 4080 4095 4110 4125 4140
GTTCCACCACCACTAAATGGCATTCTTACAAATGCAAACCCTATTGGGACTAAAAATGACATTCATTTTTCAAATGTTTTTAACAGCTAT
CAAGGTGGTGGTGATTTACCGTAAGAATGTTTACGTTTGGGATAACCCTGATTTTTACTGTAAGTAAAAAGTTTACAAAAATTGTCGATA
AluI

G P L T A F S H P S P V Y P Q G Q I W D K E L D L E H K P R
4155 4170 4185 4200 4215 4230
GGTCCACTAACTGCATTTTCACACCCAAGTCCTGTATACCCTCAAGGACAAATATGGGACAAAGAACTAGATCTTGAACACAAACCTAGA
CCAGGTGATTGACGTAAAAGTGTGGGTTCAGGACATATGGGAGTTCCTGTTTATACCCTGTTTCTTGATCTAGAACTTGTGTTTGGATCT
AvaIISau96I AccI MnlI BglIISau3A
```

FIGURE 3 (*continued*)

(*continued*)

```
L   H   I   T   A   P   F   V   C   K   N   N   A   P   G   Q   M   L   V   R   L   G   P   N   L   T   D   Q   Y   D
          4245           4260           4275           4290           4305           4320
CTTCACATAACTGCTCCATTTGTTTGTAAAAACAATGCACCTGGACAAATGTTGGTTAGATTAGGACCAAACCTAACTGACCAATATGAT
GAAGTGTATTGACGAGGTAAACAAACATTTTTGTTACGTGGACCTGTTTACAACCAATCTAATCCTGGTTTGGATTGACTGGTTATACTA
                                       BstNI                 Sau96IAvaII                 Sau3A

P   N   G   A   T   L   S   R   I   V   T   Y   G   T   F   F   W   K   G   K   L   T   M   R   A   K   L   R   A   N
          4335           4350           4365           4380           4395           4410
CCAAACGGAGCCACACTTTCTAGAATTGTTACATACGGTACATTTTTCTGGAAAGGAAAACTAACCATGAGAGCAAAACTTAGAGCTAAC
GGTTTGCCTCGGTGTGAAAGATCTTAACAATGTATGCCATGTAAAAAGACCTTTCCTTTTGATTGGTACTCTCGTTTTGAATCTCGATTG
                  XbaI EcoRI*         RsaI                                       DdeI AluI

T   T   W   N   P   V   Y   Q   V   S   A   E   D   N   G   N   S   Y   M   S   V   T   K   W   L   P   T   A   T   G
          4425           4440           4455           4470           4485           4500
ACCACTTGGAACCCAGTGTACCAAGTAAGTGCTGAAGACAATGGCAACTCATACATGAGTGTAACTAAATGGTTACCAACTGCTACTGGA
TGGTGAACCTTGGGTCACATGGTTCATTCACGACTTCTGTTACCGTTGAGTATGTACTCACATTGATTTACCAATGGTTGACGATGACCT
                  RsaI          MboII                                   BstEII

N   M   Q   S   V   P   L   I   T   R   P   V   A   R   N   T   Y   *
          4515           4530           4545           4560           4575           4590
AACATGCAGTCTGTGCCGCTTATAACAAGACCTGTTGCTAGAAATACTTACTAACTAACCATGCTTTTTCTTTCTGTACTTCATATATTA
TTGTACGTCAGACACGGCGAATATTGTTCTGGACAACGATCTTTATGAATGATTGATTGGTACGAAAAAGAAAGACATGAAGTATATAAT
               Fnu4HI                                                   RsaI

          4605           4620           4635           4650           4665           4680
TTAAGACTAATAAAGATACAACATAGAAATATAATATTACGTATAGATTTAAGAAATAGAATAATATGGTACTTAGTAACTGTTAAAAAT
AATTCTGATTATTTCTATGTTGTATCTTTATATTATAATGCATATCTAAATTCTTTATCTTATTATACCATGAATCATTGACAATTTTTA
                                      SacIII                        RsaIDdeI

          4695           4710           4725           4740           4755           4770
AATAGAACCTTTGGAATAACAAGATAGTTAGTTGGTTAATGTTAGATAGAATAAGAAGATCATGTATAATGAATAAAAGGGTGGAAGGGT
TTATCTTGGAAACCTTATTGTTCTATCAATCAACCAATTACAATCTATCTTATTCTTCTAGTACATATTACTTATTTTCCCACCTTCCCA
                                                    MboIISau3A

          4785           4800           4815           4830           4845           4860
GGTTGGTAGGTATTCCCTTAGACTTGATGTTAAGGACCAAAAAAATAATAAAACTTTTTTAAAACTCAACCAAGACTACTGTCTATTCAG
CCAACCATCCATAAGGGAATCTGAACTACAATTCCTGGTTTTTTTATTATTTTGAAAAAATTTTGAGTTGGTTCTGATGACAGATAAGTC
                 DdeI           Sau96IAvaII

          4875           4890           4905           4920           4935           4950
TGAACCAACTGAACCATTAGTATTACTATGTTTTTAGGGTGGGAGGGTGGGAGATACATGTGTTCGCTATGAGCGAACTGGTACTGGTTG
ACTTGGTTGACTTGGTAATCATAATGATACAAAAATCCCACCCTCCCACCCTCTATGTACACAAGCGATACTCGCTTGACCATGACCAAC
                                         MnlI                                   RsaI

          4965           4980           4995           5010           5025           5040
GTTGCTCTGCTCAACCAACCAGACCGGCAAAGCCGGTCTGGTTGGTTGAGCGCAACCAACCAGTACCAGTTCGCTCATAGCGAAOACATG
CAACGAGACGAGTTGGTTGGTCTGGCCGTTTCGGCCAGACCAACCAACTCGCGTTGGTTGGTCATGGTCAAGCGAGTATCGCTTGTGTAC
                       HpaII    HpaII             HhaI          RsaI

      5050      5060      5070      5080
TATCTCCCACCCTCCCACCCTAAAAACATAGTAATACTAAT
ATAGAGGGTGGGAGGGTGGGATTTTTGTATCATTATGATTA
          MnlI
```

FIGURE 3 *(continued)*

Index

MIX
Papier aus verantwortungsvollen Quellen
Paper from responsible sources
FSC® C105338

If you have any concerns about our products,
you can contact us on
ProductSafety@springernature.com

In case Publisher is established outside the EU,
the EU authorized representative is:
Springer Nature Customer Service Center GmbH
Europaplatz 3, 69115 Heidelberg, Germany

Printed by Libri Plureos GmbH
in Hamburg, Germany